Amyloid-β: Structure, Function, and Pathophysiological Significance in Neurodegenerative Diseases

Amyloid-β: Structure, Function, and Pathophysiological Significance in Neurodegenerative Diseases

Editors

Masashi Tanaka
Kenjiro Ono
Satoshi Saito

MDPI • Basel • Beijing • Wuhan • Barcelona • Belgrade • Manchester • Tokyo • Cluj • Tianjin

Editors
Masashi Tanaka
Health Science University
Yamanashi
Japan

Kenjiro Ono
Kanazawa University
Graduate School of Medical Sciences
Kanazawa
Japan

Satoshi Saito
National Cerebral and Cardiovascular Center
Osaka
Japan

Editorial Office
MDPI
St. Alban-Anlage 66
4052 Basel, Switzerland

This is a reprint of articles from the Special Issue published online in the open access journal *International Journal of Molecular Sciences* (ISSN 1422-0067) (available at: https://www.mdpi.com/journal/ijms/special_issues/Ab).

For citation purposes, cite each article independently as indicated on the article page online and as indicated below:

LastName, A.A.; LastName, B.B.; LastName, C.C. Article Title. *Journal Name* **Year**, *Volume Number*, Page Range.

ISBN 978-3-0365-6712-9 (Hbk)
ISBN 978-3-0365-6713-6 (PDF)

Contents

International Journal of
Molecular Sciences

MDPI

Editorial

Amyloid-β: Structure, Function, and Pathophysiological Significance in Neurodegenerative Diseases

Satoshi Saito [1], Kenjiro Ono [2] and Masashi Tanaka [3,*]

1 Department of Neurology, National Cerebral and Cardiovascular Center, 6-1 Kishibe-Shimmachi, Osaka 564-8565, Japan
2 Department of Neurology, Kanazawa University Graduate School of Medical Sciences, 13-1 Takara-machi, Kanazawa 920-8640, Japan
3 Department of Physical Therapy, Health Science University, 7187 Kodachi, Fujikawaguchiko-machi, Minamitsuru-gun, Yamanashi 401-0380, Japan
* Correspondence: masashi.7.tanaka@gmail.com; Tel.: +81-555-83-5200

Citation: Saito, S.; Ono, K.; Tanaka, M. Amyloid-β: Structure, Function, and Pathophysiological Significance in Neurodegenerative Diseases. *Int. J. Mol. Sci.* **2022**, *23*, 10275. https://doi.org/10.3390/ijms231810275

Received: 2 September 2022
Accepted: 5 September 2022
Published: 7 September 2022

The rate of dementia continues to increase worldwide; however, there currently exist no therapeutic strategies for this condition. Although several effective predictive markers (e.g., plasma amyloid-β [Aβ]$_{42}$/Aβ$_{40}$ ratios) for Alzheimer's disease (AD) have been identified, further studies are needed to identify more sophisticated and less expensive predictive markers for dementia.

Recent extensive studies revealed the pathological implications of Aβ in the development and progression of dementia. Aβ monomers are prone to aggregation, which forms toxic Aβ oligomers that cause neuronal and vascular injuries. Aβ is also involved in other cytotoxic mediators, such as reactive oxygen species (ROS) and neuroinflammation, thereby suggesting complex pathogenesis and leading to the development and progression of neurodegenerative diseases, including AD and cerebral amyloid angiopathy. Accordingly, there exists an urgent need to elucidate the molecular mechanisms underlying Aβ aggregation processes, ROS generation, and neuroinflammation to identify effective therapeutic targets and identify and develop drugs/bioactive molecules with preventive and/or therapeutic potentials for Aβ-related neurodegenerative diseases.

As such, this Special Issue includes 15 original manuscripts, a case report, a commentary, and review articles that contribute to the aforementioned aim and provide novel insights into the mechanisms underlying the pathogenesis of Aβ-related neurodegenerative diseases.

We have included several interesting papers that address the mechanisms of action of Aβ aggregation and oligomerization. Banerjee et al. demonstrated that the interaction between Aβ and cellular membrane triggered the on-membrane self-assembly of Aβ, thereby promoting the oligomer formation of Aβ [1]. Furthermore, Aβ aggregates bound to the cellular membrane acted as seeds for further aggregation, resulting in cell permeability and damage and subsequently inducing cell lysis, as shown by Ruiz-Arias et al. using a mouse neuroblastoma cell line [2]. These findings highlight the significance of Aβ dynamics on the cellular membrane in forming oligomers and exhibiting cytotoxicity in physiological conditions. Regarding familial AD, He et al. investigated the effects of several types of Aβ mutations on the secondary structure and subsequent aggregation and showed novel roles of these mutations in AD pathogenesis [3]. In a case report, Shimada et al. characterized a recently identified Osaka mutation in the gene amyloid precursor protein (APP) [4]. They found that patients with dementia carrying this mutation had a high tau burden and subtle Aβ accumulation in the cerebral cortex and cerebellum, thereby suggesting tau accumulation and neurodegeneration through toxic Aβ oligomers without senile plaque formation [4].

Neuroinflammation has been closely implicated in the pathogenesis of neurodegenerative diseases, which involves a shift in microglial phenotypes from homeostasis to a proinflammatory state. By analyzing the brains of patients with AD, Walker et al. showed that the expression levels of purinergic adenosine diphosphate/triphosphate receptor P2RY12 on microglia defines the boundary between the proinflammatory area, consisting of microglia interacting with and/or adjacent to Aβ plaques, and the nonaffected area away from the Aβ plaques [5]. These findings suggest that P2RY12 carries a novel pathological significance in the proinflammatory axis of microglia around Aβ plaques in AD brains [5].

A commentary paper by Friedland et al. provided a unique point of view on the pathogenesis of neurodegenerative diseases. Notably, they summarized the potential roles of gut microbiota-derived amyloid proteins in the aggregation of neuronal proteins, such as Aβ, and in neuroinflammation [6].

This Special Issue also includes papers that addressed the preventive and/or therapeutic potentials of chemical compounds or drugs for Aβ-related neurodegenerative diseases. Murakami et al., who focused on 10-Me-Aplog-1, a new protein kinase C activator, demonstrated its inhibitory effects on the intracellular formation of toxic Aβ oligomers in rat primary cerebral cortex cells [7]. In a mouse model of cerebral amyloid angiopathy, Yakushiji et al. showed that the administration of low-dose phosphodiesterase III inhibitor cilostazol improved vascular deposition of Aβ, potentially by facilitating perivascular drainage of Aβ [8]. These findings would promote research aimed at developing novel drugs and identifying repositioning drugs for Aβ-related neurodegenerative diseases.

We have also included excellent review articles in this Special Issue, which provide updates on unique topics concerning neurodegenerative diseases.

Tomiyama and Shimada, who summarized the characteristics of Osaka mutation in the APP gene, argued that the loss-of-function in APP and gain-of-function in Aβ were caused by this mutation and were implicated in the pathogenesis of AD [9]. Regarding Aβ aggregation, Watanabe-Nakayama et al. comprehensively reviewed the usefulness of high-speed atomic force microscopy to visualize the structural dynamics in the aggregation process of Aβ, which would lead to novel insights into the mechanisms underlying Aβ aggregation [10].

Regarding preventive and/or therapeutic potentials for Aβ-related neurodegenerative diseases, Ono and Tsuji reported the significance of Aβ protofibrils as a therapeutic target in AD while describing the pathological implications of Aβ protofibrils in AD [11]. Furthermore, Tadokoro et al. focused on the roles of ROS in the Aβ cascade and pathogenesis of AD. Notably, they summarized the recent findings regarding the pathological relationship between ROS and AD and described the beneficial effects of antioxidative supplements on AD pathogenesis [12]. Apart from antioxidants, several natural medicines that could beneficially affect memory decline in AD through pleiotropic mechanisms of action are emerging, as reviewed by Kuboyama et al. [13]. Conversely, there are natural compounds that would need careful consideration during administration. In line with this, Kobayashi et al. showed that some naturally occurring polyphenols suppressed Aβ aggregation and were expected to have protective effects against AD; however, some of them pose the potential risk of oxidative damage due to pro-oxidant properties [14]. As another therapeutic approach for AD, Iqbal et al. described the therapeutic potential of anti-infectious drugs based on the possibility that Aβ plaque formation is the innate immune response against microorganisms in AD brains [15].

We earnestly believe that the excellent papers included in this Special Issue improve our understanding of the pathogenesis of Aβ-related neurodegenerative diseases and help develop effective preventive and therapeutic strategies for such diseases.

Author Contributions: Conceptualization, S.S., K.O. and M.T.; writing—original draft preparation, M.T.; writing—review and editing, S.S., K.O. and M.T. All authors have read and agreed to the published version of the manuscript.

Funding: This research received no external funding.

Institutional Review Board Statement: Not applicable.

Informed Consent Statement: Not applicable.

Data Availability Statement: Not applicable.

Conflicts of Interest: The authors declare no conflict of interest.

References

1. Banerjee, S.; Hashemi, M.; Zagorski, K.; Lyubchenko, Y.L. Interaction of Abeta42 with Membranes Triggers the Self-Assembly into Oligomers. *Int. J. Mol. Sci.* **2020**, *21*, 1129. [CrossRef] [PubMed]
2. Ruiz-Arias, A.; Paredes, J.M.; Di Biase, C.; Cuerva, J.M.; Giron, M.D.; Salto, R.; Gonzalez-Vera, J.A.; Orte, A. Seeding and Growth of beta-Amyloid Aggregates upon Interaction with Neuronal Cell Membranes. *Int. J. Mol. Sci.* **2020**, *21*, 5035. [CrossRef] [PubMed]
3. He, K.C.; Chen, Y.R.; Liang, C.T.; Huang, S.J.; Tzeng, C.Y.; Chang, C.F.; Huang, S.J.; Huang, H.B.; Lin, T.H. Conformational Characterization of Native and L17A/F19A-Substituted Dutch-Type beta-Amyloid Peptides. *Int. J. Mol. Sci.* **2020**, *21*, 2571. [CrossRef] [PubMed]
4. Shimada, H.; Minatani, S.; Takeuchi, J.; Takeda, A.; Kawabe, J.; Wada, Y.; Mawatari, A.; Watanabe, Y.; Shimada, H.; Higuchi, M.; et al. Heavy Tau Burden with Subtle Amyloid beta Accumulation in the Cerebral Cortex and Cerebellum in a Case of Familial Alzheimer's Disease with APP Osaka Mutation. *Int. J. Mol. Sci.* **2020**, *21*, 4443. [CrossRef] [PubMed]
5. Walker, D.G.; Tang, T.M.; Mendsaikhan, A.; Tooyama, I.; Serrano, G.E.; Sue, L.I.; Beach, T.G.; Lue, L.F. Patterns of Expression of Purinergic Receptor P2RY12, a Putative Marker for Non-Activated Microglia, in Aged and Alzheimer's Disease Brains. *Int. J. Mol. Sci.* **2020**, *21*, 678. [CrossRef] [PubMed]
6. Friedland, R.P.; McMillan, J.D.; Kurlawala, Z. What Are the Molecular Mechanisms by Which Functional Bacterial Amyloids Influence Amyloid Beta Deposition and Neuroinflammation in Neurodegenerative Disorders? *Int. J. Mol. Sci.* **2020**, *21*, 1652. [CrossRef] [PubMed]
7. Murakami, K.; Yoshimura, M.; Nakagawa, S.; Kume, T.; Kondo, T.; Inoue, H.; Irie, K. Evaluation of Toxic Amyloid 42 Oligomers in Rat Primary Cerebral Cortex Cells and Human iPS-derived Neurons Treated with 10-Me-Aplog-1, a New PKC Activator. *Int. J. Mol. Sci.* **2020**, *21*, 1179. [CrossRef] [PubMed]
8. Yakushiji, Y.; Kawamoto, K.; Uchihashi, K.; Ihara, M.; Aoki, S.; Nagaishi, Y.; Suzuyama, K.; Tsugitomi, Y.; Hara, H. Low-Dose Phosphodiesterase III Inhibitor Reduces the Vascular Amyloid Burden in Amyloid-beta Protein Precursor Transgenic Mice. *Int. J. Mol. Sci.* **2020**, *21*, 2295. [CrossRef] [PubMed]
9. Tomiyama, T.; Shimada, H. APP Osaka Mutation in Familial Alzheimer's Disease-Its Discovery, Phenotypes, and Mechanism of Recessive Inheritance. *Int. J. Mol. Sci.* **2020**, *21*, 1413. [CrossRef] [PubMed]
10. Watanabe-Nakayama, T.; Sahoo, B.R.; Ramamoorthy, A.; Ono, K. High-Speed Atomic Force Microscopy Reveals the Structural Dynamics of the Amyloid-beta and Amylin Aggregation Pathways. *Int. J. Mol. Sci.* **2020**, *21*, 4287. [CrossRef] [PubMed]
11. Ono, K.; Tsuji, M. Protofibrils of Amyloid-beta are Important Targets of a Disease-Modifying Approach for Alzheimer's Disease. *Int. J. Mol. Sci.* **2020**, *21*, 952. [CrossRef] [PubMed]
12. Tadokoro, K.; Ohta, Y.; Inufusa, H.; Loon, A.F.N.; Abe, K. Prevention of Cognitive Decline in Alzheimer's Disease by Novel Antioxidative Supplements. *Int. J. Mol. Sci.* **2020**, *21*, 1974. [CrossRef] [PubMed]
13. Kuboyama, T.; Yang, X.; Tohda, C. Natural Medicines and Their Underlying Mechanisms of Prevention and Recovery from Amyloid Beta-Induced Axonal Degeneration in Alzheimer's Disease. *Int. J. Mol. Sci.* **2020**, *21*, 4665. [CrossRef] [PubMed]
14. Kobayashi, H.; Murata, M.; Kawanishi, S.; Oikawa, S. Polyphenols with Anti-Amyloid beta Aggregation Show Potential Risk of Toxicity Via Pro-Oxidant Properties. *Int. J. Mol. Sci.* **2020**, *21*, 3561. [CrossRef] [PubMed]
15. Iqbal, U.H.; Zeng, E.; Pasinetti, G.M. The Use of Antimicrobial and Antiviral Drugs in Alzheimer's Disease. *Int. J. Mol. Sci.* **2020**, *21*, 4920. [CrossRef] [PubMed]

International Journal of
Molecular Sciences

MDPI

Article

Seeding and Growth of β-Amyloid Aggregates upon Interaction with Neuronal Cell Membranes

Álvaro Ruiz-Arias [1], Jose M. Paredes [1], Chiara Di Biase [1,2], Juan M. Cuerva [3], María D. Giron [4], Rafael Salto [4], Juan A. González-Vera [1] and Angel Orte [1,*]

1 Departamento de Fisicoquímica, Unidad de Excelencia de Química Aplicada a Biomedicina y Medioambiente, Facultad de Farmacia, Universidad de Granada, Campus Cartuja, 18071 Granada, Spain; alvaroruizarias@ugr.es (Á.R.-A.); jmparedes@ugr.es (J.M.P.); c.dibiase@campus.uniurb.it (C.D.B.); gonzalezvera@ugr.es (J.A.G.-V.)

2 Department of Biomolecular Sciences, University of Urbino Carlo Bo, 61029 Urbino, Italy

3 Departamento de Química Orgánica, Unidad de Excelencia de Química Aplicada a Biomedicina y Medioambiente, Facultad de Ciencias, Universidad de Granada, Campus Fuentenueva, 18071 Granada, Spain; jmcuerva@ugr.es

4 Departamento de Bioquímica y Biología Molecular II, Unidad de Excelencia de Química Aplicada a Biomedicina y Medioambiente, Facultad de Farmacia, Universidad de Granada, Campus Cartuja, 18071 Granada, Spain; mgiron@ugr.es (M.D.G.); rsalto@ugr.es (R.S.)

* Correspondence: angelort@ugr.es; Tel.: +34-958243825

Received: 28 June 2020; Accepted: 15 July 2020; Published: 16 July 2020

Abstract: In recent years, the prevalence of amyloid neurodegenerative diseases such as Alzheimer's disease (AD) has significantly increased in developed countries due to increased life expectancy. This amyloid disease is characterized by the presence of accumulations and deposits of β-amyloid peptide (Aβ) in neuronal tissue, leading to the formation of oligomers, fibers, and plaques. First, oligomeric intermediates that arise during the aggregation process are currently thought to be primarily responsible for cytotoxicity in cells. This work aims to provide further insights into the mechanisms of cytotoxicity by studying the interaction of Aβ aggregates with Neuro-2a (N2a) neuronal cells and the effects caused by this interaction. For this purpose, we have exploited the advantages of advanced, multidimensional fluorescence microscopy techniques to determine whether different types of Aβ are involved in higher rates of cellular toxicity, and we measured the cellular stress caused by such aggregates by using a fluorogenic intracellular biothiol sensor. Stress provoked by the peptide is evident by N2a cells generating high levels of biothiols as a defense mechanism. In our study, we demonstrate that Aβ aggregates act as seeds for aggregate growth upon interacting with the cellular membrane, which results in cell permeability and damage and induces lysis. In parallel, these damaged cells undergo a significant increase in intracellular biothiol levels.

Keywords: neurodegenerative disease; amyloid; misfolding; Alzheimer's disease; fluorescence microscopy; FRET

1. Introduction

In recent years, neurodegenerative diseases have become a major threat to public health in industrialized countries, since the prevalence of these disorders increases as the percentage of elderly people in the population rises. Symptoms of neurodegenerative diseases begin many years after the onset of illness, making it extremely difficult to provide an effective and early diagnosis. The common feature of amyloid diseases is the aberrant aggregation of a certain protein, initially in small oligomers, which subsequently grow to form various fibrillar structures and plaques. These illnesses promote the progressive displacement of neurons along with secondary alterations due to effects on the central nervous system [1]. Approximately forty neurodegenerative diseases, including Alzheimer's disease (AD), are characterized by the presence of insoluble amyloid deposits in the affected tissues [2]. A hallmark of AD is the existence of plaques formed by β-amyloid (Aβ) fibers in the neocortex and hippocampus of the brain [3,4]. When the transmembrane amyloid precursor protein is proteolyzed by β- and γ-secretase, it is cleaved into small peptide fragments, namely, (Aβ) peptides [5–7]. The initial oligomeric structures that emerged during the Aβ aggregation process are thought to be the main sources of cytotoxicity related to AD, and they are currently one of most interesting pharmacological targets, as these oligomeric structures seem to be essential in the neurotoxicity of amyloid structures [8]. Among all Aβ isoforms, the Aβ (1–42) peptide (Aβ42) seems to be the most neurotoxic species compared to the other common 40-residue fragment. The different behaviors of these two peptides are striking, especially due to the fact that only two extra C-terminal residues are present in the Aβ42 form [9,10].

The interaction at a cellular level of such amyloid oligomers is an important step towards disease progression. During amyloidogenic misfolding, specific proteins undergo aberrant three-dimensional self-interactions. These structures expose solvent hydrophobic regions that are usually hidden in the native state. This fact facilitates protein aggregation and the subsequent nucleation into β-sheet-rich structures [11]. Thus, this process entails not only the loss of the physiological function but also the generation of abnormal protein aggregates. These filamentary oligomers rich in β-sheet structures exhibit unstable behavior and can alter the chemical composition of the extracellular matrix [12]. Recent studies have suggested that oligomeric intermediates formed throughout amyloidogenic aggregation are more toxic than mature fibrils. For instance, a slow conversion of early, easy-degradable aggregates into compact, highly structured oligomers was detected in α-synuclein fibrillization [13], which is related to the progress of Parkinson's disease. These later oligomers induce much higher levels of reactive oxygen species (ROS) in cells than do earlier oligomers [14]. While focusing on Aβ aggregation, these oligomers interact with lipid membrane surfaces through weak electrostatic interactions that promote conformational transition from α-helix structures to β-sheet conformations [15]. Moreover, it has been observed that Aβ alters membrane phospholipid fluidity through its own insertion into lipid bilayers [16]. That insertion, in form of pores, induces calcium-signaling, activating NADPH oxidase, stimulating nitric oxide production, and finally leading to increased oxidative stress and to neurodegeneration [14,17–19].

The urgent need for reliable tools that provide further information on the cellular effect of amyloid aggregates make this a very active field of research. For instance, the use of fluorescent probes to study relevant intracellular processes such as the interaction of the peptide with neuronal cells is highly advantageous due to its low invasiveness [20,21]. One of the most important steps to understand the aberrant protein aggregation process is the identification and characterization of different species formed during the misfolding process. A widely used technique to characterize amyloid fibers has been the use of thioflavin T (ThT) as a fluorophore, as its interaction with amyloid fibers enhances fluorescence. However, its binding to early aggregates is weak, so this dye cannot be used for the characterization of early oligomers [22]. Thus, new probes, such as specific labeling antibodies [23] and especially fluorescent reporters, are continually being developed for the specific and sensitive detection of the early species of aggregation, as well as advanced techniques that allow the visualization and characterization of such aggregates [22,24,25]. For instance, we have recently described a new family of highly solvatochromic and fluorogenic dyes based on the quinolimide scaffold [26], named 9-amino-quinolimides, and these dyes have shown great potential for probing β-amyloid aggregation in a zebrafish model, identifying the formation of different types of early oligomers [27]. Due to its noninvasive nature and high sensitivity, this technique allows intracellular processes to be studied. Among confocal fluorescence microscopy techniques, fluorescence lifetime imaging microscopy (FLIM) has been indicated to be an invaluable tool to characterize amyloid aggregates [28–30], provided its multidimensional capabilities, in terms of emission intensity, fluorescence lifetime, τ, and intraoligomer fluorescence resonance energy transfer (FRET) efficiency, in particular through a powerful microscopy variant called dual-color pulsed interleaved excitation FLIM (PIE-FLIM).

Once specific tools to decipher different types of amyloidogenic aggregates are in place, it is also important to be capable of studying the direct mechanisms involved in cellular toxicity. For instance, in a previous work, an ultracentrifugation gradient method has been described to separate between different types of Aβ aggregates according to their size and density [31]. Using this methodology, in another work, the toxicity of the corresponding aggregates of Aβ42 was monitored by using two methods capable of measuring the cytotoxic effect of aggregates: a quantitative assay to determine the ability of peptide aggregates to disorganize the structure of lipid bilayers and an assay that measures the production of cytokines from microglial cells as an estimation of the inflammation capacity of the different aggregates. With these methods, it was demonstrated that the larger the size of the aggregates is, the lower their capacity to disorganize lipid bilayers. However, a relatively increased ability to cause inflammation was observed as the size increases. These facts are correlated with structural changes in the different aggregates with different sizes and structures [8]. The effect of the aggregates of Aβ40 and Aβ42 peptides on neuronal cells as a function of their dose has also been studied [32], showing an increase in ROS generated by neurons as the concentration of added peptide rises.

Herein, we provide additional tools to characterize and understand the cellular effect of Aβ42 amyloid aggregates. We have used the Aβ42 peptide in our assays since it is believed to be one of the most important alloforms due to the significant differences in toxicity, aggregation mechanism, and physiological function that the Aβ42 peptide has with respect to other isoforms [33,34]. Aβ42 has a greater tendency to present more prefibrillary oligomeric states than Aβ40 [35]. We have studied the interaction process of the Aβ42 peptide with Neuro-2a (N2a) neuronal cells (a fast-growing mouse neuroblastoma cell line) by PIE-FLIM to distinguish different subpopulations of aggregates interacting with cells and causing lysis. Furthermore, we have also explored the cellular toxicity in real-time and at the single-cell level by monitoring the entire interaction process of the peptide with neurons over time and by evaluating cellular stress through the generation rate of biothiols using a fluorogenic biothiol probe.

2. Results

2.1. Interaction between Aβ42 Aggregates and N2a Cells

In previous works, we used PIE-FLIM imaging to directly detect and characterize amyloidogenic aggregates at a molecular level. Using this technique, it has been possible to unravel structural changes over time in amyloid aggregates, thereby characterizing different species of oligomers [28–30]. For instance, we unequivocally detected three different kinds of aggregates in the early stages of fibrillization of an SH3 domain [30], a validated model employed to study amyloid aggregation.

Herein, we use this technique for the first time to visualize such aggregates interacting with single cells, thus obtaining a more complete view of the process. In particular, we have focused our attention on the interaction between Aβ42 and N2a cells to analyze the effect that early aggregates and amyloid structures induce on cells and what possible interactions may exist. PIE-FLIM provides information on intraoligomer FRET efficiency as a measure of aggregate compactness. For imaging Aβ42 aggregates using PIE-FLIM, the peptide monomers must be fluorescently labeled with either a suitable energy donor or acceptor fluorophore. We employed commercial Aβ42 peptides labeled with the HiLyte Fluor 488 or HiLyte Fluor 647 (Aβ-488 and Aβ-647, respectively) fluorophore and incubated equimolar mixtures of Aβ-488 and Aβ-647 for 15 h at a concentration of 0.5 μM in SSPE buffer (sodium chloride 150 mM, sodium phosphate 10 mM, and ethylenediaminetetraacetic acid, EDTA, 1 mM), pH 7.4, and with agitation at 260 rpm. Different aliquots of the incubating mixture were taken and added to the buffer containing the N2a cells, for a final peptide concentration of 50 nM. Cells were imaged using PIE-FLIM every 10 min to follow the evolution of the interaction of Aβ42 aggregates with cells.

The combination of the PIE excitation scheme, by which one of the lasers directly excites the donor fluorophore and the second laser, temporarily delayed, excites the acceptor fluorophore, with the collection of fluorescence emission in two detection channels (one for the donor and one for the acceptor) allows for specific time gates to be set to discriminate photons and reconstruct three different FLIM images, i.e., donor, FRET, and directly excited acceptor (see Figure 1 and the Section 4 for a complete description of the instrumentation). Pixels that colocalize simultaneously in the three images are those unequivocally selected as aggregates (Figure 1). The FRET efficiency and therefore the degree of compactness of the aggregates are related to the decrease in the fluorescence lifetime of the donor fluorophore, τ_D.

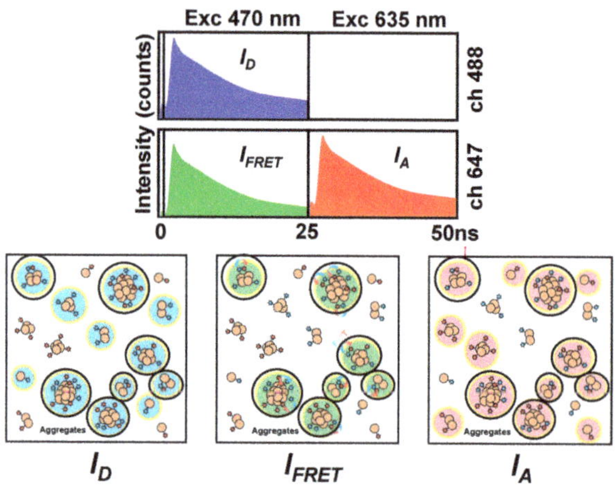

Figure 1. Dual-color pulsed interleaved excitation–fluorescence lifetime imaging microscopy (PIE-FLIM) scheme and aggregate selection criteria. The three different time gates distributed in the two detection channels define the reconstructed I_D, I_{FRET}, and I_A images. The three images are then analyzed to identify fluorescent events. Only coincident pixels in all three images are selected as aggregates (indicated with dark circles in the cartoon).

Figure 2 shows representative examples of the strong interaction of the incubated Aβ42 peptides with undifferentiated neuronal cells. As a common feature, we found that aggregates of Aβ42 surrounded cell membranes, interacting with them after 20–30 min of incubation. After 30 min of interaction, the number of Aβ42 aggregates surrounding the cell membrane increased, even inducing cell permeability, and entering the cell cytoplasm. Usually, after 60 min of interaction, N2a cells that had aggregates on the membrane were completely lysed. This behavior is particularly noticeable in Figure 2a, in which all cells appeared to be lysed after 60 min of interaction.

Interestingly, Figure 2b shows a cell in which the aforementioned effect is clearly detected. This is the accumulation of Aβ42 aggregates on the membrane, causing total lysis at the end of the studied process. However, in this experiment, it is evident that Aβ42 aggregates concentrate on one cell and leave a second cell intact. Aggregates surround the membrane and increase the size of the initial deposits within the cell over time. Surprisingly, the second cell remains intact and free from interacting aggregates, even after 60 min. The fact that aggregates grow in size and cooperatively seem to attract more aggregates within the cellular membrane holds important implications respect to cellular toxicity, and the effect of aggregated nuclei seems to act as seeds directly on the cell membrane. Figure 2c also shows interesting results, as aggregates appeared and interacted with the cell membrane of one cell and then progressively increase the size of the deposits formed to form considerable accumulations, however, these aggregates did not attack a second, nearby cell. In this case, even the cell with aggregates stuck on the membrane maintained structural integrity. This is a very important result that was consistently found across all of our repetitions. In fact, given the low amounts of peptide used (50 nM), we frequently found cells that did not show interacting aggregates. These cells remained unharmed within the time frame of the experiments.

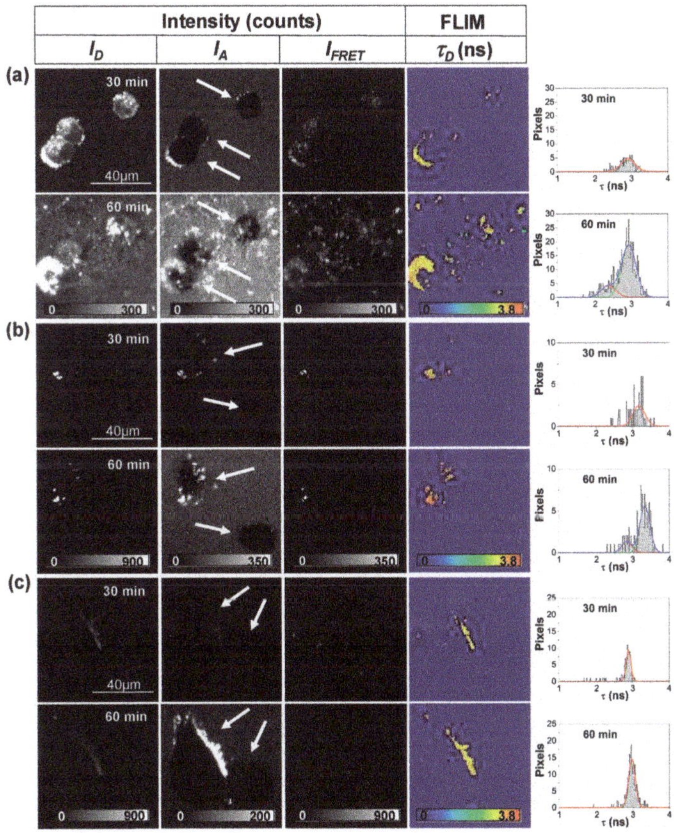

Figure 2. Fluorescence intensity images in the three detection windows (I_D, I_{FRET}, and I_A) and FLIM images of the donor fluorophore, τ_D, of equimolar mixtures of labeled Aβ-488 and Aβ-647 peptides (incubated for 15 h) interacting with Neuro-2a (N2a) cells for 30 and 60 min. Histograms in the rightmost column represent τ_D distributions for only pixels selected as aggregates. The distributions were fitted to one or two (if needed) Gaussian functions (lines). Panels (**a**–**c**) are representative images from different, repeated experiments. The arrows indicate different detected cells for identification purposes.

Importantly, the explanation for this differential behavior can be rationalized using the τ_D FLIM images and the corresponding frequency histograms. As one can see in Figure 2a,b, for the lysed cells, an earlier (first) population of aggregates surrounds cell membranes, but at the end of the process, a later (second) population of aggregates, which are characterized by a lower τ_D and hence are structurally more compact, is detected. The increasing emission intensity in the I_{FRET} images and the decreased τ_D values obtained over time indicate a larger FRET efficiency and an increased compactness of the aggregates, as previously shown in single-molecule FRET studies [13,30]. Strikingly, this second population was not detected in the cell that remained intact, as shown in Figure 2c. This suggests that such compact aggregates may play a crucial role in inducing cellular lysis, contributing to different toxicity mechanisms, specifically in the induction of membrane permeability and inflammation. This is in agreement with previous studies that show the high capability of small soluble aggregates to permeabilize the membrane, which is due to their high tendency to interact and cross the membrane, unlike what happens with larger aggregates [8]. Additional images showing more time points (each 10 min) for the experiments in Figure 2 can be found in Figures S1–S3 and an additional example of a repeated experiment is depicted in Figure S4 in the Supplementary Materials (SM).

2.2. Study of Cellular Stress Using a Biothiol Probe

In the previous experiments we confirmed how Aβ42 peptide aggregates are able to interact with neuronal cells and induce their lysis. It is thus reasonable to relate this interaction to an increase in cellular stress. Increased levels of biothiols in the cell matrix is a common physiological response as a defense mechanism associated with cellular stress [36]. Therefore, measuring the amount of intracellular biothiols generated upon interacting with pre-amyloid aggregates results in a striking way to correlate the physiological effect of cytotoxic oligomers. Generated ROS in an ensemble population of cultured cells upon addition of amyloid aggregates has commonly been reported as a measure of cellular stress and cytotoxicity [13,37,38]. However, herein, we are more interested in measuring this effect at the single-cell level.

Recently, our research group developed a fluorogenic probe to measure oxidative stress by monitoring changes in biothiol levels [38]. This probe chemically combines a highly electron-deficient 2,4-dinitrophenylsulfonyl (DNBS) group, acting as an electron sink in a photoinduced electronic transfer (PET), with a fluorescent xanthene derivative, 9-[1-(4-tert-butyl-2-methoxyphenyl)]-6-hydroxy-3H-xanthen-3-one (Granada Green, GG). The GG-DNBS probe is nonfluorescent due to PET-induced deactivation. Desulfonation of the DNBS group in the presence of thiols releases the fluorophore, resulting in an increased fluorescence emission. The GG-DNBS probe has already been successfully validated in 661W photoreceptor-derived cells [38].

In order to quantify the cellular stress caused by Aβ42 aggregates interacting with N2a cells at the single-cell level, we carried out imaging experiments using the fluorogenic biothiol sensor GG-DNBS, and determined the speed at which biothiols are generated by neuronal cells when Aβ42 aggregates are added. For these experiments, we collected aliquots of the incubating samples of Aβ-647 at different time points of incubation (0, 0.5, 1, 3, 24, and 48 h) to investigate the effect of a wide range of peptide aggregation stages. The selected incubation times are important, as evidenced by transmission electron microscopy (TEM) and dynamic light scattering (DLS) measurements (Figure 3). As the incubation proceeded, curly protofibrillar structures were detected early, between 0.5 and 3 h of incubation, in TEM images, whereas longer, straight fibrils were found after 24 and 48 h of incubation (Figure 3a). The nonincubated peptides (0 h of incubation) exhibited certain degree of amorphous aggregation (Figure 3b), usually attributed to hydrophobically-driven contacts and micellar structures [39,40]. However, as shown in Figure 3c, the size of the aggregates increases considerably after 0.3 h of the aggregation process.

Once the aliquots were collected, we added simultaneously the Aβ-647 aggregates and the GG-DNBS biothiol probe to living N2a cells. Herein, we also employed a dual-channel PIE microscopy setting; however, because the fluorescence emission of the GG-DNBS sensor is in the green spectral region, we used only the incubated red-labeled Aβ-647 peptide for the pre-amyloid aggregates. Accordingly, we used the two fluorescence intensity channels to simultaneously image the biothiol sensor response in the green channel and the Aβ-647 aggregates in the red channel (Figure 4). With this setting, we followed the kinetics of the biothiol-mediated fluorogenic reaction of GG-DNBS for 1 h by extracting the average fluorescence intensity in pixels of interest (corresponding with cell cytoplasm) as a function of time. Repetitions under same experimental conditions were carried out for each aggregation time of Aβ-647 and compared these results with those from cells in the basal state (control cells without Aβ-647 aggregates).

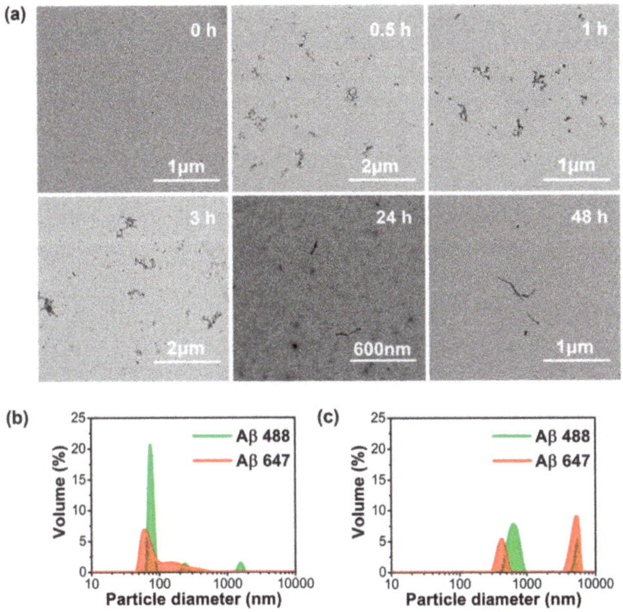

Figure 3. (**a**) Transmission electron microscopy (TEM) images of fluorescently labeled Aβ aggregates and fibers (equimolar 0.5 μM mixture of Aβ-488 and Aβ-647) at different incubation times. (**b**,**c**) dynamic light scattering (DLS) images of labeled Aβ peptides (Aβ-488: green line and Aβ-647: red line) formed at 0 h (**b**) and 0.3 h (**c**) of the incubation process, sonicated in a quartz cuvette in an NH_3 solution (pH 12) to avoid their aggregation (**b**) and in a phosphate buffered saline solution pH 7.4 after 20 min of aggregation (**c**).

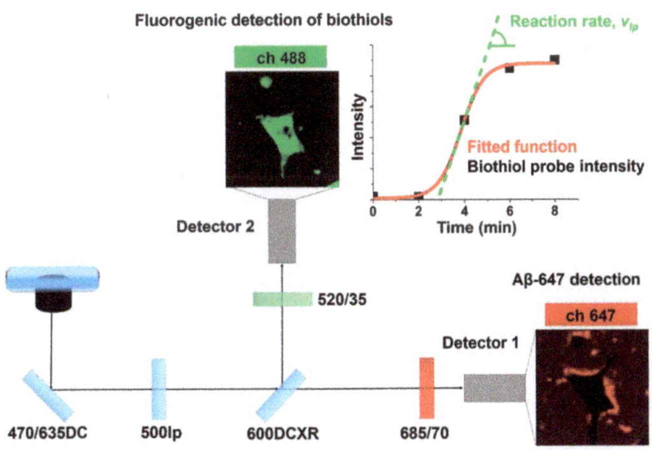

Figure 4. Scheme of the configuration of the dual-channel PIE-FLIM to simultaneously obtain Granada Green-2,4-dinitrophenylsulfonyl (GG-DNBS) biothiol probe and Aβ-647 aggregates images. For determining the rate of the fluorogenic biothiol reaction, the average pixel intensity in the regions of interest (black points) were fitted to a sigmoidal function (line red) and the slope at the inflection point (line green) was calculated.

An important novelty of the present work, compared to previous studies using fluorogenic biothiol sensors, is the method for data analysis and the parameter of interest. In previous works with biothiol sensors, we measured the total reached intensity and/or the area under the curve of the fluorogenic reaction as the parameter related to biothiol levels [38,41]. Herein, we go a step further and focus on the actual kinetics of the fluorogenic reaction by extracting the rate of the fluorescence increase. In order to compare the effect caused by different types of aggregates and the control cells, we selected the maximum rate of the fluorogenic reaction as the parameter of interest, as this is the rate in the inflection point of the intensity (I) vs. time plot (see Figure 4). To do this, we obtained the plots of average I from GG-DNBS in the selected pixels as a function of time (t), and fit these plots to a sigmoidal, dose–response function:

$$I = I_0 + \frac{I_{\max} - I_0}{1 + 10^{p(t_{ip} - t)}} \tag{1}$$

where $I_{\max}$ and I_0 represent the final and initial intensity values, respectively; t_{ip} is the time of the inflection point; and p is related to the geometry of the sigmoidal shape.

By deriving Equation (1), we obtained the speed of the fluorogenic reaction (v) at each time point:

$$v = \left(\frac{dI}{dt}\right) = \frac{\ln(10) \cdot (I_{\max} - I_0) \cdot p \cdot 10^{p(t_{ip} - t)}}{\left[1 + 10^{p(t_{ip} - t)}\right]^2} \tag{2}$$

which can be evaluated at $t = t_{ip}$ to yield the rate at the inflection point (v_{ip}):

$$v_{ip} - \left(\frac{dI}{dt}\right)_{t_{ip}} - \frac{\ln(10) \cdot (I_{\max} - I_0) \cdot p}{4} \tag{3}$$

The rate of the fluorogenic reaction at the inflection point (v_{ip}, see Figure 4) is a robust parameter that can be directly compared between different experiments and can be related to the levels of cellular biothiols. Representative examples of these experiments are shown in Figure 5. In particular, the fluorogenic biothiol reaction in a control cell in the absence of aggregates (Figure 5a) and in a cell with the addition of Aβ-647 aggregates that was incubated for 24 and 48 h (Figure 5b,c, respectively). These figures show how the probe diffuses across membranes into the cytosol and generates fluorescence when reacting with biothiols, undergoing a clear increase in the fluorescence emission as time passes. In the experiments with added Aβ-647 aggregates, we can clearly visualize how the aggregates surround the cell membrane in just a few minutes. As time proceeds, the size of these interacting aggregates considerably increased, which is consistent with the results described in the previous section. More repetitions for aggregates incubated during different times have been carried out, and representative examples are included in the Supplementary Materials (Figure S5).

As mentioned above, between three and seven repetitions for all the measurements were performed using different cell cultures and different aggregate preparations to ensure the reproducibility of the results. Figure 6 shows the average values of the v_{ip} rate as a measure of the biothiol levels upon the interaction of N2a cells with Aβ-647 aggregates and the corresponding controls. Consistently, a greater rate than the corresponding controls was found when the aggregates incubated for 0.5, 1, 3, and 48 h were added. Interestingly, nonincubated aggregates (0 h) did not cause an increase in the rate of the fluorogenic biothiol reaction, even though the interaction with the cell membrane was detected (Figure S5). These results support higher levels of biothiols in cells interacting with Aβ-647 aggregates formed after 30 min of incubation, which may be correlated with enhanced cellular stress and, hence, the notion that most cytotoxic species are formed during the early stages of aggregation.

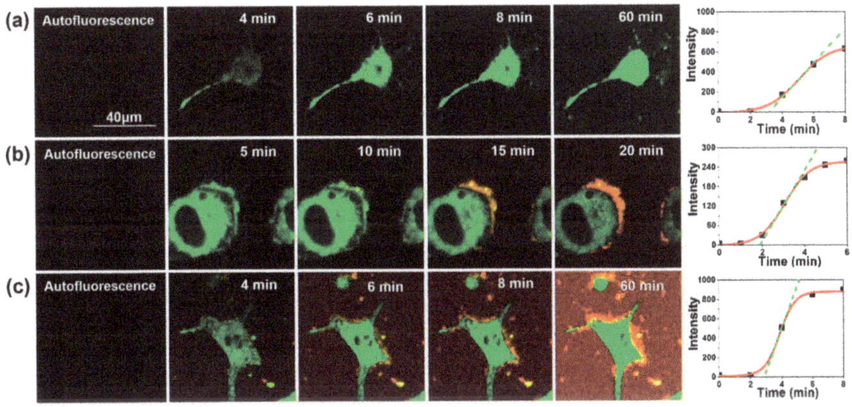

Figure 5. Representative images of cellular stress assays with the biothiol probe (green channel) at different times without control, (**a**) or with the addition of Aβ-647 aggregates (red channel) previously incubated for 24 h (**b**) and 48 h (**c**). The plots in the rightmost column represent the corresponding average intensity per pixel value of the biothiol probe as a function of time along with the fitted sigmoidal function (red curve) and the slope at the inflection point (green, dashed line).

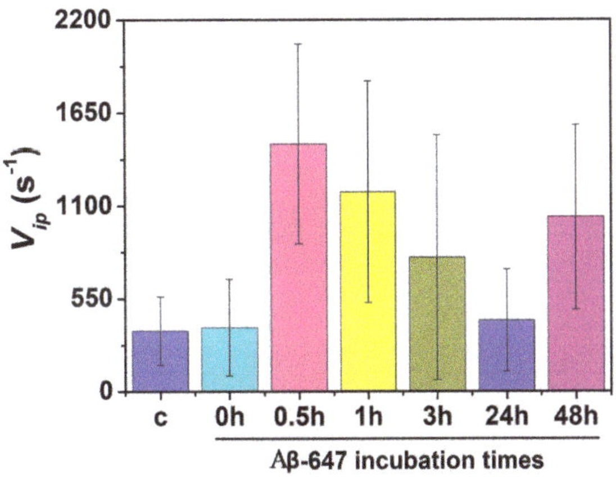

Figure 6. Biothiols sensing in cells interacting with Aβ-647 aggregates formed at different incubation times. The bar chart represents the average rate of the fluorogenic biothiol sensing reaction at the inflection point (v_{ip}) for the controls (labeled as c) and for the cells interacting Aβ-647 aggregates formed at different incubation times (0, 0.5, 1, 3, 24, and 48 h of incubation). Error bars represent standard error of the mean (s.e.m.).

To further understand the effect that the amyloid peptide can exert on neuronal cells, we also performed another assay following a different scheme, in which N2a cells were interacting with Aβ-647 aggregates for 20 min and then generated biothiols were detected by the fluorogenic probe. This experiment shows that during the time that the peptide aggregates were interacting with cells, a considerable accumulation of the peptide were formed at certain points of the membrane in accordance with the other results seen in this study. These results can be seen in the Supplementary Materials (Figure S6).

To correlate the aforementioned results with actual cytotoxicity caused by Aβ-647 aggregates to N2a cells, we carried out cell viability assays using the commercial Cell Titer-Blue reagent (see Section 4 for details). We added aliquots of Aβ-647 aggregates incubated for different times (0, 0.5, 1, 3,

24, and 48 h) to cultured N2a cells, and let them interact for 76 h. Then, the Cell Titer-Blue reagent was added, and the fluorescence emission was compared between the experimental and control cells (with and without Aβ-647 aggregates added, respectively). In all cases, the addition of Aβ-647 caused a decrease in cell viability, between 11.6 and 41.3% (Figure 7). Importantly, we performed a statistical study for significant differences in toxicity using the Holm–Bonferroni test and found that the viability of the cells interacting with aggregates incubated for 1 h was significantly lower than that of the cells interacting with aggregates incubated for 0.5, 3, and 24 h, with a 95% confidence. These results suggest that Aβ-647 aggregates are moderately toxic to N2a cells. These values are in good agreement with previous toxicity studies of the Aβ peptide interacting with N2a cells, which reported viability values of almost 50% at a much higher peptide concentration (10 μM) than that used in our assay [42]. Regarding toxicity with other cell lines, it has also been shown Aβ42 aggregates at 10 μM that were previously incubated for 24 h at 37 °C, provoked a reduction in the cell viability of SH-SY5Y neuroblastoma cells by almost 20% [40], but the effect was even larger in primary neurons from E16 rat embryos in which Aβ42 aggregates at 20 μM caused a 65% reduction in cell viability [43].

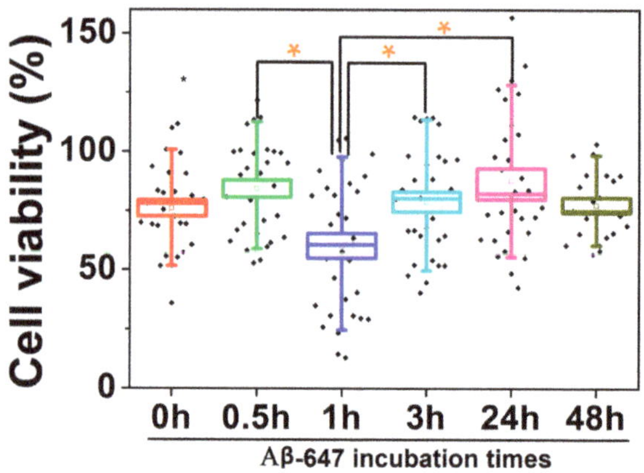

Figure 7. Box and whisker plot representing N2a cell viability upon the addition of Aβ-647 aggregates incubated for different times, expressed as the percentage of cell viability when compared with that in the control cells in the absence of aggregates. Boxes represent the standard error of the mean, and whiskers represent 90% of the results. Orange asterisks indicate populations that are significantly different from each other, with a 95% confidence interval.

In order to obtain additional information regarding the induction effect of different formed Aβ-647 aggregates on N2a neuronal cells, we took advantage of the fact that GG-DNBS is a dual sensor that is capable of not only simultaneously measuring levels of biothiols generated inside the cell but also detecting changes in the global levels of phosphate anions through the analysis of the fluorescence lifetime of the released fluorophore from the fluorogenic reaction, τ_{probe} [41,44]. The released fluorescent fragment undergoes changes in its fluorescence lifetime, influenced by the presence of phosphate anions, which promote a proton transfer reaction in the excited state of the fluorophore. Hence, we carried out an analysis of the FLIM images obtained in the previous experiments to determine appreciable changes in the intracellular phosphate levels upon interaction with Aβ-647 aggregates. Figure 8 shows FLIM images of the probe over time within the control cells (Figure 8a) and the cells treated with Aβ-647 peptide aggregates and incubated for 24 h (Figure 8b,c) and 48 h (Figure 8d), thus providing a global view of the phosphate levels present inside the cell. In the latter case, we can see that changes in the lifetime of the probe occur towards the end of the study period, decreasing its value. This finding suggests an increase in the global concentrations of phosphate ions caused by the incorporation of phosphate-enriched extracellular medium through permeabilized membranes.

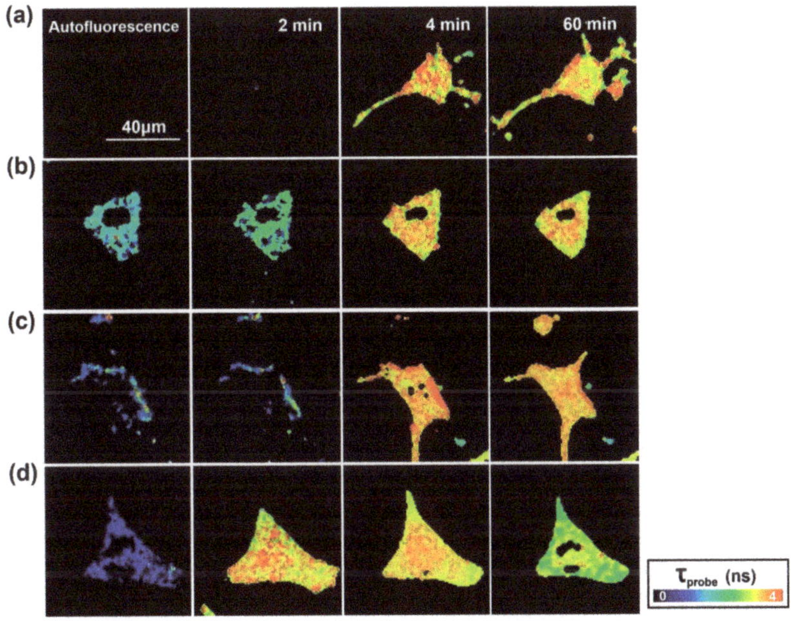

Figure 8. FLIM images of the released fluorophore after the fluorogenic reaction of GG-DNBS with biothiols in N2a cells without (**a**) and with the addition of Aβ42 aggregates, formed after incubating for 24 h (**b**) and 48 h (**c**,**d**). The pseudocolor scale represents the fluorescence lifetime of the probe, τ_{probe}, between 0 and 4 ns.

3. Discussion

In this work, we report powerful microscopy tools to understand the interaction of Aβ42 aggregates with N2a neuronal cells at the single-cell level. The direct interaction of Aβ42 oligomers with the cell membrane, fostering cellular stress, lysis and death, was probed with a variety of methods, with a particular focus on dual-color PIE-FLIM due to its capability to provide multidimensional, rich images.

We first focused on the intra-oligomer FRET as a mean to identify different types of aggregates [30] and directly imaged their interaction with N2a cells using PIE-FLIM. For these experiments, we employed an equimolar mixture of labeled peptides, Aβ-488 and Aβ-647, incubated for 15 h. After this incubation period, it is expected that a large number of heterogeneous oligomers with different aggregation capacities and toxicities and nontoxic monomers coexist in equilibrium. Our PIE-FLIM results provide evidence of several key factors in the interaction of Aβ42 amyloid aggregates with N2a cells: (i) Aβ42 aggregates interact with the cellular membrane in just a few minutes. (ii) The interaction sites can act as effective seeds for the continuous recruitment of more aggregates. Hence, in-membrane aggregate growth is more likely to occur than a new interaction with a different cell. (iii) The former behavior results in some cells interacting with Aβ42 aggregates and others remaining intact. (iv) This interaction results in cell lysis after some period of interaction as well as aggregate seeding and growth, which involves certain conformational changes, as demonstrated by the appearance of the high-FRET aggregate population. These results provide a clear and unprecedented depiction of the cellular effect of amyloid aggregates, especially demonstrating that seeding takes place within the cell membrane. This is an important conclusion, which was consistently supported by repeated experiments, and that explains how Aβ42 aggregates can be toxic to cells, even at low nanomolar concentrations, since the local concentration within the cell membrane is cooperatively increased.

These results can be rationalized by previous studies that show a similar behavior for Aβ42 aggregates on the neuronal tissue of transgenic models of Drosophila melanogaster. Cellular accumulations of Aβ42 peptide in the form of diffuse nonamyloid plaques are correlated to neurodegeneration and premature death of the animal [45,46]. Likewise, in another work, in which a transgenic mouse model that overexpresses the human mutated amyloid precursor protein was used, clear neurodegeneration caused by the peptide was observed. Interestingly, neurodegenerative activity was found to be related to intraneuronal deposits of the peptide but not to the extracellular accumulation of Aβ42 [47]. The fact that a heterogeneous interaction of Aβ42 aggregates with the membranes of some neuronal cells was observed in our results suggests that this interaction occurs selectively—only in places with a high affinity for the peptide, such as certain membrane lipids [48]. This affinity can alter the properties of the membrane and interfere with its fluidity, promoting fibrillogenesis [49]. It is also known that the formation of amyloid structures occurs in lipid rafts containing a ganglioside cluster [50]. On the other hand, the evident ability of Aβ42 to induce the permeability of some cell membranes in this work can be related to the propensity of the peptide to form Ca^{2+}-permeable channels [50]. It has been reported that some protofibers formed by Aβ behave as pore-forming structures that can alter cell activity and cause cell death [51]. Our results also agree with previous reports that describe the cellular uptake of Aβ42 aggregates, which occurs with the prerequisite of the rapid binding of β-sheet-rich aggregates to the cell membrane [52]. Therefore, our results are consistent with the idea that the entry of aggregates into the cell interior could be a crucial step in its cytotoxicity. Thus, our work provides significant information regarding the process of interaction, permeabilization and cell lysis that Aβ42 aggregates exert on cells. As membrane permeability induced by Aβ42 aggregates is observable only in certain cases here, we suggest that there are different types of aggregates that generate different levels of cellular toxicity, in line with other studies that show how certain aggregates of Aβ42 induce greater cellular permeability [8]. Among these different aggregates, those that are more stable and that compact than the initial aggregates could be the most harmful to cells [13].

Importantly, a common limitation of previous biophysical studies in terms of their physiological significance is that the formation of Aβ peptide amyloid fibrils was performed at high in vitro concentrations, whereas physiologically relevant Aβ concentrations lie in the low nanomolar range. Very recently, Lyubchenko and colleagues studied the interaction of Aβ42 peptides with membrane bilayers using AFM and computational experiments, concluding that aggregation at low concentrations of the peptide is triggered by interaction with the membrane [53]. Our results clearly support this model; we added nanomolar concentrations of incubated aggregates to the cell buffer, and we clearly detected the interaction with the membrane and subsequent seeding and growth.

In the second set of experiments, we focused on the cellular stress caused by Aβ42 aggregates using a fluorogenic sensor for biothiols and a powerful dual-channel microscopy to narrow our study to the single-cell level. The production of biothiols is associated with a complex metabolic response elicited by cellular stress that depends biologically on each individual cell and on how these individual cells respond to an adverse factor. Importantly, the dual-channel microscopy method allows us to directly visualize how Aβ-647 aggregates interact with N2a cells while simultaneously measuring biothiol levels. This is important because it allows for the direct elucidation of whether aggregates are interacting with the studied cell, thereby providing a deeper understanding on the actual effect of this interaction. We also detailed our investigation into the search for differences in the effect caused by aggregates formed in different stages of the aggregation process. Our results show that the interaction of aggregates with the cellular membrane plays a role in the enhanced cellular stress leading to toxicity. We found that aggregates formed in the initial 0.5–1 h of incubation cause the most cellular stress (Figure 6), which correlates well with our cell viability and toxicity experiments (Figure 7) as well as with previous toxicity studies [40,42,43] suggesting that aggregates formed at early stages of the aggregation process are the most powerful for inducing membrane permeability [8]. However, importantly, during the very early stages, the large majority of aggregates are nonfibrillar, but a dynamic population usually returns to monomers [54]. This oligomer kinetics reinforce the idea of a dynamic and critical seeding process within the cell membrane to foster cellular stress and cell lysis. The key link between Aβ42 toxicity and cellular stress is still obscure. However, Butterfield and colleagues suggested a close relationship with lipid peroxidation that takes place in the cell membranes of neurons promoted by the peptide [55], which may support the catalytic role of aggregates upon binding to the cell membrane.

Taken together, our results offer a broad picture of the relationship between Aβ aggregates interacting with neuronal cells, cellular stress, and neuronal toxicity (Figure 9). Amyloid aggregates formed in the early stages are the most harmful to N2a cells, which occur via interaction between the membrane and aggregates of a suitable size and structure, followed by subsequent aggregate seeding and growth, resulting in induced permeability until cellular lysis occurs in some cases. In summary, the tools that we presented herein are fully validated and have general applicability to further explore the cellular mechanisms underlying neurodegenerative diseases.

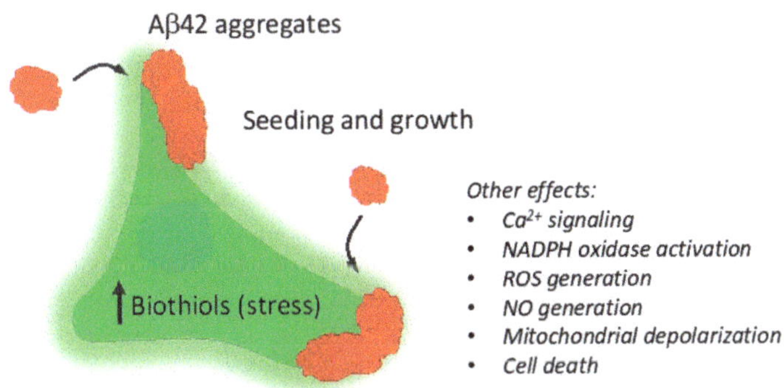

Figure 9. Main findings in this work. Aβ42 aggregates are bound to the cell membrane, acting as seeds for further growth. This membrane interaction is accompanied by increased cellular oxidative stress, finally leading to lysis and cell death.

4. Materials and Methods

4.1. Materials

Lyophilized Aβ42 peptides labeled with HiLyte 488 dye (Aβ-488) and HiLyte 647 dye (Aβ-647), were obtained from Anaspec Peptide (Seraing, Belgium). A total of 0.1 mg of each peptide was dissolved in NH_3 (1%) at a total concentration of 66 μM, sonicated in ice in an ultrasound bath for 30 min, distributed in aliquots and frozen immediately in liquid nitrogen to avoid aggregation. SSPE buffer (150 mM NaCl, 10 mM phosphate, and 1 mM EDTA) was acquired from Sigma-Aldrich (Madrid, Spain). For neuronal cells experiments, and Krebs buffer (130 mM NaCl, 2.5 mM KCl, 25 mM $NaHCO_3$, 1.2 mM NaH_2PO_4, 1.2 mM $MgCl_2$, and 2.5 mM $CaCl_2$, pH 7.2) was freshly prepared. To adjust the pH of the buffers, NaOH and HCl (both from Sigma-Aldrich) were used. All chemical compounds were used without any further purification. Cell viability assays were carried out using a CellTiter-Blue™ viability assay (Promega Biotech, Madrid, Spain).

The N2a (ATCC® CCL-131™) cell line used in these experiments is a cell line that comes from Mus musculus brain neuroblasts and presents neuronal stem cell and ameboid morphology. N2a cells were grown at 37 °C in Dulbecco's modified Eagle's medium (DMEM) from Sigma-Aldrich supplemented with 10% fetal bovine serum (FBS), 2 mM glutamine and 100 U/mL penicillin and 0.1 mg/mL streptomycin.

For the preparation of the biothiol probe, appropriate amounts of the powdered GG-DNBS (2,4-dinitrobenzenesulfinate derivative of 9-[1-(4-tert-butyl-2-methoxyphenyl)]-6-hydroxy-3H-xanthen-3-one) were dissolved in dimethyl sulfoxide (Sigma-Aldrich) to prepare a 0.36 mM stock solution. For imaging experiments, the fluorogenic probe was diluted in the buffer containing the cells, down to a final concentration of 0.25 μM.

4.2. Instruments

Dual-color PIE-FLIM experiments were performed on a MicroTime 200 system (PicoQuant GmbH, Berlin, Germany), as previously described [30]. As excitation sources, we used two pulsed laser diode heads (PicoQuant) at λ_{ex} = 470 and 635 nm, with a repetition rate of 20 MHz, and alternating in the nanosecond time regime for achieving PIE excitation. The excitation light was focused into the sample throughout a 100×, 1.4 numerical aperture (NA) oil immersion objective, which collected the fluorescence emission and focused it to the 75-μm confocal aperture. Fluorescence was then separated into two detection channels through a 600DCXR dichroic mirror, focusing the light on two single photon avalanche detectors (Perkin Elmer, Hopkinton, MA, USA). A 520/35 or 685/70 bandpass filter was employed to define the blue or red channel, respectively. A TimeHarp 200 time-correlated single-photon counting (TCSPC) module (PicoQuant, Berlin, Germany) was used for photon counting, data acquisition and imaging reconstruction.

We obtained raw images at a resolution of 0.26 μm/pixel and a temporal resolution of 232 ps/channel at the micro-time scale. Analysis of the images (separation and reconstruction of the different images and FLIM imaging) was performed in the SymPhoTime 32 software (PicoQuant). For FLIM imaging of the donor fluorophore, we fit the data to a single exponential decay function in the fluorescence decay traces obtained in each pixel after a 5 × 5 spatial binning using the maximum likelihood estimator (MLE) for parameter optimization.

We used in-house coded scripts in FiJi (distribution of ImageJ) [56] for the selection of colocalized pixels in the three images (I_D, I_{FRET}, and I_A images) in the intra-oligomeric FRET experiments to selectively pick pixels corresponding to aggregates and scripts to quantify the average intensity per pixel in the regions of interest of the fluorogenic biothiol probe.

For the cell viability assays, an FP-8500 spectrofluorometer (Jasco), equipped with a microplate reader, was used to record the fluorescence emission spectra (λ_{ex} = 550 nm) of the CellTiter-Blue reagent in each well.

TEM images were obtained using a Libra 120 Plus TEM microscope (Carl Zeiss SMT, Oberkochen, Germany). It was operated at 120 kV and equipped with a LaB6 filament and an SSCCD 2 k × 2 k direct coupling camera. TEM images of samples incubated at different times were collected by adding aliquots on Formvar 300-mesh copper grids, washed twice with Milli-Q water and stained with uranyl acetate 1% (*w/v*).

DLS measurements were carried out on a Malvern Zetasizer μV, equipped with an 850 nm laser to avoid fluorescence interference with the 647-labeled peptides. DLS traces of the aggregates of Aβ-647 and Aβ-488 formed in NH_3, pH 12, and in SSPE buffer, pH 7.4, after 20 min of incubation were collected using a 2 μL quartz cuvette.

4.3. Aβ Amyloid Aggregation

Aβ42 aggregates containing equimolar amounts of Aβ-488 and Aβ-647 used in the FRET assay were prepared at a total concentration of 0.5 μM for each peptide in SSPE buffer, and incubated for 15 h at a physiological temperature (37 °C) with continuous agitation (360 rpm). Aβ-647 aggregates, used in the experiments with the fluorogenic biothiol probe, were described as above at a total concentration of 1 μM. Aliquots were collected at different incubation times (0, 0.5, 1, 3, 24, and 48 h) and immediate frozen in liquid nitrogen until use.

4.4. Cell Viability Assays

CellTiter-Blue® (Promega) viability assay, based on the resazurin → resorufin fluorogenic reaction, was performed to study the cytotoxicity of Aβ42 aggregates. Cell viability of the samples treated with Aβ42 aggregates was evaluated by the comparison with untreated controls, for which a cell viability of

100% is assumed. Cells were grown, up to a population density of 10^3 cells/well, in black 96-well plates with 100 μL of DMEM plus 10% FBS per well. The medium was removed after 24 h of cell culture at 37 °C. Then, 100 μL aliquots of Aβ-647 incubated for different times were added to cells for 76 h. Finally, 20 μL (20% *v/v*) of CellTiter-Blue reagent was added to each well and incubated for 20 min at 37 °C before fluorescence emission was measured.

Supplementary Materials: The following are available online at http://www.mdpi.com/1422-0067/21/14/5035/s1. Figures S1–S4. Supplementary figures of the interaction of equimolar mixtures of labeled Aβ-488 and Aβ-647 peptides (incubated for 15 h) with N2a cells. Figure S5. Additional examples of cell stress assays with the biothiol probe (green channel), upon the addition of previously incubated Aβ-647 aggregates (red channel). Figure S6. Second strategy for the assay of cellular stress induced by Aβ-647 aggregates.

Author Contributions: Conceptualization, J.A.G.Z-V. and A.O.; Formal analysis, Á.R.-A., J.M.P., C.D.B., J.A.G.-V., and A.O.; Funding acquisition, M.D.G. and A.O.; Investigation, Á.R.-A., J.M.P., C.D.B., M.D.G., and R.S.; Methodology, J.M.P., J.M.C., R.S., J.A.G.-V., and A.O.; Project administration, A.O.; Resources, J.M.C., M.D.G., and R.S.; Supervision, J.M.C. and J.A.G.-V.; Validation, M.D.G.; Visualization, Á.R.-A., J.M.P., and A.O.; Writing—Original draft, Á.R.-A., and A.O.; Writing—Review and editing, Á.R.-A., J.M.P., C.D.B., J.M.C., M.D.G., R.S., J.A.G.-V., and A.O. All authors have read and agreed to the published version of the manuscript.

Funding: This work was funded by the grants CTQ2014-56370-R, CTQ2017-86568-R, and CTQ2017-86125-P from the Spanish Ministerio de Ciencia, Innovación y Universidades, the Spanish Agencia Estatal de Investigación, and the European Regional Development Fund (ERDF).

Acknowledgments: We thank the Centro de Instrumentación Científica (CIC) of the Universidad de Granada for use of the TEM facilities and Jose Mata del Pino for helping with the DLS measurements. Á.R.-A. thanks the Spanish Ministerio de Educación y Formación Profesional for the FPU Ph.D. scholarship.

Conflicts of Interest: The authors declare that they have no conflicts of interest.

Abbreviations

FLIM	Fluorescence lifetime imaging microscopy
FRET	Fluorescence resonance energy transfer
PIE	Pulsed interleaved excitation
Aβ42	β-Amyloid (1-42) peptide
Aβ-488	β-Amyloid (1-42) peptide labeled with HiLyte Fluor 488
Aβ-647	β-Amyloid (1-42) peptide labeled with HiLyte Fluor 647
GG-DNBS	Granada Green dinitrobenzene sulfonate
TEM	Transmission electron microscopy
DLS	Dynamic light scattering
TCSPC	Time-correlated single-photon counting

References

1. Temussi, P.A.; Masino, L.; Pastore, A. From Alzheimer to Huntington: Why is a structural understanding so difficult? *EMBO J.* **2003**, *22*, 355–361. [CrossRef]
2. Relja, M. Pathophysiology and Classification of Neurodegenerative Diseases. *EJIFCC* **2004**, *15*, 97–99. [PubMed]
3. Giannakopoulos, P.; Silhol, S.; Jallageas, V.; Mallet, J.; Bons, N.; Bouras, C.; Delaere, P. Quantitative analysis of tau protein-immunoreactive accumulations and beta amyloid protein deposits in the cerebral cortex of the mouse lemur, *Microcebus murinus*. *Acta Neuropathol.* **1997**, *94*, 131–139. [CrossRef] [PubMed]
4. Naslund, J.; Haroutunian, V.; Mohs, R.; Davis, K.L.; Davies, P.; Greengard, P.; Buxbaum, J.D. Correlation between elevated levels of amyloid beta-peptide in the brain and cognitive decline. *JAMA* **2000**, *283*, 1571–1577. [CrossRef]
5. Bucciantini, M.; Giannoni, E.; Chiti, F.; Baroni, F.; Formigli, L.; Zurdo, J.; Taddei, N.; Ramponi, G.; Dobson, C.M.; Stefani, M. Inherent toxicity of aggregates implies a common mechanism for protein misfolding diseases. *Nature* **2002**, *416*, 507–511. [CrossRef] [PubMed]

6. Shankar, G.M.; Li, S.; Mehta, T.H.; Garcia-Munoz, A.; Shepardson, N.E.; Smith, I.; Brett, F.M.; Farrell, M.A.; Rowan, M.J.; Lemere, C.A.; et al. Amyloid-β protein dimers isolated directly from Alzheimer's brains impair synaptic plasticity and memory. *Nat. Med.* **2008**, *14*, 837–842. [CrossRef]
7. Chiti, F.; Dobson, C.M. Protein Misfolding, Functional Amyloid, and Human Disease. *Annu. Rev. Biochem.* **2006**, *75*, 333–366. [CrossRef] [PubMed]
8. De, S.; Wirthensohn, D.C.; Flagmeier, P.; Hughes, C.; Aprile, F.A.; Ruggeri, F.S.; Whiten, D.R.; Emin, D.; Xia, Z.; Varela, J.A.; et al. Different soluble aggregates of Aβ42 can give rise to cellular toxicity through different mechanisms. *Nat. Commun.* **2019**, *10*, 1541. [CrossRef]
9. Bitan, G.; Kirkitadze, M.D.; Lomakin, A.; Vollers, S.S.; Benedek, G.B.; Teplow, D.B. Amyloid β-protein (Aβ) assembly: Aβ40 and Aβ42 oligomerize through distinct pathways. *Proc. Natl. Acad. Sci. USA* **2003**, *100*, 330–335. [CrossRef]
10. Lv, Z.; Roychaudhuri, R.; Condron, M.M.; Teplow, D.B.; Lyubchenko, Y.L. Mechanism of amyloid β−protein dimerization determined using single−molecule AFM force spectroscopy. *Sci. Rep.* **2013**, *3*, 2880. [CrossRef]
11. Bucciantini, M.; Calloni, G.; Chiti, F.; Formigli, L.; Nosi, D.; Dobson, C.M.; Stefani, M. Prefibrillar amyloid protein aggregates share common features of cytotoxicity. *J. Biol. Chem.* **2004**, *279*, 31374–31382. [CrossRef] [PubMed]
12. Bronfman, F.C.; Soto, C.; Tapia, L.; Tapia, V.; Inestrosa, N.C. Extracellular matrix regulates the amount of the β-amyloid precursor protein and its amyloidogenic fragments. *J. Cell. Physiol.* **1996**, *166*, 360–369. [CrossRef]
13. Cremades, N.; Cohen, S.I.A.; Deas, E.; Abramov, A.Y.; Chen, A.Y.; Orte, A.; Sandal, M.; Clarke, R.W.; Dunne, P.; Aprile, F.A.; et al. Direct Observation of the Interconversion of Normal and Toxic Forms of α-Synuclein. *Cell* **2012**, *149*, 1048–1059. [CrossRef] [PubMed]
14. Abramov, A.Y.; Potapova, E.V.; Dremin, V.V.; Dunaev, A.V. Interaction of Oxidative Stress and Misfolded Proteins in the Mechanism of Neurodegeneration. *Life* **2020**, *10*, 101. [CrossRef]
15. Stine, W.B., Jr.; LaDu, M.J. Apolipoprotein E: Structural and Functional Interactions with Amyloid β. *Amyloid Proteins* **2005**, 211–243. [CrossRef]
16. Relini, A.; Marano, N.; Gliozzi, A. Misfolding of amyloidogenic proteins and their interactions with membranes. *Biomolecules* **2013**, *4*, 20–55. [CrossRef]
17. Dyall, S.C. Amyloid-Beta Peptide, Oxidative Stress and Inflammation in Alzheimer's Disease: Potential Neuroprotective Effects of Omega-3 Polyunsaturated Fatty Acids. *Int. J. Alzheimer's Dis.* **2010**, *2010*, 274128. [CrossRef]
18. Di Scala, C.; Yahi, N.; Boutemeur, S.; Flores, A.; Rodriguez, L.; Chahinian, H.; Fantini, J. Common molecular mechanism of amyloid pore formation by Alzheimer's β-amyloid peptide and α-synuclein. *Sci. Rep.* **2016**, *6*, 28781. [CrossRef]
19. Kadowaki, H.; Nishitoh, H.; Urano, F.; Sadamitsu, C.; Matsuzawa, A.; Takeda, K.; Masutani, H.; Yodoi, J.; Urano, Y.; Nagano, T.; et al. Amyloid β induces neuronal cell death through ROS-mediated ASK1 activation. *Cell Death Differ.* **2005**, *12*, 19–24. [CrossRef]
20. Berezin, M.Y.; Achilefu, S. Fluorescence Lifetime Measurements and Biological Imaging. *Chem. Rev.* **2010**, *110*, 2641–2684. [CrossRef]
21. Ruedas-Rama, M.J.; Alvarez-Pez, J.M.; Crovetto, L.; Paredes, J.M.; Orte, A. *FLIM Strategies for Intracellular Sensing BT-Advanced Photon Counting: Applications, Methods, Instrumentation*; Kapusta, P., Wahl, M., Erdmann, R., Eds.; Springer International Publishing: Cham, Switzerland, 2015; pp. 191–223. ISBN 978-3-319-15636-1.
22. Aliyan, A.; Cook, N.P.; Martí, A.A. Interrogating Amyloid Aggregates using Fluorescent Probes. *Chem. Rev.* **2019**, *119*, 11819–11856. [CrossRef] [PubMed]
23. Aprile, F.A.; Sormanni, P.; Podpolny, M.; Chhangur, S.; Needham, L.-M.; Ruggeri, F.S.; Perni, M.; Limbocker, R.; Heller, G.T.; Sneideris, T.; et al. Rational design of a conformation-specific antibody for the quantification of Aβ oligomers. *Proc. Natl. Acad. Sci. USA* **2020**, *117*, 13509–13518. [CrossRef] [PubMed]
24. Jiang, B.; Aliyan, A.; Cook, N.P.; Augustine, A.; Bhak, G.; Maldonado, R.; Smith McWilliams, A.D.; Flores, E.M.; Mendez, N.; Shahnawaz, M.; et al. Monitoring the Formation of Amyloid Oligomers Using Photoluminescence Anisotropy. *J. Am. Chem. Soc.* **2019**, *141*, 15605–15610. [CrossRef] [PubMed]
25. Lv, G.; Sun, A.; Wang, M.; Wei, P.; Li, R.; Yi, T. A novel near-infrared fluorescent probe for detection of early-stage Aβ protofibrils in Alzheimer's disease. *Chem. Commun.* **2020**, *56*, 1625–1628. [CrossRef] [PubMed]

26. González-Vera, J.A.; Fueyo-González, F.; Alkorta, I.; Peyressatre, M.; Morris, M.C.; Herranz, R. Highly solvatochromic and tunable fluorophores based on a 4,5-quinolimide scaffold: Novel CDK5 probes. *Chem. Commun.* **2016**, *52*, 9652–9655. [CrossRef] [PubMed]
27. Fueyo-González, F.; González-Vera, J.A.; Alkorta, I.; Infantes, L.; Jimeno, M.L.; Aranda, P.; Acuña-Castroviejo, D.; Ruiz-Arias, A.; Orte, A.; Herranz, R. Environment-Sensitive Probes for Illuminating Amyloid Aggregation in vitro and in Zebrafish. *ACS Sens.* **2020**. [CrossRef]
28. Jurado, R.; Castello, F.; Bondia, P.; Casado, S.; Flors, C.; Cuesta, R.; Domínguez-Vera, J.M.; Orte, A.; Gálvez, N. Apoferritin fibers: A new template for 1D fluorescent hybrid nanostructures. *Nanoscale* **2016**, *8*, 9648–9656. [CrossRef]
29. Castello, F.; Casares, S.; Ruedas-Rama, M.J.; Orte, A. The First Step of Amyloidogenic Aggregation. *J. Phys. Chem. B* **2015**, *119*, 8260–8267. [CrossRef]
30. Castello, F.; Paredes, J.M.; Ruedas-rama, M.J.; Martin, M.; Roldan, M.; Casares, S.; Orte, A. Two-Step Amyloid Aggregation: Sequential Lag Phase Intermediates. *Sci. Rep.* **2017**, *7*. [CrossRef]
31. Ward, R.V.; Jennings, K.H.; Jepras, R.; Neville, W.; Owen, D.E.; Hawkins, J.; Christie, G.; Davis, J.B.; George, A.; Karran, E.H.; et al. Fractionation and characterization of oligomeric, protofibrillar and fibrillar forms of beta-amyloid peptide. *Biochem. J.* **2000**, *348 Pt. 1*, 137–144. [CrossRef]
32. Narayan, P.; Holmstrom, K.M.; Kim, D.-H.; Whitcomb, D.J.; Wilson, M.R.; St George-Hyslop, P.; Wood, N.W.; Dobson, C.M.; Cho, K.; Abramov, A.Y.; et al. Rare individual amyloid-β oligomers act on astrocytes to initiate neuronal damage. *Biochemistry* **2014**, *53*, 2442–2453. [CrossRef] [PubMed]
33. Pauwels, K.; Williams, T.L.; Morris, K.L.; Jonckheere, W.; Vandersteen, A.; Kelly, G.; Schymkowitz, J.; Rousseau, F.; Pastore, A.; Serpell, L.C.; et al. Structural basis for increased toxicity of pathological aβ42:aβ40 ratios in Alzheimer disease. *J. Biol. Chem.* **2012**, *287*, 5650–5660. [CrossRef]
34. Qiu, T.; Liu, Q.; Chen, Y. X.; Zhao, Y. F.; Li, Y. M. Aβ42 and Aβ40: Similarities and differences. *J. Pept. Sci.* **2015**, *21*, 522–529. [CrossRef] [PubMed]
35. Fisher, C.K.; Ullman, O.; Stultz, C.M. Comparative studies of disordered proteins with similar sequences: Application to Aβ40 and Aβ. *Biophys. J.* **2013**, *104*, 1546–1555. [CrossRef] [PubMed]
36. Baba, S.P.; Bhatnagar, A. Role of Thiols in Oxidative Stress. *Curr. Opin. Toxicol.* **2018**, *7*, 133–139. [CrossRef]
37. Lévy, E.; El Banna, N.; Baïlle, D.; Heneman-Masurel, A.; Truchet, S.; Rezaei, H.; Huang, M.-E.; Béringue, V.; Martin, D.; Vernis, L. Causative Links between Protein Aggregation and Oxidative Stress: A Review. *Int. J. Mol. Sci.* **2019**, *20*, 3896. [CrossRef]
38. Puente-Muñoz, V.; Paredes, J.M.; Resa, S.; Vílchez, J.D.; Zitnan, M.; Miguel, D.; Girón, M.D.; Cuerva, J.M.; Salto, R.; Crovetto, L. New Thiol-Sensitive Dye Application for Measuring Oxidative Stress in Cell Cultures. *Sci. Rep.* **2019**, *9*, 1659. [CrossRef]
39. Novo, M.; Freire, S.; Al-Soufi, W. Critical aggregation concentration for the formation of early Amyloid-β (1–42) oligomers. *Sci. Rep.* **2018**, *8*, 1783. [CrossRef]
40. Morel, B.; Carrasco, M.P.; Jurado, S.; Marco, C.; Conejero-Lara, F. Dynamic micellar oligomers of amyloid beta peptides play a crucial role in their aggregation mechanisms. *Phys. Chem. Chem. Phys.* **2018**, *20*, 20597–20614. [CrossRef]
41. Herrero-Foncubierta, P.; Paredes, J.M.; Giron, M.D.; Salto, R.; Cuerva, J.M.; Miguel, D.; Orte, A. A Red-Emitting, Multidimensional Sensor for the Simultaneous Cellular Imaging of Biothiols and Phosphate Ions. *Sensors* **2018**, *18*, 161. [CrossRef]
42. Daniels, W.; Hendricks, J.; Salie, R.; Taljaard, J. The Role of the MAP-Kinase Superfamily in β-Amyloid Toxicity. *Metab. Brain Dis.* **2001**, *16*, 175–185. [CrossRef]
43. Castelletto, V.; Ryumin, P.; Cramer, R.; Hamley, I.W.; Taylor, M.; Allsop, D.; Reza, M.; Ruokolainen, J.; Arnold, T.; Hermida-Merino, D.; et al. Self-Assembly and Anti-Amyloid Cytotoxicity Activity of Amyloid beta Peptide Derivatives. *Sci. Rep.* **2017**, *7*, 43637. [CrossRef]
44. Resa, S.; Orte, A.; Miguel, D.; Paredes, J.M.; Puente-Muñoz, V.; Salto, R.; Giron, M.D.; Ruedas-Rama, M.J.; Cuerva, J.M.; Alvarez-Pez, J.M.; et al. New Dual Fluorescent Probe for Simultaneous Biothiol and Phosphate Bioimaging. *Chem. Eur. J.* **2015**, *21*, 14772–14779. [CrossRef] [PubMed]
45. Crowther, D.C.; Kinghorn, K.J.; Miranda, E.; Page, R.; Curry, J.A.; Duthie, F.A.I.; Gubb, D.C.; Lomas, D.A. Intraneuronal Aβ, non-amyloid aggregates and neurodegeneration in a Drosophila model of Alzheimer's disease. *Neuroscience* **2005**, *132*, 123–135. [CrossRef] [PubMed]

46. Greeve, I.; Kretzschmar, D.; Tschäpe, J.-A.; Beyn, A.; Brellinger, C.; Schweizer, M.; Nitsch, R.M.; Reifegerste, R. Age-dependent neurodegeneration and Alzheimer-amyloid plaque formation in transgenic Drosophila. *J. Neurosci.* **2004**, *24*, 3899–3906. [CrossRef] [PubMed]
47. Casas, C.; Sergeant, N.; Itier, J.-M.; Blanchard, V.; Wirths, O.; van der Kolk, N.; Vingtdeux, V.; van de Steeg, E.; Ret, G.; Canton, T.; et al. Massive CA1/2 neuronal loss with intraneuronal and N-terminal truncated Aβ42 accumulation in a novel Alzheimer transgenic model. *Am. J. Pathol.* **2004**, *165*, 1289–1300. [CrossRef]
48. Terzi, E.; Hölzemann, G.; Seelig, J. Interaction of Alzheimer β-Amyloid Peptide(1−40) with Lipid Membranes. *Biochemistry* **1997**, *36*, 14845–14852. [CrossRef]
49. Verdier, Y.; Penke, B. Binding Sites of Amyloid β-Peptide in Cell Plasma Membrane and Implications for Alzheimers Disease. *Curr. Protein Pept. Sci.* **2004**, *5*, 19–31. [CrossRef] [PubMed]
50. Fantini, J.; Yahi, N. Chapter 11—Alzheimer's Disease. In *Brain Lipids in Synaptic Function and Neurological Disease*, 1st ed.; Academic Press: San Diego, CA, USA, 2015; pp. 253–277. ISBN 978-0-12-800111-0.
51. Lashuel, H.A.; Hartley, D.; Petre, B.M.; Walz, T.; Lansbury, P.T. Amyloid pores from pathogenic mutations. *Nature* **2002**, *418*, 291. [CrossRef] [PubMed]
52. Jin, S.; Kedia, N.; Illes-Toth, E.; Haralampiev, I.; Prisner, S.; Herrmann, A.; Wanker, E.E.; Bieschke, J. Amyloid-β(1-42) Aggregation Initiates Its Cellular Uptake and Cytotoxicity. *J. Biol. Chem.* **2016**, *291*, 19590–19606. [CrossRef] [PubMed]
53. Banerjee, S.; Hashemi, M.; Zagorski, K.; Lyubchenko, Y.L. Interaction of Aβ42 with Membranes Triggers the Self-Assembly into Oligomers. *Int. J. Mol. Sci.* **2020**, *21*, 1129. [CrossRef] [PubMed]
54. Dear, A.J.; Michaels, T.C.T.; Meisl, G.; Klenerman, D.; Wu, S.; Perrett, S.; Linse, S.; Dobson, C.M.; Knowles, T.P.J. Kinetic diversity of amyloid oligomers. *Proc. Natl. Acad. Sci. USA* **2020**, *117*, 12087–12094. [CrossRef] [PubMed]
55. Butterfield, D.A.; Castegna, A.; Lauderback, C.M.; Drake, J. Evidence that amyloid beta-peptide-induced lipid peroxidation and its sequelae in Alzheimer's disease brain contribute to neuronal death. *Neurobiol. Aging* **2002**, *23*, 655–664. [CrossRef]
56. Schindelin, J.; Arganda-Carreras, I.; Frise, E.; Kaynig, V.; Longair, M.; Pietzsch, T.; Preibisch, S.; Rueden, C.; Saalfeld, S.; Schmid, B.; et al. Fiji: An open-source platform for biological-image analysis. *Nat. Methods* **2012**, *9*, 676–682. [CrossRef] [PubMed]

International Journal of
Molecular Sciences

MDPI

Article

Conformational Characterization of Native and L17A/F19A-Substituted Dutch-Type β-Amyloid Peptides

Kai-Cyuan He [1,2], Yi-Ru Chen [3,†], Chu-Ting Liang [1,4], Shi-Jie Huang [2], Chung-Ying Tzeng [4], Chi-Fon Chang [5], Shing-Jong Huang [6], Hsien-Bin Huang [7,*] and Ta-Hsien Lin [1,2,4,*]

1 Basic Research Division, Medical Research Department, Taipei Veterans General Hospital, Taipei 11217, Taiwan; kay5402@hotmail.com
2 Institute of Biochemistry & Molecular Biology, National Yang-Ming University, Taipei 11221, Taiwan; ja7yrx326159487@gmail.com
3 Department and Institute of Pharmacology, National Yang-Ming University, Taipei 11221, Taiwan; YRC_Chen@adimmune.com.tw
4 Department of Life Sciences and Institute of Genome Sciences, National Yang-Ming University, Taipei 11221, Taiwan; ayumi850@gmail.com (C.-T.L.); wetlandyuhu@gmail.com (C.-Y.T.)
5 Genomics Research Center, Academia Sinica, Taipei 11529, Taiwan; chifon@gate.sinica.edu.tw
6 Instrumentation Center, National Taiwan University, Taipei 10617, Taiwan; shingjonghuang@ntu.edu.tw
7 Department of Life Science and the Institute of Molecular Biology, National Chung Cheng University, Chia-yi 62102, Taiwan
* Correspondence: biohbh@ccu.edu.tw (H.-B.H.); thlin@vghtpe.gov.tw (T.-H.L.); Tel.: +886-5-272-0411 (ext. 53200) (H.-B.H.); +886-2-2871-2121 (ext. 2703) (T.-H.L.); Fax: +886-5-272-2871 (H.-B.H.); +886-2-2875-1562 (T.-H.L.)
† Current address: Adimmune Corporation, No. 3, Sec. 1, Tanxing Rd., Tanzi Dist., Taichung City 42743, Taiwan.

Received: 29 February 2020; Accepted: 6 April 2020; Published: 7 April 2020

Abstract: Some mutations which occur in the α/β-discordant region (resides 15 to 23) of β-amyloid peptide (Aβ) lead to familial Alzheimer's disease (FAD). In vitro studies have shown that these genetic mutations could accelerate Aβ aggregation. We recently showed that mutations in this region could alter the structural propensity, resulting in a different aggregative propensity of Aβ. Whether these genetic mutations display similar effects remains largely unknown. Here, we characterized the structural propensity and aggregation kinetics of Dutch-type $A\beta_{40}$ ($A\beta_{40}$(E22Q)) and its L17A/F19A-substituted mutant ($A\beta_{40}$(L17A/F19A/E22Q)) using circular dichroism spectroscopy, nuclear magnetic spectroscopy, and thioflavin T fluorescence assay. In comparison with wild-type $A\beta_{40}$, we found that Dutch-type mutation, unlike Artic-type mutation (E22G), does not reduce the α-helical propensity of the α/β-discordant region in sodium dodecyl sulfate micellar solution. Moreover, we found that $A\beta_{40}$(L17A/F19A/E22Q) displays a higher α-helical propensity of the α/β-discordant region and a slower aggregation rate than $A\beta_{40}$(E22Q), suggesting that the inhibition of aggregation might be via increasing the α-helical propensity of the α/β-discordant region, similar to that observed in wild-type and Artic-type $A\beta_{40}$. Taken together, Dutch-type and Artic-type mutations adopt different mechanisms to promote Aβ aggregation, however, the L17A/F19A mutation could increase the α-helical propensities of both Dutch-type and Artic-type $A\beta_{40}$ and inhibit their aggregation.

Keywords: NMR; CD; Aβ; β-amyloid peptide; α/β-discordant; Dutch-type mutation; E22Q; familial Alzheimer's disease; FAD

1. Introduction

On the basis of the amyloid cascade hypothesis [1,2], aggregation of β-amyloid peptide (Aβ) is a crucial factor for the neuronal damage that leads to Alzheimer's disease (AD). The clinical hallmarks of AD are neurofibrillary tangles and senile plaques within AD patients' brains. The major components of these two hallmarks are tau protein and Aβ, respectively. Aβ, about 38–42 residues in length, is a derivative from sequentially enzymatic processing of transmembrane protein, called β-amyloid precursor protein (βAPP). It has been reported that increased Aβ production resulting from mutations in the processing enzymes of βAPP (such as β- and γ-secretase) [3] or βAPP mutations close to the cutting site of the processing enzymes [4,5] would cause family Alzheimer's disease (FAD). Point mutations within the Aβ region of βAPP have also been shown to cause family Alzheimer's disease (FAD), such as mutations occurring at A21 [6], E22 [7,8], and D23 [9,10] of Aβ. Several studies have shown that E22G (Arctic-type mutation), E22Q (Dutch-type mutation), and D23N (Iowa-type mutation) mutations would alter the aggregation behavior [11] and structure property [12–17] of Aβ.

Structures of wild-type Aβ in different environments have been reported. They adopted a mainly random coil conformation [18] or a short α-helical structure in aqueous solution [19]. In SDS micellar solution, two short α-helices were contained [20–22]. In the presence of large unilamellar vesicles (zwitterionic lipid bilayers), a partially folded structure was shown [23]. In vitro experiments have shown that Aβ would aggregate into fibrils whose secondary structure was mainly β-sheets [24–26]. Similar β-sheet conformations were also observed for the Aβ fibrils purified from AD brain tissue [27,28]. These findings suggested that Aβ would undergo conformational transitions from random coil or α-helix conformation into β-sheet structure during the process of aggregation. However, the detailed mechanism of the aggregation process of Aβ remains unclear. The aggregative propensity of Aβ is linked to its structural conversion tendency which depends on its intrinsic structural propensity and the local environments where it exists.

Previously, we reported that mutations located in the α/β-discordant region (resides 15 to 23) of Aβ (E22G and L17A/F19A mutations) could either reduce or augment α-helical propensity of Aβ, leading to either an increase or a decrease of the rates of structural transition and fibril formation of Aβ [21,29–31]. The results of these studies support the view that the α-helical and aggregative propensities of Aβ tend to be inversely correlated. It remains uncertain whether other FAD-related mutations located in the Aβ sequence would promote Aβ aggregation by reducing the α-helical propensity of Aβ or not. We have been focusing on investigating the effects of FAD-related mutations in the α/β-discordant region of Aβ on the structural propensity of Aβ. The effect of Arctic-type mutation (E22G) on the structural propensity of Aβ has been reported [16,30], however, the effects of other FAD-related mutations on the structural propensity of Aβ remain largely unknown. In the present study, we applied nuclear magnetic resonance (NMR) and circular dichroism (CD) spectroscopies to characterize the structural conformation of Dutch-type $A\beta_{40}$ ($A\beta_{40}$(E22Q) in SDS micellar solution. Moreover, the effects of Ala replacements at L17 and F19, which have been shown to increase the α-helical propensity and decrease the rate of aggregation of wild-type $A\beta_{40}$ and Arctic-type $A\beta_{40}$ ($A\beta_{40}$(E22G)), on the structure and aggregation kinetics of Dutch-type $A\beta_{40,}$ were also characterized. Our data suggested that the structural conformation of $A\beta_{40}$(E22Q) in SDS micellar solution is very similar to that of wild-type $A\beta_{40}$. There is only a slight difference between these two structures. However, there is a more significant difference in α-helical propensity between $A\beta_{40}$(E22Q) and $A\beta_{40}$(L17A/F19A/E22Q). These results are discussed in terms of the relation between the structural and aggregative propensities of Aβ mutants.

2. Results

2.1. Comparison of the Secondary Structures of Wild-Type $A\beta_{40}$ and $A\beta_{40}$(E22Q)

In our recent study, we reported the effect of Arctic-type mutation (E22G) on the structure of Aβ in SDS micellar solution. To gain more insight into the effect of FAD-related mutation at position 22 on the structure of Aβ, we characterized the structure of $A\beta_{40}$(E22Q) in the present study. $A\beta_{40}$(E22Q) had been found in FAD patients with severe cerebral amyloid angiopathy (CAA). To examine the effect of E22Q mutation on the structure of Aβ, we first analyzed the secondary structures of wild-type $A\beta_{40}$ and $A\beta_{40}$(E22Q) in SDS micellar solution using circular dichroism (CD) spectroscopy. It can be seen from Figure 1 that the CD spectrum of wild-type $A\beta_{40}$ shows a band with positive ellipticity at around 192 nm and two bands with negative ellipticity at 207 nm and 221 nm which are CD spectral characteristics of α-helix, suggesting that the secondary structure content of wild-type $A\beta_{40}$ in micellar solution is mainly α-helix. The result is consistent with that obtained in the previous studies [30,31]. The CD spectrum of $A\beta_{40}$(E22Q) displays a similar spectral pattern to that of wild-type $A\beta_{40}$ with a more positive ellipticity at around 192 nm and a slightly more negative ellipticity at 207 nm and at 221 nm, suggesting that $A\beta_{40}$(E22Q) adopts mainly α-helical conformation as well, and the α-helix content of $A\beta_{40}$(E22Q) might be slightly higher than that of wild-type $A\beta_{40}$.

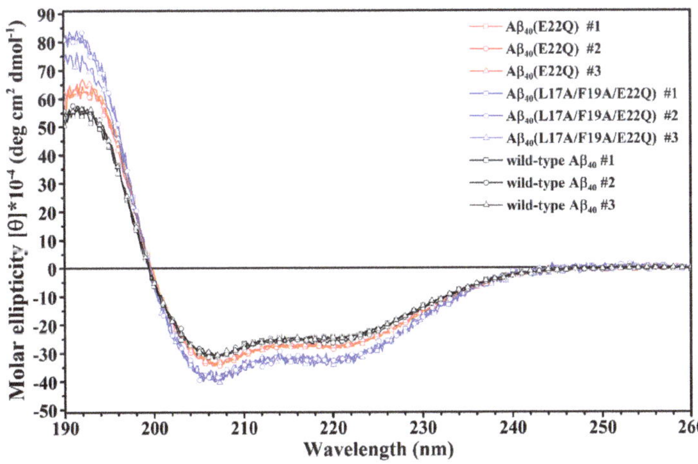

Figure 1. Overlay of CD spectra of wild-type $A\beta_{40}$ (**black**), $A\beta_{40}$(E22Q) (**red**) and $A\beta_{40}$(L17A/F19A/E22Q) (**blue**) in 100 mM SDS micellar solution.

We further applied NMR spectroscopy to characterize the secondary structure of $A\beta_{40}$(E22Q) in SDS micellar solution. In order to derive the secondary structure from the backbone atom chemical shifts, we first accomplished the sequential backbone assignment of $A\beta_{40}$(E22Q). Figure 2A shows the two-dimensional ^{1}H-^{15}N-HSQC spectrum of ^{15}N-labeled $A\beta_{40}$(E22Q) in SDS micellar solution. The result of residue assignment is shown in the figure. By comparison of the two-dimensional ^{1}H-^{15}N-HSQC spectrum of $A\beta_{40}$(E22Q) with that of wild-type $A\beta_{40}$, we obtained the effect of E22Q mutation on the two-dimensional ^{1}H-^{15}N-HSQC spectrum of wild-type $A\beta_{40}$. Figure 2B showed the superimposed two-dimensional ^{1}H-^{15}N-HSQC spectra of wild-type and Dutch-type $A\beta_{40}$. It is evident that these two spectra look almost the same except for some amide proton and nitrogen cross-peaks which displayed chemical shift changes as a result of E22Q mutation. According to the previously assigned two-dimensional ^{1}H-^{15}N-HSQC spectrum of wild-type $A\beta_{40}$ [30], some cross-peaks which displayed relatively significant chemical shift changes on account of E22Q mutation were readily assigned to L17, V18, F20, A21, and D23 (excluding E22). In general, there are three major factors which contribute to the observed chemical shift perturbations of nitrogen (^{15}N) and amide proton (^{1}HN), including the sequence effect caused by E22Q mutation, the conformational change induced by E22Q mutation, and the interaction with SDS micelle altered by E22Q mutation. Further analysis revealed that the chemical shift perturbations are very small (less than 0.05) as shown in Figure 2C, suggesting that the effects of E22Q mutation on these three factors which cause chemical shift perturbations are very small. It can also be seen from Figure 2C that residues which displayed relatively significant chemical shift perturbations resulting from E22Q mutation were located in the α/β-discordant region (resides 15 to 23). This observation suggested that E22Q mutation might slightly affect the structural conformation of the α/β-discordant region of Aβ and/or the interaction of the α/β-discordant region of Aβ with SDS micelle.

In order to confirm the inference that the effect of E22Q mutation on the structural conformation is small, we used secondary chemical shifts of $^{13}C^{\alpha}$ and $^{13}C^{\beta}$ which are mainly affected by the backbone conformation of the amino acid itself instead of any direct through-space interaction, slightly affected by the sequence [32], to estimate the secondary structure of $A\beta_{40}$(E22Q) [33–35]. Figure 3A shows the comparison of $^{13}C^{\alpha}$ secondary chemical shifts of wild-type $A\beta_{40}$ and $A\beta_{40}$(E22Q). It is apparent that the $^{13}C^{\alpha}$ secondary chemical shifts of wild-type $A\beta_{40}$ and $A\beta_{40}$(E22Q) look almost the same except for a few residues in the α/β-discordant region which displayed slightly more positive $^{13}C^{\alpha}$ secondary chemical shifts resulting from E22Q mutation. This result suggested that both wild-type $A\beta_{40}$ and $A\beta_{40}$(E22Q) adopted two short α-helices from residues 15 to 26 and residues 28 to 34 [35] and a few residues in the α/β-discordant region might have adopted slightly higher α-helical propensities (α-helicity) [33] as a result of E22Q mutation. By taking the $^{13}C^{\beta}$ secondary chemical shift into account, we further analyzed the effect of the E22Q mutation on the secondary structure of wild-type $A\beta_{40}$. The results are shown in Figure 3B. As expected, the differences between $^{13}C^{\alpha}$ and $^{13}C^{\beta}$ secondary chemical shifts of wild-type $A\beta_{40}$ and $A\beta_{40}$(E22Q) look almost the same. It can be seen from Figure 3B that a few residues in the α/β-discordant region also displayed slightly more positive values of $\Delta\delta^{13}C^{\alpha}$ $-\Delta\delta^{13}C^{\beta}$ for $A\beta_{40}$(E22Q) than for wild-type $A\beta_{40}$. This observation suggested that the E22Q mutation might result in a slight increase in the α-helical propensities of a few residues in the α/β-discordant region as well [34]. These findings were consistent with those observed from CD spectroscopy. Since the slight differences in $^{13}C^{\alpha}$ secondary chemical shifts (or $\Delta\delta^{13}C^{\alpha}$–$\Delta\delta^{13}C^{\beta}$) between wild-type $A\beta_{40}$ and $A\beta_{40}$(E22Q) are within the error limits of chemical shift measurements using three-dimensional NMR spectra, one may argue that these relatively small differences might be overinterpreted. These differences might merely come from sequence effect. At any rate, we may speculate that the effects of E22Q mutation on the secondary structure of Aβ and the interaction of Aβ with SDS micelle are insignificant. Even though it exists, it is very small according to our NMR and CD data.

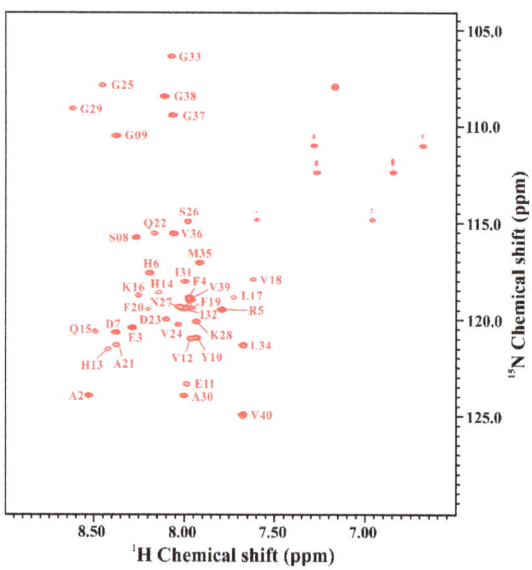

(a)

Figure 2. *Cont.*

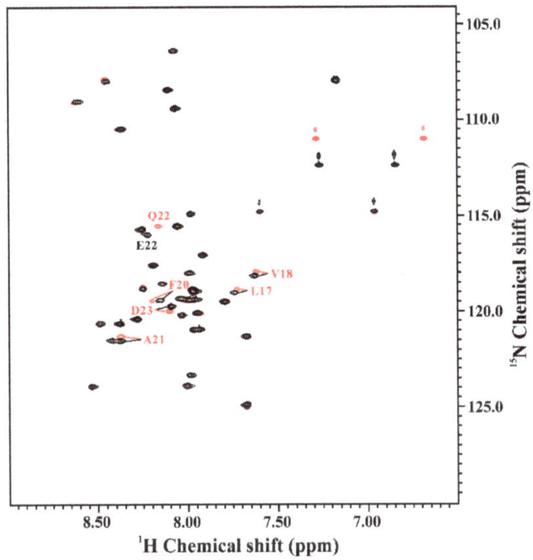

(b)

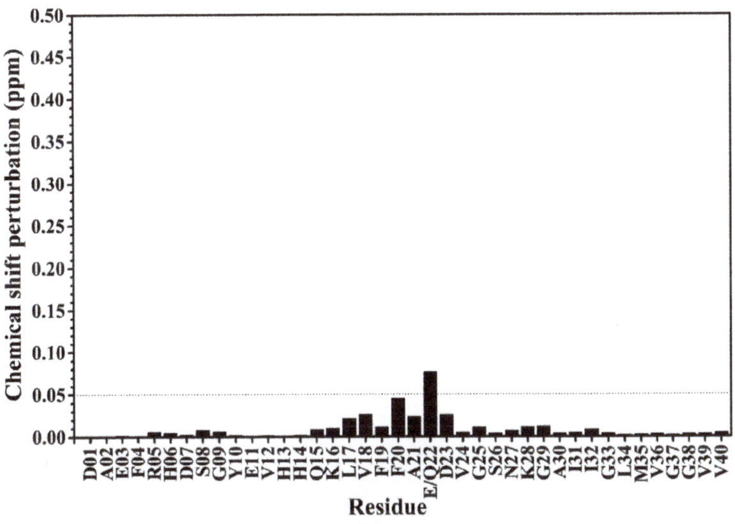

(c)

Figure 2. (**a**) Two-dimensional ^{1}H-^{15}N-HSQC spectrum of ^{15}N-labeled Aβ_{40}(E22Q) in 100 mM SDS micellar solution at 296 K; (**b**) Overlay of two-dimensional ^{1}H-^{15}N-HSQC spectra of ^{15}N-labeled wild-type Aβ_{40} (black) and Aβ_{40}(E22Q) (red) in 100 mM SDS micellar solution at 296 K. Residues which display chemical shift perturbations were labeled; (**c**) Chemical shift perturbation plotted as a function of residue number. Chemical shift perturbation was calculated using the equation $[(^{HN}\Delta ppm)^2 + (^{N}\Delta ppm/10)^2]^{1/2}$, where $^{HN}\Delta$ppm and $^{N}\Delta$ppm were equal to $^{1}H^{N}$ and ^{15}N chemical shift differences between wild-type Aβ_{40} and Aβ_{40}(E22Q), respectively [31].

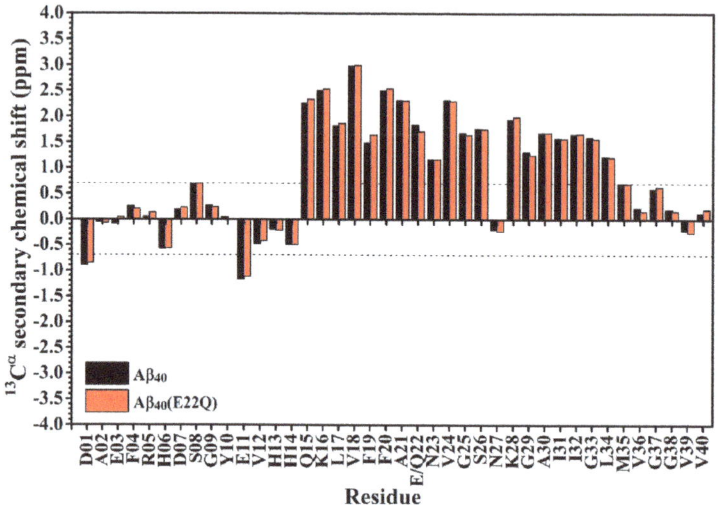

(a)

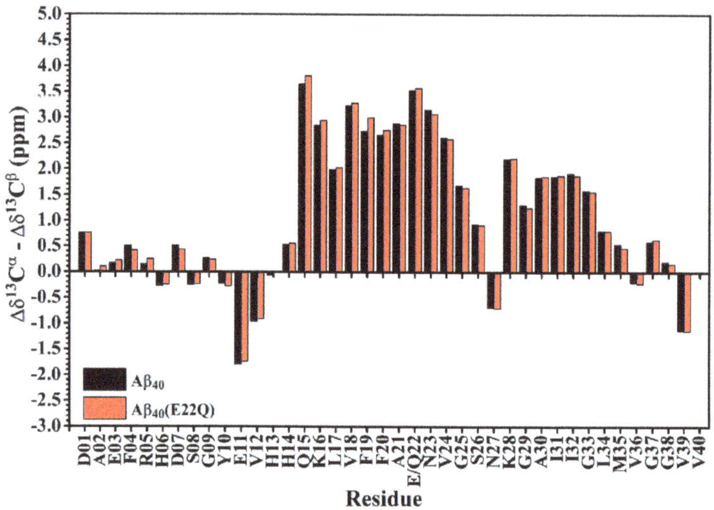

(b)

Figure 3. (**a**) $^{13}C^{\alpha}$ secondary chemical shifts of wild-type $A\beta_{40}$ (balck) and $A\beta_{40}$(E22Q) (red) plotted as a function of residue. In principle, if the $^{13}C^{\alpha}$ secondary chemical shift of an amino acid residue is greater than 0.7 ppm, its conformation would be α-helical [35]; (**b**) Differences between $\Delta\delta^{13}C^{\alpha}$ ($^{13}C^{\alpha}$ secondary chemical shift) and $\Delta\delta^{13}C^{\beta}$ ($^{13}C^{\beta}$ secondary chemical shift) of wild-type $A\beta_{40}$ (black) and $A\beta_{40}$(E22Q) (red) plotted as a function of residue. $\Delta\delta^{13}C^{\alpha}$ (or $\Delta\delta^{13}C^{\beta}$) was defined as the difference between the observed $^{13}C^{\alpha}$ (or $^{13}C^{\beta}$) chemical shift of an amino acid residue and its $^{13}C\alpha$ (or $^{13}C\beta$) chemical shift in a random coil conformation. If $\Delta\delta^{13}C^{\alpha}$–$\Delta\delta^{13}C^{\beta}$ for an amino acid residue is positive, its conformation would be α-helical. For a more detailed description of the relationship between the value of $\Delta\delta^{13}C^{\alpha}$–$\Delta\delta^{13}C^{\beta}$ and secondary structure of an amino acid residue please see the reference [34].

2.2. Comparison of the Secondary Structures of $A\beta_{40}$(E22Q) and $A\beta_{40}$(L17A/F19A/E22Q)

In our recent study, we showed that residues L17 and F19 of Aβ played an important role in the structural and aggregative propensities of wild-type $A\beta_{40}$ and $A\beta_{40}$(E22G) [21,29,31]. To examine whether the effects of Ala replacements at L17 and F19 on the structure and aggregative property of Aβ_{40}(E22Q) are similar to those observed for wild-type Aβ_{40} and Aβ_{40}(E22G) or not, we performed structural characterization and aggregation kinetic study on Aβ_{40}(L17A/F19A/E22Q). Prior to the experimental structural characterization of Aβ_{40}(L17A/F19A/E22Q), we applied propensity-based prediction to the analyzed effects of E22Q and L17A/F19A mutations on the structural propensity of the α/β-discordant region of wild-type Aβ_{40} and Aβ_{40}(E22Q), respectively [31,36]. The results obtained from in silico studies implied that wild-type Aβ_{40} and Aβ_{40}(E22Q) adopt the same structural propensity in their α/β-discordant region. Unlike the E22G mutation which would alter the structural propensity of D23 from α-helix to β-strand, the E22Q mutation has no effect on the structural propensity of wild-type Aβ_{40}. It can also be seen that the L17A/F19A mutation would alter the structural propensities of residues 15 to 21 in the α/β-discordant region of Aβ_{40}(E22Q) from β-strand to α-helix as shown in Figure 4. The same effect has also been observed on wild-type Aβ_{40} and Aβ_{40}(E22G) [31].

wild-type Aβ_{40}	**Aβ_{40}(E22Q)**	**Aβ_{40}(L17A/F19A/E22Q)**
EEEEEEEEh	EEEEEEEEh	HHHHHHHEh
QKLVFFAED	QKLVFFAQD	QKAVAFAQD
15 23	15 23	15 23

Figure 4. The predicted secondary structures of the α/β-discordant regions of wild-type Aβ_{40} Aβ_{40}(E22Q) and Aβ_{40}(L17A/F19A/E22Q). β-strands predicted with high and low probability were denoted by the symbols E and e, respectively. α-helical structures predicted with high and low probability were denoted by the symbols H and h, respectively [31,36].

We next applied CD spectroscopy to examine the effect of L17A/F19A mutation on the overall secondary structure of Dutch-type Aβ_{40}. The CD spectra of Aβ_{40}(L17A/F19A/E22Q) are shown in Figure 1. It is apparent that Aβ_{40}(E22Q) and Aβ_{40}(L17A/F19A/E22Q) exhibited similar spectral patterns in their CD spectra, suggesting that the overall secondary structure of Aβ_{40}(L17A/F19A/E22Q) is similar to that of Aβ_{40}(E22Q). They both adopt mainly α-helical structures in SDS micellar solution. However, it can be seen from Figure 1 that Aβ_{40}(L17A/F19A/E22Q) displayed more positive ellipticity at around 192 nm and more negative ellipticity at 207 nm and 221 nm than Aβ_{40}(E22Q), suggesting that the L17A/F19A mutation would result in an increase of the α-helix content of Aβ_{40}(E22Q). The difference between the CD spectra of Aβ_{40}(E22Q) and Aβ_{40}(L17A/F19A/E22Q) is more significant than that between wild-type Aβ_{40} and Aβ_{40}(E22Q), indicating that the effect of L17A/F19A mutation on the overall secondary structure of Dutch-type Aβ_{40} is more prominent than that of E22Q mutation on the overall secondary structure of wild-type Aβ_{40}.

We also applied NMR spectroscopy to characterize the secondary structure of Aβ_{40}(L17A/F19A/E22Q) in SDS micellar solution and used the same approach as that employed for analyzing the effect of E22Q mutation on the structural conformation of Aβ to analyze the effect of L17A/F19A mutation on the structural conformation of Dutch-type Aβ_{40}. A two-dimensional ^{1}H-^{15}N-HSQC spectrum of ^{15}N-labeled Aβ_{40}(L17A/F19A/E22Q) in SDS micellar solution with the result of residue assignment is shown in Figure 5A. Figure 5B shows the comparison of the two-dimensional ^{1}H-^{15}N-HSQC spectra of Aβ_{40}(E22Q) and Aβ_{40}(L17A/F19A/E22Q). It is quite obvious that many amide proton and nitrogen cross-peaks of Aβ_{40}(E22Q) display significant chemical shift changes because of L17A/F19A mutation. Cross-peaks which display significant chemical shift changes are indicated in the figure. Calculations of chemical shift perturbations were also performed for further analysis of the effect of the L17A/F19A mutation on the chemical shifts of the amide proton and nitrogen cross-peaks of Aβ_{40}(E22Q). The results are shown in Figure 5C. Residues which exhibited significant chemical shift perturbations (greater than 0.05) were readily identified as E11, H13-F20 (excluding L17 and F19), Q22, D23, and G25. These residues are mainly located in the α/β-discordant region of Aβ_{40}(E22Q), suggesting that the increases of α-helical content observed from CD spectra are mainly from the residues in the α/β-discordant region of Aβ$_{40}$(L17A/F19A/E22Q). Similar effects have also been observed on wild-type Aβ$_{40}$ and Aβ$_{40}$(E22G) [31]. This finding implied that the L17A/F19A mutation would affect the structural conformation of the α/β-discordant region of Aβ$_{40}$(E22Q) and the interaction of the α/β-discordant region of Aβ$_{40}$(E22Q) with SDS micelle.

The effect of the L17A/F19A mutation on the secondary structure of Aβ$_{40}$(E22Q) was also analyzed in terms of the changes of secondary chemical shifts of $^{13}C^{\alpha}$ and $^{13}C^{\beta}$. Figure 6A,B shows the plots of $^{13}C^{\alpha}$ secondary chemical shifts and the values of $\Delta\delta^{13}C^{\alpha}$–$\Delta\delta^{13}C^{\beta}$ of Aβ$_{40}$(E22Q) and Aβ$_{40}$(L17A/F19A/E22Q) as a function of residue, respectively. It can be seen from Figure 6A,B that residues which displayed significant changes in the $^{13}C^{\alpha}$ secondary chemical shifts and the values of $\Delta\delta^{13}C^{\alpha}$–$\Delta\delta^{13}C^{\beta}$ as a result of L17A/F19A mutation were mainly located in the α/β-discordant region of Aβ$_{40}$(E22Q). Moreover, both the $^{13}C\alpha$ secondary chemical shifts and the values of $\Delta\delta^{13}C^{\alpha}$–$\Delta\delta^{13}C^{\beta}$ for the residues in the α/β-discordant region are significantly more positive for Aβ$_{40}$(L17A/F19A/E22Q) than for Aβ$_{40}$(E22Q). These findings suggested that Aβ$_{40}$(L17A/F19A/E22Q) adopted two short α-helices from residues 15 to 26 and residues 28 to 34, and residues 15–26 of Aβ$_{40}$(L17A/F19A/E22Q) adopted higher α-helical propensities than those of Aβ$_{40}$(E22Q). It has to be noted that changes of these secondary chemical shifts are primarily contributed by structural conformational changes induced by the L17A/F19A mutation. Alternation of interaction with SDS micelle would result in changes of these secondary chemical shifts as well. We cannot rule out the possibility that interaction of the α/β-discordant region of Aβ$_{40}$(E22Q) with SDS micelle would be altered due to the L17A/F19A mutation. However, whether interaction with SDS is strong or not, its effect on the changes of these secondary chemical shifts is small.

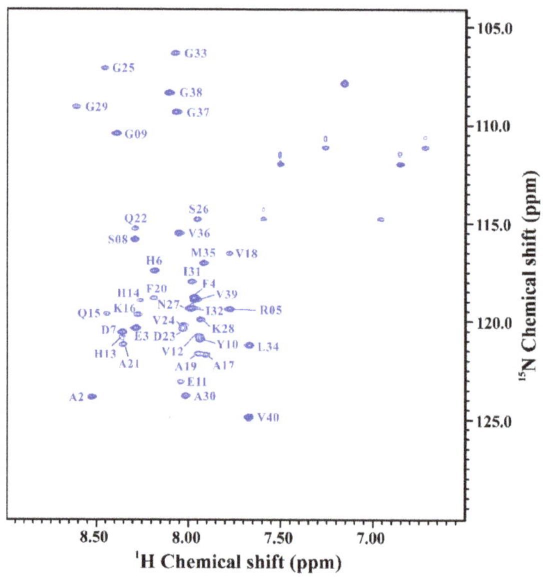

(a)

Figure 5. *Cont.*

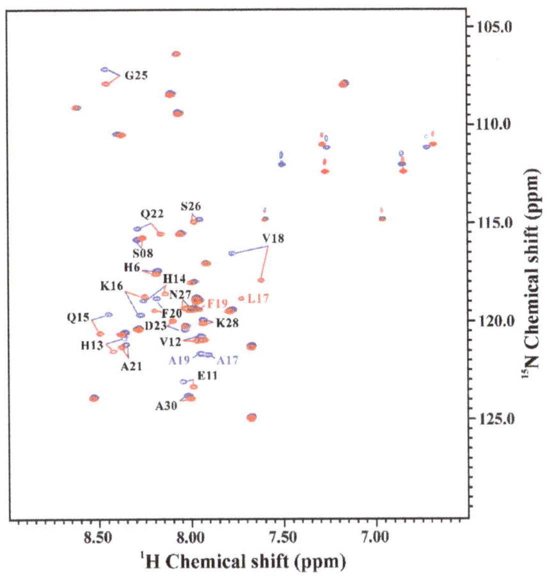

(b)

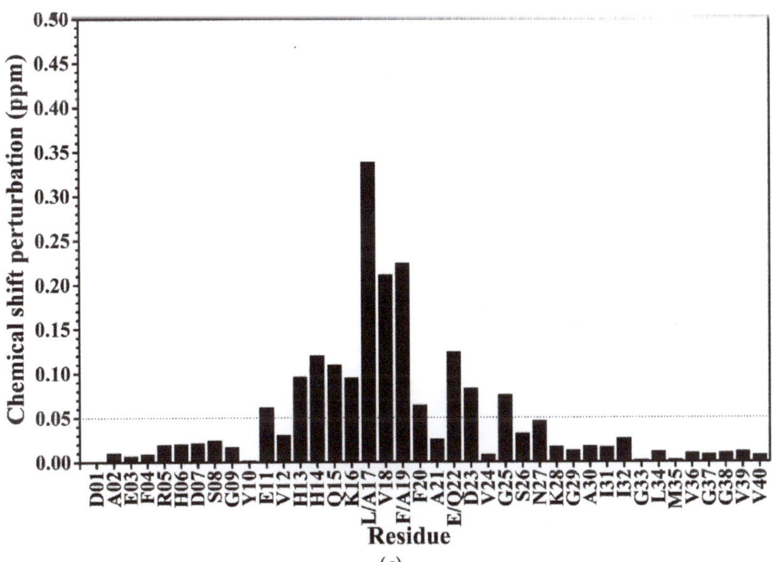

(c)

Figure 5. (**a**) Two-dimensional ^{1}H-^{15}N-HSQC spectrum of ^{15}N-labeled Aβ_{40}(L17A/F19A/E22Q) in 100 mM SDS micellar solution at 296 K; (**b**) Overlay of Two-dimensional ^{1}H-^{15}N-HSQC spectra of ^{15}N-labeled Aβ_{40}(E22Q) (red) and Aβ_{40}(L17A/F19A/E22Q) (blue) in 100 mM SDS micellar solution at 296 K. Residues which display chemical shift perturbations were labeled; (**c**) Chemical shift perturbation plotted as a function of residue number. Chemical shift perturbation was calculated using the equation $[(^{HN}\Delta ppm)^2 + (^{N}\Delta ppm/10)^2]^{1/2}$, where $^{HN}\Delta$ppm and $^{N}\Delta$ppm were equal to ^{1}HN and ^{15}N chemical shift differences between Aβ_{40}(E22Q) and Aβ_{40}(L17A/F19A/E22Q), respectively [31].

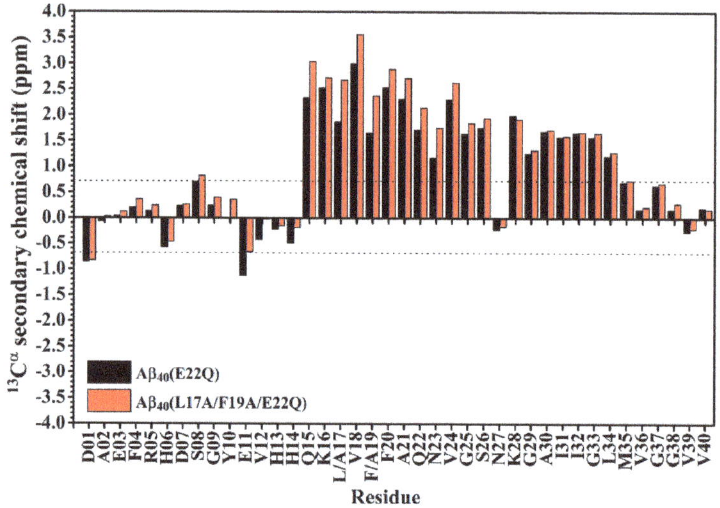

(**a**)

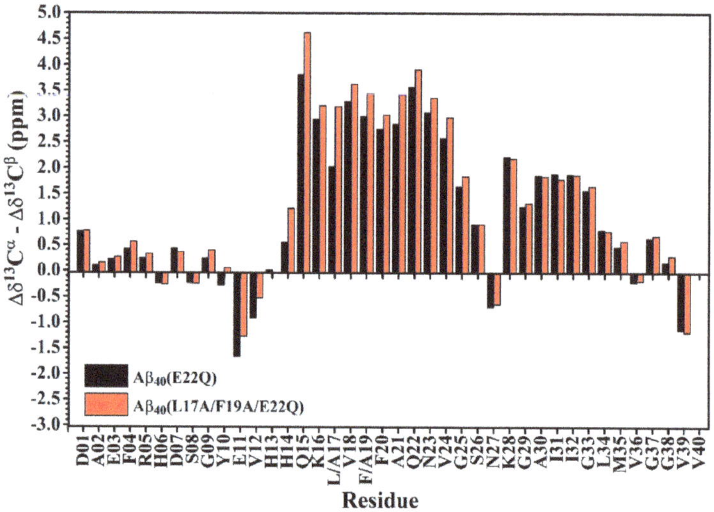

(**b**)

Figure 6. (**a**) $^{13}C\alpha$ secondary chemical shifts of $A\beta_{40}$(E22Q) (black) and $A\beta_{40}$(L17A/F19A/E22Q) (red) plotted as a function of residue. In principle, if the $^{13}C^{\alpha}$ secondary chemical shift of an amino acid residue is greater than 0.7 ppm, its conformation would be α-helical [35]; (**b**) Differences between $\Delta\delta^{13}C^{\alpha}$ ($^{13}C^{\alpha}$ secondary chemical shift) and $\Delta\delta^{13}C^{\beta}$ ($^{13}C^{\beta}$ secondary chemical shift) of $A\beta_{40}$(E22Q) (black) and $A\beta_{40}$(L17A/F19A/E22Q) (red) plotted as a function of residue. $\Delta\delta^{13}C^{\alpha}$ (or $\Delta\delta^{13}C^{\beta}$) was defined as the difference between the observed $^{13}C^{\alpha}$ (or $^{13}C^{\beta}$) chemical shift of an amino acid residue and its $^{13}C^{\alpha}$ (or $^{13}C^{\beta}$) chemical shift in a random coil conformation. If $\Delta\delta^{13}C^{\alpha}$–$\Delta\delta^{13}C^{\beta}$ for an amino acid residue is positive, its conformation would be α-helical. For a more detailed description of the relationship between the value of $\Delta\delta^{13}C^{\alpha}$–$\Delta\delta^{13}C^{\beta}$ and secondary structure of an amino acid residue please see the reference [34].

2.3. L17A/F19A Mutation Inhibits the Aggregation of Aβ_{40}(E22Q)

We characterized the effect of L17A/F19A mutation on the structural propensity of Aβ_{40}(E22Q). However, the effect of L17A/F19A mutation on the aggregative property of Aβ_{40}(E22Q) remained unclear. To investigate this issue, we applied thioflavin-T (Th-T) fluorescence assay and transmission electron microscopy (TEM) to monitor the aggregation processes of Aβ_{40}(E22Q) and Aβ_{40}(L17A/F19A/E22Q) in aqueous solution. The results of Th-T assay and TEM are shown in Figures 7 and 8, respectively. It can be seen from Figure 7 that the shapes of the aggregation profiles of Aβ_{40}(E22Q) and Aβ_{40}(L17A/F19A/E22Q) in aqueous solution are sigmoidal, suggesting that both peptides aggregated in a nucleation-dependent polymerization manner. Furthermore, the two aggregation profiles shown in Figure 7 displayed two distinct lag phases (nucleation phases) whose durations are 12 and 27 h for Aβ_{40}(E22Q) and Aβ_{40}(L17A/F19A/E22Q), respectively. This result revealed that Aβ_{40}(E22Q) aggregated more rapidly than Aβ_{40}(L17A/F19A/E22Q). Figure 8 shows the TEM images of Aβ_{40}(E22Q) and Aβ_{40}(L17A/F19A/E22Q) in aqueous solution acquired at different time points. Fibrils were observed at Day 1 and Day3 for Aβ_{40}(E22Q) and Aβ_{40}(L17A/F19A/E22Q), respectively. This observation indicated that the rate of fibril formation is more rapid for Aβ_{40}(E22Q) than for Aβ_{40}(L17A/F19A/E22Q). Taken together, these findings suggested that the L17A/F19A mutation would reduce the aggregation rate of Aβ_{40}(E22Q). From a kinetic point of view, the free energy of activation for conformational change from α-helix to β-strand would be higher for a peptide which adopts a higher α-helical propensity. Since the conformational change from the α-helix to the β-strand of Aβ is one of the key factors in governing its aggregative propensity, it is reasonable to infer that L17A/F19A mutation inhibits the aggregation of Aβ_{40}(E22Q). This might be through increasing the α-helical propensity of its α/β-discordant region, which in turn reduces its rate of conformational change from the α-helix to the β-strand.

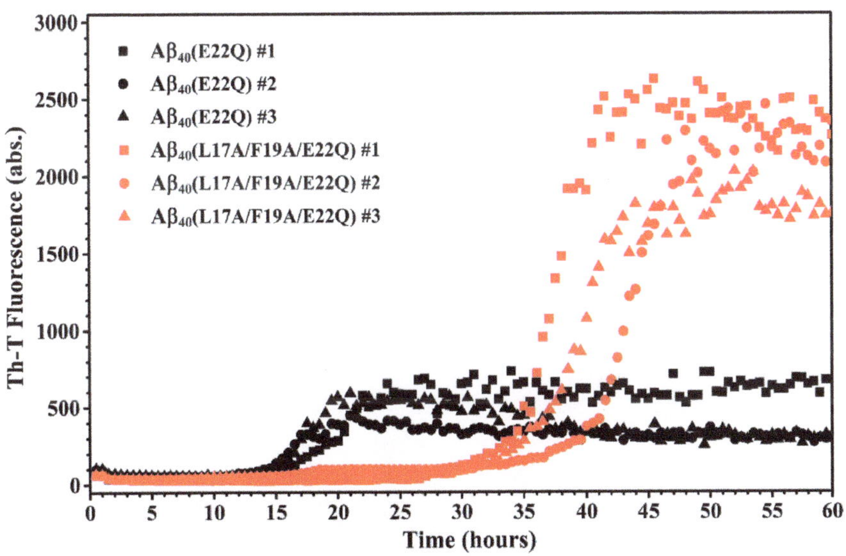

Figure 7. Aggregation kinetics of Aβ_{40}(E22Q) (red) and Aβ_{40}(L17A/F19A/E22Q) (black).

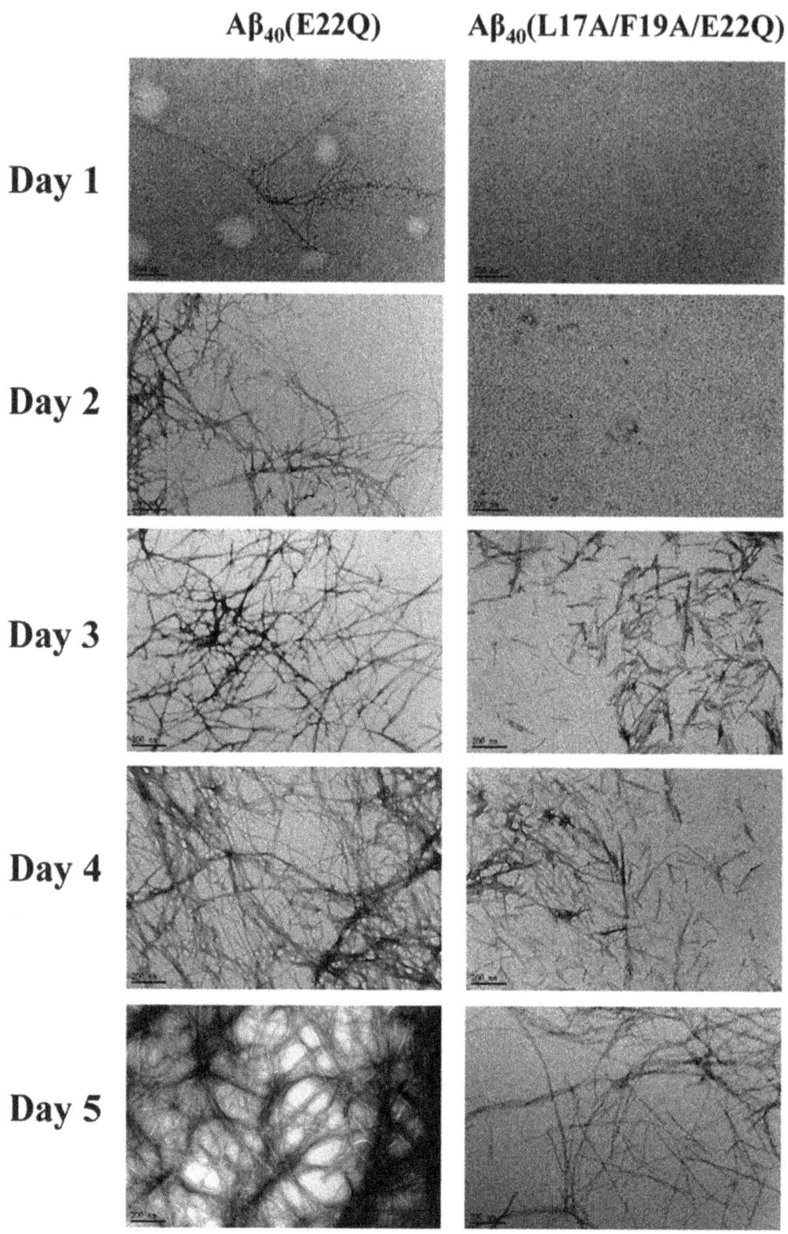

Figure 8. TEM images of Aβ_{40}(E22Q) and Aβ_{40}(L17A/F19A/E22Q). The scale bar is 200 nm.

3. Discussion

Recently, Hatami et al. reported the effects of FAD-related mutations within the Aβ sequence on the fibrils morphology and aggregation kinetics of Aβ using TEM and Th-T assay [11]. They found that most FAD-related Aβ mutants exhibited faster rates of aggregation. They also observed that Th-T fluorescence profiles of these FAD-related Aβ mutants displayed shorter times of lag phase with higher intensities of Th-T fluorescence and higher amounts of fibrils as compared to wild-type $A\beta_{40}$, however, not all FAD-related Aβ mutants displayed the same patterns. Several FAD-related Aβ mutants showed a lower intensity of Th-T fluorescence with a higher amount of fibrils. This phenomenon can be explained by the binding ability of Th-T with Aβ aggregates or fibrils, since Th-T would bind to aggregates or fibrils of different structural conformations with distinct binding abilities, resulting in different fluorescence intensities. Hatami et al. reported that the intensity of Th-T fluorescence is not correlated with the amyloid fibril content. It can also be applied to explain our data shown in Figures 7 and 8 in which the Th-T fluorescence intensity of $A\beta_{40}$(L17A/F19A/E22Q) after 45 h was higher than that of $A\beta_{40}$(E22Q) and the TEM images showed a smaller amount of aggregates and/or fibrils of $A\beta_{40}$(L17A/F19A/E22Q). These observations also suggested that the fibril conformation of $A\beta_{40}$(L17A/F19A/E22Q) should be different from that of $A\beta_{40}$(E22Q).

Many studies have reported that the FAD-related mutations, Dutch-type and Arctic-type mutations, both of which are located at position 22 within the Aβ sequence, would result in an increase of the aggregation rate of Aβ [11,12,37,38]. However, the underlying mechanisms by which these two FAD-related mutations accelerate the aggregation process of Aβ remain elusive. In general, the aggregation process of Aβ would involve conformational changes and self-association which are closely related to the intrinsic structural propensity, the intramolecular interactions within the Aβ molecule, and intermolecular interactions between Aβ molecules. Thus, any factor which varies these properties would alter its aggregation behavior as we discussed in the previous paper [30,31]. In the previous study, we investigated the mechanism of why Arctic-type mutation accelerates Aβ aggregation from a structural point of view and proposed that Arctic-type mutation would reduce the α-helical propensity of the α/β-discordant region of Aβ, resulting in an acceleration of Aβ aggregation [30]. However, it remains unclear whether or not Arctic-type mutation would enhance or reduce the intramolecular and/or intermolecular interactions of Aβ, since it is difficult to measure these interactions. In this study, we applied the same approach to investigate the underlying mechanism of how Dutch-type mutation promotes Aβ aggregation. Our data indicated that Dutch-type mutation, unlike Arctic-type mutation, has no significance on the structural propensity of Aβ. According to our data, the structural propensity of Dutch-type $A\beta_{40}$ and its interaction with SDS micelle are almost the same as those of wild-type $A\beta_{40}$. Thus, we speculated that Dutch-type mutation might alter the intramolecular and/or intermolecular interactions of Aβ, leading to an increase of the aggregation rate of Aβ. This is a very likely inference, even though these effects were not directly observed. A single mutation at the same position (position 22) within the Aβ sequence with a different amino-acid would result in a distinct mechanism by which it promotes Aβ aggregation. This might be the reason why different FAD-related mutations within the Aβ sequence displayed different clinical characteristics, such as cerebral amyloid angiopathy (CAA) for Dutch-type $A\beta_{40}$.

For L17A/F19A mutation, our data suggested that one of the factors in determining its inhibition of the aggregation of $A\beta_{40}$(E22Q) is through increasing the α-helical propensity of the α/β-discordant region of $A\beta_{40}$(E22Q). This effect was also observed on wild-type $A\beta_{40}$ and $A\beta_{40}$(E22G) [21,29]. Whether the L17A/F19A mutation could inhibit the aggregation of other FAD-related Aβ mutants through the same effect which is exerted on wild-type $A\beta_{40}$, $A\beta_{40}$(E22Q) and $A\beta_{40}$(E22G) remains to be investigated. The possibility that the intramolecular and/or intermolecular interactions of Aβ would be altered by the L17A/F19A mutation cannot be ruled out. We characterized the effects of L17A/F19A mutation on the structural propensity and aggregation kinetics of wild-type, Arctic-type and Dutch-type $A\beta_{40}$, however, the effects of L17A/F19A mutation on the structural propensity and aggregation kinetics of the more amyloidogenic $A\beta_{42}$ and its FAD-related mutants remain unclear. Since the fibril structures of $A\beta_{40}$ [25,26,28,39] and $A\beta_{42}$ [40–45] have been solved at the atomic resolution, the intramolecular interactions within Aβ molecule and intermolecular interactions between Aβ molecules can be grasped to some extent based on this structural information. The intramolecular interactions within $A\beta_{40}$ were located at K16-D23 and G29-M35 segments, which correspond to the α/β-discordant region and c-terminal α-helix, respectively. According this structural information, L17A/F19A mutation would disrupt the intramolecular interaction within Aβ_{40}. It can be seen from the fibril structure of Aβ_{42} that residues Ile41 and Ala42 are involved in the intramolecular and intermolecular interactions, suggesting that these two residues would affect the aggregation kinetics of Aβ. According to the Aβ_{42} fibril structure, we may also speculate that the L17A/F19A mutation would also disrupt the intramolecular interaction within Aβ_{42}, leading to an alternation of the aggregative propensity of Aβ_{42}. Knowing the effects of the mutations within the Aβ sequence may help us in developing agents for inhibition of the aggregation of Aβ.

4. Materials and Methods

4.1. Preparation of Aβ Peptides

The protocols for production of Aβ peptides were the same as those described in the previous studies [29–31]. All peptide samples were dissolved in 70% TFE (trifluoroethanol) and then lyophilized. For NMR studies, peptides were dissolved in 0.25 mL 100 mM SDS-d_{25} (sodium dodecyl sulfate-d_{25}) with 10% (*v*/*v*) D_2O/H_2O containing 10 mM phosphate buffer at pH 6.0. TSP (3-(trimethylsilyl)propionic-2,2,3,3,-d_4 acid) was used for internal chemical shift standard. The sample solutions were put into the 5 mm Shigemi tubes (Shigemi Co., Allison Park, PA, USA) for NMR spectra recording.

4.2. Nuclear Magnetic Resonance (NMR) Spectroscopy

NMR experiments were performed at 296 K on the Bruker AVANCE-500, 600, or 800 spectrometer equipped with a 5-mm inverse triple resonance ($^1H/^{13}C/^{15}N$), Z-axis gradient cyroprobe. NMR data were processed and analyzed using the TopSpin and AURELIA programs (Bruker BioSpin GmbH, Rheinstetten, Germany). Linear predictions were used in the indirectly detected dimensions to improve digital resolution. ^1H chemical shifts were referenced to the ^1H frequency of the methyl resonances of TSP at 0 ppm. The ^{15}N and ^{13}C chemical shifts were indirectly referenced using the following consensus ratios of the zero-point frequencies: 0.101329118 for $^{15}N/^1H$ and 0.251449530 for $^{13}C/^1H$ [46]. Backbone sequential assignments were accomplished using the three-dimensional spectra: HNCOCA, HNCO, HNCA, and CBCA(CO)NH [30,31].

4.3. Circular Dichroism (CD) Spectroscopy

Pretreated Aβ peptides (50 μM) were dissolved in 0.160 mL 100 mM SDS containing 5 mM phosphate buffer at pH 6.0. CD measurements were performed on an AVIV CD spectrometer (Aviv 410 spectropolarimeter, Aviv Biomedical, Inc., Lakewood, NJ USA) at 296 K [30,31]. The measurement was carried out three times.

4.4. Thioflavin T (Th-T) Fluorescence Assay

Pretreated Aβ peptides (30 μM) were incubated in aqueous solution (5 mM phosphate buffer, pH 7.2). The molar ratio of Aβ and thioflovin T (Th-T) (Sigma) was 1:1. Fluorescence signals were acquired (SpectraMax M5, Molecular Device, San Jose, CA, USA) every 30 min at 37 °C. The excitation and emission wavelengths of fluorescence were 450 nm and 482 nm, respectively [21,29].

4.5. Transmission Electron Microscopy (TEM)

TEM images of Aβ peptides were acquired using JEOL JEM-2100 EXII TEM (JEOL, Tokyo, Japan). Pretreated Aβ peptides (60 μM) were dissolved in 10 mM phosphate buffer, pH 7.0 and incubated at 37 °C for different time periods. Sample preparation of TEM followed the procedures as described [29,30].

5. Conclusions

In the present study, we characterized the effects of E22Q and L17A/F19A mutations on the structural propensities of wild-type Aβ_{40} and Aβ_{40}(E22Q), respectively, by CD and NMR spectroscopy. We found that the E22Q mutation has no significant effect on the structural propensity of wild-type Aβ_{40}, indicating that it does not promote aggregation by altering the α-helical propensity of the α/β-discordant region. This finding supported the view that it is not necessary for FAD-related mutations in the α/β-discordant region to promote aggregation by altering the structural propensity of Aβ. Besides wild-type and Arctic-type Aβ_{40}, the L17A/F19A mutation would increase the α-helical propensity of the α/β-discordant region of Dutch-type Aβ_{40}, resulting in inhibition of Aβ_{40}(E22Q) aggregation. It is possible that the L17A/F19A mutation can be applied to inhibit Aβ aggregation in vivo. This study provides the information for a clearer understanding of how mutations within the α/β-discordant region of Aβ affect aggregation.

Author Contributions: Conceptualization, K.-C.H., H.-B.H., and T.-H.L.; Data curation, K.-C.H., Y.-R.C., S.-J.H., and C.-Y.T.; Formal analysis, K.-C.H., Y.-R.C., C.-T.L., S.-J.H., and C.-Y.T.; Investigation, K.-C.H., Y.-R.C., S.-J.H., and C.-Y.T.; Methodology, K.-C.H., C.-F.C., S.-J.H., H.-B.H., and T.-H.L.; Project administration, H.-B.H. and T.-H.L.; Software, C.-F.C. and S.-J.H.; Validation, K.-C.H., C.-T.L., and S.-J.H.; Visualization, K.-C.H.; Writing—original draft, K.-C.H.; Writing—review & editing, H.-B.H. and T.-H.L. All authors have read and agreed to the published version of the manuscript.

Funding: This research received no external funding.

Acknowledgments: This work was supported by grants from the Ministry of Science and Technology, ROC (MOST 106-2113-M-010-003, MOST 107-2113-M-010-004, MOST 108-2113-M-010-003) and from Taipei Veterans General Hospital, Taiwan, ROC (V106C-052, V107C-051 and V108C-023) The NMR spectra were obtained at the High-Field NMR Center at Academia Sinica, Instrumentation Center, National Taiwan University and Taipei Veterans General Hospital.

Conflicts of Interest: The authors declare no conflicts of interest.

Abbreviations

Aβ	β-amyloid peptide
βAPP	β-amyloid precursor protein
AD	Alzheimer's disease
FAD	familial Alzheimer's disease
CAA	cerebral amyloid angiopathy
NMR	nuclear magnetic resonance
CD	circular dichroism
TEM	transmission electron microscopy
SDS	sodium dodecyl sulfate
TFE	trifluoroethanol
Th-T	thioflavin-T

References

1. Hardy, J.A.; Higgins, G.A. Alzheimer's disease: The amyloid cascade hypothesis. *Science* **1992**, *256*, 184–185. [CrossRef]
2. Hardy, J.; Selkoe, D.J. The amyloid hypothesis of Alzheimer's disease: Progress and problems on the road to therapeutics. *Science* **2002**, *297*, 353–356. [CrossRef] [PubMed]
3. Weggen, S.; Beher, D. Molecular consequences of amyloid precursor protein and presenilin mutations causing autosomal-dominant Alzheimer's disease. *Alzheimer's Res. Ther.* **2012**, *4*, 9. [CrossRef] [PubMed]
4. Cai, X.D.; Golde, T.E.; Younkin, S.G. Release of excess amyloid beta protein from a mutant amyloid beta protein precursor. *Science* **1993**, *259*, 514–516. [CrossRef] [PubMed]
5. Mullan, M.; Crawford, F.; Axelman, K.; Houlden, H.; Lilius, L.; Winblad, B.; Lannfelt, L. A pathogenic mutation for probable Alzheimer's disease in the APP gene at the N-terminus of beta-amyloid. *Nat. Genet.* **1992**, *1*, 345–347. [CrossRef]

6. Hendriks, L.; van Duijn, C.M.; Cras, P.; Cruts, M.; Van Hul, W.; van Harskamp, F.; Warren, A.; McInnis, M.G.; Antonarakis, S.E.; Martin, J.J.; et al. Presenile dementia and cerebral haemorrhage linked to a mutation at codon 692 of the beta-amyloid precursor protein gene. *Nat. Genet.* **1992**, *1*, 218–221. [CrossRef]
7. Levy, E.; Carman, M.D.; Fernandez-Madrid, I.J.; Power, M.D.; Lieberburg, I.; van Duinen, S.G.; Bots, G.T.; Luyendijk, W.; Frangione, B. Mutation of the Alzheimer's disease amyloid gene in hereditary cerebral hemorrhage, Dutch type. *Science* **1990**, *248*, 1124–1126. [CrossRef]
8. Nilsberth, C.; Westlind-Danielsson, A.; Eckman, C.B.; Condron, M.M.; Axelman, K.; Forsell, C.; Stenh, C.; Luthman, J.; Teplow, D.B.; Younkin, S.G.; et al. The 'Arctic' APP mutation (E693G) causes Alzheimer's disease by enhanced Abeta protofibril formation. *Nat. Neurosci.* **2001**, *4*, 887–893. [CrossRef]
9. Grabowski, T.J.; Cho, H.S.; Vonsattel, J.P.; Rebeck, G.W.; Greenberg, S.M. Novel amyloid precursor protein mutation in an Iowa family with dementia and severe cerebral amyloid angiopathy. *Ann. Neurol.* **2001**, *49*, 697–705. [CrossRef]
10. Bugiani, O.; Giaccone, G.; Rossi, G.; Mangieri, M.; Capobianco, R.; Morbin, M.; Mazzoleni, G.; Cupidi, C.; Marcon, G.; Giovagnoli, A.; et al. Hereditary cerebral hemorrhage with amyloidosis associated with the E693K mutation of APP. *Arch. Neurol.* **2010**, *67*, 987–995. [CrossRef]
11. Hatami, A.; Monjazeb, S.; Milton, S.; Glabe, C.G. Familial Alzheimer's Disease Mutations within the Amyloid Precursor Protein Alter the Aggregation and Conformation of the Amyloid-beta Peptide. *J. Biol. Chem.* **2017**, *292*, 3172–3185. [CrossRef] [PubMed]
12. Miravalle, L.; Tokuda, T.; Chiarle, R.; Giaccone, G.; Bugiani, O.; Tagliavini, F.; Frangione, B.; Ghiso, J. Substitutions at codon 22 of Alzheimer's abeta peptide induce diverse conformational changes and apoptotic effects in human cerebral endothelial cells. *J. Biol. Chem.* **2000**, *275*, 27110–27116.
13. Qiang, W.; Yau, W.M.; Luo, Y.; Mattson, M.P.; Tycko, R. Antiparallel beta-sheet architecture in Iowa-mutant beta-amyloid fibrils. *Proc. Natl. Acad. Sci. USA* **2012**, *109*, 4443–4448. [CrossRef] [PubMed]
14. Okamoto, A.; Yano, A.; Nomura, K.; Higai, S.; Kurita, N. Effect of D23N mutation on the dimer conformation of amyloid beta-proteins: Ab initio molecular simulations in water. *J. Mol. Graph. Model.* **2014**, *50*, 113–124. [CrossRef] [PubMed]
15. Sgourakis, N.G.; Yau, W.M.; Qiang, W. Modeling an in-register, parallel "iowa" abeta fibril structure using solid-state NMR data from labeled samples with rosetta. *Structure* **2015**, *23*, 216–227. [CrossRef] [PubMed]
16. Usachev, K.S.; Filippov, A.V.; Khairutdinov, B.I.; Antzutkin, O.N.; Klochkov, V.V. NMR structure of the Arctic mutation of the Alzheimer's Aβ(1–40) peptide docked to SDS micelles. *J. Mol. Struct.* **2014**, *1076* (Supplement C), 518–523. [CrossRef]
17. Glenner, G.G.; Wong, C.W. Alzheimer's disease: Initial report of the purification and characterization of a novel cerebrovascular amyloid protein. *Biochem. Biophys. Res. Commun.* **1984**, *120*, 885–890. [CrossRef]
18. Hou, L.; Shao, H.; Zhang, Y.; Li, H.; Menon, N.K.; Neuhaus, E.B.; Brewer, J.M.; Byeon, I.J.; Ray, D.G.; Vitek, M.P.; et al. Solution NMR studies of the A beta(1–40) and A beta(1–42) peptides establish that the Met35 oxidation state affects the mechanism of amyloid formation. *J. Am. Chem. Soc.* **2004**, *126*, 1992–2005. [CrossRef]
19. Vivekanandan, S.; Brender, J.R.; Lee, S.Y.; Ramamoorthy, A. A partially folded structure of amyloid-beta(1–40) in an aqueous environment. *Biochem. Biophys. Res. Commun.* **2011**, *411*, 312–316. [CrossRef]
20. Coles, M.; Bicknell, W.; Watson, A.A.; Fairlie, D.P.; Craik, D.J. Solution structure of amyloid beta-peptide(1–40) in a water-micelle environment. Is the membrane-spanning domain where we think it is? *Biochemistry* **1998**, *37*, 11064–11077. [CrossRef]
21. Chen, Y.R.; Huang, H.B.; Lo, C.J.; Wang, C.C.; Su, C.L.; Liu, H.T.; Shiao, M.S.; Lin, T.H.; Chen, Y.C. Abeta40(L17A/F19A) mutant diminishes the aggregation and neurotoxicity of Abeta40. *Biochem. Biophys. Res. Commun.* **2011**, *405*, 91–95. [CrossRef] [PubMed]
22. Jarvet, J.; Danielsson, J.; Damberg, P.; Oleszczuk, M.; Graslund, A. Positioning of the Alzheimer Abeta(1–40) peptide in SDS micelles using NMR and paramagnetic probes. *J. Biomol. Nmr* **2007**, *39*, 63–72. [CrossRef] [PubMed]
23. Korshavn, K.J.; Bhunia, A.; Lim, M.H.; Ramamoorthy, A. Amyloid-beta adopts a conserved, partially folded structure upon binding to zwitterionic lipid bilayers prior to amyloid formation. *Chem. Commun. (Camb)* **2016**, *52*, 882–885. [CrossRef] [PubMed]

24. Petkova, A.T.; Ishii, Y.; Balbach, J.J.; Antzutkin, O.N.; Leapman, R.D.; Delaglio, F.; Tycko, R. A structural model for Alzheimer's beta -amyloid fibrils based on experimental constraints from solid state NMR. *Proc. Natl. Acad. Sci. USA* **2002**, *99*, 16742–16747. [CrossRef]
25. Bertini, I.; Gonnelli, L.; Luchinat, C.; Mao, J.; Nesi, A. A new structural model of Abeta40 fibrils. *J. Am. Chem. Soc.* **2011**, *133*, 16013–16022. [CrossRef]
26. Parthasarathy, S.; Inoue, M.; Xiao, Y.; Matsumura, Y.; Nabeshima, Y.; Hoshi, M.; Ishii, Y. Structural Insight into an Alzheimer's Brain-Derived Spherical Assembly of Amyloid beta by Solid-State NMR. *J. Am. Chem. Soc.* **2015**, *137*, 6480–6483. [CrossRef]
27. Kollmer, M.; Close, W.; Funk, L.; Rasmussen, J.; Bsoul, A.; Schierhorn, A.; Schmidt, M.; Sigurdson, C.J.; Jucker, M.; Fandrich, M. Cryo-EM structure and polymorphism of Abeta amyloid fibrils purified from Alzheimer's brain tissue. *Nat. Commun.* **2019**, *10*, 4760. [CrossRef]
28. Lu, J.X.; Qiang, W.; Yau, W.M.; Schwieters, C.D.; Meredith, S.C.; Tycko, R. Molecular structure of beta-amyloid fibrils in Alzheimer's disease brain tissue. *Cell* **2013**, *154*, 1257–1268. [CrossRef]
29. Chen, Y.R.; Huang, H.B.; Lo, C.J.; Wang, C.C.; Ho, L.K.; Liu, H.T.; Shiao, M.S.; Lin, T.H.; Chen, Y.C. Effect of alanine replacement of l17 and f19 on the aggregation and neurotoxicity of arctic-type abeta40. *PloS ONE* **2013**, *8*, e61874.
30. Lo, C.J.; Wang, C.C.; Huang, H.B.; Chang, C.F.; Shiao, M.S.; Chen, Y.C.; Lin, T.H. The Arctic mutation accelerates Abeta aggregation in SDS through reducing the helical propensity of residues 15–25. *Amyloid Int. J. Exp. Clin. Investig. Off. J. Int. Soc. Amyloidosis* **2015**, *22*, 8–18. [CrossRef]
31. Liang, C.-T.; Huang, H.-B.; Wang, C.-C.; Chen, Y.-R.; Chang, C.-F.; Shiao, M.-S.; Chen, Y.-C.; Lin, T.-H. L17A/F19A Substitutions Augment the α-Helicity of β-Amyloid Peptide Discordant Segment. *PloS ONE* **2016**, *11*, e0154327. [CrossRef] [PubMed]
32. Williamson, M.P. Using chemical shift perturbation to characterise ligand binding. *Prog. Nucl. Magn. Reson. Spectrosc.* **2013**, *73*, 1–16. [CrossRef] [PubMed]
33. Weinstock, D.S.; Narayanan, C.; Baum, J.; Levy, R.M. Correlation between 13Calpha chemical shifts and helix content of peptide ensembles. *Protein Sci.* **2008**, *17*, 950–954. [CrossRef]
34. Marsh, J.A.; Singh, V.K.; Jia, Z.; Forman-Kay, J.D. Sensitivity of secondary structure propensities to sequence differences between alpha- and gamma-synuclein: Implications for fibrillation. *Protein Sci.* **2006**, *15*, 2795–2804. [CrossRef] [PubMed]
35. Wishart, D.S.; Sykes, B.D. The 13C chemical-shift index: A simple method for the identification of protein secondary structure using 13C chemical-shift data. *J. Biomol. Nmr* **1994**, *4*, 171–180. [CrossRef] [PubMed]
36. Kallberg, Y.; Gustafsson, M.; Persson, B.; Thyberg, J.; Johansson, J. Prediction of amyloid fibril-forming proteins. *J. Biol. Chem.* **2001**, *276*, 12945–12950. [CrossRef] [PubMed]
37. Murakami, K.; Irie, K.; Morimoto, A.; Ohigashi, H.; Shindo, M.; Nagao, M.; Shimizu, T.; Shirasawa, T. Synthesis, aggregation, neurotoxicity, and secondary structure of various A beta 1–42 mutants of familial Alzheimer's disease at positions 21–23. *Biochem. Biophys. Res. Commun.* **2002**, *294*, 5–10. [CrossRef]
38. Paivio, A.; Jarvet, J.; Graslund, A.; Lannfelt, L.; Westlind-Danielsson, A. Unique physicochemical profile of beta-amyloid peptide variant Abeta1–40E22G protofibrils: Conceivable neuropathogen in arctic mutant carriers. *J. Mol. Biol* **2004**, *339*, 145–159. [CrossRef]
39. Stroud, J.C.; Liu, C.; Teng, P.K.; Eisenberg, D. Toxic fibrillar oligomers of amyloid-beta have cross-beta structure. *Proc. Natl. Acad. Sci. USA* **2012**, *109*, 7717–7722. [CrossRef]
40. Schmidt, M.; Rohou, A.; Lasker, K.; Yadav, J.K.; Schiene-Fischer, C.; Fandrich, M.; Grigorieff, N. Peptide dimer structure in an Abeta(1–42) fibril visualized with cryo-EM. *Proc. Natl. Acad. Sci. USA* **2015**, *112*, 11858–11863. [CrossRef]
41. Xiao, Y.; Matsuda, I.; Inoue, M.; Sasahara, T.; Hoshi, M.; Ishii, Y. NMR-based site-resolved profiling of beta-amyloid misfolding reveals structural transitions from pathologically relevant spherical oligomer to fibril. *J. Biol. Chem.* **2020**, *295*, 458–467. [CrossRef] [PubMed]
42. Colvin, M.T.; Silvers, R.; Ni, Q.Z.; Can, T.V.; Sergeyev, I.; Rosay, M.; Donovan, K.J.; Michael, B.; Wall, J.; Linse, S.; et al. Atomic Resolution Structure of Monomorphic Abeta42 Amyloid Fibrils. *J. Am. Chem. Soc.* **2016**, *138*, 9663–9674. [CrossRef] [PubMed]
43. Gu, L.; Tran, J.; Jiang, L.; Guo, Z. A new structural model of Alzheimer's Abeta42 fibrils based on electron paramagnetic resonance data and Rosetta modeling. *J. Struct. Biol.* **2016**, *194*, 61–67. [CrossRef] [PubMed]

44. Ahmed, M.; Davis, J.; Aucoin, D.; Sato, T.; Ahuja, S.; Aimoto, S.; Elliott, J.I.; Van Nostrand, W.E.; Smith, S.O. Structural conversion of neurotoxic amyloid-beta(1–42) oligomers to fibrils. *Nat. Struct. Mol. Biol.* **2010**, *17*, 561–567. [CrossRef]
45. Xiao, Y.; Ma, B.; McElheny, D.; Parthasarathy, S.; Long, F.; Hoshi, M.; Nussinov, R.; Ishii, Y. Abeta(1–42) fibril structure illuminates self-recognition and replication of amyloid in Alzheimer's disease. *Nat. Struct. Mol. Biol.* **2015**, *22*, 499–505. [CrossRef]
46. Wishart, D.S.; Bigam, C.G.; Yao, J.; Abildgaard, F.; Dyson, H.J.; Oldfield, E.; Markley, J.L.; Sykes, B.D. 1H, 13C and 15N chemical shift referencing in biomolecular NMR. *J. Biomol. Nmr.* **1995**, *6*, 135–140. [CrossRef]

International Journal of
Molecular Sciences

MDPI

Article

Low-Dose Phosphodiesterase III Inhibitor Reduces the Vascular Amyloid Burden in Amyloid-β Protein Precursor Transgenic Mice

Yusuke Yakushiji [1,*], Kazuhiro Kawamoto [1], Kazuyoshi Uchihashi [2], Masafumi Ihara [3], Shigehisa Aoki [2], Yukiko Nagaishi [1], Kohei Suzuyama [1], Yumiko Tsugitomi [4] and Hideo Hara [1]

1 Division of Neurology, Department of Internal Medicine, Saga University Faculty of Medicine, Saga 849-8501, Japan; kawamotokazu@icloud.com (K.K.); suzusuzu61@gmail.com (Y.N.); sj8817@cc.saga-u.ac.jp (K.S.); hihara@cc.saga-u.ac.jp (H.H.)
2 Department of Pathology and Microbiology, Saga University Faculty of Medicine, Saga 849-8501, Japan; uchihashi.kazuyoshi.du@mail.hosp.go.jp (K.U.); aokis@cc.saga-u.ac.jp (S.A.)
3 Department of Neurology, National Cerebral and Cardiovascular Center, Suita 564-8565, Japan; ihara@ncvc.go.jp
4 Department of Internal Medicine, Saga University Faculty of Medicine, Saga 849-8501, Japan; y.tsugitomi@gmail.com
* Correspondence: yakushij@cc.saga-u.ac.jp; Tel.: +81-952-31-6511

Received: 28 February 2020; Accepted: 24 March 2020; Published: 26 March 2020

Abstract: A previous study reported that relatively high-dose cilostazol (0.3%) promoted the drainage of cerebrovascular amyloid-β (Aβ) protein in Aβ Precursor Protein (APP) transgenic mice overexpressing vasculotropic Aβ. We investigated whether lower-dose cilostazol can decrease micro-hemorrhages and Aβ deposition in the brain using APP transgenic mice. At baseline, 14-month-old female Tg2576 mice were randomly assigned to a control group (vehicle), aspirin group (0.01% aspirin), or cilostazol group (0.01% cilostazol). The severity of cerebral micro-hemorrhages (i.e., number), area of senile plaque, and severity of vascular amyloid burden (quantified with cerebral amyloid angiopathy (CAA) score (=number of Aβ-positive vessels × severity of amyloid burden of Aβ-positive vessels) were evaluated in the brain of mice aged 15 and 21–23 months. At 15 months, no differences were shown in each pathological change among the three groups. At 21–23 months, there were no differences in the severity of cerebral micro-hemorrhages or area of senile plaque among the three groups. However, the CAA score was significantly lower in the cilostazol compared to the control group ($p = 0.046$, Mann–Whitney U test), although no difference was seen between the control and aspirin group. Our study showed that lower-dose cilostazol could reduce the vascular amyloid burden without increasing cerebral micro-hemorrhages in APP transgenic mice.

Keywords: phosphodiesterase III inhibitor; cerebral micro-hemorrhage(s); cerebral amyloid angiopathy; amyloid-β protein; transgenic mice

1. Introduction

Sporadic cerebral amyloid angiopathy (CAA) is characterized by the progressive deposition of amyloid-β (Aβ) protein in the walls of small- to medium-sized arteries, arterioles, and capillaries in the cerebral cortex and overlying leptomeninges [1,2]. CAA is a common age-related cerebral small vessel disease (SVD) in the elderly [2,3] (especially those with Alzheimer's disease [4]), but it is most often recognized clinically by symptomatic intracerebral hemorrhage (ICH) restricted to the lobar areas of the brain [5,6]. Long-term antiplatelet therapy, which is widely used for the secondary prevention of cerebral infarction, myocardial infarction, and peripheral artery diseases, could increase the incidence of ICH. In consideration of the high incidence of cerebral bleeding in patients with

CAA and/or Alzheimer's disease, especially in antiplatelet drug users [7,8], the safety of antiplatelet therapies for those patients should be explored.

Cyclic nucleotide phosphodiesterases (PDEs) play critical roles in regulating intracellular cyclic nucleotides (cyclic adenosine monophosphate (cAMP) and cyclic guanosine monophosphate), which are important secondary messengers involved in intracellular signal transduction in all tissues. PDE III is the major cAMP-hydrolyzing PDE (a negative regulator of cAMP) uniquely expressed in vascular smooth muscle cells. A selective inhibitor of PDE III, cilostazol, has multiple effects on the vasculature including vasodilatation, antioxidation, anti-inflammation, the regulation of smooth muscle cell, and an increase in cerebral hemodynamics, pulse duration time, and arterial elasticity with the maintenance of microvascular integrity [9]. Cilostazol is known as a unique antiplatelet drug, which is superior to aspirin in terms of safety for reducing ICH after an ischemic stroke [10]. Cilostazol ameliorates collagenase-induced cerebral hemorrhage by protecting the blood–brain barrier in mice [11]. In a CAA mice model (Tg-SwDI mice), cilostazol improved cognitive performance, which may be associated with reduced Aβ deposition by cilostazol (0.3% cilostazol) [12]. However, it is unknown whether the safety and efficacy of cilostazol could be replicated in different settings (i.e., drug dose or experimental transgenic mice). Thus, we sought to determine whether lower-dose cilostazol could reduce the incidence of cerebral (micro-)hemorrhages or cerebrovascular Aβ depositions using different transgenic mice as a CAA model.

2. Results

2.1. Survival Rate, Feed Consumption, and Drug Intake

Sixty Tg2576 mice aged 3 months initially received the vehicle, but 10 mice died due to unknown reasons before 14 months. Thus, 50 mice aged 14 months were divided into three groups: control group (n = 14), aspirin group (n = 18), and cilostazol group (n = 18) (Figure 1). Within a month after the grouping, two mice died due to unknown reasons (control group, n = 1; cilostazol group, n = 1). Of the 48 survivors aged 15 months, 14 mice (control group, n = 4; aspirin group, n = 5; cilostazol group, n = 5) were randomly selected for the first evaluation (i.e., the first specimen). Among the remaining mice (n= 34), four mice died due to unknown reasons (control group, n = 1; cilostazol group, n = 3). Finally, 30 surviving Tg2576 mice aged 21–23 months were evaluated as the second specimens. There was no significant difference in the survival rate after the grouping among the three groups when the 14 mice evaluated for the first specimen were excluded from the analyses (p = 0.109, log-rank test: Figure A1: please see appendix). Even in the two group comparisons, no differences were seen in the survival rate between the control group and the aspirin or cilostazol group (p = 0.098 and p = 0.550, log-rank test). Table 1 shows estimated individual food consumption and drug intake (per mouse) of the three groups. There was no significant difference in food consumption between the control group (mean, standard deviation [SD]: 3.57 ± 0.40 g/day) and the aspirin group (3.67 ± 0.47 g/day) or the cilostazol group (3.67 ± 0.44 g/day) (p = 0.289 and p = 0.543, Mann–Whitney U test). Individual daily drug intake in the aspirin group (14.7 ± 1.9 mg/kg/day) was similar to that in the cilostazol group (14.7 ± 1.8 mg/kg/day).

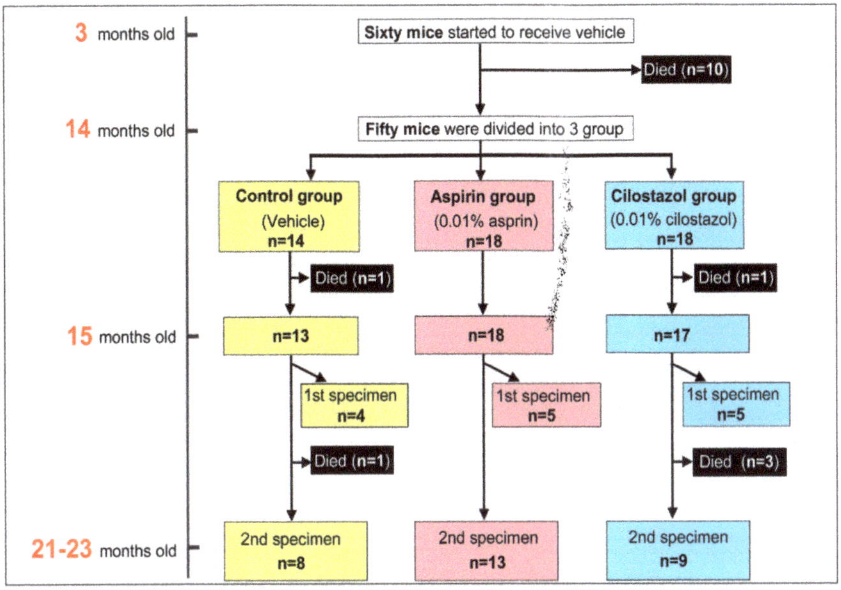

Figure 1. The flow diagram of study schedule and grouping.

Table 1. Estimated individual feed consumption and drug intake (per mouse) of the three groups.

	Control Group	Aspirin Group	Cilostazol Group
Measurements, times	75	77	75
Estimated individual food consumption, g/day (SD)	3.57 (0.40)	3.67 (0.47) [a]	3.67 (0.44) [b]
Estimated individual drug intake, mg/kg/day (SD)	NA	14.7 (1.9)	14.7 (1.8)

[a] $p = 0.289$ vs. control; [b] $p = 0.324$ vs. control (Mann–Whitney U test); NA = not applicable; SD = standard deviation.

2.2. *Confirmation of Age-Related Cerebrovascular Amyloid Burden and Smooth Muscle Cell Loss*

Confocal microscopic observation of double-immunolabeled vessels in different Tg2576 mice (15 months old, and 23 months old) fed with standard pelleted chow (i.e., control group) confirmed the age-related progression of amyloid burden and loss of smooth muscle cells (Figure 2a–d). The findings of the negative controls for Aβ show only a faint background stain in the vessel walls (Figure 2e–h).

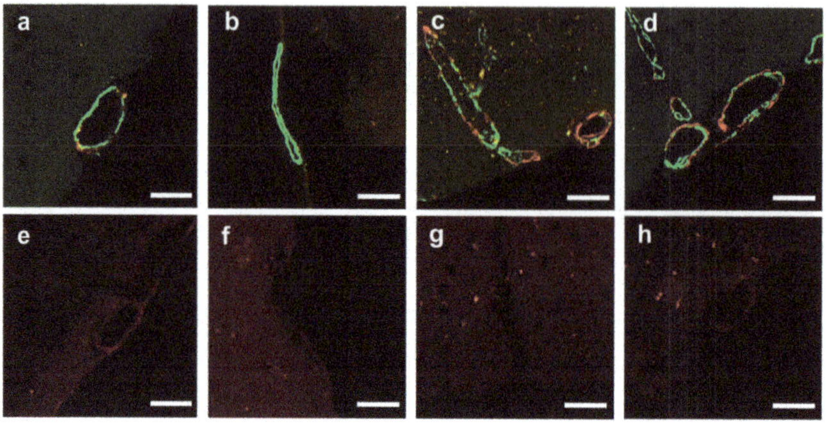

Figure 2. Upper line (**a–d**): Serial confocal microscopic changes of double-immunolabeled vessels (green, smooth muscle actin; red, amyloid). Lower line (**e–h**): Confocal microscopic findings represent negative controls without primary antibody for Aβ. a, b, Leptomeningeal vessel in a 15-month-old mouse shows small amyloid deposition and focal loss of smooth muscle cells at the site of cerebrovascular amyloid; c, d, In a 23-month-old mouse, smooth muscle cells are lost, and a thick sheet of amyloid covers the wall of a leptomeningeal vessel. e–f, Photos of the negative controls for Aβ (each section adjacent to a, b, c, d, respectively) show only a faint background stain in the vessel walls. Scale bar: 50 μm.

2.3. Acute Subdural or Cerebral Bleeding(S)

In both the first specimens (evaluated in mice aged 15 months) and the second specimens (evaluated in mice aged 21–23 months), no acute subdural or cerebral bleeding was found. However, in the second specimens in the aspirin group, small accumulations of erythrocytes were found in one mouse around a leptomeningeal artery with lymphocytic infiltration (Figure A2a). Hemosiderin depositions (arrow heads), representing old bleeding, were also seen around a leptomeningeal artery with vasculitis in the same mouse (Figure A2b). No accumulations of erythrocyte were found in either the control and the cilostazol group.

2.4. Cerebral Micro-Hemorrhages

The representative images of cerebral micro-hemorrhages are shown in Figure 3. In the first specimens (evaluated in mice aged 15 months), there were no differences in the number of cerebral micro-hemorrhages between the control group (median, interquartile range (IQR): 1, 0–2) and the aspirin group (1, 0.5–3.5) or the cilostazol group (0, 0–3.5) ($p = 0.730$ and $p = 1.000$, Mann–Whitney *U* test: Figure 4a). In the second specimens (evaluated in mice aged 21–23 months), there were also no differences in the number of cerebral micro-hemorrhages between the control group (3.5, 1.5–4) and the aspirin group (4, 1–7.5) or the cilostazol group (5, 2.5–8) ($p = 0.804$, and $p = 0.277$, Mann–Whitney *U* test, respectively: Figure 4b).

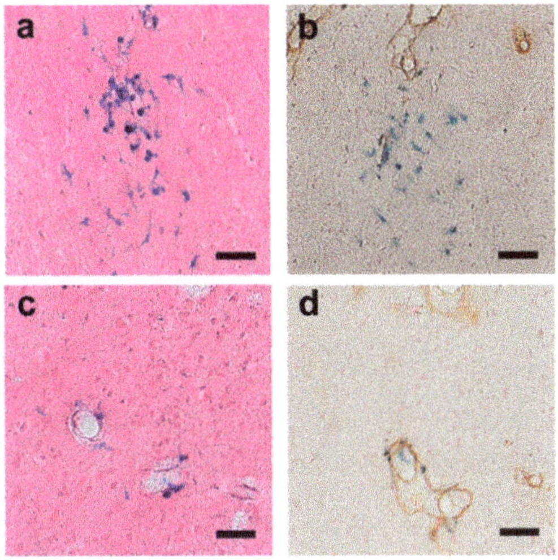

Figure 3. Representative images of cerebral micro-hemorrhage findings (a, b, Perls's Berlin blue stain with Nuclear Fast Red (Kernechtrot stain solution)); b, d, double-labeled for amyloid (brown) and hemosiderin (blue)). (**a**) Clusters of hemosiderin staining are shown in the brain parenchyma (cortex at 1.35 mm behind Bregma). (**b**) In an adjacent section to a, some of them are in contact with amyloid-β (Aβ)-positive vessels. (**c**) Localized hemosiderin shown around the vessel wall (cortex at 0.85 mm behind Bregma). (**d**) In an adjacent section to c, localized bleeding to amyloid-laden vessels is shown. Scale bars indicate 10 μm.

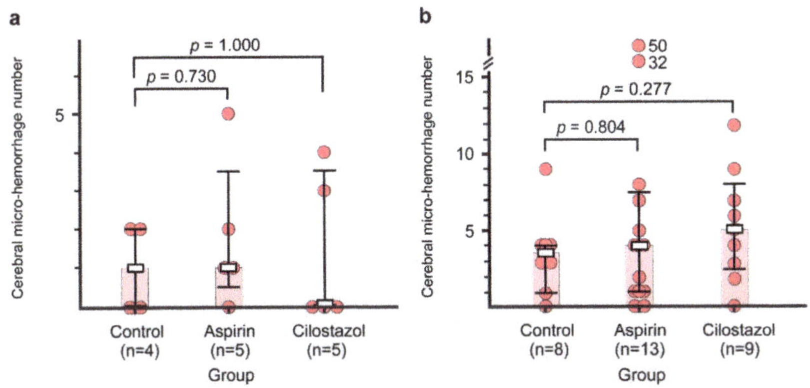

Figure 4. The number of cerebral micro-hemorrhages in the control, aspirin, and cilostazol groups evaluated at 15 months (**a**) and 21–23 months (**b**). Light-red-colored bar graphs indicate median. Boxes and bars indicate median and interquartile range, respectively. Red circles represent the number of cerebral micro-hemorrhages of each mouse. The numbers next to the red circles indicate the number of cerebral micro-hemorrhages, which are out of range of the vertical axis scale in two mice (b, Aspirin group).

2.5. CAA Burden

The representative images of CAA burden are shown in Figure 5. In the first specimens, there were no differences in the CAA score between the control group (median, IQR: 2, 1–4) and the aspirin group (6, 1–7) or the cilostazol group (2, 1–8) ($p = 0.556$ and $p = 1.000$, Mann–Whitney *U* test: Figure 6a). In the second specimens, there was no difference ($p = 0.750$, Mann–Whitney *U* test) in the CAA score between the control group (28.5, 24.5–58) and the aspirin group (28, 23–47). Even after an outlier of the CAA score in the aspirin group (168) was excluded, non-significant results still remained (median, IQR: the aspirin group, 27, 18.5–40.5, $p = 0.521$, Mann–Whitney *U* test). In contrast, the CAA score of the cilostazol group (14, 11–31) was significantly lower in the control group ($p = 0.046$, Mann–Whitney *U* test: Figure 6b).

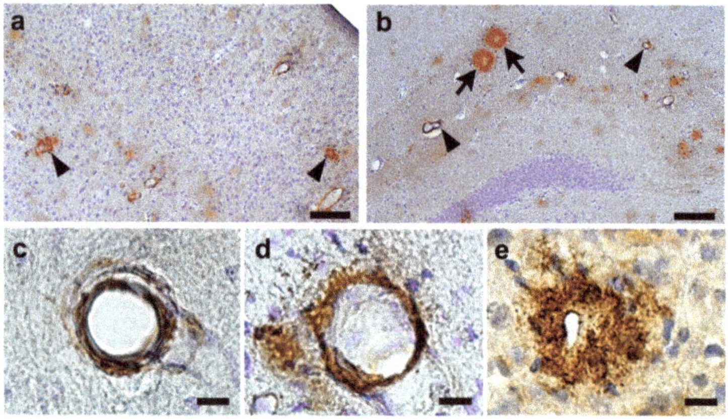

Figure 5. Representative images of cerebral amyloid angiopathy (CAA) in a Tg2576 mouse (aged 23 months old). Pan-Aβ immunostained sections show significant CAA in the cortex at 1.08 mm behind Bregma (a: arrowheads) and mild to moderate CAA (b: arrowheads) in the hippocampus at 1.33 mm behind Bregma. Arrows show senile plaques in the hippocampus (**b**). Vessel with a thin rim of amyloid in the vessel wall (c; severity grade, 1); vascular amyloid with amyloid infiltrating the surrounding neuropil (d; severity grade, 2); dysphoric amyloid with amyloid deposition within the vessel wall and with a thick and complete amyloid coat around the vessel wall (e; severity grade, 3). Scale bars indicate 100 μm (**a**,**b**) and 10 μm (**c**–**e**).

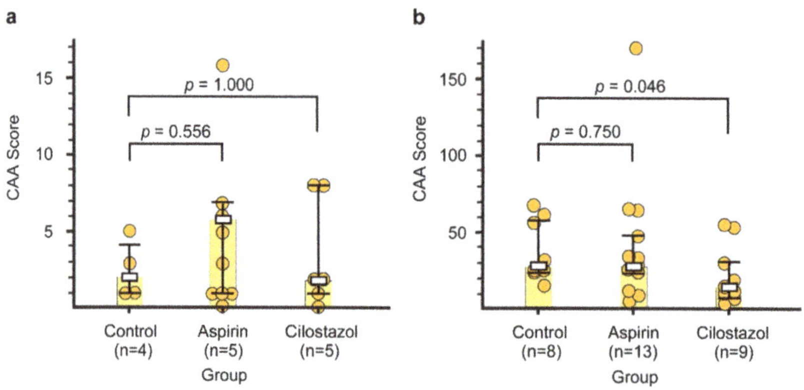

Figure 6. The cerebral amyloid angiopathy (CAA) scores in the control, aspirin, and cilostazol groups evaluated at 15 months (**a**) and 21–23 months (**b**). Light-yellow-colored bar graphs indicate median. Boxes and bars indicate median and interquartile range, respectively. Yellow circles represent the CAA score of each mouse.

2.6. Senile Plaque

In the first specimens, there were no differences in the percent area of senile plaque between the control group (median, IQR: 0.05, 0.03–0.40) and the aspirin group (0.24, 0.08–0.83) or the cilostazol group (0.73, 0.00–0.12) (p = 0.286 and p = 0.905, Mann–Whitney U test: Figure 7a). In the second specimens, there were also no differences in the percent area of senile plaque between the control group (median, IQR: 0.27, 0.19–0.54) and the aspirin group (0.28, 0.22–0.45) or the cilostazol group (0.50, 0.16–0.84) (p = 0.750 and p = 0.888, Mann–Whitney U test: Figure 7b).

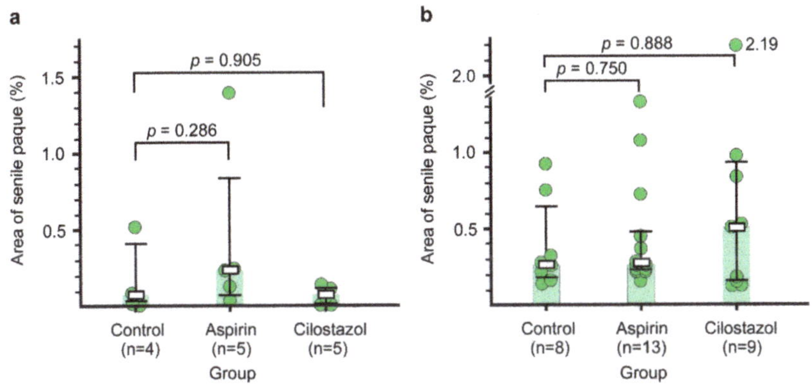

Figure 7. The percent area of senile plaque in the control, aspirin, and cilostazol groups evaluated at 15 months (**a**) and 21–23 months (**b**). Light-green-colored bar graphs indicate median. Boxes and bars indicate median and interquartile range, respectively. Green circles represent the value of percent area of senile plaque of each mouse. The numbers next to the green circle (2.19) indicate percent area of senile plaque out of the range of the vertical axis scale (b, Cilostazol group).

3. Discussion

The main finding of this study is that the CAA burden in Tg2576 mice could be reduced by half with long-term cilostazol therapy, but not with aspirin. In particular, such a long-term treatment effect on the CAA burden of APP transgenic mice was firstly archived with a lower dose of cilostazol (conducted to intake 20 mg/kg/day; resulting intake 14.7 mg/kg/day) compared to a previous study (600 mg/kg/day) [12]. However, we found no long-term treatment effect of either cilostazol or aspirin on the severity of cerebral micro-hemorrhages or on the expansion of senile plaque.

PDE family proteins, mostly expressed in the brain, have attracted attention as a source of new targets for the treatment of psychiatric and neurodegenerative disorders [13,14]. Previous studies of animal models have shown that the phosphodiesterase III inhibitor, cilostazol, decreases cerebral amyloid-β accumulation [12,15]. Regarding the long-term effects of cilostazol on CAA, one study using Tg-SwDI mice mainly expressing vasculotropic Aβ demonstrated that relatively high-dose cilostazol (equivalent to 600 mg/kg/day intake in 25 g weight mouse) decreased the Aβ accumulation of the brain, resulting in improved cognitive performance [12]. Regarding the short-term effects of cilostazol, one study using C57BL/6J mice reported that the oral administration of cilostazol (20 mg/kg/day) for 6 weeks around the injection of Aβ into the cerebral ventricle almost completely prevented Aβ accumulation in the brain [15]. Thus, our finding has added one line of evidence that the efficacy of lower-dose cilostazol was replicated in a different experimental setting. Given the dose-dependent side effects (e.g., headache and dizziness) of cilostazol [16], long-term administration of the lower dose of cilostazol could be an optional strategy of the treatment for CAA burden.

The protective role of cilostazol against Aβ burden in the CAA model mice (i.e., Tg-SwDI mice or Tg2576 mice) appeared to be its vasculotropic effects, as long-term aspirin treatment did not reverse Aβ deposition. Our result of the reduced CAA score and non-reduced senile plaque suggests that the promotion of Aβ metabolism by vasculotropic cilostazol was achieved by its easy access to the perivascular area but not to the brain parenchyma. This is consistent with the finding that cilostazol poorly penetrates the blood–brain barrier [12]. The main mechanism promoting Aβ metabolism by cilostazol appears to be the increase in perivascular drainage of Aβ, followed by the decrease in degenerative changes in vascular walls with Aβ deposits [12]. Since the motive force for perivascular Aβdrainage appears to be generated by arterial pulsations [17,18], the direct action of cilostazol on the vascular smooth muscle cells to increase pulse duration time [19] and arterial elasticity [16] may have contributed to facilitating the perivascular drainage of Aβ.

None of the Aβ-targeted phase 3 clinical trials in Alzheimer's disease has shown statistically significant benefits on its pre-specified clinical endpoints. Several of these trials, however, were mis-designed in terms of patient selection, choice of agent, target engagement, and/or dose, or they had to be halted because of the off-target side effects [20]. A recent phase 2 clinical trial in patients with CAA has shown that immunotherapy using the anti-Aβ40 antibody (Ponezumab) also did not show the prespecified efficacy (improvement in cerebrovascular reactivity measured by functional magnetic resonance imaging (MRI)) [21]. The Aβ-targeted immunotherapy could cause amyloid-related imaging abnormalities (ARIA) representing vasogenic edema, micro-hemorrhages, or cortical, superficial siderosis on MRI. ARIA, which appeared to be a dose-dependent phenomenon, sometimes causes transient symptoms of headaches, confusion, and visual disturbances [22]. Thus, it might not be realistic to increase the dose of the anti-Aβ antibody to enhance the effectiveness. To explore treatment to minimize Aβ accumulation, it might to be crucial to consider not only the suppression of Aβ over-production, but also the promotion of Aβ clearance in CAA and/or Alzheimer's disease patients. A Japanese retrospective study reported that the combination therapy group (using donepezil plus cilostazol) was more effective for cognitive decline in patients with mild dementia compared to the donepezil-only group [23]. Thus, given that the balance between Aβ synthesis and clearance determines brain Aβ accumulation, a multidrug combination (e.g., low-dose anti-Aβ antibody, and low-dose cilostazol) therapy could provide a mainstream cure in the early stages of CAA and/or Alzheimer's disease [24].

Our result that aspirin had no influence on the severity of cerebral micro-hemorrhages was in line with a previous study [25]. A characteristic feature of cilostazol is that it has weaker hemorrhagic side effects than other antiplatelet drugs [26] and does not increase the bleeding time [27]. In fact, a previous study demonstrated that the short-term administration of cilostazol (30 mg/kg/day) reduced the intracranial hemorrhage volume along with sufficient inhibition of platelet aggregation in non-transgenic mice [11]. However, the present study did not suggest the expected safety benefits of cilostazol for reducing micro-hemorrhages over aspirin. In other words, this study offered evidence that low-dose cilostazol could improve the vascular amyloid burden without increasing cerebral micro-hemorrhages in a mouse model of CAA.

We also acknowledge a limitation of this study. We did not evaluate the effects of lower-dose cilostazol on cognitive performance in Tg2576 mice. Although we had discussed this issue at the planning state of the study, we decided to just focus on whether lower-dose cilostazol could reduce the pathological burden of the brain in Tg2576 mice, because of our funding limitations to develop a research environment to fully examine the cognitive function of the mice (i.e., to buy additional Tg2576 mice or appropriate experimental devices). However, our current results allow us to believe that further investigation would be meaningful to calcify whether lower-dose cilostazol could improve cognitive performance in Tg2576 mice.

4. Materials and Methods

4.1. Standard Protocol Approval

All animal procedures were performed according to the guidelines of the Animal Use and Care Committee of the Saga University (Saga, Japan). All protocols were approved by the Animal Use and Care Committee and the Genetic Recombination Experimental Committee (ethical approval code: 23-024-2), as well as Animal Research: Reporting of In Vivo Experiments guidelines [28]. The experimental data is available from the supplemental materials: Data S1 and Data S2).

4.2. Animals

As a CAA animal model, we used female B6, SJL-Tg 2576Kha (APPSWE) transgenic mice (Tg2576 mice: Taonic Bioscience, Inc., NY, USA). The Tg2576 mice were housed in a room with a 12-h light/dark cycle (light on at 7:00 a.m.) with access to food and water ad libitum. A flow diagram of the study schedule and grouping is shown in Figure 1. Sixty Tg2576 mice aged 3 months were started on standard pelleted chow (vehicle). Eleven months after, mice aged 14 months were divided into three groups with adjustment for body weight: control group (mice fed with standard pelleted chow only), aspirin group (mice fed with the pelleted chow containing 0.01% aspirin), and cilostazol group (mice were fed with pelleted chow containing 0.01% cilostazol). The dose setting for each drug is described in the next section (please see 4.3, Drugs). The Tg2576 mice were randomly assigned to each group (in considering drug-associated death, we a priori planned to allocate more mice to the aspirin or the cilostazol group).

4.3. Drugs

In the brain of Tg2576 mice, Aβ deposits developed after 8 months, and dissemination of Aβ plaque progressed from 15 months to 23 months [29]. Therefore, we planned to evaluate the long-term effects of the drugs using mice aged 15 months or 21–23 months. For long-term administration of drugs, we selected oral drug intake using a pelleted chow containing each drug. The dose of cilostazol to be administered in mice varied because of the different bioavailability between humans and mice. A previous study reported that, in transgenic C57BL/6-Tg(Thy1-APPSwDutIowa) BWevn/J mice fed with pelleted chow containing 0.3% cilostazol (equivalent to 600 mg/kg/day intake in 25 g weight mouse), phosphodiesterase III inhibitor promoted the drainage of cerebrovascular Aβ [12], while a previous study demonstrated that in Aβ-injected wild-type mice, cilostazol administration of 10–20 mg/kg/day for 2 weeks exerted a protective effect against Aβ-induced cognitive deficits along with decreased Aβ accumulation [15]. To evaluate the effect of a lower dose of cilostazol on the CAA mice model, we used pelleted chow containing 0.01% cilostazol (equivalent to 20 mg/kg/day intake in 25 g weight mouse, donated by Otsuka Pharmaceutical, Tokyo, Japan) for treatment of the cilostazol group. Regarding the dose of aspirin, a previous study reported that, in spontaneous hypertensive rats, aspirin-attenuated collagen-induced platelet aggregation at 10–100 mg/kg in rats [30]. Clinically, daily intake of low-dose aspirin in humans is almost similar to cilostazol. Therefore, similar to the drug concentration of cilostazol, we used the pelleted chow containing 0.01% aspirin (equivalent to 20 mg/kg/day intake in 25 g weight mouse: donated by Otsuka Pharmaceutical, Tokyo, Japan) for treatment of the aspirin group.

4.4. Measurements of Estimated Individual Food Consumption and Drug Intake by the Groups

Food consumption per cage was measured once a week. Daily estimated individual food consumption (g/day) was calculated by the following formula: food consumption during a week per cage (g) / 7 (days) / the number of surviving mice in the cage. Daily estimated individual drug intake per weight of a mouse (mg/kg/day), which was also estimated per cage per week, was calculated by the following formula: the daily individual food consumption of the cage (g/day × 1000) × drug concentration/mean weight of a surviving mouse in the cage (kg).

4.5. Histology and Immunohistochemistry

To evaluate the serial effects of aspirin or cilostazol on Aβ deposition, pathological examinations were performed 1 month after the grouping (the first specimen, 15 months old, (n = 14)) and at 7–9 months after the grouping (the second specimen, 21–23 months old, (n = 30)). Mice were overdosed with sodium pentobarbital (50 mg/kg, intraperitoneal) and perfused transcardially with phosphate-buffered saline (PBS), followed by 4% paraformaldehyde in 0.1 M PBS. The brains were immediately removed, immersion fixed for 1 d in 4% paraformaldehyde, followed by 2 d in 10% sucrose in 0.01 M PBS, and 2 d in 30% sucrose in 0.01 M PBS. Post-fixed brains were cryoprotected, frozen, and sectioned at 25 μm with a freezing–sliding microtome [31]. Hematoxylin and eosin (H&E) and Congo red staining were done according to standard protocols [32]. Perls's Berlin Blue method was used to visualize ferric iron in hemosiderin (with Nuclear Fast Red (Kernechtrot stain solution: Lot number, 130312: Muto Pure Chemicals Co., LTD., Tokyo, Japan)) [32,33]. According to previously published protocols [31,34], the pan-Aβ stain was performed with the following primary and secondary antibodies: the primary antibody, rabbit polyclonal antibody to Aβ (1–40) (catalog number, 44–136: Invitrogen by Thermo Fisher Scientific Inc., CA, USA); the secondary antibody, biotinylated goat anti-rabbit IgG antibody (catalog number, BA-1000: Vector Laboratories, CA, USA). For confocal microscopy, double-labeling for Aβ and smooth muscle cells was achieved simultaneously using frozen sections as follows: for Aβ, the primary antibody with rabbit polyclonal antibody to Aβ (1–40) (catalog number, 44–136: Invitrogen by Thermo Fisher Scientific Inc., CA, USA); the secondary antibody with Donkey anti-Rabbit IgG antibody Cy3 conjugate (1:500; product number, AP182C: Millipore Co., CA, USA); for smooth muscle cells, monoclonal anti-alpha-smooth muscle actin conjugate FITC (product number, F3777: Sigma-Aldrich Co. LLC, MO, USA). Sections were mounted with Fluoromount (catalog number: K 024: Diagnostic BioSystems, Hague, Netherlands) and analyzed with a Confocal Laser Scanning Microscope LSM880+Airyscan Fast (Zeiss, Oberkochen, Germany).

4.6. Pathological Evaluations

4.6.1. Observation of Natural Changes of Cerebrovascular Amyloid Burden and Smooth Muscle Cell Loss

To confirm the findings of the age-related progression of cerebrovascular amyloid burden and accompanying smooth muscle cell loss in Tg2576 mice, mice fed with standard pelleted chow (i.e., control group) were evaluated at different times (15 months old and 23 months old) using confocal microscopy with double-labeling for Aβ and smooth muscle cell actin. We also evaluated with negative controls without primary antibody for Aβ (follow the same staining protocol without the addition of a primary antibody) to dismiss a possible age-dependent nonspecific stain for the secondary antibody.

4.6.2. Specimens and Raters

All findings were evaluated through the cortex and the hippocampus of the right hemisphere. The ratings of hemorrhagic findings of the brain and any Aβ-positive vessels were evaluated by two raters (K.K. and K.U.), who were blinded to our hypothesis and information of the food content on each mouse. If results were different between the raters, the final decision was made after discussions by the two raters. CAA severity, described below, was classified by a single rater (K.K.) after the determinations of Aβ-positive vessels. Regarding senile plaque, quantitative analysis was performed with a semiautomatic computer-assisted processing system, as mentioned below, by a single rater (K.K.).

4.6.3. Quantitation of Cerebral Hemorrhage(s)

Any acute subdural or cerebral bleeding(s) was defined as a large accumulation of erythrocytes in the intracranial space observed on the H&E stains with sets of systematically sampled sections (every 10th section throughout the cortex and the hippocampus (right hemisphere only)). Cerebral micro-hemorrhages, defined as clusters of hemosiderin staining on Perls's Berlin blue stain with a delayed appearance of hemosiderin-positive microglia [35], located in the brain parenchyma and the around the vessel walls (Figure 3), were quantified on additional sets of every 10th section (right hemisphere only). The ratings of these findings were evaluated by two raters (K.K. and K.U.), who were blinded to our hypothesis and information of the food content on each mouse.

4.6.4. Quantitative Analysis of CAA Burden

All quantification of CAA burden was done as previously published [31]. The frequency and severity of CAA were quantified on systematically sampled serial pan-Aβ immunostained sections throughout the region of interest (every 10th section through the cortex, and the hippocampus). Severity of CAA was classified by a single rater (K.K.), who were blinded to our hypothesis and the information of the food content on each mouse. "CAA frequency" was calculated by counting the total number of any Aβ-positive vessels in the entire set of systematically sampled sections. Regarding the CAA severity, all Aβ-positive vessels were classified into three grades (Figure 5) with a rating scale as described previously [36,37]: severity grade 1 = vessels with a thin rim of amyloid in the vessel wall; severity grade 2 = vascular amyloid with amyloid infiltrating the surrounding neuropil; severity grade 3 = dysphoric amyloid with amyloid deposition within the vessel wall and with a thick and complete amyloid coat around the vessel wall. The mean for all Aβ-positive vessels was taken as CAA severity. To evaluate comprehensive CAA burden, a "CAA score" was calculated by multiplying CAA frequency with CAA severity [31].

4.6.5. Quantitative Analysis of Senile Plaque

Using a computer-assisted processing system (Image J version 1.49 for Mac; National Institutes of Health, Bethesda, MD, USA), the area of pan-Aβ stained lesions in the cortex and the hippocampus, corresponding to senile plaques, was quantified semi-automatically [38] (Scheme A1: please see appendix) by a single rater (K.K.), who was blinded to our hypothesis and the information of the food content on each mouse. Every section of the right hemisphere with Aβ stained electrically was converted to Joint Photographic Experts Group (JPEG) images with the same scale. These JPEG images were analyzed with Image J (version 1.49 for Mac: National Institutes of Health, Bethesda, MD, USA) with appropriate calibrations, as follows. Measurement of area of the section: (1) fill section with red color; (2) dichotomization of color for black and white using a semiautomatic method with an appropriate color threshold; (3) measurement of the black area. Measurement of total area of senile plaque: (1) digital stripping of Aβ-stained lesions located out of the regions of interest; (2) change color of the remaining Aβ-stained area to red using manual calibration with appropriate color threshold; (3) dichotomization of color for black (Aβ-stained lesions in regions of interest (i.e., cortex or hippocampus)) and white (other) using a semiautomatic method with an appropriate color threshold; (4) measurement of the black area. Thus, the total areas of the section, as well as total areas of Aβ-stained lesions, in the right hemisphere were quantitatively measured to count the pixels with a given intensity. To evaluate the degree of the senile plaque in brain parenchyma (including cortex and hippocampus), the percentage of senile plaque area in the brain (% area of senile plaque) was calculated with the following formula:

$$\% \text{ senile plaque} = \text{total area of senile plaque} / \text{total area of the sections} \quad (1)$$

4.6.6. Statistical Analyse

All statistical analyses were performed with two group comparisons (control group vs. aspirin group or cilostazol group) using the IBM SPSS statistics software program, version 21.0 (IBM, Armonk, NY, USA). As our variables (including cerebral micro-hemorrhages number, CAA score, and percent area of senile plaque) did not follow a normal distribution, a non-parametric test (Mann–Whitney *U* test) was used for the two group comparisons. Log-rank test was used to compare survival rates. p values < 0.05 were considered statistically significant.

5. Conclusions

The present study shows pathological evidence that CAA burden is reduced by cilostazol, even at a low dose. Cilostazol may provide a novel, promising therapeutic target for patients with CAA and/or Alzheimer's disease, potentially in combination with early Aβ immunization therapy.

Supplementary Materials: Supplementary materials can be found at http://www.mdpi.com/1422-0067/21/7/2295/s1.

Author Contributions: Conceptualization, Y.Y. and H.H.; methodology, Y.Y., S.A. and Y.T.; software, Y.Y., K.K.; validation, K.K. and K.U.; formal analysis, K.K. and K.U.; investigation, Y.Y. and K.K.; resources, Y.Y. and H.H.; data curation, Y.Y., Y.N. and K.S.; writing—original draft preparation, Y.Y.; writing—review and editing, M.I. and H.H.; visualization, Y.Y. and K.K.; supervision, H.H.; project administration, Y.Y.; funding acquisition, Y.Y. All authors have read and agreed to the published version of the manuscript.

Funding: This study was funded by Otsuka Pharmaceuticals, which provided the experimental animals purchase cost and the feed.

Acknowledgments: We wish to express thanks to Tetsuro Ago (Department of Medical and Clinical Science, Graduate School of Medical Science, Kyusyu University), Ataru Nishihara (Department of Neurosurgery, Graduate School of Medical Science, Kyusyu University), and Naoaki Oyama (Department of Stroke Medicine, Kawasaki Medical School) for their technical advice.

Conflicts of Interest: The authors declare no conflicts of interest. The funders had no role in the design of the study; in the collection, analyses, or interpretation of data; in the writing of the manuscript, or in the decision to publish the results.

Abbreviations

Aβ	amyloid-β
APP	amyloid-β Protein Precursor
CAA	cerebral amyloid angiopathy
H&E	hematoxylin and eosin
IQR	interquartile range
JPEG	joint photographic experts group
PBS	phosphate-buffered saline
PDE	phosphodiesterase

Appendix A

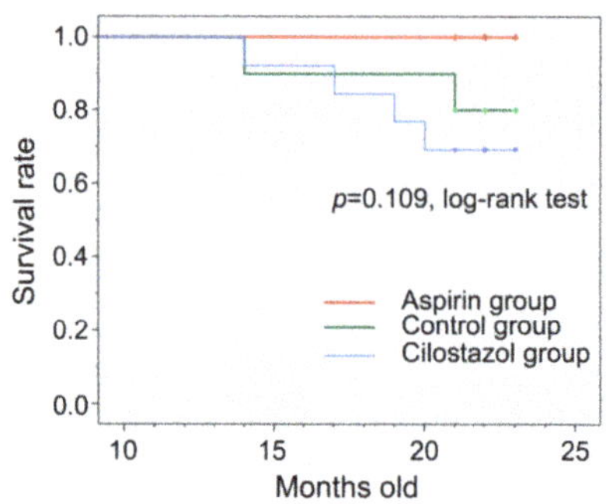

Figure A1. The survival rates after the grouping. There is no significant difference in the survival rate after the grouping among the three groups (control group, 80%; aspirin group, 100%; cilostazol group, 69% ($p = 0.109$, log-rank test: 14 mice evaluated for the first specimen are excluded from the analyses). Even in the two group comparisons, no differences are seen in the survival rate between the control group and the aspirin or cilostazol group ($p = 0.098$ and $p = 0.550$, log-rank test).

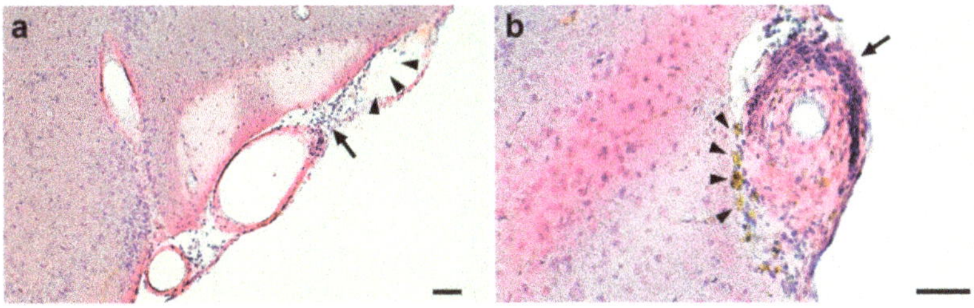

Figure A2. A representative image of accumulation of erythrocytes or hemosiderin. Accumulation of erythrocytes or hemosiderin shown in a Tg2576 mouse (aged 21 months old) fed with aspirin (i.e., aspirin group) (H&E stain). (**a**) Small accumulations of erythrocytes (arrow heads) are found around a leptomeningeal artery with lymphocytic infiltration (arrow). (**b**) Hemosiderin depositions (arrow heads) are seen around a leptomeningeal artery with vasculitis (arrow). Scale bars indicate 50 μm.

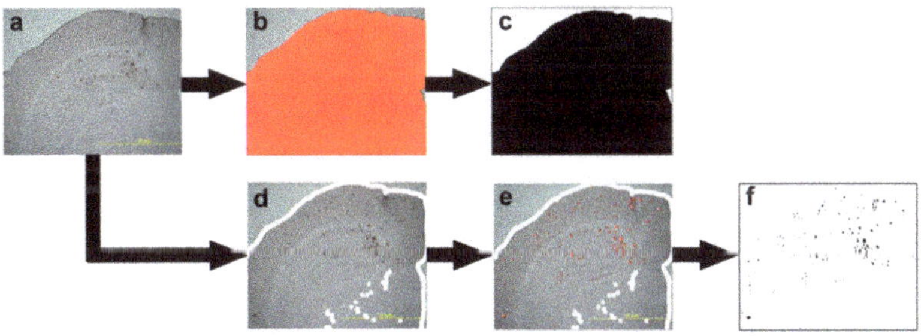

Scheme A1. Semi-automatic measurement of senile plaques in the regions of interest (cortex and the hippocampus). Upper line: Measurement of the area of the section: (1) Filling of the section (**a**) with red color (**b**); (2) dichotomization of color into black and white using a semiautomatic method with an appropriate color threshold (**c**). Lower line: Measurement of the total area of the senile plaque: (1) digital stripping of Aβ-stained lesions located out of the regions of interest (**d**); (2) change of color of the remaining Aβ-stained area to red using manual calibration with an appropriate color threshold (**e**); (3) dichotomization of color to black (Aβ-stained lesions in regions of interest) and white (other) using a semiautomatic method with an appropriate color threshold (**f**).

References

1. Masuda, J.; Tanaka, K.; Ueda, K.; Omae, T. Autopsy study of incidence and distribution of cerebral amyloid angiopathy in Hisayama, Japan. *Stroke* **1988**, *19*, 205–210. [CrossRef] [PubMed]
2. Xuereb, J.H.; Brayne, C.; Dufouil, C.; Gertz, H.; Wischik, C.; Harrington, C.; Mukaetova-Ladinska, E.; McGee, M.A.; O'Sullivan, A.; O'Connor, D.; et al. Neuropathological findings in the very old. Results from the first 101 brains of a population-based longitudinal study of dementing disorders. *Ann. N. Y. Acad. Sci.* **2000**, *903*, 490–496. [CrossRef] [PubMed]
3. Neuropathology Group of the Medical Research Council Cognitive Function and Ageing Study (MRC CFAS). Pathological Correlates of Late-Onset Dementia in a Multicentre, Community-Based Population in England and Wales. Neuropathology Group of the Medical Research Council Cognitive Function and Ageing Study (MRC CFAS). *Lancet* **2001**, *357*, 169–175. [CrossRef]
4. Pfeifer, L.A.; White, L.R.; Ross, G.W.; Petrovitch, H.; Launer, L.J. Cerebral amyloid angiopathy and cognitive function: The HAAS autopsy study. *Neurology* **2002**, *58*, 1629–1634. [CrossRef] [PubMed]
5. Charidimou, A.; Gang, Q.; Werring, D.J. Sporadic cerebral amyloid angiopathy revisited: Recent insights into pathophysiology and clinical spectrum. *J. Neurol. Neurosurg. Psychiatry* **2012**, *83*, 124–137. [CrossRef] [PubMed]
6. Yakushiji, Y. Cerebral microbleeds: Detection, associations, and clinical implications. *Front. Neurol. Neurosci.* **2016**, *37*, 78–92. [CrossRef]

7. Biffi, A.; Halpin, A.; Towfighi, A.; Gilson, A.; Busl, K.; Rost, N.; Smith, E.E.; Greenberg, M.S.; Rosand, J.; Viswanathan, A. Aspirin and recurrent intracerebral hemorrhage in cerebral amyloid angiopathy. *Neurology* **2010**, *75*, 693–698. [CrossRef]
8. Thoonsen, H.; Richard, E.; Bentham, P.; Gray, R.; van Geloven, N.; De Haan, R.J.; Van Gool, W.A.; Nederkoorn, P.J. Aspirin in Alzheimer's disease: Increased risk of intracerebral hemorrhage: Cause for concern? *Stroke* **2010**, *41*, 2690–2692. [CrossRef]
9. Begum, N.; Shen, W.; Manganiello, V. Role of PDE3A in regulation of cell cycle progression in mouse vascular smooth muscle cells and oocytes: Implications in cardiovascular diseases and infertility. *Curr. Opin. Pharmacol.* **2011**, *11*, 725–729. [CrossRef]
10. Shinohara, Y.; Katayama, Y.; Uchiyama, S.; Yamaguchi, T.; Handa, S.; Matsuoka, K.; Ohashi, Y.; Tanahashi, N.; Yamamoto, H.; Genka, C.; et al. Cilostazol for prevention of secondary stroke (CSPS 2): An aspirin-controlled, double-blind, randomised non-inferiority trial. *Lancet Neurol.* **2010**, *9*, 959–968. [CrossRef]
11. Takagi, T.; Imai, T.; Mishiro, K.; Ishisaka, M.; Tsujimoto, M.; Ito, H.; Nagashima, K.; Matsukawa, H.; Tsuruma, K.; Shimazawa, M.; et al. Cilostazol ameliorates collagenase-induced cerebral hemorrhage by protecting the blood-brain barrier. *J. Cereb. Blood Flow Metab.* **2017**, *37*, 123–139. [CrossRef] [PubMed]
12. Maki, T.; Okamoto, Y.; Carare, R.O.; Hase, Y.; Hattori, Y.; Hawkes, C.A.; Saito, S.; Yamamoto, Y.; Terasaki, Y.; Ishibashi-Ueda, H.; et al. Phosphodiesterase III inhibitor promotes drainage of cerebrovascular beta-amyloid. *Ann. Clin. Transl. Neurol.* **2014**, *1*, 519–533. [CrossRef] [PubMed]
13. Garcia-Osta, A.; Cuadrado-Tejedor, M.; Garcia-Barroso, C.; Oyarzabal, J.; Franco, R. Phosphodiesterases as therapeutic targets for Alzheimer's disease. *ACS Chem. Neurosci.* **2012**, *3*, 832–844. [CrossRef]
14. Menniti, F.S.; Faraci, W.S.; Schmidt, C.J. Phosphodiesterases in the CNS: Targets for drug development. *Nat. Rev. Drug Discov.* **2006**, *5*, 660–670. [CrossRef] [PubMed]
15. Park, S.H.; Kim, J.H.; Bae, S.S.; Hong, K.W.; Lee, D.S.; Leem, J.Y.; Choi, B.T.; Shin, H.K. Protective effect of the phosphodiesterase III inhibitor cilostazol on amyloid beta-induced cognitive deficits associated with decreased amyloid beta accumulation. *Biochem. Biophys. Res. Commun.* **2011**, *408*, 602–608. [CrossRef]
16. Han, S.W.; Lee, S.S.; Kim, S.H.; Lee, J.H.; Kim, G.S.; Kim, O.J.; Koh, I.S.; Lee, J.Y.; Suk, S.H.; Lee, S.I.; et al. Effect of cilostazol in acute lacunar infarction based on pulsatility index of transcranial Doppler (ECLIPse): A multicenter, randomized, double-blind, placebo-controlled trial. *Eur. Neurol.* **2013**, *69*, 33–40. [CrossRef]
17. Weller, R.O.; Djuanda, E.; Yow, H.Y.; Carare, R.O. Lymphatic drainage of the brain and the pathophysiology of neurological disease. *Acta Neuropathol.* **2009**, *117*, 1–14. [CrossRef]
18. Schley, D.; Carare-Nnadi, R.; Please, C.P.; Perry, V.H.; Weller, R.O. Mechanisms to explain the reverse perivascular transport of solutes out of the brain. *J. Theor. Biol.* **2006**, *238*, 962–974. [CrossRef]
19. Aruna, D.; Naidu, M.U. Pharmacodynamic interaction studies of Ginkgo biloba with cilostazol and clopidogrel in healthy human subjects. *Br. J. Clin. Pharmacol.* **2007**, *63*, 333–338. [CrossRef]
20. Selkoe, D.J.; Hardy, J. The amyloid hypothesis of Alzheimer's disease at 25 years. *EMBO Mol. Med.* **2016**, *8*, 595–608. [CrossRef]
21. Leurent, C.; Goodman, J.A.; Zhang, Y.; He, P.; Polimeni, J.R.; Gurol, M.E.; Lindsay, M.; Frattura, L.; Sohur, U.S.; Viswanathan, A.; et al. Immunotherapy with ponezumab for probable cerebral amyloid angiopathy. *Ann. Clin. Transl. Neurol.* **2019**, *6*, 795–806. [CrossRef] [PubMed]
22. Sperling, R.A.; Jack, C.R., Jr.; Black, S.E.; Frosch, M.P.; Greenberg, S.M.; Hyman, B.T.; Scheltens, P.; Carrillo, M.C.; Thies, W.; Bednar, M.M.; et al. Amyloid-related imaging abnormalities in amyloid-modifying therapeutic trials: Recommendations from the Alzheimer's Association Research Roundtable Workgroup. *Alzheimers Dement. J. Alzheimers Assoc.* **2011**, *7*, 367–385. [CrossRef]
23. Ihara, M.; Nishino, M.; Taguchi, A.; Yamamoto, Y.; Hattori, Y.; Saito, S.; Takahashi, Y.; Tsuji, M.; Kasahara, Y.; Takata, Y.; et al. Cilostazol add-on therapy in patients with mild dementia receiving donepezil: A retrospective study. *PLoS ONE* **2014**, *9*, e89516. [CrossRef]
24. Saito, S.; Ihara, M. New therapeutic approaches for Alzheimer's disease and cerebral amyloid angiopathy. *Front. Aging Neurosci.* **2014**, *6*, 290. [CrossRef]
25. Hattori, Y.; Maki, T.; Saito, S.; Yamamoto, Y.; Nagatsuka, K.; Ihara, M. Influence of Low-Dose Aspirin on Cerebral Amyloid Angiopathy in Mice. *J. Alzheimers Dis.* **2016**, *52*, 1037–1045. [CrossRef] [PubMed]
26. Niu, P.P.; Guo, Z.N.; Jin, H.; Xing, Y.Q.; Yang, Y. Antiplatelet regimens in the long-term secondary prevention of transient ischaemic attack and ischaemic stroke: An updated network meta-analysis. *BMJ Open* **2016**, *6*, e009013. [CrossRef] [PubMed]

27. Wilhite, D.B.; Comerota, A.J.; Schmieder, F.A.; Throm, R.C.; Gaughan, J.P.; Rao, A.K. Managing PAD with multiple platelet inhibitors: The effect of combination therapy on bleeding time. *J. Vasc. Surg.* **2003**, *38*, 710–713. [CrossRef]
28. Kilkenny, C.; Browne, W.J.; Cuthill, I.C.; Emerson, M.; Altman, D.G. Improving bioscience research reporting: The ARRIVE guidelines for reporting animal research. *PLoS Biol.* **2010**, *8*, e1000412. [CrossRef] [PubMed]
29. Kawarabayashi, T.; Younkin, L.H.; Saido, T.C.; Shoji, M.; Ashe, K.H.; Younkin, S.G. Age-dependent changes in brain, CSF, and plasma amyloid (beta) protein in the Tg2576 transgenic mouse model of Alzheimer's disease. *J. Neurosci.* **2001**, *21*, 372–381. [CrossRef]
30. Ito, H.; Hashimoto, A.; Matsumoto, Y.; Yao, H.; Miyakoda, G. Cilostazol, a phosphodiesterase inhibitor, attenuates photothrombotic focal ischemic brain injury in hypertensive rats. *J. Cereb. Blood Flow Metab.* **2010**, *30*, 343–351. [CrossRef]
31. Winkler, D.T.; Bondolfi, L.; Herzig, M.C.; Jann, L.; Calhoun, M.E.; Wiederhold, K.H.; Tolnay, M.; Staufenbiel, M.; Jucker, M. Spontaneous hemorrhagic stroke in a mouse model of cerebral amyloid angiopathy. *J. Neurosci.* **2001**, *21*, 1619–1627. [CrossRef] [PubMed]
32. Carson, F.L.; Cappellano, C.H. *Histotechnology, A Self-Instructional Text*, 4th ed.; American Society for Clinical Pathology: Chicago, IL, USA, 2014.
33. Gomori, G. Microtechnical Demonstration of Iron: A Criticism of its Methods. *Am. J. Pathol.* **1936**, *12*, 655. [PubMed]
34. Hara, H.; Kataoka, S.; Anan, M.; Ueda, A.; Mutoh, T.; Tabira, T. The therapeutic effects of the herbal medicine, Juzen-taiho-to, on amyloid-beta burden in a mouse model of Alzheimer's disease. *J. Alzheimers Dis.* **2010**, *20*, 427–439. [CrossRef] [PubMed]
35. Koeppen, A.H.; Dickson, A.C.; McEvoy, J.A. The cellular reactions to experimental intracerebral hemorrhage. *J. Neurol. Sci.* **1995**, *134*, 102–112. [CrossRef]
36. Olichney, J.M.; Hansen, L.A.; Galasko, D.; Saitoh, T.; Hofstetter, C.R.; Katzman, R.; Thal, L.J. The apolipoprotein E epsilon 4 allele is associated with increased neuritic plaques and cerebral amyloid angiopathy in Alzheimer's disease and Lewy body variant. *Neurology* **1996**, *47*, 190–196. [CrossRef]
37. Calhoun, M.E.; Burgermeister, P.; Phinney, A.L.; Stalder, M.; Tolnay, M.; Wiederhold, K.H.; Abramowski, D.; Sturchler-Pierrat, C.; Sommer, B.; Staufenbiel, M.; et al. Neuronal overexpression of mutant amyloid precursor protein results in prominent deposition of cerebrovascular amyloid. *Proc. Natl. Acad. Sci. USA* **1999**, *96*, 14088–14093. [CrossRef]
38. Qu, B.; Boyer, P.J.; Johnston, S.A.; Hynan, L.S.; Rosenberg, R.N. Abeta42 gene vaccination reduces brain amyloid plaque burden in transgenic mice. *J. Neurol. Sci.* **2006**, *244*, 151–158. [CrossRef]

International Journal of
Molecular Sciences

MDPI

Article

Evaluation of Toxic Amyloid β42 Oligomers in Rat Primary Cerebral Cortex Cells and Human iPS-derived Neurons Treated with 10-Me-Aplog-1, a New PKC Activator

Kazuma Murakami [1,*], Mayuko Yoshimura [1], Shota Nakagawa [2], Toshiaki Kume [2,3], Takayuki Kondo [4,5,6], Haruhisa Inoue [4,5,6] and Kazuhiro Irie [1,*]

1 Division of Food Science and Biotechnology, Graduate School of Agriculture, Kyoto University, Kyoto 606-8502, Japan; myk0128dw@gmail.com
2 Department of Pharmacology, Graduate School of Pharmaceutical Sciences, Kyoto University, Kyoto 606-8501, Japan; nakagawa.shota.0812@gmail.com (S.N.); tkume@pha.u-toyama.ac.jp (T.K.)
3 Department of Applied Pharmacology, Graduate School of Medicine and Pharmaceutical Sciences, University of Toyama, Toyama 930-0194, Japan
4 Center for iPS Cell Research and Application (CiRA), Kyoto University, Kyoto 606-8507, Japan; takayuki.kondo@cira.kyoto-u.ac.jp (T.K.); haruhisa@cira.kyoto-u.ac.jp (H.I.)
5 iPSC-based Drug Discovery and Development Team, RIKEN BioResource Research Center (BRC), Kyoto 619-0237, Japan
6 Medical-risk Avoidance based on iPS Cells Team, RIKEN Center for Advanced Intelligence Project (AIP), Kyoto 606-8507, Japan
* Correspondence: alzkazu@kais.kyoto-u.ac.jp (K.M.); irie@kais.kyoto-u.ac.jp (K.I.); Tel.: +81-75-753-6282 (K.M.); +81-75-753-6281 (K.I.)

Received: 15 January 2020; Accepted: 6 February 2020; Published: 11 February 2020

Abstract: Amyloid β42 (Aβ42), a causative agent of Alzheimer's disease (AD), is derived extracellularly from Aβ precursor protein (APP) following the latter's cleavage by β-secretase, but not α-secretase. Protein kinase Cα (PKCα) activation is known to increase α-secretase activity, thereby suppressing Aβ production. Since Aβ42 oligomer formation causes potent neurotoxicity, APP modulation by PKC ligands is a promising strategy for AD treatment. Although bryostatin-1 (bryo-1) is a leading compound for this strategy, its limited natural availability and the difficulty of its total synthesis impedes further research. To address this limitation, Irie and colleagues have developed a new PKC activator with few side effects, 10-Me-Aplog-1, (**1**), which decreased Aβ42 in the conditioned medium of rat primary cerebral cortex cells. These results are associated with increased α-secretase but not PKCε-dependent Aβ-degrading enzyme. The amount of neuronal embryonic lethal abnormal vision (nELAV), a known β-secretase stabilizer, was reduced by treatment with **1**. Notably, **1** prevented the formation of intracellular toxic oligomers. Furthermore, **1** suppressed toxic oligomerization within human iPS-derived neurons such as bryo-1. Given that **1** was not neurotoxic toward either cell line, these findings suggest that **1** is a potential drug lead for AD therapy.

Keywords: alzheimer's disease; amyloid β; bryostatin-1; ECE1; iPS; nELAV; neurotoxicity; oligomer; protein kinase C; α-secretase

---

1. Introduction

The 40-mer and 42-mer amyloid β-proteins (Aβ40 and Aβ42) are considered causative agents of Alzheimer's disease (AD) [1,2]. Aβ40 and Aβ42 are known to be produced from Aβ precursor protein (APP) following cleavage of the latter by β-secretase, but not α-secretase. APP proteolysis may be more complex, given the recent discovery of APP proteolysis by η- and δ-secretases, for example, in [3]. The ability of Aβ42 to aggregate and exhibit neurotoxicity is higher than that of Aβ40 despite the lower in vivo amounts of Aβ42 [4]. Aβ42 oligomer formation causes synaptic dysfunction and neuronal death in AD pathology, whereas the contribution of end-stage mature fibrils of Aβ42 to AD is lower than that of oligomers [5]. Higher-order toxic oligomers that show potent synaptotoxicity and neurotoxicity have been reported, such as protofibrils (PFs), Aβ-derived diffusible ligands, and amylospheroids [6]. Therefore, suppressing toxic oligomerization of Aβ42 is a favorable strategy for developing AD therapies. This suppression can also be achieved by simultaneously decreasing Aβ production while inducing Aβ degradation.

Protein kinase C (PKC) is a family of serine/threonine kinases that plays a pivotal role in various biological events such as signal transduction, proliferation, and apoptosis mediated by the second messenger 1,2-diacyl-*sn*-glycerol [7]. The PKC family, which contains at least 10 isozymes, is divided into three groups, namely conventional (α, βI, βII, and γ), novel (δ, ε, η, and θ), and atypical (μ, ξ, and ι) [7]. PKC activity is related to memory formation and learning [8], while PKC downregulation may induce cognitive impairment and memory loss in AD [9]. Regarding Aβ-driven molecular events, PKCα reportedly upregulates α-secretase activity either directly or indirectly through the mitogen-activated protein kinase (MAPK) pathway [10]. PKCα activation in a mouse model of AD has beneficial effects on AD pathology, including the disruption of Aβ production and reduction of toxic Aβ oligomer formation [11]. Neuronal embryonic lethal abnormal vision (nELAV), also known as HuD protein, may contribute to mRNA stability through a PKCα-dependent mechanism due to adenine- and uridine-rich elements (AREs) [12]. PKCε may also be a target beneficial for preventing AD. A mouse study demonstrated that PKCε activation reduces senile plaque formation, although its effect on oligomer generation was not determined [13]. Similarly, the stimulator specific for PKCε (DCP-LA) rescued synaptic dysfunction and cognitive deficits as well as senile plaques in another mouse study [14]. PKCε stimulates the degradation of Aβ42 and Aβ40 by activating endothelin converting enzyme 1 (ECE1) [15]. These reports indicate that PKC activation may offer a promising strategy for AD treatment.

Bryostatin-1 (bryo-1), which was isolated from the marine bryozoan *Bugula neritina* [16], is a potent PKC activator with few side effects such as tumor-promoting and proinflammatory activities. Bryo-1 was found to activate both PKCα and PKCε, and to restore loss of hippocampal synapses and memory impairment by suppressing the levels of Aβ oligomers detected by the A11 antibody [14]. Bryo-1 may have beneficial effects against Aβ-induced abnormality in human fibroblasts [17]. These findings indicate that bryo-1 is a potential drug lead for AD [18]. However, its limited availability from natural sources and the difficulty of total synthesis both hamper further development, despite scalable synthetic routes reported by the Wender [19] and Trost groups [20]. Taking an alternative approach, Irie and colleagues developed 10-Me-Aplog-1 (**1**; Figure 1a), a simplified analog of aplysiatoxin [21], which is a potent PKC activator with tumor-promoting activity. It should be noted that **1** exhibited anti-proliferative activity towards cancer cell lines without significant tumor-promoting or proinflammatory activities [22,23].

The ratio of Aβ42 to Aβ40 (Aβ42/Aβ40) is a known biomarker for predicting AD onset in cerebrospinal fluid (CSF) and plasma [24]. However, such a biomarker could correlate with senile plaque depositions containing less toxic fibrils according to brain imaging of Aβ deposition with positron emission tomography (Aβ-PET) [25–27]. Furthermore, the PKC activation strategy is not expected to modulate Aβ42/Aβ40, since the proteolysis of APP by γ-secretase can predominantly determine the length of secreted Aβ.

Irie and colleagues identified a toxic Aβ42 conformer with a turn at positions 22–23 (toxic turn) [28], and proposed the ratio of the toxic conformer to total Aβ42 as a possible biomarker for AD progression in CSF using sandwich ELISA specific for Aβ42 toxic oligomers based on the anti-toxic turn antibody (24B3) [29]. A change in Aβ42 toxic conformer ratio may be a good predictor for long-term cognitive outcomes in idiopathic normal pressure hydrocephalus (iNPH) [30]. Toxic conformers can easily form toxic oligomers [31]. Here, we offer a novel, direct evaluation platform that determines the ratio of toxic oligomers to Aβ42 (toxic oligomers/Aβ42) in rat primary cerebral cortex cells and human induced pluripotent stem (iPS)-derived neurons using 24B3-based ELISA [29], which were treated with **1**. The therapeutic potential of **1** and its mechanism of action in AD prevention were also investigated.

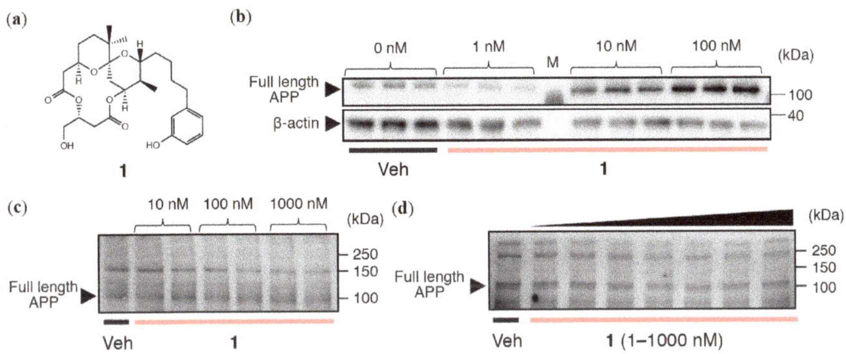

Figure 1. (**a**) Structure of 10-Me-Aplog-1 (**1**). APP expression levels in (**b**) HEK293-APPwt, (**c**) SH-SY5Y, and (**d**) rat primary cerebral cortex cells treated with **1** at the indicated concentrations for 24 h. M indicates marker. In (**d**), 1, 10, 50, 100, 500, and 1000 nM (from left to right) of **1** were used. Veh: vehicle.

2. Results

2.1. APP Expression Levels in Cultured Neuronal Cell Lines Treated with 1

The reason why research on PKC modulators faces difficulties in the AD field is the abnormal enhancement of APP itself upon addition of PKC ligand to cultured animal cells, including rat PC12 cells [32] and human HeLa cells [33], resulting in unwanted Aβ overproduction. Alternatively, APP secreted after α-secretase processing (sAPPα) or AD-index calculated from Erk1/2 phosphorylation have been used as evaluation criteria for PKC modulators [17]; however, there are very few reports concerning the direct quantification of Aβ in cell-based experiments. As expected, **1** enhanced APP levels in HEK293 cells overexpressing wild-type APP (HEK293-APPwt) in a dose-dependent manner (Figure 1b). By contrast, APP levels in both SH-SY5Y cells (Figure 1c) and rat primary cerebral cortex cells (Figure 1d) were largely unaltered.

2.2. Effects of 1 on Extracellular Aβ42/Aβ40 and Aβ Oligomerization in Rat Primary Cerebral Cortex Cells

Since the amount of Aβ42 secreted by SH-SY5Y cells was near to the detection limit of specific ELISA (#27711 Human Amyloid β 1-42 Assay Kit—IBL), we selected rat primary cerebral cortex cells for evaluating PKC modulators in the following study. After a 24 h incubation, **1** did not reduce Aβ42/Aβ40 as expected above, because the amounts of both Aβ42 and Aβ40 were lowered (Figure 2a). 12-*O*-Tetradecanoylphorbol 13-acetate (TPA) is a PKC ligand that exerts a similar effect [34]. Because the extracellular levels of toxic oligomers after a 24 h incubation were under the detection limit for specific ELISA (#27709 Human Amyloid β Toxic Oligomer Assay Kit—IBL) and Aβ42 easily aggregates to form amyloid fibrils after a 24 h incubation in vitro [35,36], we sampled at an earlier time point, 6 h, to determine the formation of toxic Aβ oligomers. As shown in Figure 2b, the ratio of toxic oligomers to Aβ42 (toxic oligomers/Aβ42) in cerebral cortex cells did not increase following treatment with **1** even at a higher concentration range than that in Figure 2a. However, the toxic oligomer levels were unchanged by **1** (Figure 2b).

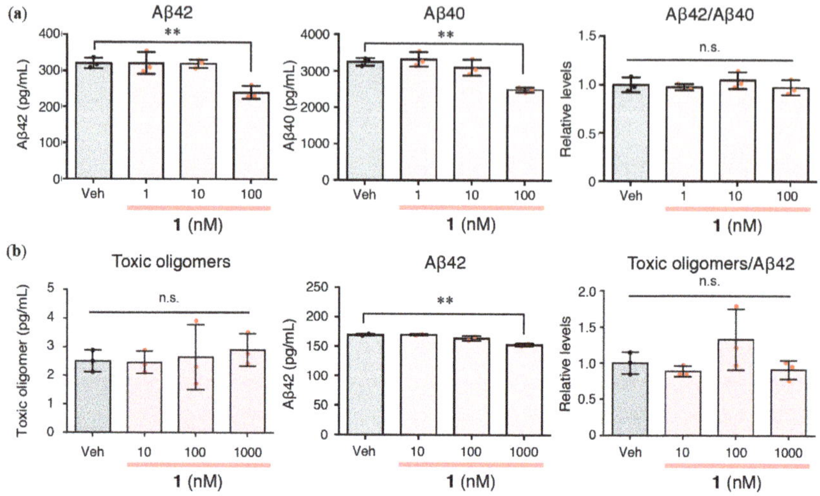

Figure 2. (**a**) Monomeric Aβ42, Aβ40, and their ratio (Aβ42/Aβ40) in the conditioned medium of rat primary cerebral cortex cells treated with **1** at the indicated concentrations for 24 h. (**b**) Toxic Aβ oligomers, monomeric Aβ42, and their ratio (toxic oligomers/Aβ42) in the conditioned medium of rat primary cerebral cortex cells treated with **1** at the indicated concentrations for 6 h. The data are presented as mean ± SD ($n = 3$). ** $p < 0.01$ versus Veh (vehicle). n.s.: not significant. Red or black dots represent each value.

2.3. Effects of 1 on Aβ Production and Degradation in Rat Primary Cerebral Cortex Cells

Given the moderate reduction in Aβ42 secretion to the extracellular space caused by **1** treatment (Figure 2b), we investigated the contribution of **1** to Aβ production and degradation in cultured cells. The concentration of **1** was set to 10–1000 nM in the following study of primary cultured cells. The amount of disintegrin and metalloproteinase 10 (ADAM10), as one of the α-secretases, was increased in Western blotting, using the ratio of the processed to active form of ADAM10 in the case of **1** (Figure 3a). nELAV proteins are known to act as PKCα-dependent Aβ modulators via α-secretase [12,37] or β-secretase [38]. As shown in Figure 3b, the amounts of nELAV were decreased by **1**.

Next, ECE1 levels were also measured. ECE1 levels were almost unchanged in cells treated with **1** (Figure 3c). These results indicate that the decrease in Aβ42 caused by **1** could be due to enhanced α-secretase expression, but not Aβ degradation.

2.4. Effects of 1 on Intracellular Aβ Oligomerization in Rat Primary Cerebral Cortex Cells

Intracellular Aβ accumulation appears to be an early event in AD pathogenesis. In particular, Aβ oligomerization may begin to induce mitochondrial toxicity, proteasome impairment, and synaptic damage [39]. To elucidate the intracellular mechanism, lysates were prepared from cells after 6 h of incubation with **1** and subjected to Western blotting using 24B3 antibody [29]. Notably, the formation of intracellular toxic oligomers, which are 20–30-mers according to synthetic studies [40,41] of Aβ oligomer models that inhibited long-term potentiation (LTP) in mouse hippocampal slices (T. Kume, personal communication, unpublished results), was significantly decreased by **1** (Figure 4). These results suggest that **1** may modulate toxic Aβ oligomerization.

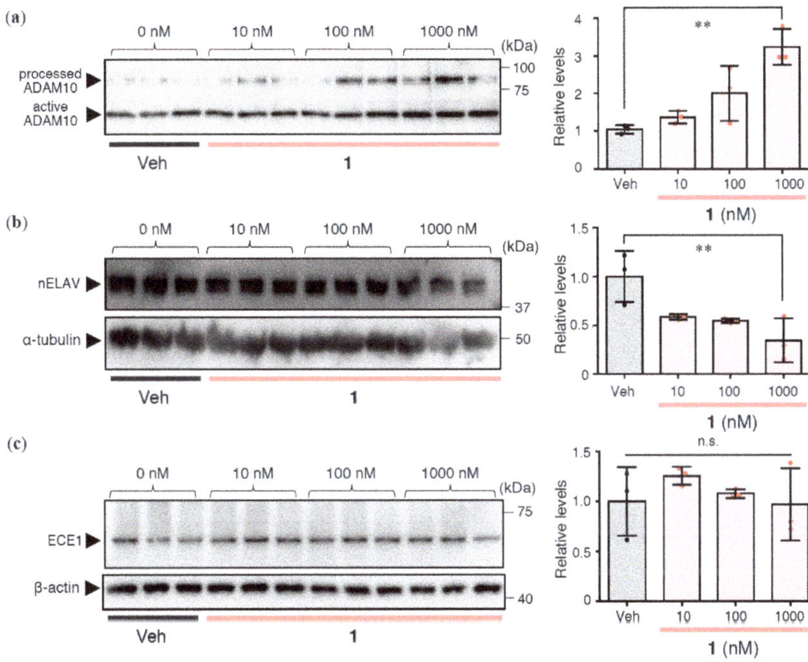

Figure 3. (**a**) Processed ADAM10, (**b**) nELAV, and (**c**) ECE1 in the cell lysate prepared from rat primary cerebral cortex cells treated with **1** at indicated concentrations for 24 h. The relative levels of (**a**) active ADAM10, (**b**) α-tubulin, and (**c**) β-actin are presented as mean ± SD ($n = 3$). $^{**}p < 0.01$ versus Veh (vehicle). n.s.: not significant. Red or black dots represent each value.

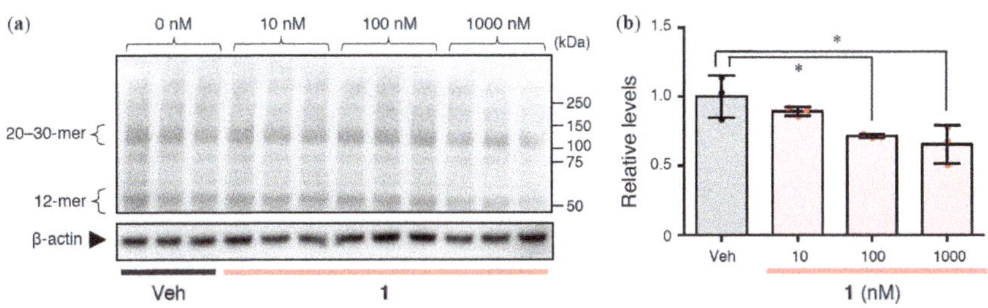

Figure 4. (**a**,**b**) Toxic oligomer formation in lysate from rat primary cerebral cortex cells treated with **1** at the indicated concentration for 6 h. (**a**) The representative Western blot shown was probed with anti-Aβ42 toxic turn (24B3) antibody. (**b**) Band intensities corresponding to 20–30-mers relative to β-actin in (**a**) are presented as mean ± SD ($n = 3$). $^{*}p < 0.05$ versus Veh (vehicle). Red or black dots represent each value.

2.5. Effects of 1 on the Cytotoxicity of Rat Primary Cerebral Cortex Cells

To examine the neurotoxicity of **1**, a 3-(4,5-dimethylthiazol-2-yl)-2,5-diphenyltetrazolium bromide (MTT) assay was performed on rat primary cerebral cortex cells. As shown in Figure 5, it was confirmed that **1** did not exhibit neurotoxicity at the concentrations used in the above tests (Figure 2b,3,4). This finding suggests that **1** is potentially as safe as bryo-1, with few side effects.

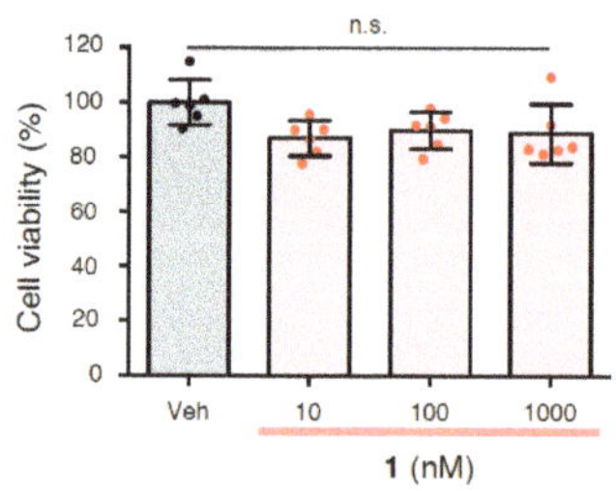

Figure 5. Neurotoxicity of rat primary cerebral cortex cells treated with **1** at the indicated concentrations for 24 h. The data are presented as mean ± SD (n = 6). n.s.: not significant. Veh: vehicle. Red or black dots represent each value.

2.6. *Effects of 1 on Aβ42/Aβ40, Aβ Oligomerization, and Neurotoxicity in Human iPS-Derived Neurons*

To further verify the preventative effects of **1** against AD, human iPS-derived neurons were adopted for this experiment because of a slight difference in Aβ sequence between rat and human. Recent studies also imply a large gap in the effectiveness of drug discovery studies between iPS-derived neurons and cultured cell lines [42]. Recently, Inoue and colleagues developed a reliable and robust iPS-based screening system for anti-Aβ drugs [43]. After incubating the differentiated neurons from iPS with PKC ligands for 24 h, Aβ42 and Aβ40 levels in the conditioned medium were calculated using electrochemiluminescence assays. Bryo-1 was used as a positive control, which significantly decreased the amount of Aβ42 and Aβ40 in a dose-dependent manner. Bryo-1 therefore suppressed the Aβ42/Aβ40 ratio (Figure 6d). Treatment with **1** lowered Aβ42 and Aβ40 levels to almost the same extent, resulting in almost no alternation of Aβ42/Aβ40 (Figure 6a). **1** failed to show cytotoxicity such as bryo-1 (Figure 6b,e) measured by the ToxiLight assay that reflects the release of adenylate kinase from damaged cells [44].

Lysate prepared from iPS-derived neurons was subjected to ELISA measurement for toxic oligomers (Figure 6c,f). In Figure 6c, the amount of Aβ42 toxic oligomers following **1** treatment showed a tendency to decrease, like bryo-1, in a dose-dependent manner (Figure 6f), in spite of one anomalous value at 30 nM, which might originate from a technical issue. These findings suggest that **1** may also prevent toxic oligomer formation in iPS-derived neurons.

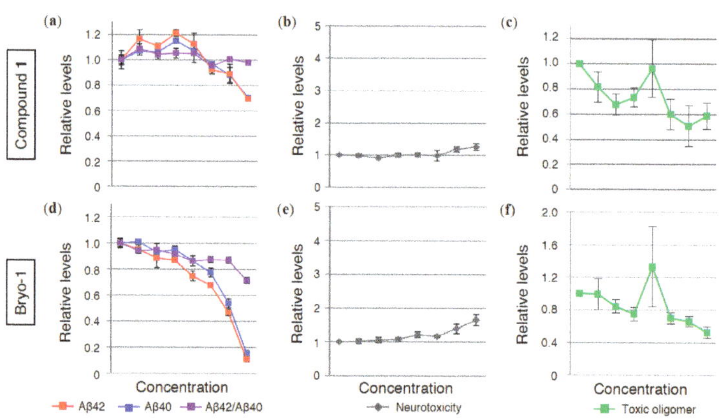

Figure 6. (**a**,**d**) Monomeric Aβ42, Aβ40, and Aβ42/Aβ40 in conditioned medium, (**b**,**e**) neurotoxicity, and (**c**,**f**) toxic Aβ42 oligomers in lysate from human iPS-derived neurons treated with (**a**,**b**,**c**) **1** and (**d**,**e**,**f**) bryo-1 at the indicated concentration (0, 1, 3, 10, 30, 100, 300, and 1000 nM from left to right) for 24 h. The data are presented as mean ± SEM (n = 3).

3. Discussion

Alkon and colleagues hypothesized that deficits in PKC signaling are involved in AD symptoms [18]. PKCα and PKCε are thought to induce Aβ diminution, leading to beneficial effects for AD. Indeed, the results of several clinical trials provide encouragement for bryo-1 as a potential drug against AD [45]. It is worth noting that **1** prevented nELAV accumulation within the cell (Figure 3b). nELAV levels were higher in AD patients compared with non-AD controls [38]. The nELAV-driven stabilization of β-secretase mRNA (β-site amyloid β precursor protein cleaving enzyme, BACE1) [38] and tau mRNA [46] may be involved in AD progression. On the other hand, experiments using SH-SY5Y cells suggest that the stabilization of ADAM10 by the binding of nELAV may contribute to beneficial effects against AD via the PKCα pathway [12]. Bryo-1 counteracted the deficit in ADAM10 in SH-SY5Y cells in which HuD expression had been silenced [47]. Although the involvement of nELAV in AD remains controversial, nELAV is a novel putative target for anti-AD therapies. Furthermore, Jarosz-Griffiths et al. reported that ADAM10-modulated shedding of cellular prion protein reduced the neurotoxicity of Aβ oligomers [48]. The present findings illustrate that **1** prevented the formation of intracellular Aβ42 oligomers as well as extracellular Aβ42, which is associated with enhanced α-secretase cleavage of APP. Further studies will be required to clarify whether **1** might affect toxic oligomerization directly, and if so, how. Given that the parent analogue (Aplog-1) of **1** can activate PKCδ [22] and **1** binds potently to the PKCα-C1A and PKCε-C1B domains [23], **1** is a promising substitute for bryo-1 as a therapeutic drug lead for AD.

Recently, Yanagisawa and colleagues identified plasma APP669-711/Aβ42 [49] in addition to Aβ42/Aβ40 as an alternative biomarker using Japanese and Australian cohorts [50]. Using a composite biomarker calculated from APP669-711/Aβ42 and Aβ42/Aβ40 may enhance the accuracy of diagnosis during disease progression from mild cognitive impairment (MCI) to AD. However, in their work, the potential of Aβ oligomerization in CSF or plasma as a biomarker was not fully addressed. Recently, a detection method for Aβ oligomers using single molecule arrays (Simoa) as a highly sensitive platform was reported using the same anti-Aβ N-terminal antibody (bapineuzumab), both for antigen capture and detection [51]. However, this strategy cannot exclude the possibility of detecting mature fibrils, resulting in lower specificity for Aβ oligomers [52]. The use of the anti-N-terminal antibody (82E1) [53] may address the problem by using the same antibodies for capture and detection [54]. Alternatively, the development of highly specific antibodies for toxic oligomeric species with synaptotoxicity would be most ideal for finding biomarkers. In the Aβ42 toxic oligomer ELISA used in this study, the 24B3 antibody against the Aβ toxic turn and 82E1 antibody against the Aβ N-terminus are used for detection and capture, respectively [29].

Ohshima et al. reported that familial mutations of AD increase oligomer formation of Aβ in the conditioned medium of wild-type APP-transfected cells, but intracellular levels of Aβ oligomer in these mutant APP-transfected cells were unaltered compared with wild-type APP-transfected cells [55]. These results may be due to the stronger ability of Aβ either to be formed or to aggregate due to these mutations in the precursor APP, namely Swedish, Dutch, and London mutations. The Osaka mutant (E693Δ) of Aβ tends to be found as oligomers within cell bodies in both cultured cells [56] and human iPS-derived neurons [57]. It was therefore difficult to determine oligomer levels in non-mutated APP cell models with precision. Regarding the intracellular accumulation of Aβ, the key question of how intracellular Aβ accumulates remains unanswered, thereby invoking the involvement of tau pathology; that is, the possible interaction of intraneuronal Aβ with neurofibrillary tangles [58]. The relevance of liquid–liquid phase separation to intracellular accumulation of amyloidogenic proteins (tau [59,60] and TDP-43 [61]) should also be considered.

In conclusion, to the best of our knowledge, we have developed the first direct evaluation system not only for Aβ monomers, but also for their assembly into toxic oligomers in small amounts using two reliable and prevalent cell models of AD. Compared with bryo-1, whose efficiency has been recognized in several clinical trials for AD and cancer, **1** may play a pivotal role in AD prevention as a promising drug lead.

4. Materials and Methods

4.1. Rat Primary Cerebral Cortex Cells

Animals were treated according to guidelines issued by the Kyoto University Animal Experimentation Committee and by the Japanese Pharmacological Society. The experimental procedures were approved by the Kyoto University Animal Experimentation Committee [#16-12-1 (14 Mar 2016), #16-12-2 (21 Mar 2017)]. Primary cultures were obtained from the cerebral cortex of fetal Wistar rats (Nihon SLC; 17–19 d of gestation) as previously described [62]. Briefly, single cells dissociated from whole cerebral cortices of fetal rats were plated on 0.1% polyethyleneimine-coated plastic 12-well plates (10^6 cells/well, 1 mL). Cells were incubated in Eagle's minimal essential medium (E-MEM) supplemented with 10% heat-inactivated fetal bovine serum (FBS) before half the medium was exchanged for fresh medium 2 and 4 d after plating. Subsequently, half the medium was exchanged for fresh medium containing 20 nM cytosine arabinoside 6 d after plating and again with fresh medium containing 10% heat-inactivated horse serum (HS) 8 d after plating. The cultures were maintained at 37 °C under a humidified 5% CO_2 atmosphere. Mature cerebral cortex cell cultures (10 d after plating) were used for all experiments.

DMSO stock of **1** was dissolved in E-MEM with 10% heat-inactivated HS (the concentration of DMSO in the medium was under 0.1%). After 6 or 24 h incubation, 100 μL of cell lysis buffer (RIPA buffer, Wako, Tokyo, Japan) containing a phosphatase inhibitor cocktail (Roche, Mannheim, Germany) and protease inhibitor cocktail (Roche) was added to prepare cell lysates. Supernatant was obtained by centrifugation (17,860 *g*, 4 °C) and stored at –80 °C until use.

4.2. ELISA

To determine the amounts of Aβ42 [#290-62601 Human/Rat β Amyloid(42) ELISA Kit Wako (Osaka, Japan) or #27711 Human Amyloid β 1-42 Assay Kit—IBL (Gunma, Japan)] and Aβ toxic oligomers (#27709 Human Amyloid β Toxic Oligomer Assay Kit—IBL), 100 μL of cell lysate was applied to the corresponding sandwich ELISA plate.

4.3. Western Blotting

Total protein concentration of the brain was determined using the Bradford protein assay (Bio-Rad; Hercules, CA, USA). Brain proteins diluted to 1 μg/μL were treated with 4× LDS sample buffer (Invitrogen; Carlsbad, CA, USA) and 5 mM dithiothreitol before heating at 70 °C for 10 min. The denatured sample solution was subjected to Western blotting, following SDS-PAGE on a 10% Bis-Tris gel (Invitrogen) and subsequent transfer to PVDF (0.22 μm pore size, Bio-Rad). PVDF membranes were blocked in 2.5% ECL prime blocking (GE Healthcare; Madison, WI, USA) dissolved in phosphate-buffered saline (PBS) containing 0.5% Tween-20 (PBS-T), and incubated with primary antibody at the following dilutions: 1:1000 anti-Aβ (4G8) (Signet; Dedham, MA, USA), 1:1000 anti-APP(N) (IBL), 1:1000 anti-Aβ42 toxic turn (24B3) (IBL, Gunma, Japan), 1:500 anti-ADAM10 (B-3) (IBL, Gunma, Japan), 1:140 anti-nELAV(HuD+HuC) (Santa Cruz; Santa Cruz, CA, USA), or 1:1000 anti-ECE1 (abcam; Cambridge, MA, USA). Following primary antibody incubation, blots were washed before being incubated with the appropriate secondary antibody. Blots were developed with enhanced chemiluminescence and quantified using Lumino Graph II (ATTO; Tokyo, Japan).

4.4. MTT Assay

Neurotoxicity was assessed by the MTT assay according to a previously reported protocol [63]. In brief, mature cerebral cortex cultures were moved to Neurobasal Medium with 2% B-27 supplement, 25 μM sodium glutamate, and penicillin/streptomycin before plating on 96-well plates (1.5×10^6 cells/well, 100 μL). Four days after plating, the medium was replaced with sodium glutamate-free Neurobasal Medium. Half the medium was exchanged for fresh medium 7 or 8 d after plating, and all the medium was exchanged for Neurobasal Medium with 2% B-27 supplement minus AO (Gibco; Grand Island, NY,

USA), and penicillin/streptomycin containing DMSO stock of **1** (10 d after plating). After incubation for 24 h, the culture medium was replaced with a medium containing 0.5 mg/mL MTT, and cells were incubated for 15 min at 37 °C. After removing the medium, 2-propanol (100 μL) was added to lyse the cells, and absorbance was measured at 595 nm with an absorption spectrophotometer (MultiScan JX, Thermo Scientific; Waltham, MA, USA). The absorbance measured following vehicle treatment (DMSO final concentration = 0.1%) was fixed as 100% for comparison.

4.5. Generation and Characterization of Human iPS-Derived Neurons

Human iPS cells of Alzheimer's patients were generated as described from skin fibroblasts [43,57] and maintained using StemFit AK02N medium (Ajinomoto, Tokyo, Japan) [64] and expanded for neural differentiation. To establish a robust and rapid differentiation method, we utilized direct conversion technology. We differentiated iPS cells into neurons by using a direct conversion method, as previously described [43]. Briefly, we transduced human NGN2 cDNA by using the piggyBac transposon system, transiently under tetracycline-inducible promoter (tetO), and converted iPS cells into neuronal cells with more than 96% purity.

4.6. Electrochemiluminescence Assays

Aβ species in culture media after 24 h cultivation with PKC ligands were measured by human (6E10) Aβ 3-VPlex Kit (Meso Scale Discovery; Rockville, MD, USA). This assay uses the 6E10 anti-β-amyloid antibody to capture Aβ peptides and SULFO-TAG-labeled C-terminus specific anti-Aβ antibodies for detection by electrochemiluminescence with Sector Imager 2400 (Meso Scale Discovery). Quantified Aβ values were adjusted using total protein in neurons and compared among conditions.

4.7. ToxiLight Assay

Cytotoxicity was determined to measure the release of the enzyme adenylate kinase from damaged cells [44] using The ToxiLight™ Non-destructive Cytotoxicity BioAssay Kit (Lonza; Walkersville, MD, USA). In brief, the cultured medium was collected after 24 h incubation, and applied to the assay.

4.8. Statistical Analyses

The differences were subjected to one-way analysis of variance (ANOVA) followed by Bonferroni's test; p values < 0.05 versus vehicle were considered significant.

Author Contributions: Conceptualization and supervision, K.M. and K.I.; formal analysis, K.M., M.Y., S.N., and T.K. (Takayuki Kondo); investigation, K.M., M.Y., S.N., and T.K. (Takayuki Kondo); resources, T.K. (Toshiaki Kume) and H.I.; writing—original draft preparation, K.M.; writing—review and editing, K.I.; project administration, K.I. All authors have read and agreed to the published version of the manuscript.

Acknowledgments: We are thankful to Masahiro Maeda in IBL for providing ELISA kits. This study was in part supported by the Japanese government (Monbukagakusho: MEXT), grant number 17H06405 to K.I. and K.M. for Scientific Research on Innovative Areas 'Frontier Research on Chemical Communications', a grant for Core Center for iPS Cell Research of Research Center Network for Realization of Regenerative Medicine from AMED to H.I., and the JSPS KAKENHI, grant number 18K18452 to T.Ko. and H.I., grant number 17K16121 to T.Ko.

Conflicts of Interest: The authors declare no conflict of interest.

Abbreviations

Aβ	Amyloid β-protein
AD	Alzheimer's disease
ADAM10	A disintegrin and metalloproteinase 10
APP	Amyloid β precursor protein
BACE1	β-site amyloid β precursor protein cleaving enzyme 1
DMSO	Dimethyl sulfoxide
ECE1	Endothelin converting enzyme 1

ELISA	Enzyme-linked immunosorbent assay
E-MEM	Eagle's minimal essential medium
FBS	Fetal bovine serum
HFIP	1,1,1,3,3,3-hexafluoro-2-propanol
HRP	Horseradish peroxidase
HS	Horse serum
iPS	Induced pluripotent stem
MCI	Mild cognitive impairment
MTT	3-(4,5-dimethylthiazol-2-yl)-2,5-diphenyltetrazolium bromide
PBS	Phosphate-buffered saline
PKC	Protein kinase C
sAPP	Secreted amyloid β precursor protein
WB	Western blotting

References

1. Glenner, G.G.; Wong, C.W. Alzheimer's disease: initial report of the purification and characterization of a novel cerebrovascular amyloid protein. *Biochem. Biophys. Res. Commun.* **1984**, *120*, 885–890. [CrossRef]
2. Masters, C.L.; Simms, G.; Weinman, N.A.; Multhaup, G.; McDonald, B.L.; Beyreuther, K. Amyloid plaque core protein in Alzheimer disease and Down syndrome. *Proc. Natl. Acad. Sci. USA* **1985**, *82*, 4245–4249. [CrossRef] [PubMed]
3. Andrew, R.J.; Kellett, K.A.; Thinakaran, G.; Hooper, N.M. A Greek Tragedy: The growing complexity of Alzheimer amyloid precursor protein proteolysis. *J. Biol. Chem.* **2016**, *291*, 19235–19244. [CrossRef]
4. Karran, E.; Mercken, M.; De Strooper, B. The amyloid cascade hypothesis for Alzheimer's disease: an appraisal for the development of therapeutics. *Nat. Rev. Drug Discov.* **2011**, *10*, 698–712. [CrossRef] [PubMed]
5. Roychaudhuri, R.; Yang, M.; Hoshi, M.M.; Teplow, D.B. Amyloid β-protein assembly and Alzheimer disease. *J. Biol. Chem.* **2009**, *284*, 4749–4753. [CrossRef] [PubMed]
6. Benilova, I.; Karran, E.; De Strooper, B. The toxic Aβ oligomer and Alzheimer's disease: an emperor in need of clothes. *Nat. Neurosci.* **2012**, *15*, 349–357. [CrossRef] [PubMed]
7. Nishizuka, Y. Protein kinase C and lipid signaling for sustained cellular responses. *FASEB J.* **1995**, *9*, 484–496. [CrossRef]
8. Sun, M.K.; Alkon, D.L. The "memory kinases": roles of PKC isoforms in signal processing and memory formation. *Prog. Mol. Biol. Transl. Sci.* **2014**, *122*, 31–59.
9. Govoni, S.; Bergamaschi, S.; Racchi, M.; Battaini, F.; Binetti, G.; Bianchetti, A.; Trabucchi, M. Cytosol protein kinase C downregulation in fibroblasts from Alzheimer's disease patients. *Neurology* **1993**, *43*, 2581–2586. [CrossRef]
10. Yang, H.Q.; Pan, J.; Ba, M.W.; Sun, Z.K.; Ma, G.Z.; Lu, G.Q.; Xiao, Q.; Chen, S.D. New protein kinase C activator regulates amyloid precursor protein processing in vitro by increasing α-secretase activity. *Eur. J. Neurosci.* **2007**, *26*, 381–391. [CrossRef]
11. Etcheberrigaray, R.; Tan, M.; Dewachter, I.; Kuiperi, C.; Van der Auwera, I.; Wera, S.; Qiao, L.; Bank, B.; Nelson, T.J.; Kozikowski, A.P.; et al. Therapeutic effects of PKC activators in Alzheimer's disease transgenic mice. *Proc. Natl. Acad. Sci. USA* **2004**, *101*, 11141–11146. [CrossRef] [PubMed]
12. Pascale, A.; Amadio, M.; Scapagnini, G.; Lanni, C.; Racchi, M.; Provenzani, A.; Govoni, S.; Alkon, D.L.; Quattrone, A. Neuronal ELAV proteins enhance mRNA stability by a PKCα-dependent pathway. *Proc. Natl. Acad. Sci. USA* **2005**, *102*, 12065–12070. [CrossRef] [PubMed]
13. Choi, D.S.; Wang, D.; Yu, G.Q.; Zhu, G.; Kharazia, V.N.; Paredes, J.P.; Chang, W.S.; Deitchman, J.K.; Mucke, L.; Messing, R.O. PKCε increases endothelin converting enzyme activity and reduces amyloid plaque pathology in transgenic mice. *Proc. Natl. Acad. Sci. USA* **2006**, *103*, 8215–8220. [CrossRef] [PubMed]
14. Hongpaisan, J.; Sun, M.K.; Alkon, D.L. PKCε activation prevents synaptic loss, Aβ elevation, and cognitive deficits in Alzheimer's disease transgenic mice. *J. Neurosci.* **2011**, *31*, 630–643. [CrossRef]
15. Pacheco-Quinto, J.; Eckman, E.A. Endothelin-converting enzymes degrade intracellular β-amyloid produced within the endosomal/lysosomal pathway and autophagosomes. *J. Biol. Chem.* **2013**, *288*, 5606–5615. [CrossRef]

16. Pettit, G.R.; Herald, C.L.; Doubek, D.L.; Herald, D.L.; Arnold, E.; Clardy, J. Isolation and structure of bryostatin 1. *J. Am. Chem. Soc.* **1982**, *104*, 6846–6848. [CrossRef]
17. Khan, T.K.; Nelson, T.J.; Verma, V.A.; Wender, P.A.; Alkon, D.L. A cellular model of Alzheimer's disease therapeutic efficacy: PKC activation reverses Aβ-induced biomarker abnormality on cultured fibroblasts. *Neurobiol. Dis.* **2009**, *34*, 332–339. [CrossRef]
18. Alkon, D.L.; Sun, M.K.; Nelson, T.J. PKC signaling deficits: a mechanistic hypothesis for the origins of Alzheimer's disease. *Trends Pharmacol. Sci.* **2007**, *28*, 51–60. [CrossRef]
19. Wender, P.A.; Hardman, C.T.; Ho, S.; Jeffreys, M.S.; Maclaren, J.K.; Quiroz, R.V.; Ryckbosch, S.M.; Shimizu, A.J.; Sloane, J.L.; Stevens, M.C. Scalable synthesis of bryostatin 1 and analogs, adjuvant leads against latent HIV. *Science* **2017**, *358*, 218–223. [CrossRef]
20. Trost, B.M.; Dong, G. Total synthesis of bryostatin 16 using atom-economical and chemoselective approaches. *Nature* **2008**, *456*, 485–488. [CrossRef]
21. Kato, Y.; Scheuer, P.J. Aplysiatoxin and debromoaplysiatoxin, constituents of the marine mollusk Stylocheilus longicauda (Quoy and Gaimard, 1824). *J. Am. Chem. Soc.* **1974**, *96*, 2245–2246. [CrossRef] [PubMed]
22. Nakagawa, Y.; Yanagita, R.C.; Hamada, N.; Murakami, A.; Takahashi, H.; Saito, N.; Nagai, H.; Irie, K. A simple analogue of tumor-promoting aplysiatoxin is an antineoplastic agent rather than a tumor promoter: development of a synthetically accessible protein kinase C activator with bryostatin-like activity. *Am. Chem. Soc.* **2009**, *131*, 7573–7579. [CrossRef] [PubMed]
23. Kikumori, M.; Yanagita, R.C.; Tokuda, H.; Suzuki, N.; Nagai, H.; Suenaga, K.; Irie, K. Structure-activity studies on the spiroketal moiety of a simplified analogue of debromoaplysiatoxin with antiproliferative activity. *J. Med. Chem.* **2012**, *55*, 5614–5626. [CrossRef]
24. Blennow, K.; Mattsson, N.; Scholl, M.; Hansson, O.; Zetterberg, H. Amyloid biomarkers in Alzheimer's disease. *Trends Pharmacol. Sci.* **2015**, *36*, 297–309. [CrossRef]
25. Pike, K.E.; Savage, G.; Villemagne, V.L.; Ng, S.; Moss, S.A.; Maruff, P.; Mathis, C.A.; Klunk, W.E.; Masters, C.L.; Rowe, C.C. β-Amyloid imaging and memory in non-demented individuals: evidence for preclinical Alzheimer's disease. *Brain* **2007**, *130*, 2837–2844. [CrossRef]
26. Rowe, C.C.; Bourgeat, P.; Ellis, K.A.; Brown, B.; Lim, Y.Y.; Mulligan, R.; Jones, G.; Maruff, P.; Woodward, M.; Price, R.; et al. Predicting Alzheimer disease with β-amyloid imaging: results from the Australian imaging, biomarkers, and lifestyle study of ageing. *Ann. Neurol.* **2013**, *74*, 905–913. [CrossRef]
27. Shaw, L.M.; Vanderstichele, H.; Knapik-Czajka, M.; Figurski, M.; Coart, E.; Blennow, K.; Soares, H.; Simon, A.J.; Lewczuk, P.; Dean, R.A.; et al. Qualification of the analytical and clinical performance of CSF biomarker analyses in ADNI. *Acta Neuropathol.* **2011**, *121*, 597–609. [CrossRef]
28. Irie, K. New diagnostic method for Alzheimer's disease based on the toxic conformation theory of amyloid β. *Biosci. Biotechnol. Biochem.* **2020**, *84*, 1–16. [CrossRef]
29. Murakami, K.; Tokuda, M.; Suzuki, T.; Irie, Y.; Hanaki, M.; Izuo, N.; Monobe, Y.; Akagi, K.; Ishii, R.; Tatebe, H.; et al. Monoclonal antibody with conformational specificity for a toxic conformer of amyloid β42 and its application toward the Alzheimer's disease diagnosis. *Sci. Rep.* **2016**, *6*, 29038. [CrossRef]
30. Akiba, C.; Nakajima, M.; Miyajima, M.; Ogino, I.; Motoi, Y.; Kawamura, K.; Adachi, S.; Kondo, A.; Sugano, H.; Tokuda, T.; et al. Change of amyloid-β1-42 toxic conformer ratio after cerebrospinal fluid diversion predicts long-term cognitive outcome in patients with idiopathic normal pressure hydrocephalus. *J. Alzheimers Dis.* **2018**, *63*, 989–1002. [CrossRef]
31. Masuda, Y.; Uemura, S.; Ohashi, R.; Nakanishi, A.; Takegoshi, K.; Shimizu, T.; Shirasawa, T.; Irie, K. Identification of physiological and toxic conformations in Aβ42 aggregates. *ChemBioChem.* **2009**, *10*, 287–295. [CrossRef] [PubMed]
32. Trejo, J.; Massamiri, T.; Deng, T.; Dewji, N.N.; Bayney, R.M.; Brown, J.H. A direct role for protein kinase C and the transcription factor Jun/AP-1 in the regulation of the Alzheimer's β-amyloid precursor protein gene. *J. Biol. Chem.* **1994**, *269*, 21682–21690. [PubMed]
33. Lahiri, D.K.; Nall, C. Promoter activity of the gene encoding the β-amyloid precursor protein is up-regulated by growth factors, phorbol ester, retinoic acid and interleukin-1. *Brain Res. Mol. Brain Res.* **1995**, *32*, 233–240. [CrossRef]
34. Sarajarvi, T.; Jantti, M.; Paldanius, K.M.A.; Natunen, T.; Wu, J.C.; Makinen, P.; Tarvainen, I.; Tuominen, R.K.; Talman, V.; Hiltunen, M. Protein kinase C-activating isophthalate derivatives mitigate Alzheimer's disease-related cellular alterations. *Neuropharmacology* **2018**, *141*, 76–88. [CrossRef]

35. Sato, M.; Murakami, K.; Uno, M.; Nakagawa, Y.; Katayama, S.; Akagi, K.; Masuda, Y.; Takegoshi, K.; Irie, K. Site-specific inhibitory mechanism for amyloid β42 aggregation by catechol-type flavonoids targeting the Lys residues. *J. Biol. Chem.* **2013**, *288*, 23212–23224. [CrossRef]
36. Yoshioka, T.; Murakami, K.; Ido, K.; Hanaki, M.; Yamaguchi, K.; Midorikawa, S.; Taniwaki, S.; Gunji, H.; Irie, K. Semisynthesis and structure-activity studies of uncarinic acid C isolated from *Uncaria rhynchophylla* as a specific inhibitor of the nucleation phase in amyloid β42 aggregation. *J. Nat. Prod.* **2016**, *79*, 2521–2529. [CrossRef]
37. Amadio, M.; Pascale, A.; Wang, J.; Ho, L.; Quattrone, A.; Gandy, S.; Haroutunian, V.; Racchi, M.; Pasinetti, G.M. nELAV proteins alteration in Alzheimer's disease brain: a novel putative target for amyloid-β reverberating on AβPP processing. *J. Alzheimers Dis.* **2009**, *16*, 409–419. [CrossRef]
38. Kang, M.J.; Abdelmohsen, K.; Hutchison, E.R.; Mitchell, S.J.; Grammatikakis, I.; Guo, R.; Noh, J.H.; Martindale, J.L.; Yang, X.; Lee, E.K.; et al. HuD regulates coding and noncoding RNA to induce APP–>Aβ processing. *Cell Rep.* **2014**, *7*, 1401–1409. [CrossRef]
39. LaFerla, F.M.; Green, K.N.; Oddo, S. Intracellular amyloid-β in Alzheimer's disease. *Nat. Rev. Neurosci.* **2007**, *8*, 499–509. [CrossRef]
40. Irie, Y.; Murakami, K.; Hanaki, M.; Hanaki, Y.; Suzuki, T.; Monobe, Y.; Takai, T.; Akagi, K.I.; Kawase, T.; Hirose, K.; et al. Synthetic models of quasi-stable amyloid β40 oligomers with significant neurotoxicity. *ACS Chem. Neurosci.* **2017**, *8*, 807–816. [CrossRef]
41. Irie, Y.; Hanaki, M.; Murakami, K.; Imamoto, T.; Furuta, T.; Kawabata, T.; Kawase, T.; Hirose, K.; Monobe, Y.; Akagi, K.I.; et al. Synthesis and biochemical characterization of quasi-stable trimer models of full-length amyloid β40 with a toxic conformation. *Chem. Commun.* **2018**, *55*, 182–185. [CrossRef] [PubMed]
42. Paul, S.M.; Mytelka, D.S.; Dunwiddie, C.T.; Persinger, C.C.; Munos, B.H.; Lindborg, S.R.; Schacht, A.L. How to improve R&D productivity: the pharmaceutical industry's grand challenge. *Nat. Rev. Drug Discov.* **2010**, *9*, 203–214. [PubMed]
43. Kondo, T.; Imamura, K.; Funayama, M.; Tsukita, K.; Miyake, M.; Ohta, A.; Woltjen, K.; Nakagawa, M.; Asada, T.; Arai, T.; et al. iPSC-based compound screening and in vitro trials identify a synergistic anti-amyloid β combination for Alzheimer's disease. *Cell Rep.* **2017**, *21*, 2304–2312. [CrossRef] [PubMed]
44. Crouch, S.P.; Kozlowski, R.; Slater, K.J.; Fletcher, J. The use of ATP bioluminescence as a measure of cell proliferation and cytotoxicity. *J. Immunol. Methods* **1993**, *160*, 81–88. [CrossRef]
45. Nelson, T.J.; Sun, M.K.; Lim, C.; Sen, A.; Khan, T.; Chirila, F.V.; Alkon, D.L. Bryostatin effects on cognitive function and PKCε in Alzheimer's disease phase IIa and expanded access trials. *J. Alzheimers Dis.* **2017**, *58*, 521–535. [CrossRef]
46. Aranda-Abreu, G.E.; Behar, L.; Chung, S.; Furneaux, H.; Ginzburg, I. Embryonic lethal abnormal vision-like RNA-binding proteins regulate neurite outgrowth and tau expression in PC12 cells. *J. Neurosci.* **1999**, *19*, 6907–6917. [CrossRef]
47. Marchesi, N.; Amadio, M.; Colombrita, C.; Govoni, S.; Ratti, A.; Pascale, A. PKC activation counteracts ADAM10 deficit in HuD-silenced neuroblastoma cells. *J. Alzheimers Dis.* **2016**, *54*, 535–547. [CrossRef]
48. Jarosz-Griffiths, H.H.; Corbett, N.J.; Rowland, H.A.; Fisher, K.; Jones, A.C.; Baron, J.; Howell, G.J.; Cowley, S.A.; Chintawar, S.; Cader, M.Z.; et al. Proteolytic shedding of the prion protein via activation of metallopeptidase ADAM10 reduces cellular binding and toxicity of amyloid-β oligomers. *J. Biol. Chem.* **2019**, *294*, 7085–7097. [CrossRef]
49. Kaneko, N.; Yamamoto, R.; Sato, T.A.; Tanaka, K. Identification and quantification of amyloid β-related peptides in human plasma using matrix-assisted laser desorption/ionization time-of-flight mass spectrometry. *Proc. Jpn. Acad. Ser. B. Phys. Biol. Sci.* **2014**, *90*, 104–117. [CrossRef]
50. Nakamura, A.; Kaneko, N.; Villemagne, V.L.; Kato, T.; Doecke, J.; Dore, V.; Fowler, C.; Li, Q.X.; Martins, R.; Rowe, C.; et al. High performance plasma amyloid-β biomarkers for Alzheimer's disease. *Nature* **2018**, *554*, 249–254. [CrossRef]
51. Hwang, S.S.; Chan, H.; Sorci, M.; Van Deventer, J.; Wittrup, D.; Belfort, G.; Walt, D. Detection of amyloid β oligomers toward early diagnosis of Alzheimer's disease. *Anal. Biochem.* **2019**, *566*, 40–45. [CrossRef]
52. Klaver, A.C.; Patrias, L.M.; Finke, J.M.; Loeffler, D.A. Specificity and sensitivity of the Aβ oligomer ELISA. *J. Neurosci. Methods* **2011**, *195*, 249–254. [CrossRef]

53. Horikoshi, Y.; Mori, T.; Maeda, M.; Kinoshita, N.; Sato, K.; Yamaguchi, H. Aβ N-terminal-end specific antibody reduced β-amyloid in Alzheimer-model mice. *Biochem. Biophys. Res. Commun.* **2004**, *325*, 384–387. [CrossRef]
54. Xia, W.; Yang, T.; Shankar, G.; Smith, I.M.; Shen, Y.; Walsh, D.M.; Selkoe, D.J. A specific enzyme-linked immunosorbent assay for measuring β-amyloid protein oligomers in human plasma and brain tissue of patients with Alzheimer disease. *Arch. Neurol.* **2009**, *66*, 190–199. [CrossRef]
55. Ohshima, Y.; Taguchi, K.; Mizuta, I.; Tanaka, M.; Tomiyama, T.; Kametani, F.; Yabe-Nishimura, C.; Mizuno, T.; Tokuda, T. Mutations in the β-amyloid precursor protein in familial Alzheimer's disease increase Aβ oligomer production in cellular models. *Heliyon* **2018**, *4*, e00511. [CrossRef]
56. Umeda, T.; Ramser, E.M.; Yamashita, M.; Nakajima, K.; Mori, H.; Silverman, M.A.; Tomiyama, T. Intracellular amyloid β oligomers impair organelle transport and induce dendritic spine loss in primary neurons. *Acta Neuropathol. Commun.* **2015**, *3*, 51. [CrossRef]
57. Kondo, T.; Asai, M.; Tsukita, K.; Kutoku, Y.; Ohsawa, Y.; Sunada, Y.; Imamura, K.; Egawa, N.; Yahata, N.; Okita, K.; et al. Modeling Alzheimer's disease with iPSCs reveals stress phenotypes associated with intracellular Aβ and differential drug responsiveness. *Cell Stem Cell* **2013**, *12*, 487–496. [CrossRef]
58. Guo, J.P.; Arai, T.; Miklossy, J.; McGeer, P.L. Aβ and tau form soluble complexes that may promote self aggregation of both into the insoluble forms observed in Alzheimer's disease. *Proc. Natl. Acad. Sci. USA* **2006**, *103*, 1953–1958. [CrossRef]
59. Ambadipudi, S.; Biernat, J.; Riedel, D.; Mandelkow, E.; Zweckstetter, M. Liquid-liquid phase separation of the microtubule-binding repeats of the Alzheimer-related protein Tau. *Nat. Commun.* **2017**, *8*, 275. [CrossRef]
60. Boyko, S.; Qi, X.; Chen, T.H.; Surewicz, K.; Surewicz, W.K. Liquid-liquid phase separation of tau protein: The crucial role of electrostatic interactions. *J. Biol. Chem.* **2019**, *294*, 11054–11059. [CrossRef]
61. Vogler, T.O.; Wheeler, J.R.; Nguyen, E.D.; Hughes, M.P.; Britson, K.A.; Lester, E.; Rao, B.; Betta, N.D.; Whitney, O.N.; Ewachiw, T.E.; et al. TDP-43 and RNA form amyloid-like myo-granules in regenerating muscle. *Nature* **2018**, *563*, 508–513. [CrossRef]
62. Kume, T.; Kouchiyama, H.; Kaneko, S.; Maeda, T.; Akaike, A.; Shimohama, S.; Kihara, T.; Kimura, J.; Wada, K.; Koizumi, S. BDNF prevents NO mediated glutamate cytotoxicity in cultured cortical neurons. *Brain Res* **1997**, *756*, 200–204. [CrossRef]
63. Izuo, N.; Kume, T.; Sato, M.; Murakami, K.; Irie, K.; Izumi, Y.; Akaike, A. Toxicity in rat primary neurons through the cellular oxidative stress induced by the turn formation at positions 22 and 23 of Aβ42. *ACS Chem. Neurosci.* **2012**, *3*, 674–681. [CrossRef]
64. Nakagawa, M.; Taniguchi, Y.; Senda, S.; Takizawa, N.; Ichisaka, T.; Asano, K.; Morizane, A.; Doi, D.; Takahashi, J.; Nishizawa, M.; et al. A novel efficient feeder-free culture system for the derivation of human induced pluripotent stem cells. *Sci. Rep.* **2014**, *4*, 3594. [CrossRef]

International Journal of
Molecular Sciences

MDPI

Article

Interaction of Aβ42 with Membranes Triggers the Self-Assembly into Oligomers

Siddhartha Banerjee †, Mohtadin Hashemi †, Karen Zagorski and Yuri L. Lyubchenko *

Department of Pharmaceutical Sciences, University of Nebraska Medical Center, 986025 Nebraska Medical Center, Omaha, NE 68198-6025, USA; siddhartha.banerjee@unmc.edu (S.B.); mohtadin.hashemi@unmc.edu (M.H.); karen.zagorski@unmc.edu (K.Z.)

* Correspondence: ylyubchenko@unmc.edu; Tel.: +1-402-559-1971

† These authors contributed equally to this work.

Received: 29 January 2020; Accepted: 6 February 2020; Published: 8 February 2020

Abstract: The self-assembly of amyloid β (Aβ) proteins into oligomers is the major pathogenic event leading to Alzheimer's disease (AD). Typical in vitro experiments require high protein concentrations, whereas the physiological concentration of Aβ is in the picomolar to low nanomolar range. This complicates the translation of results obtained in vitro to understanding the aggregation process in vivo. Here, we demonstrate that Aβ42 self-assembles into aggregates on membrane bilayers at low nanomolar concentrations - a pathway in which the membrane plays the role of a catalyst. Additionally, physiological ionic conditions (150 mM NaCl) significantly enhance on-membrane aggregation, leading to the rapid formation of oligomers. The self-assembly process is reversible, so assembled aggregates can dissociate from the membrane surface into the bulk solution to further participate in the aggregation process. Molecular dynamics simulations demonstrate that the transient membrane-Aβ interaction dramatically changes the protein conformation, facilitating the assembly of dimers. The results indicate peptide–membrane interaction is the critical step towards oligomer formation at physiologically low protein concentrations.

Keywords: amyloid aggregation; supported lipid bilayers; time-lapse AFM; Alzheimer's disease; molecular dynamics simulation

1. Introduction

Growing evidence suggests the involvement of protein oligomers in the development of protein misfolding diseases, including Alzheimer's disease (AD) and Parkinson's disease (PD); however, limited knowledge exists regarding the molecular mechanisms behind the aggregation processes for these oligomers [1–4]. The amyloid cascade hypothesis (ACH), proposed more than a quarter-century ago [5], is the major model used to describe the pathology of AD and other neurodegenerative diseases [5–9]. ACH posits that the onset of diseases involves the spontaneous assembly of an amyloidogenic polypeptide. In turn, the accumulation of aggregates defines the disease state. Translational studies in the framework of ACH are focused on decreasing the concentration of amyloid proteins to decelerate the aggregation process (reviewed in [6,7,10]). However, drug development efforts based on decreasing the amyloid β (Aβ) concentration, as well as disaggregating the plaques, have failed [11,12], which challenges the validity of ACH [13]. In fact, there are a number of problems with ACH [7], one being the concentration of Aβ. Specifically, in vitro aggregation experiments require Aβ concentrations in the micromolar range, whereas Aβ levels in brain and cerebral spinal fluid (CSF) are frequently in the low nanomolar range [10,14–16]; this value remains in the same range regardless of the disease state [17,18]. Therefore, an explanation is needed as to how Aβ proteins at such low concentrations can assemble into aggregates. It is also unknown what causes the amyloid protein to aggregate. These problems with ACH have driven alternative models for the development of AD [10].

We have recently discovered that amyloid proteins, including amyloid β 42 (Aβ42) and its Aβ (14–23) segment, are capable of assembling into aggregates in the nanomolar concentration range at the surface–liquid interface [19–21]. We developed a model, according to which monomers that were transiently immobilized on surfaces increased the local protein monomer concentration, and thus worked as nuclei to dramatically accelerate the entire aggregation process [21]. This theoretical model was verified by experimental AFM studies, and we termed this aggregation pathway the surface-catalyzed aggregation process. We have recently examined whether the surface catalysis process can be extended to membrane surfaces. We selected α-synuclein (α-syn) and performed time-lapse AFM studies [20,22]. We used supported lipid bilayers (SLBs) formed by 1-palmitoyl-2-oleoyl-glycero-3-phosphocholine (POPC) and 1-palmitoyl-2-oleoyl-sn-glycero-3-phospho-L-serine (POPS), which are major components of neuronal membranes. We demonstrated that SLBs catalyze the aggregation of α-syn at concentrations as low as 10 nM, which corresponds to the concentration range in the CSF [23]. The aggregation kinetics were also found to be dependent on the SLB composition, being considerably greater for the POPS bilayer compared to POPC. Computational modeling suggested that α-syn monomers changed conformation upon interaction with the bilayers, and these interactions were dependent on the composition of the bilayer. The conformations of α-syn after binding to POPS dramatically facilitated assembly into the dimer, a property that is in contrast with POPC and in line with experimental data. Note that a similar phenomenon has been reported for another amyloid protein—transthyretin—which causes systemic amyloidosis and leads to a variety of tissue damage [24,25].

In the present study, we applied the same combined experimental and computational approaches to characterize the aggregation of Aβ42 at its physiologically relevant low concentration (10 nM). The time-lapse AFM imaging revealed Aβ42 aggregation on the bilayer surfaces, while no self-assembly of Aβ42 was detected in bulk solution. The assembled aggregates were not strongly bound to the surface and were capable of spontaneous dissociation into solution. Importantly, the self-assembly process did not cause damage to the surface, as no defects were detected after the aggregates were dissociated. We performed computational modeling of the aggregation process on the membrane surfaces and demonstrated that interaction with the membrane dramatically changes the conformation of Aβ42 monomers. Moreover, membrane-bound Aβ42 proteins trigger the assembly of dimers, propagating the misfolded states of the Aβ42 molecules. Thus, interaction with membranes results in the transition of Aβ42 into the aggregation-prone, misfolded conformations. Such conformations have not been reported in simulations or experiments performed in bulk solutions. Given that the membrane-assembled aggregates can dissociate into solution, the recently discovered on-membrane aggregation can be the mechanism by which amyloid oligomers, or their disease-prone seeds, are produced and spread over the organism.

2. Results

2.1. Experimental Studies

2.1.1. Aggregation on Supported Lipid Bilayers (SLB)

We used time-lapse AFM to directly visualize the aggregation process of the Aβ42 protein on SLBs formed from POPC and POPS. The key in these studies was the assembly of a topographically smooth bilayer that remains stable over the entire observation period, which typically was over several hours. Freshly cleaved mica was used as the support substrate. We developed a method which is described in papers [19,20,22,26]. Typical examples of AFM images of POPC, POPS and POPC:POPS bilayers are shown in Figure S1. The surface was smooth over several microns, with a few defects that did not change over time. These defects were used to measure the thickness of the bilayer. Cross-section profiles in this figure reveal the height value of 4.2–4.5 nm, validating the assembly of the bilayer. The smooth surface topography remains stable during hours of observation, regardless of the bilayer composition.

2.1.2. Aggregation on the POPC Bilayer

After the assembly of the bilayer and its inspection with AFM, a solution of 10 nM Aβ42 in 10 mM sodium phosphate (pH 7.4) was placed on top of the SLBs and imaged at room temperature. Figure 1a shows the surface topography taken just after the injection of protein sample. The surface remains smooth; however, we cannot exclude the binding of Aβ42 monomers, as these are too small (~4 kDa) for reliable visualization with AFM [19]. The time-dependent aggregation of 10 nM Aβ42 on the POPC bilayer is shown in Figure 1a–c, with Aβ42 aggregates appearing on the SLB after 6 h. The time-lapse experiment was continued up to 9 h. The quantitative analysis of the number and sizes of aggregates, shown in Figure 1c and Figure S2a, demonstrate that these parameters increase gradually over time, whereas no aggregation in the bulk solution was observed, even after 48 h [19].

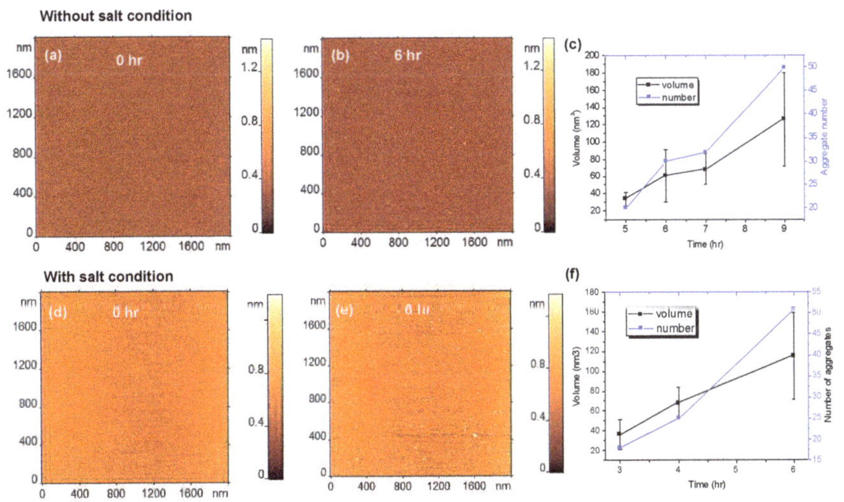

Figure 1. Aggregation of 10 nM amyloid β 42 (Aβ42) on 1-palmitoyl-2-oleoyl-sn-glycero-3-phosphocholine (POPC) supported lipid bilayer (SLB). (**a–c**) Show the results obtained in the absence of 150 mM NaCl in the buffer (without salt condition) and (**d–f**) show the aggregation results with 150 mM NaCl (with salt condition). (**a**) AFM topographic image of POPC SLB just after the addition of Aβ42 onto the SLB surface. (**b**) The same SLB surface after incubating for 6 h, without 150 mM NaCl. The small globular features indicate the presence of aggregates. (**c**) The plot shows the increase in number and volume of the aggregates with time. (**d–f**) show similar experiments but with 150 mM NaCl added to the buffer.

We also tested how ionic conditions close to physiological ones affect the aggregation propensity on the POPC bilayer by performing time-lapse AFM experiments in the presence of 150 mM NaCl. The results are shown in Figure 1d–f. Frame d shows the clean surface just after the addition of Aβ42, and frame e shows a substantial amount of aggregates after 6 h of incubation. Figure 1f shows the gradual increase in both the number and volume of the aggregates in presence of NaCl. Quantitative analysis (Figure 1c,f and Figure S2a,b) demonstrates that aggregates with volumes ~120 nm^3 appear after 6 h in the presence of 150 mM NaCl, while in the absence of NaCl similar aggregates are formed after 9 h. These data suggest that physiological ionic conditions further increase the aggregation propensity on the POPC bilayer.

2.1.3. Aggregation on the POPS Bilayer

Next, we tested how critical the composition of the bilayer is for the on-surface catalysis of aggregation of Aβ42. We performed time-lapse aggregation studies for the negatively charged POPS

SLB in the presence of 150 mM NaCl. The results are shown in Figure 2a–c. Frames a-b represent typical time-lapse images of the aggregates assembled by 10 nM Aβ42 on the POPS SLB. After 5 h of time-lapse imaging, a significant amount of aggregates are observed on the SLB surface (Figure 2b). The time-dependent values of the number and volumes of aggregates are plotted in Figure 2c and Figure S2d; they demonstrate that both values increase gradually with time. A comparison of the data obtained for POPC and POPS bilayers (Figures 1e and 2b) indicates that the aggregates are large on the negatively charged POPS surface compared to those on POPC. To identify the effect of NaCl on the aggregation kinetics of Aβ42 on POPS SLB, we then performed a similar experiment but with salt excluded from the buffer. Similar to POPC, the aggregation process is slower without 150 mM NaCl (Figure 2e and Figure S2c). Figure 2e shows an AFM image of the POPS surface after incubation of 10 nM Aβ42 without salt. The surface contains a substantially smaller number of aggregates compared to the salt condition (Figure 2b). Figure 2f shows the increase in number and volume of aggregates with time.

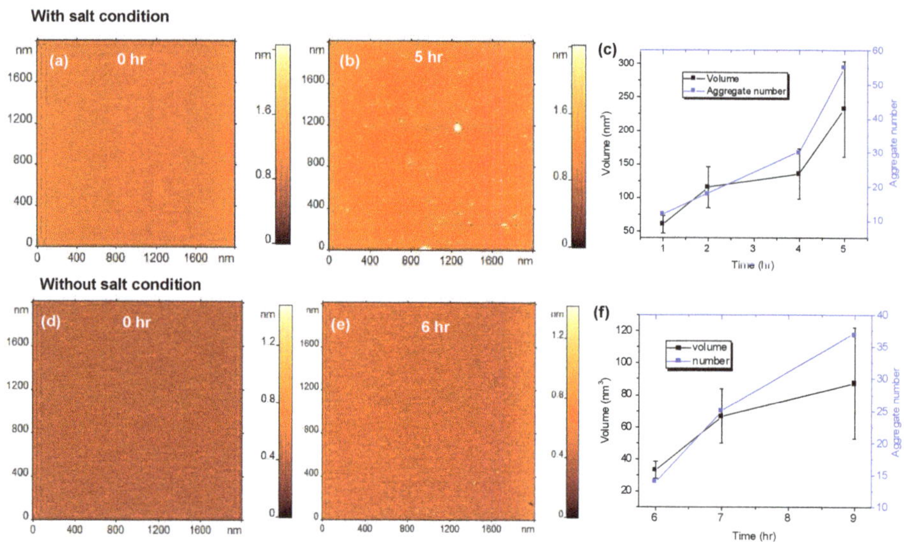

Figure 2. Aggregation of 10 nM amyloid β 42 (Aβ42) on 1-palmitoyl-2-oleoyl-sn-glycero-3-phospho-L-serine (POPS) supported lipid bilayer (SLB). (**a–c**) The results obtained in the presence of 150 mM NaCl in the buffer (with salt condition) and (**d–f**) shows the aggregation results without 150 mM NaCl (without salt condition). (**a**) AFM topographic image of POPS SLB just after the addition of 10 nM Aβ42 onto the SLB surface. (**b**) The same SLB surface having Aβ42 aggregates after incubating for 5 h with 150 mM NaCl (with salt condition). The small globular features indicate the presence of aggregates. (**c**) The plot shows the increase in number and volume of the aggregates with increase in time. (**d–f**) show similar aggregation experiments but without 150 mM NaCl in the buffer.

2.1.4. Aggregation on Binary SLB Mixture (POPC:POPS)

To further characterize the effect of the bilayer composition on the aggregation catalysis, similar time-lapse AFM experiments were performed for the SLB formed by an equimolar mixture of POPC and POPS in presence of 150 mM NaCl. The results are shown in Figure 3a–c. As observed in the previous experiments, aggregates appeared on the surface, and after 5 h the surface contained a significantly higher number of aggregates (Figure 3b) compared to the same surface without salt present (Figure 3e). The sizes of the aggregates in the presence of NaCl, are also larger compared to the no salt condition. The aggregate size after 5 h of incubation without salt was ~60 nm^3 (Figure 3f), whereas similar sized aggregates were observed within 1 h of incubation in the presence of NaCl

(Figure 3c). These results clearly indicate that the presence of NaCl strongly influences the acceleration of aggregation and becomes an important factor allowing aggregates to grow faster.

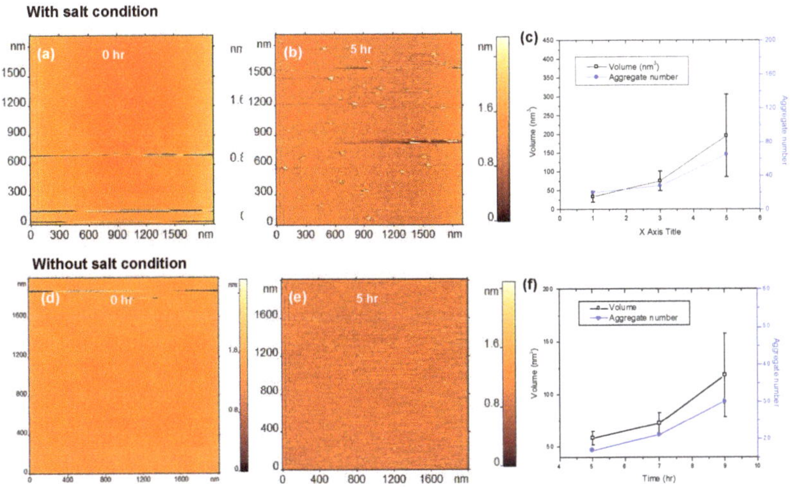

Figure 3. Aggregation of 10 nM amyloid β 42 (Aβ42) on 1-palmitoyl-2-oleoyl-sn-glycero-3-phosphocholine:1-palmitoyl-2-oleoyl-sn-glycero-3-phospho-L-serine (POPC-POPS) supported lipid bilayer (SLB). (**a**–**c**) Show the results obtained in the presence of 150 mM NaCl in the buffer (with salt condition) and (**d**–**f**) show the aggregation results without 150 mM NaCl (without salt condition). (**a**) AFM topographic image of POPC-POPS SLB just after the addition of Aβ42 onto the SLB surface. (**b**) The same SLB surface having Aβ42 aggregates after incubating for 5 h with 150 mM NaCl in the buffer (with salt condition). The small globular features indicate the presence of aggregates. (**c**) The plot shows the increase in number and volume of the aggregates with increase in time. (**d**–**f**) show results from similar experiment but without 150 mM NaCl in the buffer.

A quantitative comparison of aggregation on the three bilayer surfaces has been done and the data is assembled as histograms, shown in Figure 4. The cumulative plot in Figure 4a shows the number of aggregates observed on the three different SLB surfaces after 1, 3, and 5 h of incubation with 10 nM Aβ42. The grey bars represent the condition with 150 mM NaCl in the buffer and black bars shows the no salt condition. The missing black bars for some time-points indicate that no substantial number of aggregates were observed at that time-point. For example, no aggregates were visible after 1 h and 3 h on POPC without salt (missing black bars), but aggregates were present on the POPC surface (gray bars) with NaCl at those time-points. Overall, aggregates were visible on all three surfaces after 1 h in presence of NaCl, but the aggregates appeared later without the salt condition. Figure 4b summarizes and compares the size of the aggregates formed on the three SLB surfaces after 5 h of incubation with and without salt. In all three cases, the presence of NaCl promotes the formation of larger aggregates, indicating the role of NaCl in the on-membrane aggregation of Aβ42.

2.1.5. Dissociation of Aggregates from the Bilayer Surface

Time-lapse AFM experiments allowed us to directly monitor the assembly-disassembly process for each aggregate (Figure 5a–d). The data show that aggregates formed on the surface are able to dissociate from the surface. The black square in frames "a–d" indicates a feature that is present in all frames and acts as a fiducial marker. This allows us to detect the relative position of newly formed aggregates and desorption of already formed aggregates. The aggregates marked with yellow circles are absent in the next frame, while aggregates marked with green circles have appeared. For example, the aggregate circled in yellow in frame "b" is gone in frame "c", whereas, the aggregate circled in green in frame "c" has appeared and was absent in frame "b". These images demonstrate that assembled Aβ aggregates can dissociate from the surface and go into the solution.

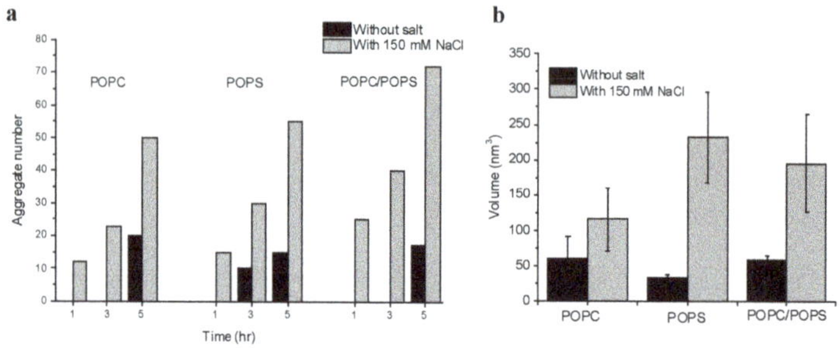

Figure 4. Volume and number of aggregates formed on different supported lipid bilayer (SLB) surfaces. (**a**) Comparison of number of aggregates formed by 10 nM Aβ42 after 1 h, 3 h, and 5 h of incubation with 150 mM NaCl (gray bars) and without NaCl (black bars) on the three SLB surfaces. The missing black bars at different time-points indicate no aggregates were visible under those conditions. (**b**) Comparison of the volume of aggregates formed on 1-palmitoyl-2-oleoyl-sn-glycero-3-phosphocholine (POPC), 1-palmitoyl-2-oleoyl-sn-glycero-3-phospho-L-serine (POPS), and POPC-POPS SLBs with and without salt condition after 5 h of incubation. The volume of the aggregates formed after 5 h of incubation is much larger in presence of NaCl on all three surfaces.

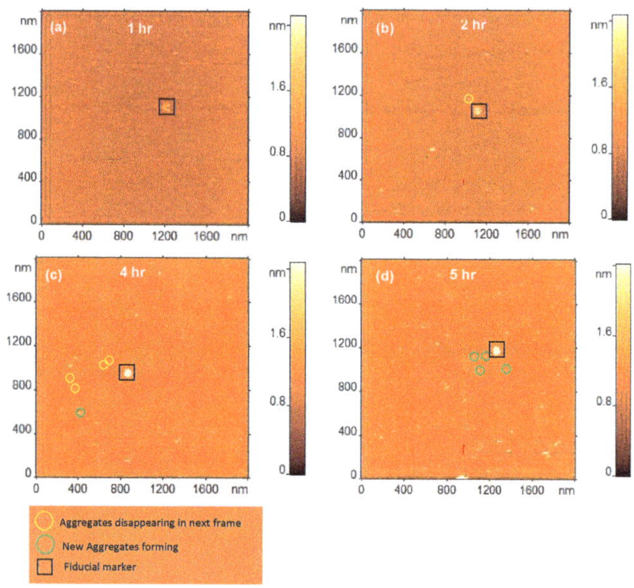

Figure 5. Dynamics of aggregate formation and dissociation from 1-palmitoyl-2-oleoyl-sn-glycero-3-phospho-L-serine (POPS) supported lipid bilayer (SLB) surface. (**a**) Time-lapse AFM imaging of 10 nM amyloid β 42 (Aβ42) on POPS SLB after 1 h incubation. (**b–d**) Same area of the surface scanned at the noted time-points. Black square depicts an aggregate that is present in all frames and acts as a fiducial marker. Aggregates circled in yellow are desorbed from the surface between frames, whereas those circled in green are newly formed on the surface.

In case the dissociation of aggregates can lead to their accumulation in the solution, the concentration of aggregates above the bilayer should grow. To test this theory, aliquots from the solution above the SLB were taken, deposited on mica, imaged, and analyzed. A set of images corresponding to aliquots from different time-points demonstrates the accumulation of aggregates in the solution above the SLB, shown in Figure S3. Frames shown in Figure S3a depict representative images of aliquots taken from the solution above the SLB surface, whereas frames in Figure S3b depict the control experiment without SLB surface. The plot (Figure S3c) shows a substantial amount of aggregate accumulation in 6 h and 24 h in the solution above the SLB, whereas only a few aggregates appear in the control.

2.2. Computer Simulations of Aβ42 Interacting with Bilayers

To determine the underlying mechanism of interaction between Aβ42 and lipid bilayers, we employed extensive microseconds-long molecular dynamics (MD) simulations. Briefly, POPC and POPS bilayers were assembled using 512 lipids, solvated with 40:1 water:lipid, and neutralized using NaCl. After energy minimization, each bilayer system was then simulated for 150 ns to obtain a relaxed bilayer. Aβ42 monomer or dimer, obtained from [27], were then added to the relaxed bilayers at a distance of 5 nm from the bilayer centers. The systems were then neutralized and kept at 150 mM ion concentration using NaCl counterions. Following short preparatory simulations, each system was then run on the special purpose Anton 2 supercomputer for 5 μs each and the interaction of the Aβ molecules with the bilayers was recorded. All simulation system configurations and durations are presented in Table 1.

Table 1. Description and nomenclature of simulated bilayer systems.

Name	Bilayer	Aβ42 Species	Duration [μs]
PC	POPC′	-	0.150
PS	POPS″	-	0.150
PCm	POPC	Monomer	5
PSm	POPS	Monomer	5
PCmm	POPC	Bound monomer + monomer	5
PSmm	POPS	Bound monomer + monomer	5
PCd	POPC	Dimer	5
PSd	POPS	Dimer	5

′1-Palmitoyl-2-oleoyl-sn-glycero-3-phosphocholine, ″1-palmitoyl-2-oleoyl-sn-glycero-3-phospho-L-serine.

2.2.1. Interactions of Aβ42 Monomer with POPC

Upon and during interaction with the POPC bilayer, the Aβ42 monomer experiences structural and conformational changes. Figure 6a shows snapshots illustrating the transition of the Aβ monomer induced by interaction with the POPC bilayer. The initial conformation of the Aβ monomer (snapshot (i)), taken from our recent paper [27], is a compact collapsed conformation with two small helical regions, shown as blue and purple ribbons, and an otherwise essentially random coil. The conformation gradually transitions to an elongated shape after ~ 500 ns, and two β-strands (yellow arrows) appear at the C- and N-termini of the molecule (snapshot (ii)). This elongated conformation remains stable for approximately 2 μs; however, the β-strand segments increase in size. This is then followed by the formation of two β-strands in the central region of the molecule (snapshot (iii)). A cooperative transition within the monomer then occurs, accompanied by the formation of a β-hairpin structure (snapshot (iv)); this structure remains stable until the end of the 5 μs-long simulation. The structural transition of the Aβ monomer induced by the interaction with the POPC bilayer is shown graphically in Figure 6b and can be viewed in Movie S1. The monomer starts with negligible β-structure content and ~12% helical content, but rapidly transitions to conformations with 5–10% β-content (after approximately 500 ns), followed by an abrupt transition at around ~2.5 μs to ~17% β-content that remains for the remainder of the simulation. Importantly, such dramatic conformational transitions within the Aβ monomer are induced by transient interactions between the POPC surface and peptide residues, with the primary interactions being through residues 7–12 and 27–29; these segments also experience the largest change in structure and adopt β-strands during the simulation (Figure 6c).

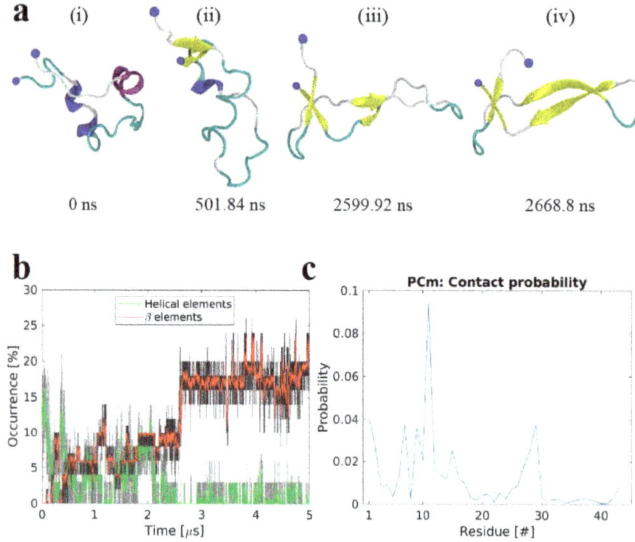

Figure 6. Molecular dynamics simulations of the interaction of amyloid β 42 (Aβ42) monomer with 1-palmitoyl-2-oleoyl-sn-glycero-3-phosphocholine (POPC) bilayer. (**a**) Snapshots showing the change of Aβ42 conformation and secondary structure at different time-points while interacting with the POPC bilayer. Protein is depicted as a cartoon following VMD coloring scheme (yellow β-strands and purple α-helices), N- and C-terminal Cα are presented as a large and a small blue sphere, respectively. (**b**) The evolution of secondary β-structure (β-sheet and β-bridge, red), and helical (α-, π-, and 3/10-helices, green), elements of the Aβ42 protein as determined by DSSP. The graphs are moving averages using a 1 ns window; raw data is presented as dark and light grey graphs, respectively. (**c**) Per residue contact probability plot between residues of the Aβ42 monomer and the P atoms of the POPC headgroups.

2.2.2. Interactions of Preformed Aβ42 Dimer with POPC

We then investigated whether interactions with the POPC bilayer have an effect on the structure of Aβ42 dimer assembled in bulk solution, and if these interactions are capable of changing the conformation of the dimer. Aβ42 dimer, with the structure identified in [27], was placed above the POPC bilayer and the results of MD simulations are summarized in Figure 7. Snapshots in Figure 7a demonstrate a rapid formation of β-sheet within the central and C-terminal segments of one monomer (i), followed by the β-strand formation in C-terminal segment in the other monomer (ii). This strand is further extended and becomes a nine-residue strand that remains stable until the end of the simulation (iii–iv). Graphically, this is depicted in Figure 7b, which shows the evolution of secondary structure elements versus time. Initially the dimer has <10% β-structure content, which over time gradually increases to ~20%; on the other hand, the helical content decreases with time.

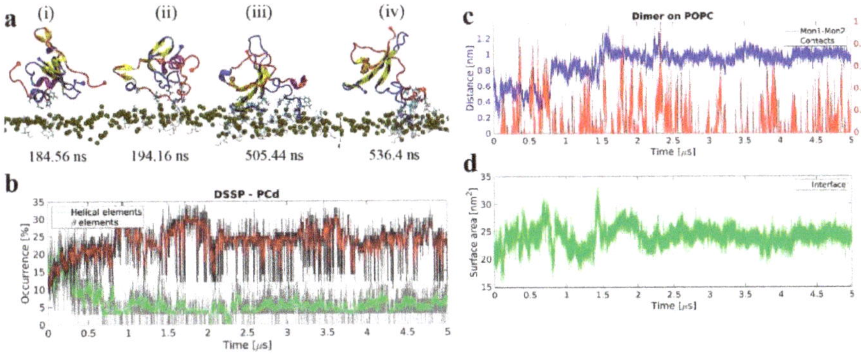

Figure 7. The dynamics of interactions between preformed dimer and 1-palmitoyl-2-oleoyl-sn-glycero-3-phosphocholine (POPC) bilayer. (**a**) Snapshots showing conformational change of preformed amyloid β 42 (Aβ42) dimer upon interaction with the POPC bilayer. Protein side-chains and POPC head groups are shown for interacting residues. Protein is depicted as cartoon following VMD coloring scheme, N- and C-terminal Cα are represented as large and small spheres in blue and red for Mon 1 and Mon 2, respectively, POPC P atoms are depicted as gold spheres. (**b**) Evolution of β-structure (β-sheet and β-bridge, red), and helical (α-, π-, and 3/10, green), elements of the preformed dimer over time. The graphs are moving averages using a 1 ns window; raw data is presented as dark and light grey graphs, respectively. (**c**) Center of mass (CoM) distance between the two monomers within the preformed dimer, blue, and normalized contacts between the dimer and the lipid headgroups, red. (**d**) Time evolution of the dimer interface surface area depicting the total area of interaction between the monomers.

The preformed Aβ42 dimer remains stable during the entire simulation, but experiences conformational change during and following interactions with the bilayer, as seen on Figure 7c. This is clearly observed from the change in center of mass (CoM) distance between the two monomers within the dimer, seen in Figure 7c. The figure also shows that the greatest change occurs when the dimer is in contact with or immediately after being in contact with the bilayer. The reorganization of the dimer also changes the interfacial area between the two monomers; initially ~20 nm^2, it increases to approximately 25 nm^2, shown in Figure 7d. Furthermore, as the dimer interacts with the bilayer, the residues forming the dimer interface also change, seen in Figure S4a. Initially, the dimer is stabilized by contacts between the N-termini and the central regions (i); however, as the dimer conformation changes, the interactions shift to also include residues of the C-termini (ii) before finally shifting to interactions between the C–C and C–N termini (iii). The probability of a residue to interact with the POPC membrane is very similar for the two monomers within the dimer, with the N-terminal and residues 29–31 having the highest probability to interact with the membrane, shown in Figure S4b.

2.2.3. Dimerization on POPC Bilayer

To elucidate the role of conformational transitions within the Aβ monomer in the aggregation process, we modeled the assembly of a dimer by the interaction of an Aβ monomer at the surface with a free monomer. Briefly, we selected the last frame of the monomer-POPC simulation (Figure 6) and added another monomer at 4 nm CoM distance with respect to the membrane-bound monomer and at 5 nm from the bilayer center. The results in Figure 8 and Movie S2 reveal a number of features of the dimer assembly process. First, the dimer on the surface assembles rapidly- in a time span below 100 ns- as is shown by the rapid drop in distance between the monomers in Figure 8a. Second, both monomers within the dimer undergo conformational transitions (Figure 8b). Third, the conformational transitions are accompanied by the formation of multiple contacts between the two monomers and the increase in interfacial surface area. The central part of the surface bound monomer, Mon 1, and the C-terminal segment of the free monomer, Mon 2, play the major role in this transition (Figure 8c and Figure S5a). Fourth, similar to the results for the monomer and preformed dimer (Figures 6 and 7), transient interactions of the dimer with the bilayer, primarily via Mon 1, are accompanied by conformational changes (Figure 8c and Figure S5b).

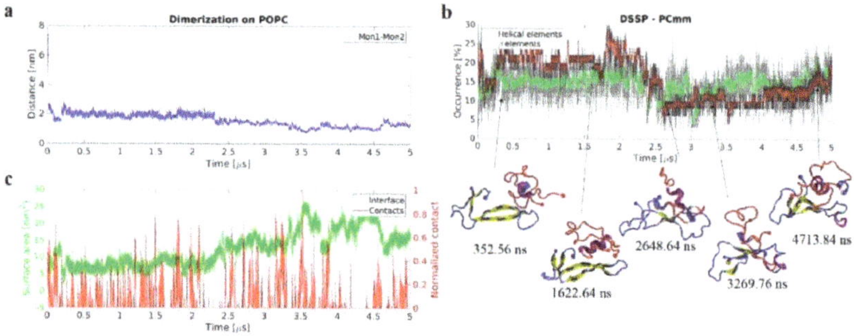

Figure 8. Dynamics of on-surface dimerization of amyloid β42 (Aβ42) monomers in presence of 1-palmitoyl-2-oleoyl-sn-glycero-3-phosphocholine (POPC) bilayer. (**a**) Time-dependent center of mass (CoM) distance between the two Aβ42 monomers revealing that the dimer rapidly forms and remains stable. (**b**) Evolution of secondary structure β-structure (β-sheet and β-bridge, red), and helical (α-, π-, and 3/10, green), elements, as determined by DSSP. The graphs are moving averages using a 1 ns window; raw data is presented as dark and light grey graphs, respectively. Key transition points are highlighted with cartoon representation of the protein structures. Initial surface bound monomer is depicted in blue and the initially free Aβ42 monomer in red. N- and C-terminal Cα are large and small spheres, respectively. Secondary structure elements are depicted using VMD color scheme (yellow β-strands and purple α-helices). (**c**) Dimer interface surface area, green, and normalized number of contacts between Aβ42 and lipid headgroups, red, showing evolution and correlation between these parameters.

Compared to the preformed dimer, the on-surface dimer shows similar stability; however, it has a smaller interfacial area, is overall less structured with the β-structure content being <20%, and shows a dramatically different interaction pattern. In particular, the interaction pattern shows that while some of the same regions are responsible for stability in the two dimer species, the on-surface dimer is more prone to interactions between the central region and the C-terminal, shown in Figure S5a. The probability of surface interactions is also skewed in favor of the initial surface-bound monomer, whereas the preformed dimer showed no significant difference in preference between the two monomers.

2.2.4. Interactions with POPS Bilayer

Simulations were also performed with POPS bilayers to identify the role of bilayer composition on the Aβ42 interaction and aggregation. The Aβ42 monomer rapidly interacts with POPS and undergoes structural transition into a conformation with β-structure, depicted in Figure S6a. The monomer interacts transiently with the bilayer; however, adsorption (around 2.4 μs for ~600 ns duration) to the bilayer is also observed, Figure S6b. Bilayer interactions are primarily through the N-terminal and residues 23–33, shown in Figure S6c.

Interaction between a POPS-bound (Mon1) and a free (Mon2) Aβ42 monomer happens after ~520 ns and results in a stable dimer, seen in Figure S7a. The formation of the dimer causes a dramatic increase in the β-structure content and a decrease in the helical content, depicted in Figure S7b. This change is further enhanced around ~2.6 μs and results in a final β- structure content of ~20%. The dimer interacts with the POPS bilayer through the N- and central regions of the monomers, shown in Figure S8a. The dimer is stabilized primarily by N-C and C-C terminal interactions, depicted in Figure S8b.

An already formed dimer interacts with the POPS bilayer and undergoes structural transitions, with both monomers within the dimer experiencing an increase in β-structure, shown in Figure S9a. This leads to a dimer interface with the inter-peptide β-sheet. Interactions between the monomers within the dimer are primarily focused around the residues of N–C termini and the central-central region, depicted in Figure S9b. Surface interactions are transient and primarily involve residues in the N-terminal and central regions of the proteins, seen in Figure S9c.

3. Discussion

The interaction of Aβ proteins with membranes is widely recognized as a problem of great importance. One of the major foci in previous studies was the ability of Aβ aggregates to make pores within membranes. The formation of these aberrant membrane channels is considered to be among the neurotoxic effects of Aβ aggregates [28,29]. Another interest in the interaction of Aβ proteins with membranes was the elevated rate of assembly of aggregates- primarily fibrils [30,31]; however, the role of membranes in the assembly of oligomers remains unexplored. We have recently discovered a novel property of surfaces, including phospholipid bilayers, in which surface plays the role of a catalyst enabling the assembly of oligomers to occur at nanomolar concentrations [19–22]. This finding is of great importance, as concentrations of Aβ proteins in vivo are in the low nanomolar concentration range. The model developed in [21] and validated by AFM experiments suggests that aggregation propensity of the on-surface aggregation is defined by the affinity of the protein to the surface and can exceed the bulk aggregation rate by several orders of magnitude. Recent combined experimental and computational studies of the on-membrane aggregation of α-syn revealed a number of properties of the α-syn oligomers assembly on membrane bilayers, including the critical role of the bilayer composition [20,22].

In the present study, we have used time-lapse AFM imaging along with all-atom MD simulations to characterize the self-assembly process of Aβ42 on POPC and POPS bilayers. Experimental studies demonstrate that Aβ42 aggregation occurs at concentrations as low as 10 nM, which is in the range of Aβ42 in vivo concentration and more than three orders of magnitude less than the Aβ42 concentration required for the spontaneous self-assembly of aggregates in test tubes. All previously obtained experimental data were limited to very high concentrations of Aβ42, and this issue was the major concern of the ACH [13,32], which is currently the main model for the development of AD. The on-surface aggregation pathway eliminates this fundamental problem with the ACH.

We then characterized the effect of chemical milieu on the self-assembly process and how the sizes and number of aggregates assembled on the surface increased over time. At ambient conditions and neutral pH values, ionic conditions contribute to the aggregation process, in such a way that aggregation is faster at salt concentrations close to physiological, 150 mM NaCl (Figure 4a,b). A comparison of Aβ42 oligomers assembly on the POPC and POPS bilayers revealed an elevated propensity of POPS compared to POPC, but the strongest effect was observed for the bilayer composed of the two phospholipids in an equimolar ratio (Figure 3). These data lead to the observation that the propensity of Aβ42 to the membrane bilayer surface plays a major role during the self-assembly.

The on-surface aggregation process is dynamic. Oligomers assembled on the surface can dissociate into the solution; moreover, their concentration increases over time (Figure 5 and Figure S3), suggesting that on-surface aggregation is the mechanism by which Aβ42 oligomeric species in solution are produced. Aβ oligomers are widely regarded as the most neurotoxic species, initiating neuron damage and eventually leading to Alzheimer's disease [32,33] via various pathways, as reviewed in [34]. The disease initiation and development require the accumulation of Aβ oligomers in the brain, and we posit that on-membrane self-assembly is the molecular mechanism for such an accumulation process.

Computational modeling provides important details of the molecular mechanism of the Aβ42 oligomerization catalyzed by the membrane bilayer. According to the MD simulation, interaction with the membrane surface rapidly changes the conformation of the Aβ42 monomer by the formation of extended β-structure motifs (Figure 6). Which are considered major building blocks within Aβ proteins facilitating their assembly into amyloid aggregates (e.g., as reviewed in [35]) and this conformation is termed as a misfolded state. This model was directly confirmed by our simulations of the assembly of Aβ42 dimer through the interaction of unstructured Aβ42 monomer with the surface-bound monomer. The data in Figure 8 demonstrate that misfolded Aβ42 monomer rapidly dimerizes when another monomer appears in proximity to the misfolded one. In turn, as is illustrated in Figure 7, the pre-formed dimer undergoes conformation changes after interaction with the bilayer. Importantly, neither monomers nor dimers stay firmly bound to the bilayer surface; rather, they are dynamic and mobile, able to dissociate, associate, and tumble over the surface, exposing different segments of the molecules to the surface. Evidently, this dynamic behavior further facilitates the misfolding process of either the monomer or dimer.

As we mentioned above, damage to the membranes caused by Aβ42 (including the formation of pores) is considered a major mechanism of the neurotoxicity of Aβ42 oligomers [36]. However, our simulations did not reveal changes in the bilayer surface, neither by monomers nor dimers of Aβ42. We also did not observe any damage to the bilayers during incubation with Aβ42 over a long time period. Moreover, no changes to the surface morphology were identified after the oligomers dissociated from the surface. These findings are in line with results presented in [36], in which no pores were observed for Aβ42 monomers. Nor did they observe pores when 50 nM Aβ42 oligomers were used; however, they were able to detect pores when Aβ42 oligomers at concentrations above 0.5 mM were used.

Based on these data, we propose the following model for the assembly of oligomers from monomers, shown in Figure 9, in which cellular membranes play a key role in initiating the aggregation process. In this model, the amyloid monomer changes its conformation upon interaction with the membrane. Therefore, another monomer can interact after the induced conformational change and assemble a dimer. The aggregate grows upon further on-surface docking of additional monomers. Collectively, we termed this process aggregation, catalyzed by interaction with the surfaces. Aggregates can dissociate from the surface and initiate disease-related effects via different pathways [7,36–42], including interference with recently identified mitochondrial protein biogenesis [43,44] and neuronal hyperactivation at the prodromal state of the disease [4]. Notably, our model is focused on the very early stages of amyloid aggregation, which represent the early onset of disease.

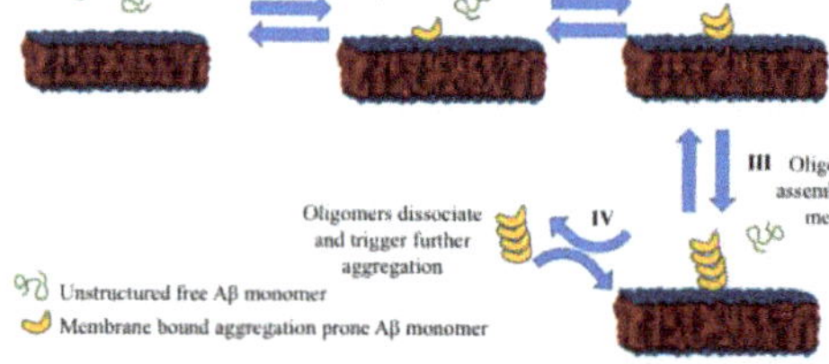

Figure 9. Schematic for the surface-mediated amyloid aggregation model. Amyloid monomers misfold upon interaction with the membrane. In turn, the misfolded dimer assembles by docking to the misfolded monomer. The process continues resulting in the formation of the oligomer, which can dissociate from the membrane to start oligomer-mediated neurotoxic events, defining the disease state at the organism level.

The proposed model is a significant departure from the current ACH model and has three key features. First, it does not require an increase in protein synthesis to the level required for the spontaneous assembly of in vitro aggregates (orders of magnitude higher compared to the in vivo concentrations of amyloids). It explains why attempts at lowering the protein level did not succeed [45,46]- aggregation on membrane surfaces can occur in the physiological range of concentrations, so aggregates can assemble even at very low amyloid concentrations. Second, Aβ, α-syn, and other amyloidogenic proteins are actively involved in important physiological processes such as signal transduction in neuron synapses [42,47,48]. Reducing their levels can impair these important processes [13,16,32,48–50]. Third, the composition of the membrane contributes to its interaction with amyloid proteins [51]. Thus, the membrane composition can be the key factor that triggers the aggregation process, and therefore it defines the disease state [52–54]. We suggest that preventative and treatment strategies should focus on controlling the membrane composition and interaction of amyloids with membranes.

4. Materials and Methods

4.1. Materials

Amyloid β (1-42) (Aβ42) was purchased from AnaSpec, Fremont, CA, USA. 1-Palmitoyl-2-oleoyl-sn-glycero-3-phosphocholine (POPC) and 1-palmitoyl-2-oleoyl-sn-glycero-3-phospho-L-serine (POPS) were obtained from Avanti Polar Lipids, Inc, Alabama, USA; Chloroform (Sigma-Aldrich Inc., St. Louis, MO, USA); sonicator (Branson 1210, Branson Ultrasonics, Danbury, CT, USA). Mainly two buffer solutions were used: a 10 mM sodium phosphate buffer with a pH of 7.4 (for without salt condition), and a 10 mM sodium phosphate, 150 mM NaCl, with a pH of 7.4 (for with salt condition). Deionized water (18.2 MΩ, filter pore size: 0.22 μm; APS Water Services Corp., Van Nuys, CA, USA) was used in all the experiments wherever required. Glass vial and glass pipettes (Fisher Scientific, Waltham, MA, USA) were used to handle the lipid solution.

4.2. Preparation of Aβ42 Protein Solution

Protein solution was prepared as described previously [19]. Briefly, a measured amount of Aβ42 was dissolved and sonicated for 5 min in 100 μL of 1,1,1,3,3,3-hexafluoroisopropanol (HFIP, Sigma-Aldrich Inc.) to remove any preformed oligomers. The tubes were then put into vacufuge until the solvent is completely evaporated. The stock solution was prepared in DMSO (Sigma-Aldrich Inc.) and kept at −20 °C. Then, 10 nM solution was prepared from the stock in 10 mM sodium phosphate, pH 7.4 just before the experiment.

4.3. Preparation of Supported Lipid Bilayers (SLBs)

A similar methodology was implemented to that described previously [20]. Briefly, a 25 mg/mL stock solution of POPC and POPS was prepared in chloroform. Glass vial and glass pipettes were used to handle the lipid solution. Stock solution was stored in −20 °C. An aliquot of 20 μL was taken in another glass vial and dried with a flow of Ar and kept overnight in a vacuum chamber to remove any trace of chloroform. Then, the dried lipid was resuspended in 1 mL 10 mM sodium phosphate, pH 7.4 buffer solution to prepare a 0.5 mg/mL solution which was used for SLB preparation. The solution was gently vortexed for 1 min and then sonicated until the solution became clear. This solution was deposited onto a freshly cleaved mica surface, which was then attached to a glass slide and incubated for 1 h at 60 °C. After the incubation, the slide was cooled to room temperature, the excess of the lipid solution was removed, and the substrate was rinsed with the buffer gently. The prepared SLB was never allowed to dry by keeping ~300 μL buffer on top of it.

4.4. AFM Imaging and Data Analysis

Time-lapse AFM imaging was performed in MFP-3D (Asylum Research, Santa Barbara, CA) instrument in tapping mode. MSNL AFM probe (cantilever 'E') was used for all the experiments. The nominal resonance frequency and the spring constant of the probe were 7–9 kHz and 0.1 N/m, respectively. The typical scan speed was kept at 1–2 Hz. For time-lapse imaging, images of the same area of the surface were acquired at different time-points. Typically, continuous scanning was avoided. Between each time-point, the cantilever was electronically retracted by the software and engaged again to record the image for the next time-point.

All the AFM images shown were subjected to minimum processing. Only flattening (1st order polynomial) was performed by Femtoscan software (Advanced Technologies Center, Moscow, Russia). The volume of the aggregates was measured by the 'Grain analysis' tool of the software and then the distribution of the volume was obtained by plotting them in histogram and fitting the histogram with Gaussian distribution using Origin Pro software (OriginLab, Northampton, MA, USA). Standard deviation was obtained as a half-width of the distribution.

4.5. Computational Methods

4.5.1. Molecular Dynamics Simulation of Bilayers

To generate the initial bilayers of POPC and POPS, we employed CHARMM-GUI [55] to produce bilayers consisting of 512 lipid molecules with 40 TIP3P waters [56] per lipid. The lipid systems were neutralized and kept at a 150 mM ionic concentration using Na and Cl counterions and converted to AMBER format using the lipid17 force field (an extension and refinement of lipid14 [57]). The systems then underwent steepest-descent energy minimization, following which 150 ns isothermal-isobaric ensemble (NPT) MD simulations were performed using a 2 fs integration time step. The simulations employed periodic boundary conditions with a semi-isotropic pressure coupling at 1 bar, a constant temperature of 300 K, non-bonded interactions truncated at 10 Å, and electrostatic interactions treated using particle-mesh Ewald [58]. Simulations were performed using the Amber16 package [59].

4.5.2. Molecular Dynamics Simulation of Aβ42 Interactions with Bilayers

To investigate the interaction of Aβ42 monomer with the bilayers, we extracted the POPC and POPS bilayers from the final frame of the pure bilayer simulations, added Aβ42 molecules (monomer or dimer conformations adopted from [27]) at 5 nm center-of-mass (CoM) from the bilayer center, solvated in TIP3P water (in an orthorhombic box [$a = b \neq c$] with side:height ratio 0.75), neutralized with NaCl counter ions, and maintained a final NaCl concentration of 150 mM. Proteins were described using the Amber ff99SB-ILDN force field [60]. Each system then underwent H-mass repartitioning to increase the H mass to 3.024 Da allowing for 4 fs time steps [61], following which the systems were simulated as an NPT ensemble for 5 ns (using the same parameters as pure bilayer simulations) before being submitted to the special purpose supercomputer Anton2 for long production runs. Simulations on Anton2 employed the multigrator algorithm and treated electrostatics using the Gaussian split Ewald method.

4.5.3. Interaction between Membrane-Bound and free Aβ42 Monomer

To investigate the interaction between membrane-bound and free Aβ42 monomers, we used the last membrane-bound conformation of the previous simulation systems and added a monomer to the simulation systems following the same procedure as used initially to add Aβ42 molecules to the bilayer systems. Newly added molecules were placed at 4 nm CoM with respect to the membrane-bound molecule and at 5 nm distance to the membrane core. Simulation parameters and steps were the same as the initial Aβ42-bilayer simulations.

4.5.4. Analysis of MD Trajectories

Gromacs suite of programs (v2016) [62] was used to analyze the obtained simulation trajectories.

Supplementary Materials: Supplementary materials can be found at http://www.mdpi.com/1422-0067/21/3/1129/s1.

Author Contributions: Y.L.L., S.B. and M.H. designed the project. S.B. and K.Z. performed the AFM experiments; M.H. performed the molecular dynamics simulations. All authors wrote and edited the manuscript. All authors have read and agreed to the published version of the manuscript.

Funding: This research was funded by National Institutes of Health, grants GM096039 and GM118006 to Y.L.L. Anton 2 computer time was provided by the Pittsburgh Supercomputing Center (PSC) through Grant R01GM116961 from the National Institutes of Health.

Acknowledgments: The Anton 2 machine at PSC was generously made available by D.E. Shaw Research. This work was completed utilizing the Holland Computing Center of the University of Nebraska, which receives support from the Nebraska Research Initiative.

Conflicts of Interest: The authors declare no conflict of interest. The funders had no role in the design of the study; in the collection, analyses, or interpretation of data; in the writing of the manuscript, or in the decision to publish the results.

Abbreviations

Aβ	Amyloid β
ACH	Amyloid cascade hypothesis
AFM	Atomic force microscopy
CSF	Cerebrospinal fluid
CoM	Center of mass
MD	Moleuclar dynamics
POPC	1-palmitoyl-2-oleoyl-glycero-3-phosphocholine
POPS	1-palmitoyl-2-oleoyl-sn-glycero-3-phospho-L-serine
SLB	Supported lipid bilayer

References

1. Festa, G.; Mallamace, F.; Sancesario, G.M.; Corsaro, C.; Mallamace, D.; Fazio, E.; Arcidiacono, L.; Garcia Sakai, V.; Senesi, R.; Preziosi, E.; et al. Aggregation states of Aβ1-40, Aβ1-42 and Aβp3-42 amyloid beta peptides: A SANS study. *Int. J. Mol. Sci.* **2019**, *20*, 4126. [CrossRef]
2. Sancesario, G.M.; Cencioni, M.T.; Esposito, Z.; Borsellino, G.; Nuccetelli, M.; Martorana, A.; Battistini, L.; Sorge, R.; Spalletta, G.; Ferrazzoli, D.; et al. The load of amyloid-β oligomers is decreased in the cerebrospinal fluid of Alzheimer's disease patients. *J. Alzheimer's Dis.* **2012**, *31*, 865–878. [CrossRef]
3. Zhao, Y.; Sivaji, S.; Chiang, M.C.; Ali, H.; Zukowski, M.; Ali, S.; Kennedy, B.; Sklyar, A.; Cheng, A.; Guo, Z.; et al. Amyloid beta peptides block new synapse assembly by nogo receptor-mediated inhibition of T-type calcium channels. *Neuron* **2017**, *96*, 355–372. [CrossRef]
4. Zott, B.; Simon, M.M.; Hong, W.; Unger, F.; Chen-Engerer, H.-J.; Frosch, M.P.; Sakmann, B.; Walsh, D.M.; Konnerth, A. A vicious cycle of β amyloid-dependent neuronal hyperactivation. *Science* **2019**, *365*, 559–565. [CrossRef]
5. Hardy, J.A.; Higgins, G.A. Alzheimer's disease: The amyloid cascade hypothesis. *Science* **1992**, *256*, 184–185. [CrossRef] [PubMed]
6. Mohamed, T.; Shakeri, A.; Rao, P.P.N. Amyloid cascade in Alzheimer's disease: Recent advances in medicinal chemistry. *Eur. J. Med. Chem.* **2016**, *113* (Suppl. C), 258–272. [CrossRef] [PubMed]
7. Armstrong, R. A critical analysis of the 'amyloid cascade hypothesis'. *Folia Neuropathol.* **2014**, *52*, 211–225. [CrossRef]
8. Hardy, J. Has the amyloid cascade hypothesis for Alzheimer's disease been proved? *Curr. Alzheimer Res.* **2006**, *3*, 71–73. [CrossRef] [PubMed]
9. Hardy, J. Alzheimer's disease: The amyloid cascade hypothesis: An update and reappraisal. *J. Alzheimer's Dis.* **2006**, *9* (Suppl. 3), 151–153. [CrossRef] [PubMed]
10. McGeer, P.L.; McGeer, E.G. The amyloid cascade-inflammatory hypothesis of Alzheimer disease: Implications for therapy. *Acta Neuropathol.* **2013**, *126*, 479–497. [CrossRef]
11. Abbott, A.; Dolgin, E. Leading Alzheimer's theory survives drug failure. *Nature* **2016**, *540*, 15–16. [CrossRef] [PubMed]
12. Hardy, J.; De Strooper, B. Alzheimer's disease: Where next for anti-amyloid therapies? *Brain* **2017**, *140*, 853–855. [CrossRef] [PubMed]
13. Hillen, H. The Beta Amyloid Dysfunction (BAD) hypothesis for Alzheimer's disease. *Front. Neurosci.* **2019**, *13*, 1154. [CrossRef] [PubMed]

14. An, S.S.A.; Lee, B.S.; Yu, J.S.; Lim, K.; Kim, G.J.; Lee, R.; Kim, S.; Kang, S.; Park, Y.H.; Wang, M.J.; et al. Dynamic changes of oligomeric amyloid beta levels in plasma induced by spiked synthetic Abeta42. *Alzheimer's Res. Ther.* **2017**, *9*, 86. [CrossRef]
15. Bjerke, M.; Portelius, E.; Minthon, L.; Wallin, A.; Anckarsater, H.; Anckarsater, R.; Andreasen, N.; Zetterberg, H.; Andreasson, U.; Blennow, K. Confounding factors influencing amyloid Beta concentration in cerebrospinal fluid. *Int. J. Alzheimer's Dis.* **2010**. [CrossRef]
16. Copani, A. The underexplored question of β-amyloid monomers. *Eur. J. Pharmacol.* **2017**, *817* (Suppl. C), 71–75. [CrossRef]
17. Potter, R.; Patterson, B.W.; Elbert, D.L.; Ovod, V.; Kasten, T.; Sigurdson, W.; Mawuenyega, K.; Blazey, T.; Goate, A.; Chott, R.; et al. Increased in vivo amyloid-beta42 production, exchange, and loss in presenilin mutation carriers. *Sci. Transl. Med.* **2013**, *5*. [CrossRef]
18. Palmqvist, S.; Scholl, M.; Strandberg, O.; Mattsson, N.; Stomrud, E.; Zetterberg, H.; Blennow, K.; Landau, S.; Jagust, W.; Hansson, O. Earliest accumulation of beta-amyloid occurs within the default-mode network and concurrently affects brain connectivity. *Nat. Commun.* **2017**, *8*, 1214. [CrossRef]
19. Banerjee, S.; Hashemi, M.; Lv, Z.; Maity, S.; Rochet, J.C.; Lyubchenko, Y.L. A novel pathway for amyloids self-assembly in aggregates at nanomolar concentration mediated by the interaction with surfaces. *Sci. Rep.* **2017**, *7*, 45592. [CrossRef]
20. Lv, Z.; Hashemi, M.; Banerjee, S.; Zagorski, K.; Rochet, J.C.; Lyubchenko, Y.L. Assembly of alpha-synuclein aggregates on phospholipid bilayers. *Biochim. Biophys. Acta Proteins Proteom.* **2019**, *1867*, 802–812. [CrossRef]
21. Pan, Y.; Banerjee, S.; Zagorski, K.; Shlyakhtenko, L.S.; Kolomeisky, A.B.; Lyubchenko, Y.L. Molecular model for the surface-catalyzed protein self-assembly. *J. Phys. Chem. B* **2020**, *124*, 366–372. [CrossRef] [PubMed]
22. Lv, Z.; Banerjee, S.; Zagorski, K.; Lyubchenko, Y.L. Supported lipid bilayers for atomic force microscopy studies. *Methods Mol. Biol.* **2018**, *1814*, 129–143. [PubMed]
23. Wang, Y.; Shi, M.; Chung, K.A.; Zabetian, C.P.; Leverenz, J.B.; Berg, D.; Srulijes, K.; Trojanowski, J.Q.; Lee, V.M.; Siderowf, A.D.; et al. Phosphorylated alpha-synuclein in Parkinson's disease. *Sci. Transl. Med.* **2012**, *4*. [CrossRef] [PubMed]
24. Hou, X.; Mechler, A.; Martin, L.L.; Aguilar, M.-I.; Small, D.H. Cholesterol and anionic phospholipids increase the binding of amyloidogenic transthyretin to lipid membranes. *Biochim. Biophys. Acta Biomembr.* **2008**, *1778*, 198–205. [CrossRef]
25. Koike, H.; Katsuno, M. Ultrastructure in transthyretin amyloidosis: From pathophysiology to therapeutic insights. *Biomedicines* **2019**, *7*, 11. [CrossRef]
26. Lyubchenko, Y.L. Direct AFM visualization of the nanoscale dynamics of biomolecular complexes. *J. Phys. D Appl. Phys.* **2018**, *51*, 403001. [CrossRef]
27. Zhang, Y.; Hashemi, M.; Lv, Z.; Lyubchenko, Y.L. Self-assembly of the full-length amyloid Abeta42 protein in dimers. *Nanoscale* **2016**, *8*, 18928–18937. [CrossRef]
28. Lee, J.; Kim, Y.H.; Arce, F.; Gillman, A.L.; Jang, H.; Kagan, B.L.; Nussinov, R.; Yang, J.; Lal, R. Amyloid β Ion channels in a membrane comprising brain total lipid extracts. *ACS Chem. Neurosci.* **2017**, *8*, 1348–1357. [CrossRef]
29. Serra-Batiste, M.; Ninot-Pedrosa, M.; Bayoumi, M.; Gairí, M.; Maglia, G.; Carulla, N. Aβ42 assembles into specific β-barrel pore-forming oligomers in membrane-mimicking environments. *Proc. Natl. Acad. Sci. USA* **2016**, *113*, 10866–10871. [CrossRef]
30. Lindberg, D.J.; Wesén, E.; Björkeroth, J.; Rocha, S.; Esbjörner, E.K. Lipid membranes catalyse the fibril formation of the amyloid-β (1–42) peptide through lipid-fibril interactions that reinforce secondary pathways. *Biochim. Biophys. Acta Biomembr.* **2017**, *1859*, 1921–1929. [CrossRef]
31. Sani, M.-A.; Gehman, J.D.; Separovic, F. Lipid matrix plays a role in Abeta fibril kinetics and morphology. *FEBS Lett.* **2011**, *585*, 749–754. [CrossRef]
32. Fagiani, F.; Lanni, C.; Racchi, M.; Pascale, A.; Govoni, S. Amyloid-beta and synaptic vesicle dynamics: A cacophonic orchestra. *J. Alzheimer's Dis.* **2019**, *72*, 1–14. [CrossRef] [PubMed]
33. Selkoe, D.J. Light at the end of the amyloid tunnel. *Biochemistry* **2018**, *57*, 5921–5922. [CrossRef] [PubMed]
34. Viola, K.L.; Klein, W.L. Amyloid beta oligomers in Alzheimer's disease pathogenesis, treatment, and diagnosis. *Acta Neuropathol.* **2015**, *129*, 183–206. [CrossRef] [PubMed]

35. Perez, C.; Miti, T.; Hasecke, F.; Meisl, G.; Hoyer, W.; Muschol, M.; Ullah, G. Mechanism of fibril and soluble oligomer formation in amyloid beta and hen egg white lysozyme proteins. *J. Phys. Chem. B* **2019**, *123*, 5678–5689. [CrossRef] [PubMed]
36. Bode, D.C.; Baker, M.D.; Viles, J.H. Ion channel formation by amyloid-β42 oligomers but not amyloid-β40 in cellular membranes. *J. Biol. Chem.* **2017**, *292*, 1404–1413. [CrossRef]
37. Wang, J.; Gu, B.J.; Masters, C.L.; Wang, Y.J. A systemic view of Alzheimer disease—Insights from amyloid-beta metabolism beyond the brain. *Nat. Rev. Neurol.* **2017**, *13*, 612–623. [CrossRef]
38. Rasmussen, J.; Mahler, J.; Beschorner, N.; Kaeser, S.A.; Häsler, L.M.; Baumann, F.; Nyström, S.; Portelius, E.; Blennow, K.; Lashley, T.; et al. Amyloid polymorphisms constitute distinct clouds of conformational variants in different etiological subtypes of Alzheimer's disease. *Proc. Natl. Acad. Sci. USA* **2017**, *114*, 13018–13023. [CrossRef]
39. Kumar, S.; Henning-Knechtel, A.; Chehade, I.; Magzoub, M.; Hamilton, A.D. Foldamer-mediated structural rearrangement attenuates Aβ oligomerization and cytotoxicity. *J. Am. Chem. Soc.* **2017**, *139*, 17098–17108. [CrossRef]
40. Eskici, G.; Axelsen, P.H. Amyloid beta peptide folding in reverse micelles. *J. Am. Chem. Soc.* **2017**, *139*, 9566–9575. [CrossRef]
41. Aleksis, R.; Oleskovs, F.; Jaudzems, K.; Pahnke, J.; Biverstål, H. Structural studies of amyloid-β peptides: Unlocking the mechanism of aggregation and the associated toxicity. *Biochimie* **2017**, *140* (Suppl. C), 176–192. [CrossRef] [PubMed]
42. Zhao, Y.; Wu, X.; Li, X.; Jiang, L.L.; Gui, X.; Liu, Y.; Sun, Y.; Zhu, B.; Pina-Crespo, J.C.; Zhang, M.; et al. TREM2 is a receptor for beta-amyloid that mediates microglial function. *Neuron* **2018**, *97*, 1023–1031. [CrossRef] [PubMed]
43. Cenini, G.; Rub, C.; Bruderek, M.; Voos, W. Amyloid beta-peptides interfere with mitochondrial preprotein import competence by a coaggregation process. *Mol. Biol. Cell* **2016**, *27*, 3257–3272. [CrossRef] [PubMed]
44. Sorrentino, V.; Romani, M.; Mouchiroud, L.; Beck, J.S.; Zhang, H.; D'Amico, D.; Moullan, N.; Potenza, F.; Schmid, A.W.; Rietsch, S.; et al. Enhancing mitochondrial proteostasis reduces amyloid-beta proteotoxicity. *Nature* **2017**, *552*, 187–193. [CrossRef] [PubMed]
45. Busche, M.A.; Grienberger, C.; Keskin, A.D.; Song, B.; Neumann, U.; Staufenbiel, M.; Forstl, H.; Konnerth, A. Decreased amyloid-beta and increased neuronal hyperactivity by immunotherapy in Alzheimer's models. *Nat. Neurosci.* **2015**, *18*, 1725–1727. [CrossRef] [PubMed]
46. Brundin, P.; Dave, K.D.; Kordower, J.H. Therapeutic approaches to target alpha-synuclein pathology. *Exp. Neurol.* **2017**, *298*, 225–235. [CrossRef]
47. Venda, L.L.; Cragg, S.J.; Buchman, V.L.; Wade-Martins, R. Alpha-synuclein and dopamine at the crossroads of Parkinson's disease. *Trends Neurosci.* **2010**, *33*, 559–568. [CrossRef]
48. Bate, C.; Williams, A. Monomeric amyloid-β reduced amyloid-β oligomer-induced synapse damage in neuronal cultures. *Neurobiol. Dis.* **2018**, *111*, 48–58. [CrossRef]
49. Drachman, D.A. The amyloid hypothesis, time to move on: Amyloid is the downstream result, not cause, of Alzheimer's disease. *Alzheimer's Dement.* **2014**, *10*, 372–380. [CrossRef]
50. Carrillo-Mora, P.; Luna, R.; Colin-Barenque, L. Amyloid beta: Multiple mechanisms of toxicity and only some protective effects? *Oxid. Med. Cell. Longev.* **2014**, *2014*, 795375. [CrossRef]
51. Parikh, N.; Klimov, D.K. Inclusion of lipopeptides into the DMPC lipid bilayers prevents Abeta peptide insertion. *Phys. Chem. Chem. Phys.* **2017**, *19*, 10087–10098. [CrossRef]
52. Gónzalez de San Román, E.; Manuel, I.; Giralt, M.T.; Ferrer, I.; Rodríguez-Puertas, R. Imaging Mass Spectrometry (IMS) of cortical lipids from preclinical to severe stages of Alzheimer's disease. *Biochim. Biophys. Acta Biomembr.* **2017**, *1859*, 1604–1614. [CrossRef] [PubMed]
53. Tong, M.; Leão, R.; Vimbela, G.V.; Yalcin, E.B.; Kay, J.; Krotow, A.; de la Monte, S.M. Altered temporal lobe white matter lipid ion profiles in an experimental model of sporadic Alzheimer's disease. *Mol. Cell. Neurosci.* **2017**, *82*, 23–34. [CrossRef] [PubMed]
54. Wong, M.W.; Braidy, N.; Poljak, A.; Pickford, R.; Thambisetty, M.; Sachdev, P.S. Dysregulation of lipids in Alzheimer's disease and their role as potential biomarkers. *Alzheimer's Dement.* **2017**, *13*, 810–827. [CrossRef] [PubMed]
55. Jo, S.; Kim, T.; Iyer, V.G.; Im, W. CHARMM-GUI: A web-based graphical user interface for CHARMM. *J. Comput. Chem.* **2008**, *29*, 1859–1865. [CrossRef] [PubMed]

56. Jorgensen, W.L.; Chandrasekhar, J.; Madura, J.D.; Impey, R.W.; Klein, M.L. Comparison of simple potential functions for simulating liquid water. *J. Chem. Phys.* **1983**, *79*, 926–935. [CrossRef]
57. Dickson, C.J.; Madej, B.D.; Skjevik, Å.A.; Betz, R.M.; Teigen, K.; Gould, I.R.; Walker, R.C. Lipid14: The amber lipid force field. *J. Chem. Theory Comput.* **2014**, *10*, 865–879. [CrossRef]
58. Darden, T.; York, D.; Pedersen, L. Particle mesh Ewald: An N·log(N) method for Ewald sums in large systems. *J. Chem. Phys.* **1993**, *98*, 10089–10092. [CrossRef]
59. Case, D.A.; Babin, V.; Berryman, J.T.; Betz, R.M.; Cai, Q.; Cerutti, D.S.; Cheatham, T.E., III; Darden, T.A.; Duke, R.E.; Gohlke, H.; et al. *AMBER 16*; University of California: San Francisco, CA, USA, 2016.
60. Lindorff-Larsen, K.; Piana, S.; Palmo, K.; Maragakis, P.; Klepeis, J.L.; Dror, R.O.; Shaw, D.E. Improved side-chain torsion potentials for the Amber ff99SB protein force field. *Proteins* **2010**, *78*, 1950–1958. [CrossRef]
61. Hopkins, C.W.; Le Grand, S.; Walker, R.C.; Roitberg, A.E. Long-time-step molecular dynamics through hydrogen mass repartitioning. *J. Chem. Theory Comput.* **2015**, *11*, 1864–1874. [CrossRef]
62. Abraham, M.J.; Murtola, T.; Schulz, R.; Páll, S.; Smith, J.C.; Hess, B.; Lindahl, E. GROMACS: High performance molecular simulations through multi-level parallelism from laptops to supercomputers. *SoftwareX* **2015**, *1–2*, 19–25. [CrossRef]

International Journal of
Molecular Sciences

MDPI

Article

Patterns of Expression of Purinergic Receptor P2RY12, a Putative Marker for Non-Activated Microglia, in Aged and Alzheimer's Disease Brains

Douglas G. Walker [1,2,*], Tiffany M. Tang [2], Anarmaa Mendsaikhan [1], Ikuo Tooyama [1], Geidy E. Serrano [3], Lucia I. Sue [3], Thomas G. Beach [3] and Lih-Fen Lue [2,3]

1 Molecular Neuroscience Research Center, Shiga University of Medical Science, Seta-Tsukinowa-cho, Otsu 520-0072, Japan; anarmaa@belle.shiga-med.ac.jp (A.M.); kinchan@belle.shiga-med.ac.jp (I.T.)
2 Neurodegenerative Disease Research Center, Arizona State University, Tempe, AZ 85287, USA; ttang@pennstatehealth.psu.edu (T.M.T.); lih-fen.lue@bannerhealth.com (L.-F.L.)
3 Civin Neuropathology Laboratory, Banner Sun Health Research Institute, Sun City, AZ 85351, USA; geidy.serrano@bannerhealth.com (G.E.S.); lucia.sue@bannerhealth.com (L.I.S.); thomas.beach@bannerhealth.com (T.G.B.)
* Correspondence: walkerdg@gmail.com; Tel.: +81-80-3113-7049

Received: 11 December 2019; Accepted: 11 January 2020; Published: 20 January 2020

Abstract: Neuroinflammation is considered a key pathological process in neurodegenerative diseases of aging, including Alzheimer's disease (AD). Many studies have defined phenotypes of reactive microglia, the brain-resident macrophages, with different antigenic markers to identify those potentially causing inflammatory damage. We took an alternative approach with the goal of characterizing the distribution of purinergic receptor P2RY12-positive microglia, a marker previously defined as identifying homeostatic or non-activated microglia. We examined the expression of P2RY12 by dual-color light and fluorescence immunohistochemistry using sections of middle temporal gyrus from AD, high plaque and low plaque non-demented cases in relation to amyloid beta (Aβ) plaques and phosphorylated tau, markers of pathology, and HLA-DR, IBA-1, CD68, and progranulin, microglial phenotype markers. In low plaque cases, P2RY12-positive microglia mostly had non-activated morphologies, while the morphologies of P2RY12-positive microglia in AD brains were highly variable, suggesting its expression could encompass a wider range of phenotypes than originally hypothesized. P2RY12 expression by microglia differed depending on the types of plaques or tangles they were associated with. Areas of inflammation characterized by lack of P2RY12-positive microglia around mature plaques could be observed, but many diffuse plaques showed colocalization with P2RY12-positive microglia. Based on these results, P2RY12 expression by microglia should not be considered solely a marker of resting microglia as P2RY12 immunoreactivity was identifying microglia positive for CD68, progranulin and to a limited extent HLA-DR, markers of activation.

Keywords: activation phenotypes; microglia; neuroinflammation; immunohistochemistry; temporal cortex; Alzheimer's disease; amyloid

1. Introduction

Alzheimer's disease (AD) is the leading cause of dementia, currently affecting an estimated 47 million people worldwide, but this number will increase unless effective treatments are discovered [1]. Since the identification of strongly immunoreactive major histocompatibility class II HLA-DR (MHC-II)-positive microglia associated with AD pathological structures [2,3], neuroinflammation is considered a prominent feature of AD pathology [4,5]. These early studies established the hypothesis that inflammatory responses to extracellular Aβ plaques and neurofibrillary tangles might be accelerating neurodegeneration through the production of toxic inflammatory cytokines, reactive oxygen species and enzymes [6,7]. Microglia, the brain-resident macrophages, are considered the main source of these molecules. These studies suggested that anti-inflammatory agents might be effective in slowing disease progression [8], but clinical trials of anti-inflammatories have generally shown no protective effect for AD subjects [9].

There is now greater appreciation of the complexity of microglia and their many specialized functions, both pathological and reparative. Recent gene expression profiling studies of microglia isolated from human AD tissue or AD animal models have provided large amounts of data on microglial properties and identified potentially new phenotypic markers for studying microglia in disease [10–13]. These and other studies have consistently identified the purinergic adenosine diphosphate/triphosphate (ADP/ATP)) receptor P2RY12 as a significant marker for non-activated/homeostatic microglia (examples: [10,12,14–16]). Increased understanding of neuroinflammation will come from further classification of microglia for expression of these newly identified functional markers in human AD brain tissues. Microglial markers studied in human brains only represent a small number of potential targets. The most widely-used markers in neuropathology studies of human brains have been MHCII protein HLA-DR and ionized calcium-binding adaptor molecule (IBA-1). Increased expression of HLA-DR by microglia in AD gray matter has been consistently observed, but the specificity, significance or mechanism for this is unclear, while IBA-1 identifies all microglia and does not discriminate between phenotypes of microglia in human brains. Other microglial markers characterized in AD brains include CD68, a lysosomal-associated membrane protein associated with phagocytosis, CD32 and CD64, immunoglobulin Fc receptors, CD11b, colony stimulating factor-1 receptor (CSF-1R), Toll-like receptors (TLR)-2, 3 and 4, ferritin, CD163, Transmembrane Protein (TMEM)-119 [17–23] as well as Triggering receptor expressed on myeloid cells-2 (TREM-2) and CD33, microglial genes with genetic associations to AD [24,25].

P2RY12 is a member of the P2 purinergic family of receptors, a seven transmembrane-spanning G protein-coupled receptor that responds to ADP/ATP by increasing cell migration [26]. P2RY12 is mainly expressed by platelets and microglia [27]. Its function has been widely studied in relation to platelet activation and blood clotting, but its role in neuroinflammation requires further investigation. Microglia expressed significantly higher levels of P2RY12 than macrophages, in culture and in tissue, allowing discrimination between microglia and blood macrophages [11,28]. Activation of microglial P2RY12 by ADP/ATP promotes microglial chemotaxis towards sites of release [29]; these molecules are released in increased amounts by necrotic and apoptotic cells There was significantly reduced microglial chemotaxis and process formation in response to injury in P2RY12 gene-deficient mice [26]. Expression of P2RY12 by microglia (rodent and human) is downregulated after inflammatory stimulation. Injection of lipopolysaccharide (LPS) into rat brains resulted in rapid loss of P2RY12 immunoreactivity [16,22,26].

There have been limited numbers of studies of P2RY12 expression in microglia in human brains by immunohistochemistry. The most detailed previous study of P2RY12 microglia in human brains across different ages, brain regions and diseases showed expression early in brain development in all regions with limited decline with aging, while characterization of P2RY12-immunopositive microglia in 3 AD cases identified absence of P2RY12-positive microglia around Aβ plaques [30]. P2RY12-expressing microglia in astrocytomas were increased in low-grade but reduced in high-grade tumors [31]. Immunohistochemistry of human brain sections from multiple sclerosis cases confirmed loss of P2RY12 microglial immunoreactivity in areas associated with enhanced inflammation [22,30,32,33]. Another study in 2 AD cases showed that microglia positive for P2RY12 did not express TREM-2 [34].

In this report, we sought to determine if P2RY12 could be used for phenotyping the progression of human brain microglial changes in response to AD pathology by extending previous studies [30] by characterizing P2RY12 expression by microglia in detail in a staged series of AD and non-demented aged cases. The major findings showed that P2RY12 identified populations of microglia with features of resting microglia but also other populations of microglia. There was significant reduction in P2RY12 total protein levels in AD compared to ND cases, but significant amount of P2RY12 expression was present even in severe AD cases. We identified CD68 and progranulin expression in most P2RY12-positive microglia. In pathologically-involved brains, as P2RY12 expression identified microglia with many of the different morphologies associated with inflammatory activation, classifying P2RY12 expression as a marker of homeostatic (non-activated) microglia needs to be reconsidered.

2. Results

2.1. Patterns of Expression of P2RY12 by Microglia

The aim of this study was to determine if P2RY12 expression delineates populations of resting, non-activated or reparative microglia in human brains affected by different amounts of AD plaque and tangle pathology from microglia considered as pro-inflammatory and activated. If correct, areas of inflammation around pathological structures could be defined by the presence of P2RY12-positive microglia surrounding areas with P2RY12-negative (activated) microglia.

Figure 1 illustrates initial observations of P2RY12-immunopositive microglia in middle temporal gyrus (MTG) and selected hippocampal sections. Figure 1A,B illustrates MTG sections stained with antibody to P2RY12 (Novus, rabbit polyclonal) from a non-demented (ND) (A) and AD (B) case. P2RY12-immunoreactive microglia with similar morphology were in sections of hippocampus from ND (Figure 1C) and AD cases (Figure 1D) (CA 2–3 region). A feature to note in Figure 1D is that although this region of hippocampus had significant neurodegenerative pathology as tangles, there was little difference in numbers and abundance of P2RY12 microglia compared to the ND case shown (Figure 1C) The distribution of P2RY12-positive microglia was noticeable with areas showing minimal staining in high pathology and AD cases (AD case shown in Figure 1B—red arrows). Areas with reduced staining were identified by accumulations of HLA-DR-positive microglia (Figure 1E,F). However, within these clusters of HLA-DR-positive microglia were isolated P2RY12-positive microglia (Figure 1E,F, blue arrows). Absence of P2RY12-positive microglia in the area around a cored Aβ immunoreactive plaque is shown (Figure 1G). However, the observation of P2RY12-expressing microglia with activated morphologies in close association with diffuse Aβ-positive plaques (high plaque non-demented (HPND) case—Figure 1H) suggested that P2RY12 expression by microglia was not restricted to non-activated microglia. These observations were the basis for further examination of P2RY12 expression by different types of microglia in this report.

Antibody validation was carried out to confirm that the observed immunostaining represented P2RY12 expression by microglia. Firstly, absorption of the Novus antibody with its immunizing peptide was carried out. Preincubation of diluted antibody with this 40 amino acid-recombinant peptide resulted in absence of microglial staining (Figure 1I—absorbed + PEP) compared to staining with non-absorbed antibody (Figure 1J, -PEP). The second validation step was to demonstrate the same microglial staining pattern using an independent P2RY12 antibody (Alomone Labs) produced against a different immunizing sequence. Both antibodies produced similar immunoreactivity patterns in sections from a low plaque non-demented (LPND) case (Figure 1K: Novus, Figure 1L: Alomone). To demonstrate immunoreactivity with the Alomone antibody, sections required antigen retrieval (80 °C, 30 min, 1 mM EDTA, pH 8.0), which was not required for the Novus antibody. The third stage was western blot analyses using brain protein extracts to demonstrate that the Novus antibody could detect a P2RY12 polypeptide of approximately 58 kDa, (representative western blots are shown in Figure 2A). This antibody also identified a polypeptide of approximately 30 kDa, a presumptive cleavage fragment of the full-length P2RY12 polypeptide. Although a number of different molecular weights for P2RY12 have been shown depending on cell source and antibody, a 58 kDa polypeptide is consistent with previous observations.

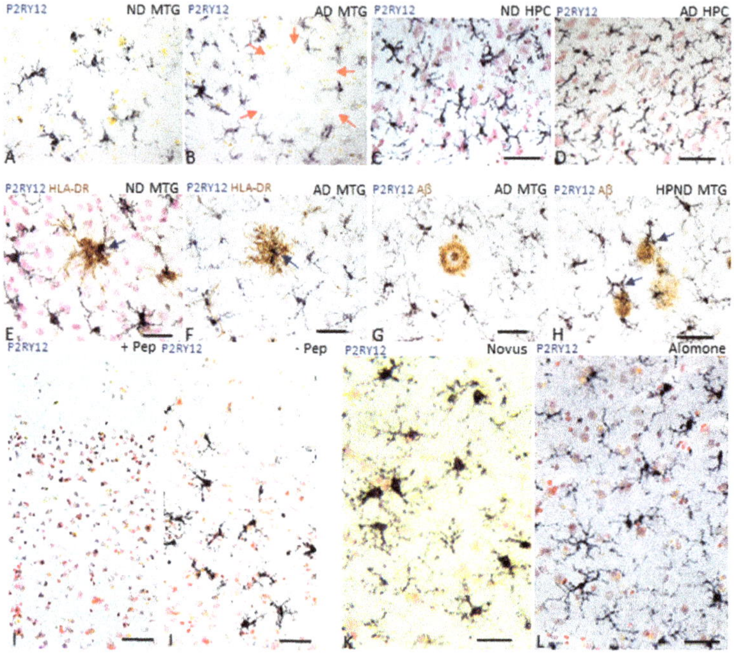

Figure 1. Features of P2RY12-immunoreactive microglia. (**A**,**B**). Morphology of P2RY12-immunoreactive microglia (purple) in a low plaque non-demented (LPND) case (**A**) and AD case (**B**). Sections of middle temporal gyrus (MTG) were single-stained with antibody to P2RY12. Red arrows in panel B illustrate the lack of P2RY12 immunoreactive cells in an area occupied by plaque. (**C**,**D**). P2RY12-immunoreactive microglia (purple) are a feature in hippocampus sections from non-demented (ND) case (**H**) and Alzheimer's disease (AD) case. Section shows staining in CA2 region of hippocampus. Continued presence of P2RY12-positive microglia in AD hippocampus (**D**) was noticeable. (**E**,**F**). Double-staining of section of ND and AD case with P2RY12 (purple) and HLA-DR (brown) showed limited overlap. HLA-DR-positive microglial clusters over plaques were P2RY12-negative except for single cells observed within the cluster (arrows). (**G**,**H**). Interaction of P2RY12-immunoreactive microglia (purple) and Aβ plaques (brown). The panels show two types of interactions of P2RY12-positive microglia with plaques. Positive microglia are not present in close association with mature cored plaque (**G**), while they are present in close association with diffuse type of plaques (**H**). Specificity controls for P2RY12 staining of microglia. (**I**,**J**). Staining of representative sections with P2RY12 (Novus) antibody preabsorbed with immunizing peptide (I, +Pep) compared to staining of matched section with P2RY12 antibody non-absorbed (**J**, -Pep). (**K**,**L**). Staining of matched sections with alternative P2RY12 antibody. Same staining pattern of microglia revealed with P2RY12 (Novus) antibody (**C**) as with P2RY12 (Alomone Labs) antibody (**D**). Sections reacted with Alomone Lab P2RY12 required antigen retrieval to obtain positive staining pattern. All sections shown had been counterstained with neutral red to identify nuclei (red color). Abbreviations: ND: non-demented. AD: Alzheimer's disease. MTG: middle temporal gyrus.—Pep: antibody without immunizing peptide. + Pep: antibody with immunizing peptide. Scale bars represent 50 µm.

2.2. Continued Expression of P2RY12 in MTG Brain Samples with Increasing Pathology and AD: Biochemical Measurements

Increased microglial activation and pro-inflammatory cytokines have long been considered a feature of AD. If widespread throughout the AD brain, one would expect levels of P2RY12 to be very low in severe AD cases if expression was restricted to non-activated microglia. We measured the levels of P2RY12 expression in MTG sections from brains with increasing amounts of plaque pathology (Table 1, set 2). Western blot measurements of levels of P2RY12 polypeptides were made in protein samples from LPND (n = 10), HPND (n = 9) and AD cases (n = 9) (representative western blot–Figure 2A). Protein extracts were not available from 6 cases used in immunohistochemistry. The Novus P2RY12 antibody detected two polypeptide bands in brain samples, one of approximately 58 kDa (considered to represent full-length P2RY12) and one of approximately 30 kDa. Semi-quantitative measurements of band intensities, normalized for β-actin levels, showed significant decreased levels of 58 kDa polypeptide in AD cases (Figure 2B) but increased levels of the 30 kDa band (Figure 2C). Spearman non-parametric correlation analysis between levels of 58 kDa polypeptide and plaque and tangle scores showed significant negative correlation (P2RY12 (58 kDa)/β-actin levels compared to plaque scores; r = −0.503, p = 0.0039 ($F_{1,29}$ = 7.618, $p < 0.001$): compared to tangle scores; r = −0.612, p = 0.0002 ($F_{1,29}$= 24.06, $p < 0.0001$). As the Novus antibody was raised against an intracellular C-terminal peptide sequence of P2RY12, the 30 kDa bands could represent accumulations of P2RY12 after proteolytic cleavage. Its functional significance is unclear. Measurement of P2RY12 mRNA expression in cDNA derived from a similar but separate group of MTG samples (Table 1, set 3) showed non-statistically significant decrease in expression in the AD cases (Figure 2D). These results showed that significant amounts of microglial expression of P2RY12 mRNA and protein were still occurring in AD brains.

Table 1. Demographic details of human brain cases used.

Set 1: Middle temporal gyrus (Immunohistochemistry)						
Disease State (n)	***Age***	***Sex***	***ApoE4***	***Plaques***	***Tangles***	***Braak***
LPND (n = 12)	85.9 ± 8.9	6M/6F	4.5%	1.3 ± 1.9	4.8 ± 2.8	I-IV
HPND (n = 12)	88 ± 8	4M/8F	13.6%	12.2 ± 1.6	5.4 ± 2.3	II-IV
AD (n = 12)	79.2 ± 5.1	7M/5F	33.3%	14.2 ± 0.8	13.8 ± 1.9	V-VI
Set 2: Middle temporal gyrus (Western blot)						
Disease State (n)	***Age***	***Sex***	***ApoE4***	***Plaques***	***Tangles***	***Braak***
LPND (n = 10)	86.3 ± 8.9	6M/4F	4.5%	1.7 ± 2.1	5.5 ± 2.3	I-IV
HPND (n = 9)	86.9 ±8.6	3M/6F	13.6%	12.2 ± 1.6	5.4 ± 1.9	II-IV
AD (n = 9)	78.2 ± 3.9	5M/4F	33.3%	14.4 ± 0.6	13.8 ± 1.9	V-VI
Set 3: Middle temporal gyrus (RNA expression)						
Disease State (n)	***Age***	***Sex***	***ApoE4***	***Plaques***	***Tangles***	***Braak***
LPND (n = 13)	85.7 ± 9.3	7M/6F	0.0%	1.2 ± 1.9	4.9 ± 2.8	I-IV
HPND (n = 12)	86.1 ± 6	6M/6F	12.5%	11.4 ± 2	4.7 ± 2.3	II-IV
AD (n = 11)	81.2 ± 3.4	8M/3F	31.8%	14.4 ± 0.7	13.0 ± 2.7	V-VI

Abbreviations: ApoE4: % ApoE4 alleles; Plaques: mean plaque score + SEM (scale 0–15); tangles: mean tangle score + SEM (scale demented. AD: Alzheimer's disease 0–15); LPND: low plaque non-demented; HPND: high plaque non- demented.

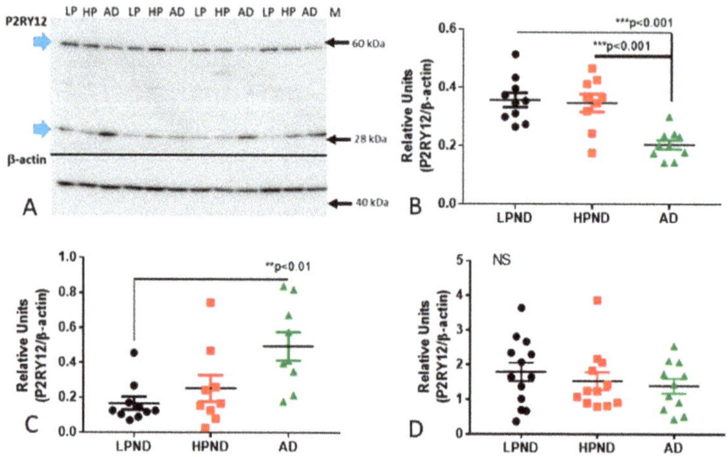

Figure 2. Quantitative biochemical measurements of P2RY12 protein and mRNA in human brains. (**A–C**). Western blot measurements of P2RY12 levels in MTG samples from LP, HP and AD brains. (**A**). Representative western blot image of P2RY12 polypeptide of MTG protein extracts identified with Novus antibody. Blots were normalized for levels of β actin. (**B**). Scatter plot showing individual P2RY12 expression levels. Significant decrease in protein levels of 58 kDa full-length P2RY12 band in AD (green shapes) compared to LPND (black) and HPND (red) cases. Chart indicates mean + Standard error of mean (SEM). Statistical analysis by one-way ANOVA with Tukey post-hoc test ($F_{2,26} = 11.54$, $p < 0.001$). (**C**). Scatter plot showing significant increase in protein levels of 30 kDa cleaved P2RY12 band in AD cases compared to LPND cases. Statistical analysis by one-way ANOVA with Tukey post-hoc test ($F_{2,26} = 5.649$, $p < 0.01$). Bar chart indicates mean + SEM. (**D**). Scatter plot showing expression levels of P2RY12 mRNA (normalized for β actin mRNA) in MTG samples. Lack of significant difference in expression of P2RY12 mRNA between LPND, HPND, and AD cases (MTG set 2). Samples measured by real time polymerase chain reaction. Statistical analysis by one-way ANOVA with Tukey post-hoc test ($F_{2,32} = 1.031$, $p > 0.05$). Abbreviations: LPND: low plaque non-demented. HPND: high plaque non-demented. AD: Alzheimer's disease. NS: non-significant. * $p < 0.05$, ** $p < 0.01$, *** $p < 0.001$.

2.3. Patterns of Expression of P2RY12-Positive Microglia and Amyloid Beta Plaques

When considering features of microglial activation in AD, one feature not generally considered is the anatomical distribution of microglia within brain structure. The features of neuronal architecture throughout the cortex results in different patterns of pathological development and different phenotypes of microglia. Figure 3A–C illustrate the distributions and morphology of P2RY12-positive microglia between disease groups, with boxed areas of layer I and II for each of these sections shown at higher magnification (Figure 3D–F). These areas appeared to have high expression of P2RY12 in AD cases. Sections were double-stained for P2RY12 (purple) and Aβ brown). In the HPND case illustrated (Figure 3B,E), diffuse-type Aβ plaques are observable, particularly in layers I and II (panel E), while in the AD case, the plaques are more consolidated. The morphologies of P2RY12-positive microglia varied, particularly in HPND and AD cases, and many of the plaques had associated P2RY12-positive microglia with activated morphologies. Due to the shrinkage of cortical layers in AD, the layers in LPND and HPND sections are slightly wider; white matter regions in Figure 3A (LPND) and Figure 3B (HPND) are not visible.

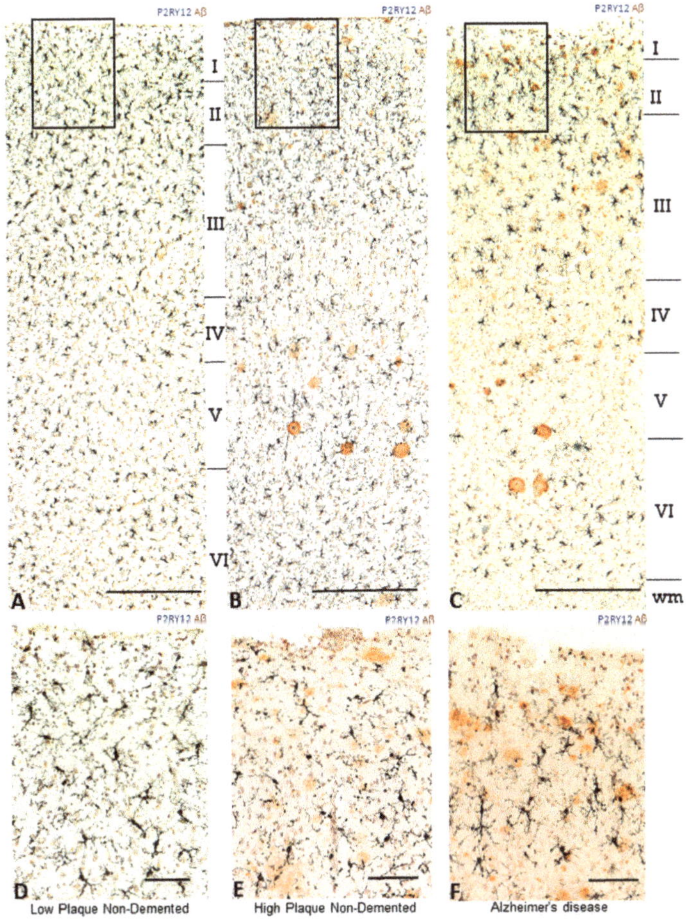

Figure 3. Distribution of P2RY12 microglia within cortical layers in relation to amyloid beta plaques in pathologically staged samples (**A–C**). Lower magnification photomicrographs showing the changes in P2RY12 microglia distribution compared to Aβ plaques within cortical layers of MTG. Sections from low plaque, high plaque and AD cases stained for P2RY12 (purple) and Aβ (brown). Sections were counterstained with neutral red to identity cellular morphology. Scale bars represent 200 μm. (**D–F**). Higher magnification photomicrographs of the areas in panels (**A–C**) indicated by frames. Scale bars represent 50 μm.

2.4. Immunohistochemical Measurements of P2RY12 Expression

Measurement of the occupied areas of P2RY12 immunoreactivity in a complete series of MTG sections was carried out. These sections were imaged at low magnification to include all cortical layers and analyzed using ImageJ software. There was a small but significant decrease in mean occupied area of P2RY12 immunoreactivity in AD cases (Figure 4A). Representative low-magnification images of sections used for measurements are shown (Figure 4B). Determining the numbers of P2RY12 immunoreactive microglia in stained sections in a defined area using a microscope eyepiece reticule was also carried out. These were counted in five individual fields through the cortical laminar columns (layer I and II, layer III, layer IV, layer V and layer VI), and three separate columns from each section. Mean total values for all laminar showed no significant difference between disease groups (Figure 4C), but we observed a redistribution of P2RY12-positive microglia into layers I and II in AD cases. As shown in Figure 3F, this area contained significant amount of Aβ. When cell counts for layer I and II were excluded from the analyses, the decreased number of P2YR12-positive microglia in the other layers in the AD cases was statistically significant ($p < 0.05$) (Figure 4D).

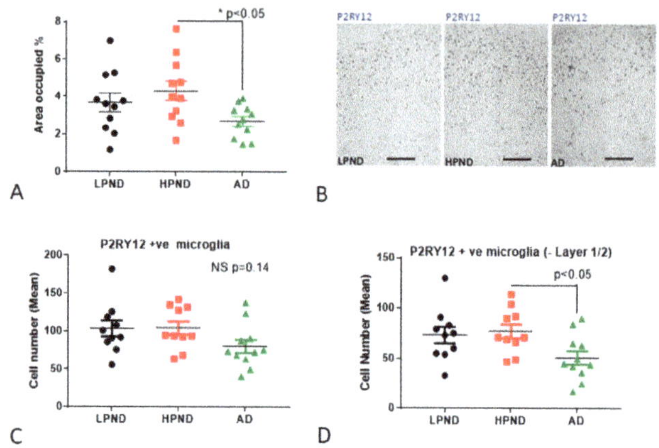

Figure 4. Quantitative measurements of P2RY12 immunoreactive structures in human brain tissue sections. (**A**,**B**). Measurements of area occupied of P2RY12 immunoreactivity in LPND (n = 11) (black), HPND (n = 11) (red) and AD (n = 12) (green) cases. (**A**). Sections were imaged at 4x magnification (three random areas/cases) and area occupied in thresholded images measured using Image J software. Results show significant decrease between HPND and AD cases. Statistical analysis by one-way ANOVA with Tukey post-hoc test ($F_{2,29}$ = 3.903, $p < 0.05$). (**B**). Representative images of low magnification images of LPND, HPND and AD used for measurements show distribution of P2RY12 immunoreactive microglia. Scale bars represent 400 µm. (**C**). Numbers of P2RY12 immunoreactive microglia in all cortical layers. Scatter plot showing the estimated total mean number of microglia in cortical layers I–VI. Individual points represent the mean of total numbers from three separate measures for each slide. The numbers of microglia/2 mm^2 field were counted. Results show insignificant decline in mean number of P2RY12-positive microglia in AD cases. Statistical analysis by one-way ANOVA with Tukey post-hoc test ($F_{2,29}$ = 1.704, p = 0.14). (**D**). Numbers of P2RY12 immunoreactive microglia in cortical layers III-VI. Scatter plot showing the estimated total mean number of microglia in cortical layers III-VI. The numbers of microglia counted in Layers I and II were subtracted from the total. Individual points represent the mean of total numbers (except Layers I and II) from three separate measures for each slide. The numbers of microglia/2 mm^2 fields were counted. Results show significant decline in mean number of P2RY12 positive microglia in AD cases. Statistical analysis by one-way ANOVA with Tukey post-hoc test ($F_{2,29}$ = 3.201, $p < 0.05$). **Abbreviations**: LPND: low plaque non-demented. HP: high plaque non-demented. AD: Alzheimer's disease. ND: non-demented. NS: non-significant.

2.5. P2RY12 Expression and Microglial Morphology

The morphology of microglia has been considered to reflect their activation states, but these types of classifications have limitations especially in aged human brains [35,36]. The typical ramified appearance of resting microglia positive for P2RY12 in layer III of LPND case is shown (Figure 5A). Other morphologies can be observed, including dystrophic (fragmented) (Figure 5B,C) and "tufted" microglia (Figure 5D). It was observed that all P2RY12-positive microglia showed immunoreactivity for IBA-1 (purple/brown colocalization), but there were IBA-1-positive/P2RY12-negative (arrowhead) or IBA-1-positive/P2RY12-weakly positive (arrow) (Figure 5E—LPND case). Figure 5F shows the appearance of a cluster of IBA-1-positive/P2RY12-negative activated microglia (arrowhead) surrounded by strongly stained P2RY12-positive microglia. Considering the interaction of P2RY12-positive microglia with Aβ plaques, variable responses were seen. Figure 5G shows that most of the diffuse plaques (brown) in an HPND case had associated P2RY12-positive microglia (purple) with extended processes. Figure 5H,I from AD cases show microglia with more activated morphologies (shorter processes and enlarged cell bodies) interacting with more mature plaques (arrowheads). Another frequent observation was the presence of large P2RY12-positive rod-shaped microglia (Figure 5J,K), with some directly interacting with plaques (Figure 5K). One other noticeable feature seen in most cases was that many P2RY12-positive microglia, especially those with ramified processes, showed close interactions of processes with neurons (Figure 5L).

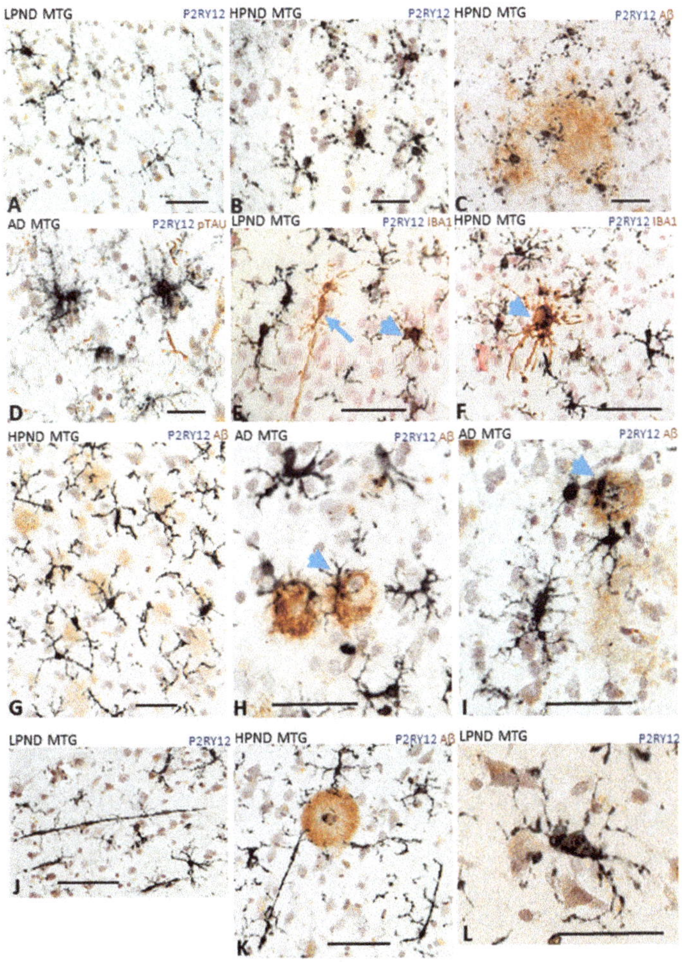

Figure 5. Different microglial morphologies associated with P2RY12 expression. Representative immunohistochemistry results of tissue sections stained to identify P2RY12 (purple) alone and Aβ (brown), IBA-1 or phosphorylated tau. (**A**). Ramified microglia in LPND case. (**B**,**C**). Microglia with fragmented morphology in HPND cases. (**C**). Fragmented microglia associated with diffuse Aβ plaques. (**D**). P2RY12 microglia with tufted morphology in AD case. (**E**,**F**). Colocalization of P2RY12 and IBA-1.

(**E**). Rod shaped IBA-1-positive microglia (brown arrow) with minimal P2RY12 immunoreactivity. P2RY12-positive, IBA-1-positive microglia (arrowheads (**E**,**F**)). All P2RY12 immunoreactive microglia showed some IBA-1 immunoreactivity. (**F**). IBA-1 positive cluster surrounded by P2RY12 microglia. (**G–I**). Different morphologies of P2RY12-positive microglia interacting with Aβ plaques. (**G**). P2RY12-positive microglia with long processes interacting with diffuse plaques in HPND case. (**H**,**I**). P2RY12-positve microglia with activated morphologies (large cell bodies, short processes) interacting with dense Aβ plaques. (**J**,**K**). P2RY12-positive rod-shaped microglia in LPND (**J**) and HPND (**K**) sections. (**L**) P2RY12-positive microglial processes show interactions with neurons. Scale bars represent 50 μm.

2.6. Confocal Microscopy Localization of P2RY12 with HLA-DR, CD68 and Progranulin in Brain Microglia

To follow up observations, further investigations of the phenotypes of P2RY12-positive microglia using antibodies to activation markers HLA-DR, CD68 and progranulin were carried out using multicolor laser confocal microscopy to demonstrate cellular colocalization.

In LPND cases with fewer HLA-DR positive microglia, those present showed colocalization with P2RY12 immunoreactivity (example: Figure 6A–C). In HPND cases, with greater numbers of HLA-DR positive microglia cases, there was less colocalization between P2RY12 (green) and HLA-DR (red) immunoreactivity (Figure 6D–F). Similarly, in AD sections (Figure 6G–I), separation of P2RY12 and HLA-DR immunoreactivity can be observed though small amounts of P2RY12 immunoreactivity with HLA-DR (Figure 6I—yellow arrow) were present in some cells.

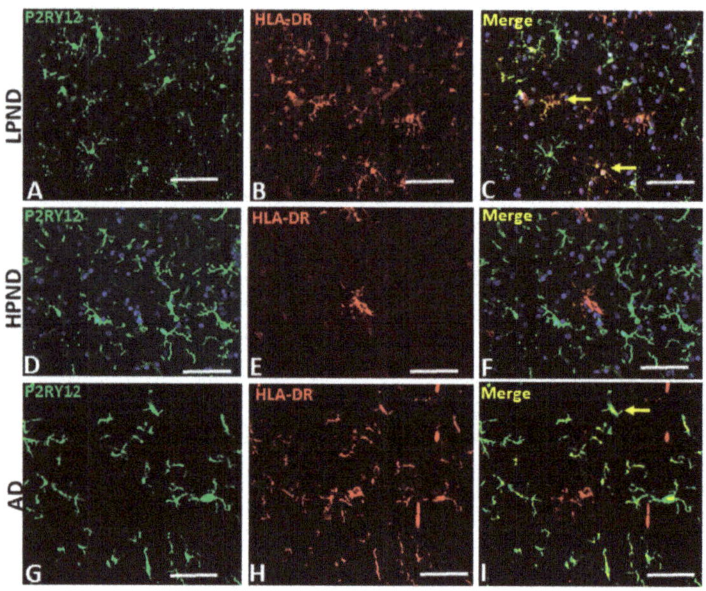

Figure 6. Confocal microscopy of P2RY12 and HLA-DR positive microglia in pathologically staged cases. (**A–I**) Images of P2RY12 (green), HLA-DR (red) and merged (yellow) with DAPI (blue) in MTG of LPND (**A–C**), HPND (**D–F**) and AD case (**G–I**) to show the distribution of P2RY12 and HLA-DR immunoreactivity. Examples of colocalization (yellow arrows) are shown in Merge image. Scale bar represents 50 μm.

Similar analyses were carried out comparing co-expression by microglia in LPND, HPND and AD MTG sections for P2YR12 and CD68, the monocyte-specific phagocytic lysosomal marker (Figure 7), and with progranulin, another lysosomal-associated marker, (Figure 8). Both proteins have been considered as markers of activated microglia. We had hypothesized that there would be a clear discrimination between P2RY12 and CD68 staining to distinguish between resting and phagocytic microglia, but this was incorrect. We observed that most P2RY12 positive microglia were also positive for CD68. The distribution of P2RY12 and CD68 in top cortical layers are shown in Figure 7A–C at low-magnification. Higher magnification images of these cases show that there was strong CD68 staining in low pathology cases (Figure 7D—red CD68 alone and Figure 7G—merged images of P2RY12 and CD68). There was no noticeable increase in CD68 intensity in HPND and AD cases (Figure 7E,F), but there were more CD68 positive cells with limited P2RY12 staining (Figure 7I) in AD cases.

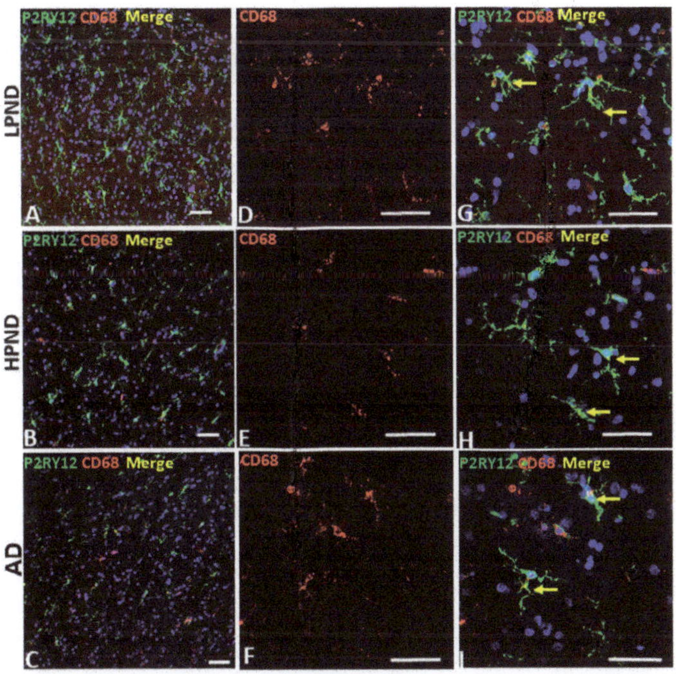

Figure 7. Confocal microscopy of P2RY12 and CD68-positive microglia in pathologically staged cases. **A–C).** Low magnification merged images of P2RY12 (green), CD68 (red) and DAPI (blue) in MTG of LPND (**A**), HPND (**B**) and AD cases (**C**) to show the distribution of P2RY12 and CD68 immunoreactivity through cortical layer. Scale bars represent 50 µm. (**D–F**). Higher magnification merged images of CD68 (red) in MTG of LPND (**D**), HPND (**E**) and AD cases (**F**) to show the distribution of CD68 immunoreactivity. Similar amounts of CD68 immunoreactivity was present in each disease group. Scale bars represent 50 µm. (**G–I**). Merged images of P2RY12 (green) with the CD68 (red) images shown in (**D**,**E**)) with DAPI (blue). Scale bars represent 50 µm.

Analyses of expression of progranulin in P2RY12-positive microglia in staged samples were also carried out. Progranulin is a multi-functional protein with anti-inflammatory, growth factor and lysosomal regulatory properties. Similar to CD68, microglial expression of progranulin has been associated with activated microglia [37]. The representative images shown in Figure 8 show that most P2RY12-positive microglia, whether in LPND, HPND or AD MTG sections, are also positive for progranulin. The lower magnification images (Figure 8A–C) show the distribution of P2RY12 and progranulin-positive microglia through sections of Layer II and III. All progranulin immunoreactivity was associated with cells showing different amounts of P2RY12 immunoreactivity. Although progranulin immunoreactivity can be detected in neurons with the antibody used, this was mainly detectable only in large pyramidal neurons in Layer V and not in the layers shown in this figure [38]. Higher magnifications images show progranulin immunoreactivity alone (Figure 8D–F), and merged images combined with P2RY12 show that all P2RY12 microglia appeared to be progranulin positive (Figure 8G–I).

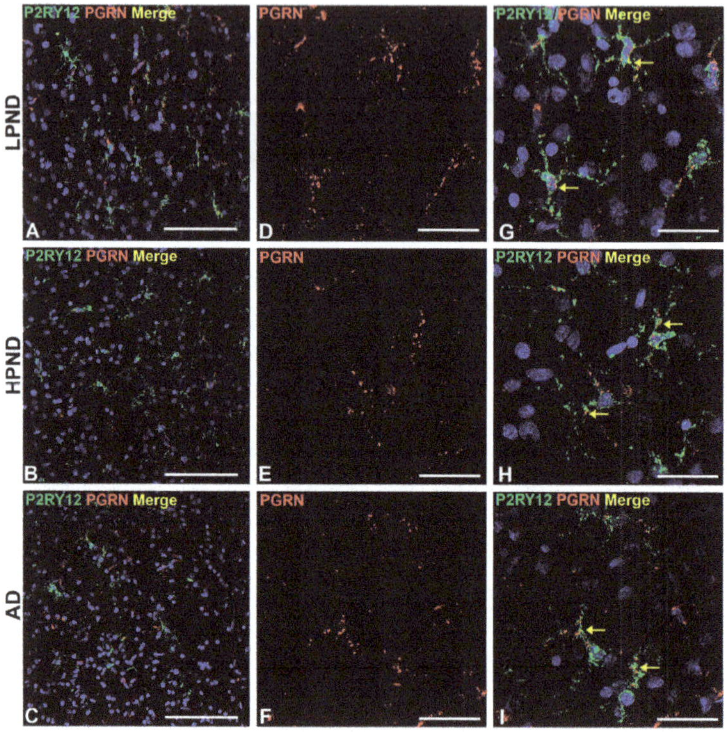

Figure 8. Confocal microscopy of P2RY12 and progranulin-positive microglia in pathologically staged cases. (**A–C**). Low magnification merged images of P2RY12 (green), progranulin (PGRN) (red) and DAPI (blue) in MTG of LPND (**A**), HPND (**B**) and AD cases (**C**) to show the distribution of P2RY12 and PGRN immunoreactivity through cortical layers. Scale bar represents 50 µm. (**D–F**). Higher magnification images of PGRN (red) in MTG of LPND (**D**), HPND (**E**) and AD cases (**F**) to show the distribution of immunoreactivity. Similar amounts of PGRN immunoreactivity was present in each disease group. Scale bar represents 50 µm. (**G–I**). Merged images of P2RY12with PGRNimages shown in (**D–F**) with DAPI (blue) showing expression in same cells (yellow arrows). Scale bar represents 25 µm.

2.7. Patterns of Expression of P2RY12-Microglia with Different Types of Plaques in Pathologically Staged Cases

Confocal microscopy with multi-layered images of P2RY12-immunoreactive microglia with Aβ provide additional information on their interactions. We had earlier observed that there were two types of P2RY12-positive microglia having interactions with Aβ plaques. Those surrounding plaques and those interacting with plaques. It might be assumed that all Aβ plaques would activate the microglia in a proinflammatory manner resulting in downregulation of P2RY12 expression. Figure 9 illustrates that P2RY12-positive microglia interact with diffuse-like, non-cored amyloid (earlier plaques) (Figure 9A—LPND case: Figure 9C—HPND case: Figure 9E—AD case), while the cored plaques had zones without P2RY12-expressing microglia (Figure 9B—LPND: Figure 9D—HPND: Figure 9F—AD).

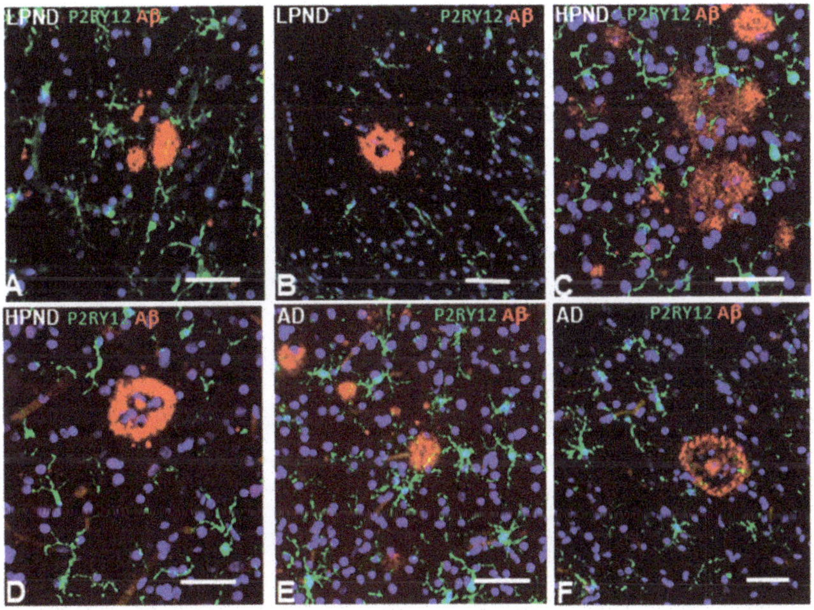

Figure 9. Interaction of P2RY12-positive microglia with different types of Aβ-positive plaques in pathologically-staged cases. (**A–F**). P2RY12-positive microglia (green) interacting with Aβ-positive diffuse-type plaques (red) (**A**,**C**,**E**), in LPND (**A**), HPND (**C**), and AD (**E**), but not with mature-type cored plaques (**B**-LPND) (**D**-HPND) (**F**-AD). Scale bars represent 50 μm.

2.8. Patterns of Expression of P2RY12-Positive Microglia and Phosphorylated Tau-Positive Tangles

Neurofibrillary tangles and dystrophic neurites are also a hallmark pathological feature of AD pathology. In this study, we identified them using antibody AT8 that recognizes phosphorylated forms of tau (serine 202/threonine 205)(pTAU) that accumulate in tangles and neurites. The cases studied had been staged based on plaque not tangle pathology. There was little difference in tangle scores between LPND and HPND cases, but a large increase in the AD cases (Table 1-set 1). AT8 staining was only prominent in the AD cases. A complete series of sections in this study were double-stained using enzyme histochemistry for P2YR12 (purple) and AT8 (brown), and a subset of these also examined by confocal microscopy. Figure 10A and 10B shows P2RY12-immunoreactive microglia interacting with sparse pTau-positive neurites (Figure 10A) and surrounding an early-stage intracellular tangle (arrow) (Figure 10B). Figure 10, panels C–E show features of P2RY12 immunoreactive microglia surrounding different AT8-immunoreactive structures. It can be seen that these microglia do not have the morphology of resting microglia. This is particular noticeable in Figure 10D with a microglia closely interacting with a tangle-containing neuron. A frequent observation in regions with heavy density of AT8 staining were the small numbers of microglia that were strongly immunoreactive for P2RY12 (Figure 10E). A rare feature observed in only one of our LPND case was the presence of AT8-positive glial cells (brown) (Figure 10F) with closely-associated P2RY12-positive microglia. Overall, the AT8-positive tangled structures did not appear to provide the inflammatory stimuli to cause downregulation of P2RY12 expression. Using confocal microscopy, Figure 10G shows an early intracellular tangle with an intact nucleus in a LPND case surrounded by P2RY12 microglia. More mature tangles and tangled neurites (Figure 10H) in AD cases did not have directly interacting P2RY12-positive microglia. Based on the characteristic morphology, the AT8-positive structure in Figure 10I is considered to represent a neuritic plaque, an accumulation of phosphorylated tau-containing neurites associated with an Aβ plaque.

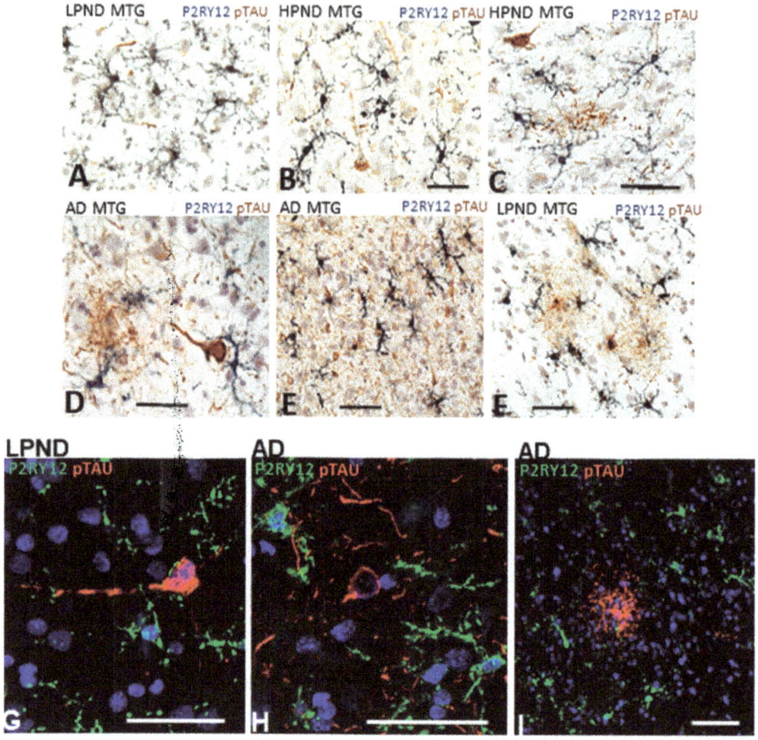

Figure 10. Features of P2RY12-positive microglia interacting with phosphorylated tau-containing structures. (**A–F**). Dual-color DAB enzyme histochemistry illustrating different features of P2RY12-positive microglia (purple) with phosphorylated tau-positive structures (brown). Abbreviations: LPND; low plaque non-demented. HPND; high plaque non-demented. AD; Alzheimer's disease. MTG; middle temporal gyrus. Scale bars represent 50 µm. (**G–I**). Dual-color laser confocal histochemistry showing interaction of P2RY12 positive microglia with early intracellular tangle in LPND case (**G**). Features of P2RY12-positive microglia interacting with mature tangle (**H**) and neuritic plaque (**I**) in AD cases. Scale bars represent 50 µm.

2.9. In Vitro Analysis of P2RY12 Expression by Human Microglia

Our brain tissue observations of P2RY12-expressing microglia have shown multiple features of these cells in relation to their activation states. To further investigate if increased P2RY12 expression represents a marker of alternative activation, cultured human brain-derived microglia were treated with interleukin-4 (IL-4), Aβ peptide and other proinflammatory agents to determine how activation affects P2RY12 mRNA expression. Samples analyzed by qPCR for P2RY12 mRNA expression showed strong induction of expression by IL-4 treatment, and downregulation to different extents with Aβ and other proinflammatory agents (Figure 11A). Induction of P2RY12 protein was also observed by western blotting in IL-4-treated microglia (Figure 11B).

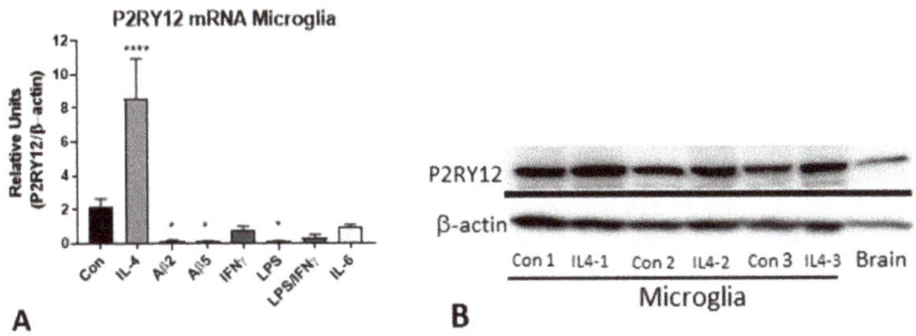

Figure 11. Expression of P2RY12 mRNA and protein by in vitro cultures of human microglia. (**A**). Interleukin-4 stimulates P2RY12 mRNA expression. Bar chart showing results real time PCR analysis for P2RY12 mRNA of human microglia stimulated with indicated agents. Results of analysis of single human microglia case (each in triplicate) and representative of other analyses. Abbreviations: Con, control unstimulated: IL-4, interleukin-4 (40 ng/mL): Aβ2 and Aββ5 (Aβ (1–42) 2 μM and 5 μM): IFNγ, interferon-γ (100 ng/mL): LPS, lipopolysaccharide (100 ng/mL): LPS/IFNγ (doses combined): IL-6, interleukin-6 (40 ng/mL). **** $p < 0.0001$. * $p < 0.05$. (**B**). Western blot of human microglia protein samples probed with antibody to P2RY12. Increased amounts of P2RY12 (58 kDa) in IL-4 treated samples.

3. Discussion

The aim of this work was to examine phenotypes of P2RY12-immunopositive microglia in aging and AD brains in relation with AD-associated pathological structures. P2RY12 has been defined as a specific marker to discriminate between microglia, with high levels of expression, and macrophages, with low levels of expression [16,30]. In addition, based on experimental findings, continued expression by microglia of P2RY12 in brain should define them as non-activated, namely those not producing proinflammatory cytokines associated with enhanced inflammation. As experimental studies have shown that proinflammatory activation of microglia resulted in significant reduction in P2RY12 expression [16,26], it was hypothesized that identifying microglia with high expression of P2RY12 compared to microglia positive for classical activation markers but with low to negative expression of P2RY12 would provide a means of identifying areas of active inflammation in brain tissue [30]. The findings of this work showed that microglial expression of P2RY12 was downregulated in AD tissue samples, but immunohistochemistry identified more complex patterns of increased P2RY12 expression associated with pathological structures than previously identified [30].

Recent gene expression profiling of single-cell microglia from rodent and human sources had confirmed that P2RY12 mRNA expression was associated with a non-activated phenotype and expression was downregulated with progression of disease [10,14,15]. However, our initial observations in human brains suggested that this might not cover all features of P2RY12 expression by microglia; for this reason, we sought to provide detailed characterization of P2RY12 microglia in human aged and AD brains. The importance of microglial phenotyping to identifying functional markers is now appreciated along with the need for greater numbers of markers [39]. For a number of years, the classification of microglia (and macrophages) into functional M1 and M2 groups was applied but it is now appreciated that this system does not account for the complexities of microglial phenotypes in diseased brains [40]. Recent findings have defined a phenotype of microglia designated "disease-associated microglia" (DAM), which describes a transcriptional signature first associated with response to neurodegeneration-associated molecular patterns (DAM stage 1) that progresses to a signature associated with a protective role to limit inflammation (DAM stage 2) that is coordinated by TREM-2 signaling. Downregulation of P2RY12 from homeostatic to stage 1 DAM confirms earlier findings but our cellular localization findings suggest that upregulation of P2RY12 may also be a feature associated with later stages of AD.

We have made some new findings on P2RY12 expression by microglia in human brains as part of this study that extend previous findings [30]. There were decreased levels of total P2RY12 protein of 58 kDa in brain extracts from AD cases compared to non-demented low and high pathology cases, as could be expected, but we also identified increased levels of 30 kDa P2RY12 polypeptide in AD cases. This indicates that downregulation of P2RY12 protein levels might be due to enhanced cleavage of this plasma membrane protein. The antibody we used for our study was prepared against a 40-amino acid recombinant peptide corresponding to the C-terminal cytoplasmic domain of P2RY12 (amino acids 303–342). P2RY12 is a G-protein-coupled receptor for ADP containing 7-transmembrane domains. Based on this structure, the 30 kDa polypeptide would represent a remaining cell-associated peptide that does not contain the N-terminal sequences. As P2RY12 has been reported to have multiple ADP-binding domains, it is unclear if this fragment will be biologically active for ADP binding and signal transduction. Downregulation of P2RY12 expression as a result of proinflammatory activation and upregulation as a response to IL-4 were regarded as features of alternatively activated microglia though the mechanisms for this to occur in brains is unclear as IL-4 expression has not been consistently detected in brain tissue [41]. We showed that downregulation of P2RY12 expression following inflammatory stimulation also occurred in cultured human microglia as did others [42]. The involvement of progranulin, a neuroprotective and anti-inflammatory molecule, in microglial function is still unclear. Progranulin positive microglia are found throughout brain but our findings showed P2RY12-positive microglia in all disease groups were positive for intracellular progranulin. It has been shown that IL-4 upregulates progranulin expression by cultured human microglia [43]. Based on current findings of gene regulation, one can speculate that P2RY12/progranulin positive microglia are protective rather than reactive, but further studies are required.

A previous study of P2RY12 microglia across different human brain regions and ages made similar observations as this study that most P2RY12-positive microglia were also CD68 positive [30]. CD68, a myeloid specific lysosomal-associated membrane protein associated with phagocytosis, has been considered as a microglial activation marker in a number of studies [23,44], but colocalization of CD68 and P2RY12 would suggest its involvement in normal microglial function. Two earlier studies have observed that P2RY12 was not expressed by microglia accumulating around plaques in AD brains. We also observed this for many plaques, but there were noticeable exceptions as many diffuse-like Aβ(6E10-immunoreactive) plaques had P2RY12-positive microglia interacting with them. We also observed varied morphologies for P2RY12-microglia interacting with plaques, including fragmented, tufted and rod-shaped. The study of Mildner et al. ([30]) employed a different antibody (Sigma-Aldrich HPA014518) than we used (Novus NBP2-33870), however both antibodies were produced against the same C-terminal 40-amino acid peptide sequence so should have the same properties. Detection differences in these studies could be due to tissue fixation conditions. Our study employed lightly-fixed free-floating brain sections for immunohistochemistry, while Mildner et al. employed harder-fixed paraffin-embedded sections [30]. For their study, antigen retrieval was required for all antibodies, while we found this not necessary for free-floating sections when using the Novus antibody. However, sensitivity of P2RY12 detection to fixation was observed as the alternate antibody (Alomone-APR-012) we used only worked when free-floating sections underwent antigen retrieval processing.

Although P2RY12 appears to be an excellent marker for microglia in brain, it is unclear whether expressing microglia can be classified as protective or proinflammatory. How P2RY12-mediated responses by microglia are involved in AD pathogenesis is unresolved. P2RY12-mediated chemotactic responses to ATP and ADP, which are released by damaged or dying cells, appears to be an early inflammatory response. The rapid downregulation of P2RY12 expression with proinflammatory activation would appear to function to anchor microglia at sites of inflammation. It has been proposed that downregulation of P2RY12 is accompanied by increased expression of adenosine A2 receptor, the breakdown products of P2RY12 ligands ADP/ATP [45]. A recent in vitro study of microglia demonstrated the proinflammatory consequences of inflammasome and NFκB activation by extracellular ADP activation of P2RY12 [46]. In a rodent ischemia model, blockade of microglial P2RY12 with ticagrelor, an antagonist, reduced ischemic damage by microglia by reducing their migration to sites of injury [47]. In another ischemia animal model, and with an in vitro model of oxygen–glucose deprivation (OGD), inhibition of microglia P2YR12 with clopidogrel, another antagonist, or by P2RY12 gene expression knock-down significantly reduced microglial migration and neurotoxicity [48].

The potential for using P2RY12-positive microglia distribution to define distribution of neuroinflammation in AD or other neurodegenerative diseases is still valid, though not as clear as originally hypothesized. There has been a widely-held concept that AD neuroinflammation is widespread through affected areas, but these findings suggest it is highly localized. If one considers the scheme outlined in Figure 12, which is a representation of Figure 1E,F,G and Figure 9D,F, that activated microglia (low-negative P2RY12 microglia) associated with mature amyloid plaques are secreting cytokines and other molecules that downregulate P2RY12 expression by microglia, these will be present in the zone around the plaques (Zones 1 and 2), but not in Zone 3 where the P2RY12-positive microglia are located. With the use of laser capture microscopy and proteomics techniques, analyses of micro-dissected regions are feasible and could identify key neuroinflammatory factors associated with AD that are not represented when larger dissected pieces of brain tissue are analyzed. The feasibility of this approach has been demonstrated in proteomic analyses of micro-dissected plaques and tangles from AD brains [49,50].

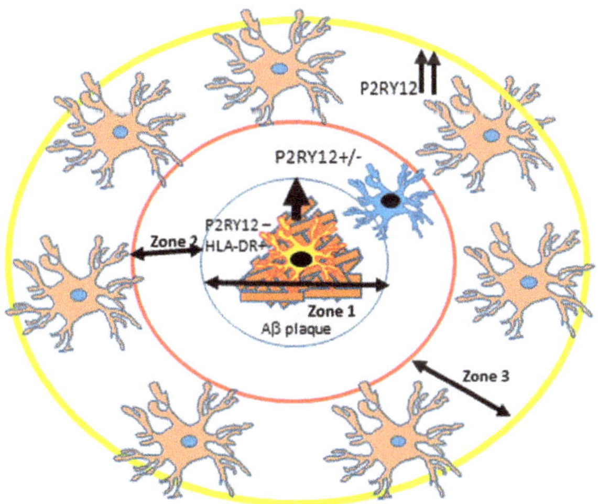

Figure 12. Proposed scheme of arrangement of different P2RY12-expressing microglia around Aβ plaques. Suggested scheme to describe localized areas of microglial inflammation around plaques. Zone 1: microglia interacting with mature plaques (HLA-DR high, P2RY12 negative) producing proinflammatory cytokines. Zone 2: Area adjacent to plaque with low or negative P2RY12 positive microglia. Zone 3. P2RY12 high expression in surrounding area defining the boundary between proinflammatory area (Zone 1 and 2) and non-affected area (Zone 3 and beyond). As described in this report, exceptions to this scheme were observed.

4. Materials and Methods

4.1. Human Brain Tissue Samples

Human brain tissue samples used in this study were obtained from the Banner Sun Health Research Institute Brain and Body Donation Program, Sun City, Arizona, U.S.A. [51,52]. The operations of the Brain and Body Donation Program (BBDP) have received continuous approval of Institutional Review Boards (IRB). Current operations have been reviewed by Western IRB (Puyallup, WA, USA).

Written informed consent for collection and use of brain and other tissues for research purposes were obtained from donors or next-of-kin. Tissue studies carried out in the U.S.A. were considered non-human subject research under federal regulations. Tissue studies carried out in Japan were approved by Shiga University of Medical Science Ethical Committee (Project Certificate no. 29-114). Demographic details of cases used in this study are summarized in Table 1.

4.2. Brain Tissue Preservation and Fixation

All brains were processed at autopsy in a standardized manner [52]. The median postmortem interval for autopsies in the BBDP was 3.8 h. After brain removal, the cerebellum and brain stem are separated from the hemispheres, then each brain is sectioned in a frame into 1cm thick coronal slabs. The hemispheres are divided, with the left hemisphere being frozen on dry ice for storage at −80 °C, and the right hemisphere being fixed for 48 h in buffered formalin solution. After fixation, the coronal pieces are rinsed and transferred to a phosphate-buffered solution of 15% glycerol/15% ethylene glycol as cryoprotectant. Brain regions used for subsequent studies are dissected from frozen or fixed coronal slices by experienced neuroanatomists.

4.3. Neuropathological Diagnosis Criteria

All donated brains received full neuropathological diagnosis including reference to pre-mortem clinical history of each case. Consensus clinical and neuropathological criteria were used to diagnose AD, Dementia with Lewy bodies (DLB) and PD in these cases [53,54]. To assess severity of AD pathology in each case, tissue sections from 5 brain regions (entorhinal cortex, hippocampus, frontal cortex, temporal cortex and parietal cortex) were stained with Thioflavin-S, Gallyas or Campbell–Switzer histological stains and assessed semi-quantitatively for the density of neurofibrillary tangles and amyloid plaques with each brain region being ranked on a scale of 0–3. By combining the measures across these 5 brain regions, assessment of AD pathology was ranked on an ordinal scale of 0–15 for plaques and tangles [55]. The two sets of cases used in this study were subdivided into non-demented low plaque (LPND) (plaque score < 6), non-demented high plaque (HPND)(plaque score 6–14) and AD with dementia (plaque score > 12).

4.4. Peroxidase/Diaminobenzidine Immunohistochemistry

Formaldehyde-fixed tissue sections (25 µm) from middle temporal gyrus (MTG) from 36 cases cut on a sliding microtome were used for cellular localization of purinergic receptor P2RY12 in relation with AD pathological markers, amyloid-beta (Aβ and phosphorylated tau, or markers of microglia activation (HLA-DR, and IBA-1). Sections of hippocampus from 4 cases (2 LP, 2 AD) were also used for comparison. A free-floating immunohistochemistry method was used. Tissue sections from a series of cases were rinsed three times with phosphate-buffered saline-Triton X100 (PBSTx) (0.1 M phosphate Buffer, pH 7.6, 0.137 M NaCl, 0.3% Triton X100) and pretreated with 1% hydrogen peroxide in PBSTx for 30 min to quench endogenous tissue peroxidases. For certain antibodies, antigen retrieval was carried out by heating sections in 10 mM EDTA (pH 8.0) at 80 °C for 30 min and then cooling to room temperature for 30 min. Sections were incubated free-floating at room temperature for 18 h with shaking in PBSTx with optimal dilution of antibodies. To identify localized antibodies, sections were sequentially washed three times in PBSTx for 10 min, incubated in biotinylated secondary antibody (1:1000), sequentially washed again three times in PBSTx and then incubated in preformed avidin-biotin horseradish peroxidase enzyme complex (ABC-Vector Laboratories, Burlingame, CA USA) solution (1:1000) for 1 h. Sections were then washed three times in PBSTx and two times in 50 mM Tris-HCl (pH 7.6) before incubation in peroxidase substrate. Most frequently used was nickel ammonium sulfate-enhanced diaminobenzidine as substrate to produce a purple reaction product (50 mM Tris-HCl, pH 7.6, 1% saturated nickel ammonium sulfate, 40 mM imidazole, 100 µg/mL diaminobenzidine-HCl (Dojindo, Kumamoto, Japan) and 0.0003% hydrogen peroxide). For two-color immunohistochemistry, reacted sections were rinses in PBSTx, treated with 1% hydrogen peroxide to remove residual peroxidase activity and then incubated for a second time in primary antibody overnight at room temperature. The detection procedure followed the above described protocol except the substrate used was diaminobenzidine without nickel ammonium sulfate as substrate to produce a brown reaction product (50 mM Tris-HCl, pH 7.6, 20 mM imidazole, 200 μg/mL diaminobenzidine-HCl and 0.0006% hydrogen peroxide). Reacted sections were mounted on slides, counterstained in most cases with 0.5% neutral red, dehydrated, cleared and coverslipped using Permount mounting media (ThermoFisher, Waltham, MA, USA Sections used for quantitative measurements were not counterstained.

4.5. Fluorescent Confocal Immunohistochemistry

Multiple-color fluorescent confocal immunohistochemistry was carried out for antibody pairs to verify cellular co-localization of antigens with P2RY12-expressing cells [19]. Tissue sections were incubated with optimal dilutions of antibodies at room temperature overnight with shaking. After three washes (10 min each) in PBSTx, sections were incubated in the dark with optimal concentrations of fluorescent-labeled secondary antibodies. Bound primary antibodies were detected with Alexa Fluor 488 (donkey anti-goat IgG), Alexa Fluor 568 (donkey anti-rabbit or anti-mouse IgG) or Alexa Fluor 647 (donkey anti-mouse IgG) (ThermoFisher, San Jose, CA, USA). After washing and mounting, sections were counterstained with Sudan Black (1% solution in 70% ethanol for 10 min) to quench tissue autofluorescence, destained with 70% ethanol, and stained with DAPI to reveal nuclei. Sections were coverslipped using antifading-hardening-fluorescent mounting media (Vector Labs, Burlingame, CA, USA). Sections were imaged using an Olympus FV1000 confocal microscope, and compiled Z-scans obtained and processed using Olympus microscope system software (Olympus Corporation, Tokyo, Japan). Some sections were imaged using a Leica SP8 confocal microscope system (Leica-Microsystems, Wetzlar, Germany).

4.6. Antibodies

The following primary antibodies were used in this study: P2RY12 (Novus Biologicals, Centennial, CO, USA) catalog no. NBP2-33870; rabbit, 1:1000–1:2000 used for immunohistochemistry (IHC) and western blot (WB). P2RY12 (Alomone Labs, Tel Aviv, Israel); catalog no. APR-012; rabbit, 1:200 used for IHC. P2RY12 (Abcam, Cambridge, UK); catalog no. AB83066; rabbit, 1:2000 used for WB. HLA-DR clone LN3 (Abcam); catalog no. AB80658; mouse, 1:750 used for IHC. CD68 (Biolegend, San Diego, CA, USA); catalog no 916104; mouse, used at 1:250 for IHC. Progranulin (R&D Systems, Minneapolis, MN, USA; catalog no. AF2420, goat used at 1:100 for IHC. IBA-1 (Wako, Richmond, VA, USA); catalog no. 019-19741; rabbit, 1:1000 used for IHC. Aβ clone 6E10 (Biolegend); catalog no. 803001; mouse, 1:2000 used for IHC. pTau clone AT8 (ThermoFisher, Waltham, MA, USA); catalog no. MN1020: mouse, 1:3000 used for IHC.

Secondary biotinylated-antibodies used for enzyme histochemistry and Avidin-Biotin-Complex (ABC) peroxidase were obtained from Vector Labs (Burlingame, CA, USA). Fluorescent-labeled secondary antibodies used for confocal microscopy, and horseradish peroxidase (HRP)-conjugated secondary antibodies used for western blots were obtained from ThermoFisher (Waltham, MA, USA).

4.7. Verification of Antibody Specificity

The majority of the immunohistochemistry studies reported in this communication were carried out with the Novus P2RY12 antibody. Peptide absorption studies were carried out using a recombinant 40-amino acid protein (NBP2-33870PEP, Novus), the immunizing peptide for the Novus antibody (NBP2-33870). P2YR12 antibody (1:1000) was mixed with 20-fold molar excess of protein for 18 h, and these materials were used to stain sections from three separate cases in parallel with non-absorbed antibody. In addition, comparisons of immunostaining patterns were carried using an independent antibody to P2RY12 (1:250, Alomone Labs), prepared against an 18-amino acid peptide sequence that did not overlap with the immunizing sequence of the Novus antibody.

4.8. Brain Sample Extraction and Western Blot

Brain tissues samples (middle temporal gyrus – MTG) were dissected frozen and then further trimmed so samples being analyzed primarily contained gray matter. Detergent-soluble extracts were prepared by gently sonicating each tissue sample in 5 volumes (weight to volume) of RIPA buffer (20 mM Tris-HCl, pH 7.5. 150 mM NaCl, 1% Triton X100, 1% sodium deoxycholate, 0.1% sodium dodecyl sulfate) supplemented with protease and phosphatase inhibitors (Nacalai-Tesque, Kyoto, Japan). After 30 min incubation on ice, samples were centrifuged at 15,000× *g* for 30 min. The supernatants were transferred to new tubes and total protein concentration of each extracted sample was determined using a Micro BCA assay kit (ThermoFisher, Rockford, IL, USA) with bovine serum albumin as standard. P2RY12 polypeptides were detected in MTG samples by western blots using the Novus P2RY12 antibody (1:1000). Protein samples were dissolved in SDS-sample buffer, adjusted to contain 1 μg/μL protein, denatured by heating to 90 °C for 10 min, centrifuged at 15,000× *g* for 10 min to remove insoluble material and then separated on 4–20% Tris-glycine precast gradient polyacrylamide gels (Nacalai-Tesque, Kyoto, Japan). Separated polypeptides were transferred to PVDF membranes (Millipore-EMD) using a semi-dry transfer apparatus. Membranes were blocked in 5% skimmed milk solution diluted in Tris-buffered saline with 0.1% Tween 20 (TBST), and incubated for 18 h in optimal dilutions of antibody solution in 2% milk in TBST. Membranes were washed in TBST and then reacted with HRP-labeled secondary antibody for 2 h. Membranes were reacted with Chemi-Lumi-One Super chemiluminescent substrate (Nacalai-Tesque, Kyoto, Japan) and sequential images captured using an ImageQuant LAS 4000 system (GE LifeSciences, Pittsburgh, PA, USA). Band intensities were measured using Image Studio Lite (LI-COR, Lincoln, NE, USA). All membranes were subsequently reacted with HRP-labeled antibody to β-actin (FujiFilm Wako Pure Chemicals, Osaka, Japan) at 1:15,000 for 1 h and imaged in the above-described manner. Band intensities were normalized for levels of β-actin in samples.

4.9. Area of P2RY12 Immunoreactivity

To measure area of P2RY12 immunoreactivity, a complete series of cases (see Table 1) were single-stained with P2RY12 antibody (Novus) using nickel-enhanced DAB to reveal dark purple reaction product. After mounting and coverslipping of sections, images at 4x magnification were captured from each section, taking four separate, random fields of gray matter for each section. Captured images were analyzed using ImageJ analysis software (NIH, Bethesda, MD, USA) (https://imageJ.net/ImageJ, version 1.52a, accessed on 5 January, 2018) to measure area of immunoreactivity occupied by reaction product on each slide. Images were converted to gray scale and thresholds adjusted to identify positively stained microglia. Mean values of area occupied between the four separate fields measured were calculated for each case, and then mean data were compiled into respective disease groups for further statistical analysis.

4.10. P2RY12 Immunoreactive Cell Counts

Further analyses were carried out to estimate numbers of P2RY12 immunoreactive microglia in each section. Using a 25-grid-square microscope eyepiece reticule (Nikon) and 20× objective, the numbers of microglia were counted in five consecutive areas through the cortex. Each field counted corresponded to 2 mm^2. These areas corresponded approximately to Layer I and II, Layer III, Layer IV, Layer V, and Layer VI. For each section, three separate areas were counted where distinct cortical layers could be detected. The patterns of cortical layers was confirmed by reference to standard text. These measures allowed sampling through all cortical layers though the AD cases had thinner cortical layers. Counting criteria required the presence of a microglial cell body to be present within the area of the reticule grid. For each case, the mean total numbers of cells were calculated, and then mean and standard error of mean for all samples from each disease group.

4.11. Quantitative Reverse Transcription Polymerase Chain Reaction (qPCR) Analysis of mRNA Expression.

RNA was prepared from human brain tissue samples and human microglia using RNAeasy Plus-Mini kits (Qiagen, Valencia, CA, USA) according to the manufacturer's protocol. RNA concentrations and purities were measured using a Nanodrop 1000 spectrophotometer and RNA integrity with an Agilent Bioanalyzer and RNA 6000 Nano kits (Agilent, Santa Clara, CA, USA). Samples used for qPCR had RIN values greater than 7.0. RNA from brain samples (0.5 μg) and cultured cell samples (0.2 μg) were reverse transcribed using the Quantitect reverse transcription kit (Qiagen) with genomic DNA elimination reagent according to manufacturer's protocol. Appropriate numbers of no reverse transcriptase controls were prepared in parallel for each batch of samples. For qPCR, cDNA samples were amplified using Perfecta Fast Mix 2x reaction mixture (Quanta Biosciences, Gaithersburg, MD, USA) supplemented with 1.25 μM of Eva Green. The primers used were as follows: P2RY12 sense: AGTCCCCAGGAAAAAGGTG; P2RY12 antisense: GTTTGGCTCAGGGTGTAAGG (reference sequence NM_022788.40). Expression results were normalized with relative levels of β-actin mRNA using primer sequences: β-actin sense: TCCTATGTGGGCGACGAG. β-actin antisense: ATGGCTGGGGTGTTGAAG. QPCR was carried out using a Stratagene Mx3000p machine and abundance of gene expression quantified relative to a standard curve of pooled samples. QPCR analyses followed recommended criteria for minimum information for publication of quantitative real-time PCR experiment (MIQE) [56].

4.12. Human Brain-Derived Microglia Isolation and Stimulation

Human brain microglia were prepared from frontal cortex from three different donor cases for this study following our published procedures [9,13]. After 10–14 days in culture, microglia were replated into wells at 10^5 cells/well in 12-well plates prior to stimulation. For these experiments, microglia were unstimulated or treated with interleukin (IL)-4 (20 ng/mL), aggregated Aβ42 (2 μM and 5 μM), interferon-γ (IFNγ) (20 ng/mL), lipopolysaccharide (LPS) (100 ng/mL), LPS and IFNγ combined, and IL-6 (20 ng/mL) [9]. After 24 h treatment, RNA was isolated from microglia using the above described methodology. Expression of P2RY12 mRNA in treated and untreated cells were carried out as described above for brain samples. Western blot analysis for P2RY12 protein expression by IL-4 stimulated cells was carried out with microglia from a single additional case.

4.13. Data Analysis

Data for relative changes in relation to disease classification were analyzed by one-way analysis of variance (ANOVA) with Newman–Keuls post-hoc test for significance between paired groups. Significant differences were assumed if p values of less than 0.05 were obtained. Statistical analyses were carried out using Graphpad Prism Version 7 software (Graphpad Software, La Jolla, CA, USA).

Author Contributions: D.G.W. conceptualized this project, planned this study and carried out the immunohistochemistry and biochemical studies with the assistance and input of T.M.T. and L.-F.L. D.G.W. supervised and carried out data analyses and writing of drafts of this manuscript. T.M.T. contributed equally to this project as D.G.W., carried out immunohistochemistry to establish the methodology and initial findings, and carried out all image analyses studies. A.M. carried out confocal immunohistochemistry and microscopy and assisted in image production. I.T. provided funding for studies in Japan and operational support. G.E.S., L.I.S., and T.G.B. selected and provided brain tissues for this study and accompanying clinicopathological data for these cases used from the Brain and Body Donation Program, Banner Sun Health Research Institute. All authors have read and agreed to the published version of the manuscript.

Funding: The studies carried out in Japan were supported by grant JP26290022 from Japan Society for Promotion of Science to IT.

Conflicts of Interest: The authors declare no conflicts of interest. The funders had no role in the design of the study; in the collection, analyses, or interpretation of data; in the writing of the manuscript, or in the decision to publish the results.

Abbreviations

P2RY12	Purinergic receptor P2Y12
LPND	Low plaque non-demented
HPND	High plaque non-demented
AD	Alzheimer's disease
Aβ	Amyloid beta
IL	Interleukin
LPS	Lipopolysaccharide

References

1. World Alzheimer Report. 2016. Available online: http://www.alz.co.uk/research/WorldAlzheimerReport2016.pdf (accessed on 10 December 2019).
2. McGeer, P.L.; Itagaki, S.; Tago, H.; McGeer, E.G. Reactive microglia in patients with senile dementia of the Alzheimer type are positive for the histocompatibility glycoprotein HLA-DR. *Neurosci. Lett.* **1987**, *79*, 195–200. [CrossRef]
3. Mattiace, L.A.; Davies, P.; Dickson, D.W. Detection of HLA-DR on microglia in the human brain is a function of both clinical and technical factors. *Am. J. Pathol.* **1990**, *136*, 1101–1114.
4. Sarlus, H.; Heneka, M.T. Microglia in Alzheimer's disease. *J. Clin. Investig.* **2017**, *127*, 3240–3249. [CrossRef]
5. Heppner, F.L.; Ransohoff, R.M.; Becher, B. Immune attack: The role of inflammation in Alzheimer disease. *Nat. Rev. Neurosci.* **2015**, *16*, 358–372. [CrossRef] [PubMed]
6. McGeer, P.L.; Kawamata, T.; Walker, D.G.; Akiyama, H.; Tooyama, I.; McGeer, E.G. Microglia in degenerative neurological disease. *Glia* **1993**, *7*, 84–92. [CrossRef] [PubMed]
7. Akiyama, H.; Barger, S.; Barnum, S.; Bradt, B.; Bauer, J.; Cole, G.M.; Cooper, N.R.; Eikelenboom, P.; Emmerling, M.; Fiebich, B.L.; et al. Inflammation and Alzheimer's disease. *Neurobiol. Aging* **2000**, *21*, 383–421. [CrossRef]
8. In'T Veld, B.A.; Ruitenberg, A.; Hofman, A.; Launer, L.J.; van Duijn, C.M.; Stijnen, T.; Breteler, M.M.; Stricker, B.H. Nonsteroidal antiinflammatory drugs and the risk of Alzheimer's disease. *N. Engl. J. Med.* **2001**, *345*, 1515–1521. [CrossRef] [PubMed]
9. Miguel-Alvarez, M.; Santos-Lozano, A.; Sanchis-Gomar, F.; Fiuza-Luces, C.; Pareja-Galeano, H.; Garatachea, N.; Lucia, A. Non-steroidal anti-inflammatory drugs as a treatment for Alzheimer's disease: A systematic review and meta-analysis of treatment effect. *Drugs Aging* **2015**, *32*, 139–147. [CrossRef] [PubMed]
10. Keren-Shaul, H.; Spinrad, A.; Weiner, A.; Matcovitch-Natan, O.; Dvir-Szternfeld, R.; Ulland, T.K.; David, E.; Baruch, K.; Lara-Astaiso, D.; Toth, B.; et al. A Unique Microglia Type Associated with Restricting Development of Alzheimer's Disease. *Cell* **2017**, *169*, 1276–1290. [CrossRef]
11. Bennett, M.L.; Bennett, F.C.; Liddelow, S.A.; Ajami, B.; Zamanian, J.L.; Fernhoff, N.B.; Mulinyawe, S.B.; Bohlen, C.J.; Adil, A.; Tucker, A.; et al. New tools for studying microglia in the mouse and human CNS. *Proc. Natl. Acad. Sci. USA* **2016**, *113*, E1738–E1746. [CrossRef] [PubMed]
12. Galatro, T.F.; Holtman, I.R.; Lerario, A.M.; Vainchtein, I.D.; Brouwer, N.; Sola, P.R.; Veras, M.M.; Pereira, T.F.; Leite, R.E.P.; Moller, T.; et al. Transcriptomic analysis of purified human cortical microglia reveals age-associated changes. *Nat. Neurosci.* **2017**, *20*, 1162–1171. [CrossRef] [PubMed]
13. Mathys, H.; Adaikkan, C.; Gao, F.; Young, J.Z.; Manet, E.; Hemberg, M.; De Jager, P.L.; Ransohoff, R.M.; Regev, A.; Tsai, L.-H. Temporal Tracking of Microglia Activation in Neurodegeneration at Single-Cell Resolution. *Cell Rep.* **2017**, *21*, 366–380. [CrossRef] [PubMed]
14. Friedman, B.A.; Srinivasan, K.; Ayalon, G.; Meilandt, W.J.; Lin, H.; Huntley, M.A.; Cao, Y.; Lee, S.-H.; Haddick, P.C.G.; Ngu, H.; et al. Diverse Brain Myeloid Expression Profiles Reveal Distinct Microglial Activation States and Aspects of Alzheimer's Disease Not Evident in Mouse Models. *Cell Rep.* **2018**, *22*, 832–847. [CrossRef] [PubMed]
15. Deczkowska, A.; Keren-Shaul, H.; Weiner, A.; Colonna, M.; Schwartz, M.; Amit, I. Disease-Associated Microglia: A Universal Immune Sensor of Neurodegeneration. *Cell* **2018**, *173*, 1073–1081. [CrossRef] [PubMed]

16. Butovsky, O.; Jedrychowski, M.P.; Moore, C.S.; Cialic, R.; Lanser, A.J.; Gabriely, G.; Koeglsperger, T.; Dake, B.; Wu, P.M.; Doykan, C.E.; et al. Identification of a unique TGF-beta-dependent molecular and functional signature in microglia. *Nat. Neurosci.* **2014**, *17*, 131–143. [CrossRef] [PubMed]
17. Akiyama, H.; McGeer, P.L. Brain microglia constitutively express β-2 integrins. *J. Neuroimmunol.* **1990**, *30*, 81–93. [CrossRef]
18. Akiyama, H.; Nishimura, T.; Kondo, H.; Ikeda, K.; Hayashi, Y.; McGeer, P.L. Expression of the receptor for macrophage colony stimulating factor by brain microglia and its upregulation in brains of patients with Alzheimer's disease and amyotrophic lateral sclerosis. *Brain Res.* **1994**, *639*, 171–174. [CrossRef]
19. Walker, D.G.; Tang, T.M.; Lue, L.-F. Increased expression of toll-like receptor 3, an anti-viral signaling molecule, and related genes in Alzheimer's disease brains. *Exp. Neurol.* **2018**, *309*, 91–106. [CrossRef]
20. Grundke-Iqbal, I.; Fleming, J.; Tung, Y.C.; Lassmann, H.; Iqbal, K.; Joshi, J.G. Ferritin is a component of the neuritic (senile) plaque in Alzheimer dementia. *Acta Neuropathol.* **1990**, *81*, 105–110. [CrossRef]
21. Pey, P.; Pearce, R.K.B.; Kalaitzakis, M.E.; Griffin, W.S.T.; Gentleman, S.M. Phenotypic profile of alternative activation marker CD163 is different in Alzheimer's and Parkinson's disease. *Acta Neuropathol. Commun.* **2014**, *2*, 21. [CrossRef]
22. Zrzavy, T.; Hametner, S.; Wimmer, I.; Butovsky, O.; Weiner, H.L.; Lassmann, H. Loss of "homeostatic" microglia and patterns of their activation in active multiple sclerosis. *Brain* **2017**, *140*, 1900–1913. [CrossRef] [PubMed]
23. Hendrickx, D.A.E.; van Eden, C.G.; Schuurman, K.G.; Hamann, J.; Huitinga, I. Staining of HLA-DR, Iba1 and CD68 in human microglia reveals partially overlapping expression depending on cellular morphology and pathology. *J. Neuroimmunol.* **2017**, *309*, 12–22. [CrossRef] [PubMed]
24. Lue, L.-F.; Schmitz, C.T.; Serrano, G.; Sue, L.I.; Beach, T.G.; Walker, D.G. TREM2 Protein Expression Changes Correlate with Alzheimer's Disease Neurodegenerative Pathologies in Post-Mortem Temporal Cortices. *Brain Pathol.* **2015**, *25*, 469–480. [CrossRef] [PubMed]
25. Walker, D.G.; Whetzel, A.M.; Serrano, G.; Sue, L.I.; Beach, T.G.; Lue, L.-F. Association of CD33 polymorphism rs3865444 with Alzheimer's disease pathology and CD33 expression in human cerebral cortex. *Neurobiol. Aging* **2015**, *36*, 571–582. [CrossRef] [PubMed]
26. Haynes, S.E.; Hollopeter, G.; Yang, G.; Kurpius, D.; Dailey, M.E.; Gan, W.-B.; Julius, D. The $P2Y_{12}$ receptor regulates microglial activation by extracellular nucleotides. *Nat. Neurosci.* **2006**, *9*, 1512–1519. [CrossRef]
27. Gachet, C. $P2Y_{12}$ receptors in platelets and other hematopoietic and non-hematopoietic cells. *Purinergic Signal.* **2012**, *8*, 609–619. [CrossRef]
28. Sasaki, Y.; Hoshi, M.; Akazawa, C.; Nakamura, Y.; Tsuzuki, H.; Inoue, K.; Kohsaka, S. Selective expression of Gi/o-coupled ATP receptor $P2Y_{12}$ in microglia in rat brain. *Glia* **2003**, *44*, 242–250. [CrossRef]
29. Franco-Bocanegra, D.K.; McAuley, C.; Nicoll, J.A.R.; Boche, D. Molecular Mechanisms of Microglial Motility: Changes in Ageing and Alzheimer's Disease. *Cells* **2019**, *8*, 639. [CrossRef]
30. Mildner, A.; Huang, H.; Radke, J.; Stenzel, W.; Priller, J. $P2Y_{12}$ receptor is expressed on human microglia under physiological conditions throughout development and is sensitive to neuroinflammatory diseases. *Glia* **2017**, *65*, 375–387. [CrossRef]
31. Zhu, C.; Kros, J.M.; van der Weiden, M.; Zheng, P.; Cheng, C.; Mustafa, D.A.M. Expression site of P2RY12 in residential microglial cells in astrocytomas correlates with M1 and M2 marker expression and tumor grade. *Acta Neuropathol. Commun.* **2017**, *5*, 4. [CrossRef]
32. Amadio, S.; Montilli, C.; Magliozzi, R.; Bernardi, G.; Reynolds, R.; Volonte, C. $P2Y_{12}$ receptor protein in cortical gray matter lesions in multiple sclerosis. *Cereb. Cortex* **2010**, *20*, 1263–1273. [CrossRef]
33. Van Wageningen, T.A.; Vlaar, E.; Kooij, G.; Jongenelen, C.A.M.; Geurts, J.J.G.; van Dam, A.-M. Regulation of microglial TMEM119 and P2RY12 immunoreactivity in multiple sclerosis white and grey matter lesions is dependent on their inflammatory environment. *Acta Neuropathol. Commun.* **2019**, *7*, 206. [CrossRef] [PubMed]
34. Jay, T.R.; Miller, C.M.; Cheng, P.J.; Graham, L.C.; Bemiller, S.; Broihier, M.L.; Xu, G.; Margevicius, D.; Karlo, J.C.; Sousa, G.L.; et al. TREM2 deficiency eliminates TREM2+ inflammatory macrophages and ameliorates pathology in Alzheimer's disease mouse models. *J. Exp. Med.* **2015**, *212*, 287–295. [CrossRef] [PubMed]
35. Davies, D.S.; Ma, J.; Jegathees, T.; Goldsbury, C. Microglia show altered morphology and reduced arborization in human brain during aging and Alzheimer's disease. *Brain Pathol.* **2017**, *27*, 795–808. [CrossRef] [PubMed]

36. Boche, D.; Perry, V.H.; Nicoll, J.A.R. Review: Activation patterns of microglia and their identification in the human brain. *Neuropathol. Appl. Neurobiol.* **2013**, *39*, 3–18. [CrossRef] [PubMed]
37. Pereson, S.; Wils, H.; Kleinberger, G.; McGowan, E.; Vandewoestyne, M.; Van Broeck, B.; Joris, G.; Cuijt, I.; Deforce, D.; Hutton, M.; et al. Progranulin expression correlates with dense-core amyloid plaque burden in Alzheimer disease mouse models. *J. Pathol.* **2009**, *219*, 173–181. [CrossRef]
38. Mendsaikhan, A.; Tooyama, I.; Bellier, J.-P.; Serrano, G.E.; Sue, L.I.; Lue, L.-F.; Beach, T.G.; Walker, D.G. Characterization of lysosomal proteins Progranulin and Prosaposin and their interactions in Alzheimer's disease and aged brains: Increased levels correlate with neuropathology. *Acta Neuropathol. Commun.* **2019**, *7*, 215. [CrossRef]
39. Minett, T.; Classey, J.; Matthews, F.E.; Fahrenhold, M.; Taga, M.; Brayne, C.; Ince, P.G.; Nicoll, J.A.R.; Boche, D. Microglial immunophenotype in dementia with Alzheimer's pathology. *J. Neuroinflammation* **2016**, *13*, 135. [CrossRef]
40. Ransohoff, R.M. A polarizing question: Do M1 and M2 microglia exist? *Nat. Neurosci.* **2016**, *19*, 987–991. [CrossRef]
41. Walker, D.G.; Dalsing-Hernandez, J.E.; Campbell, N.A.; Lue, L.-F. Decreased expression of CD200 and CD200 receptor in Alzheimer's disease: A potential mechanism leading to chronic inflammation. *Exp. Neurol.* **2009**, *215*, 5–19. [CrossRef]
42. Moore, C.S.; Ase, A.R.; Kinsara, A.; Rao, V.T.S.; Michell-Robinson, M.; Leong, S.Y.; Butovsky, O.; Ludwin, S.K.; Seguela, P.; Bar-Or, A.; et al. $P2Y_{12}$ expression and function in alternatively activated human microglia. *Neurol. Neuroimmunol. Neuroinflamm.* **2015**, *2*, e80. [CrossRef] [PubMed]
43. Suh, H.S.; Choi, N.; Tarassishin, L.; Lee, S.C. Regulation of progranulin expression in human microglia and proteolysis of progranulin by matrix metalloproteinase-12 (mmp-12). *PLoS ONE* **2012**, *7*, e35115. [CrossRef] [PubMed]
44. Paquet, C.; Amin, J.; Mouton-Liger, F.; Nasser, M.; Love, S.; Gray, F.; Pickering, R.M.; Nicoll, J.A.R.; Holmes, C.; Hugon, J.; et al. Effect of active Abeta immunotherapy on neurons in human Alzheimer's disease. *J. Pathol.* **2015**, *235*, 721–730. [CrossRef] [PubMed]
45. Orr, A.G.; Orr, A.L.; Li, X.-J.; Gross, R.E.; Traynelis, S.F. Adenosine A_{2A} receptor mediates microglial process retraction. *Nat. Neurosci.* **2009**, *12*, 872–878. [CrossRef] [PubMed]
46. Suzuki, T.; Kohyama, K.; Moriyama, K.; Ozaki, M.; Hasegawa, S.; Ueno, T.; Saitoe, M.; Morio, T.; Hayashi, M.; Sakuma, H. Extracellular ADP augments microglial inflammasome and NF-kappaB activation via the $P2Y_{12}$ receptor. *Eur. J. Immunol.* **2019**. [CrossRef] [PubMed]
47. Gelosa, P.; Lecca, D.; Fumagalli, M.; Wypych, D.; Pignieri, A.; Cimino, M.; Verderio, C.; Enerback, M.; Nikookhesal, E.; Tremoli, E.; et al. Microglia is a key player in the reduction of stroke damage promoted by the new antithrombotic agent ticagrelor. *J. Cereb. Blood Flow Metab.* **2014**, *34*, 979–988. [CrossRef]
48. Webster, C.M.; Hokari, M.; McManus, A.; Tang, X.N.; Ma, H.; Kacimi, R.; Yenari, M.A. Microglial $P2Y_{12}$ deficiency/inhibition protects against brain ischemia. *PLoS ONE* **2013**, *8*, e70927. [CrossRef]
49. Drummond, E.; Nayak, S.; Faustin, A.; Pires, G.; Hickman, R.A.; Askenazi, M.; Cohen, M.; Haldiman, T.; Kim, C.; Han, X.; et al. Proteomic differences in amyloid plaques in rapidly progressive and sporadic Alzheimer's disease. *Acta Neuropathol.* **2017**, *133*, 933–954. [CrossRef]
50. Drummond, E.; Nayak, S.; Pires, G.; Ueberheide, B.; Wisniewski, T. Isolation of Amyloid Plaques and Neurofibrillary Tangles from Archived Alzheimer's Disease Tissue Using Laser-Capture Microdissection for Downstream Proteomics. *Methods Mol. Biol.* **2018**, *1723*, 319–334. [CrossRef]
51. Beach, T.G.; Sue, L.I.; Walker, D.G.; Roher, A.E.; Lue, L.; Vedders, L.; Connor, D.J.; Sabbagh, M.N.; Rogers, J. The Sun Health Research Institute Brain Donation Program: Description and experience, 1987–2007. *Cell Tissue Bank.* **2008**, *9*, 229–245. [CrossRef]
52. Beach, T.G.; Adler, C.H.; Sue, L.I.; Serrano, G.; Shill, H.A.; Walker, D.G.; Lue, L.; Roher, A.E.; Dugger, B.N.; Maarouf, C.; et al. Arizona Study of Aging and Neurodegenerative Disorders and Brain and Body Donation Program. *Neuropathology* **2015**, *35*, 354–389. [CrossRef] [PubMed]
53. Newell, K.L.; Hyman, B.T.; Growdon, J.H.; Hedley-Whyte, E.T. Application of the National Institute on Aging (NIA)-Reagan Institute criteria for the neuropathological diagnosis of Alzheimer disease. *J. Neuropathol. Exp. Neurol.* **1999**, *58*, 1147–1155. [CrossRef] [PubMed]

54. McKeith, I.G.; Dickson, D.W.; Lowe, J.; Emre, M.; O'Brien, J.T.; Feldman, H.; Cummings, J.; Duda, J.E.; Lippa, C.; Perry, E.K.; et al. Diagnosis and management of dementia with Lewy bodies: Third report of the DLB Consortium. *Neurology* **2005**, *65*, 1863–1872. [CrossRef]
55. Beach, T.G.; Sue, L.I.; Walker, D.G.; Sabbagh, M.N.; Serrano, G.; Dugger, B.N.; Mariner, M.; Yantos, K.; Henry-Watson, J.; Chiarolanza, G.; et al. Striatal amyloid plaque density predicts Braak neurofibrillary stage and clinicopathological Alzheimer's disease: implications for amyloid imaging. *J. Alzheimer's Dis.* **2012**, *28*, 869–876. [CrossRef] [PubMed]
56. Bustin, S.A.; Benes, V.; Garson, J.A.; Hellemans, J.; Huggett, J.; Kubista, M.; Mueller, R.; Nolan, T.; Pfaffl, M.W.; Shipley, G.L.; et al. The MIQE guidelines: Minimum information for publication of quantitative real-time PCR experiments. *Clin. Chem.* **2009**, *55*, 611–622. [CrossRef] [PubMed]

International Journal of
Molecular Sciences

MDPI

Review

The Use of Antimicrobial and Antiviral Drugs in Alzheimer's Disease

Umar H. Iqbal, Emma Zeng and Giulio M. Pasinetti *

Department of Neurology, Icahn School of Medicine at Mount Sinai, New York, NY 10029, USA; umar.iqbal@mssm.edu (U.H.I.); emma.zeng@mssm.edu (E.Z.)
* Correspondence: giulio.pasinetti@mssm.edu; Tel.: +1-212-241-7938

Received: 29 May 2020; Accepted: 10 July 2020; Published: 12 July 2020

Abstract: The aggregation and accumulation of amyloid-β plaques and tau proteins in the brain have been central characteristics in the pathophysiology of Alzheimer's disease (AD), making them the focus of most of the research exploring potential therapeutics for this neurodegenerative disease. With success in interventions aimed at depleting amyloid-β peptides being limited at best, a greater understanding of the physiological role of amyloid-β peptides is needed. The development of amyloid-β plaques has been determined to occur 10–20 years prior to AD symptom manifestation, hence earlier interventions might be necessary to address presymptomatic AD. Furthermore, recent studies have suggested that amyloid-β peptides may play a role in innate immunity as an antimicrobial peptide. These findings, coupled with the evidence of pathogens such as viruses and bacteria in AD brains, suggests that the buildup of amyloid-β plaques could be a response to the presence of viruses and bacteria. This has led to the foundation of the antimicrobial hypothesis for AD. The present review will highlight the current understanding of amyloid-β, and the role of bacteria and viruses in AD, and will also explore the therapeutic potential of antimicrobial and antiviral drugs in Alzheimer's disease.

Keywords: Alzheimer's disease; amyloid-β; antimicrobial; antiviral; antimicrobial peptide

1. Alzheimer's Epidemiology

Alzheimer's disease (AD) is a progressive neurological disorder that accounts for the greatest number of dementia cases. As of 2019, 5.8 million people were living with AD, with its prevalence predicted to increase to 13.8 million by 2050 [1]. The vast majority of cases are concentrated in ages over 65, impacting 10% of people in this age group. In addition, the economic toll of AD on the United States economy is significant, estimated to be roughly USD 290 billion in 2019 [1]. As the number of cases is only expected to rise over the coming decades, research in this field is critical in order to understand the pathology of this disease, as well as potential therapeutics.

2. Timeline/Characterization of Alzheimer's Disease

The understanding and characterization of AD can be traced back over 100 years to Alois Alzheimer, from whom the disease takes its name. After completing an autopsy of a patient with progressive dementia, Alzheimer noticed a severe amount of cortical degeneration and an accumulation of protein deposits, specifically extraneuronal plaques and intraneural tangles [2]. By 1991, the buildup of extraneuronal amyloid-β (Aβ) plaques became the hallmark trait of the pathogenesis of Alzheimer's disease [3], initiating the development of the amyloid cascade hypothesis [4]. In parallel to Aβ plaques formation, the accumulation of other naturally unfolded proteins is central to AD and other cerebral proteopathies [5]. The intracellular aggregation of tau proteins in the form of neurofibrillary tangles (NFTs) is also an essential trait in the pathogenesis of Alzheimer's disease [3,6]. A recent 2020 study found that neuroinflammation could play a role in the aggregation of tau, as DNA extracted from various bacterial species promoted tau misfolding [7]. Whereas Aβ plaques are more critical to AD pathogenesis, the tau protein appears to be more responsible for subsequent cognitive impairment and dementia symptoms associated with AD [8]. Indeed, tau hyperphosphorylation and NFT levels are closely correlated with cognition, and exhibit potential as therapeutic targets for AD treatment [9]. Furthermore, tau protein production has been shown to have a positive correlation with the production of Aβ plaques [10], with the formation and lack of clearance of Aβ plaques also being proposed to induce tau protein formation into NFTs [11]. Coupling this with the bi-directional relationship between Aβ plaques and neuroinflammation [12] would cement Aβ's key role in driving AD pathology.

2.1. *Amyloid-β Generation*

Aβ formation begins with the breakdown of the amyloid precursor proteins (APP) embedded in the membranes of cells, such as neurons, as a type 1 transmembrane glycoprotein [13]. Aβ peptides are produced through a two-step cleavage process, in which APP is metabolized into smaller fragments. In the first step, APP is cleaved by β-secretase 1 into a membrane-bound CTFβ fragment (containing 99 amino acids) and an extracellular fragment sAPPβ. CTFβ is then further cleaved by γ-secretase to create the final Aβ peptide [13–15], as illustrated in Figure 1. The length of Aβ peptides is not fixed, and can consist of anywhere between 37 and 49 amino acids, depending on where the cleavage was done by β-secretase 1 and γ-secretase [16,17]. The most abundant length is A$\beta_{1\text{–}40}$, representing approximately 80–90% of Aβ peptides, whereas the least soluble of the Aβ peptides, A$\beta_{1\text{–}42}$, represents roughly 5–10% [13,15]. As a greater number of Aβ peptides form, they begin to aggregate into oligomers, which then form fibrils, and eventually the insoluble plaques characteristic of AD [13]. Of the different isoforms of Aβ peptides, A$\beta_{1\text{–}40}$ and A$\beta_{1\text{–}42}$ are the most common in plaques. Regarding the comparative role of A$\beta_{1\text{–}40}$ and A$\beta_{1\text{–}42}$ peptides in the pathogenesis of AD, A$\beta_{1\text{–}42}$ peptides have been found in higher concentrations in AD. Furthermore, A$\beta_{1\text{–}42}$ peptides have been found to be more prone to forming insoluble amyloid fibrils than A$\beta_{1\text{–}40}$ [18]. This is further supported by a study using transgenic mice that expressed either A$\beta_{1\text{–}40}$ (BRI- Aβ40) or A$\beta_{1\text{–}42}$ (BRI- Aβ42A). The authors found that the mice that selectively expressed A$\beta_{1\text{–}40}$ did not develop AD pathology at any age. However, the same did not hold true for BRI- Aβ42A mice, which had developed Aβ deposits [19]. In addition, another study found A$\beta_{1\text{–}42}$ peptides to promote Aβ plaques formation, and A$\beta_{1\text{–}40}$ to decrease Aβ deposition [20]. These findings would indicate the key role A$\beta_{1\text{–}42}$ peptides play in the pathogenesis of AD.

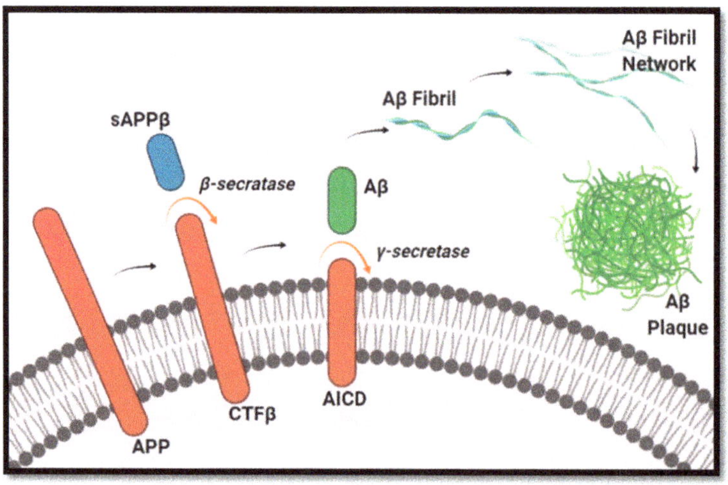

Figure 1. Generation of Aβ plaques.

2.2. Current Interventions and Limitations

Even with the tremendous effort that has been put into developing potential therapies for AD over the past few decades, there has been little success in reaching an effective therapy, with no new drug being approved in over a decade. While cholinesterase inhibitors and memantine are FDA-approved drugs for AD, and do address some of its symptoms, they lack the ability to attenuate disease progression. Over the past 20 years, a majority of the therapies have been based on the amyloid cascade hypothesis, and hence have focused on depleting Aβ peptides. Theses therapies often aim to inhibit γ-secretase or β-secretase activity in order to limit Aβ peptide production. Therapies that use such methods have, however, seen an increase in the rate of infection during clinical studies, with one study seeing 6% of participants develop meningoencephalitis [21,22]. Tarenflurbil, for instance, had been administered clinically, after it was shown to modulate γ-secretase and increase production of the less toxic $A\beta_{1-38}$ peptide, rather than $A\beta_{1-42}$ peptide [23,24]. However, in addition to not showing any significant benefit in individuals with mild AD, participants in the treatment group experienced an increase in upper respiratory infections and dizziness compared to the placebo group [25]. Additionally, in the time of COVID-19 infection, respiratory-related side effects, such as the ones related to tarenflurbil, are of growing concern. The γ-secretase inhibitor Semagacestat has not only been associated with increased levels of infection, but also with a failure to provide any cognitive improvement in patients with probable AD [26]. Furthermore, when patients with mild to moderate AD were administered ELND005, a compound that inhibits Aβ fibrils and plaque formation, it was observed that higher doses of this treatment led to serious infection. This led to lower dosage recommendations for future trials [27,28]. Lastly, the B-site ABPP cleaving enzyme 1 (BACE1)-inhibitor E2609 has also been associated with oralabial herpes relapse [29,30]. A rise in infection occurring in tandem with the reduction of Aβ peptide production could indicate these peptides' potential role in immune function.

Through these past clinical trials, it is evident that therapies largely based on the amyloid cascade hypothesis, which in turn aim to eradicate Aβ peptides in the brain, have historically been ineffective. These failures could imply that current approaches either intervene at a stage that is too late, or possess a therapeutic target that is not as relevant to disease progression [31]. This would make sense in the context of AD especially, as Aβ deposition occurs 10–20 years before the occurrence of clinical symptoms [32]. Therefore, these treatments that target Aβ peptides specifically may already be too late. To create a successful therapy, it may be necessary to consider intervention in the presymptomatic stage of the AD instead. To do so, it would be crucial to identify biomarkers for early identification of AD. In March of 2016, a meeting was convened in which international, interdisciplinary experts identified a list of biomarkers that could be used for identifying AD early on [33]. CSF levels of $A\beta_{1-42}$ and $A\beta_{1-40}$, and the ratio of $A\beta_{1-42}/A\beta_{1-40}$, were determined to be among the candidates [33,34]. In addition, plasma levels of the same biomarkers were determined to decrease in AD patients when compared to healthy subjects [35]. Other biomarkers to consider would include plasma levels of tau protein and neurofilament light [33]. Plasma levels of the latter have been shown to be able to detect neurodegeneration in presymptomatic AD [36]. Other highly sensitive methods that have shown promise in the early identification of AD include protein misfolding cyclic amplification (PMCA) and real-time quaking-induced conversion (RT-QuIC), to determine Aβ oligomers and tau protein levels in CSF [37,38]. Taking a more preventative approach to AD treatment, and understanding and addressing factors that contribute to the progression of AD, may be favorable for identifying therapeutic targets earlier in the disease. Based on evidence of the association between bacterial/viral infection and AD progression, the antimicrobial hypothesis suggests that Aβ peptides may be produced as a protective mechanism by the innate immune system, and act as an antimicrobial peptide (AMP) against foreign agents. If Aβ peptides do in fact play a beneficial role in immunity, then the aim of treatment should not be to eradicate the compound entirely. Rather, it should be to target the root cause of its over production, and reduce its deleterious effects and general neuroinflammation in AD.

3. Antimicrobial Protection Hypothesis

3.1. Role of Microorganisms and Viruses in AD and Aβ Generation

Neuroinflammation is inflammation within the brain or spinal cord due to infection, toxins or injury [39]. In the brain specifically, resident glial cells, such as microglia and astrocytes, along with endothelial cells and mast cells, all aid in defending the brain against foreign pathogens [40]. Microglia, the main immune effector cells of the central nervous system (CNS), are constantly surveying their environment for potential threats to the brain [41,42]. When an invading agent is detected, microglia change into an activated state, characterized by an enlarged soma and the production of inflammatory cytokines and chemokines [39]. Astrocytes also play a critical role in mediating neuroinflammation as they are responsible for many neuroprotective functions, such as maintaining blood brain barrier (BBB) integrity and buffering neurotransmitters [43]. Upon injury, astrocytes likewise undergo morphological changes, and exhibit increased reactivity and secretion of cytokines and chemokines [41]. While acute inflammatory responses are common to healthy individuals, chronic inflammation is damaging to the natural balance of pro- and anti-inflammatory signaling in the brain, and can lead to the development and progression of neurodegenerative diseases like AD [39].

Over the years, there has been increasing evidence of neuroinflammation's role in AD. In addition to Aβ plaques and NFTs, markers of sustained inflammation and microglial activation have repeatedly been found in AD brain samples [39]. The cytokines interleukin 1 and interleukin 6 are especially elevated [44]. One source of the neuroinflammation in AD patients could be the response to invading microorganisms and viruses. In fact, researchers have found evidence pointing to the presence of pathogens, such as viruses, bacteria and fungi, in AD brains [45–50]. This notion draws parallels with the measles virus, which can lead to the development of the neurological disease known as subacute sclerosing panencephalitis [51]. These findings include the identification of viral [52] and bacterial DNA in post-mortem brain samples, and the detection of pathogens [53] and/or their respective antibodies [54] in the serum or cerebrospinal fluid of patients. Furthermore, detection of lipopolysaccharide is commonly used by researchers to measure the presence of Gram-negative bacteria, like *P. gingivalis* specifically, as it is found in their cell walls and can stimulate an inflammatory response in the immune cells [42]. Herpes simplex virus-1 (HSV-1) [52,55–58] was the first pathogen found to be present in AD brain samples [59], and it thereafter became the most widely-researched pathogen regarding the linkage between viral infection and AD. Since then, other viruses have been identified in leading to the progression of AD, including human cytomegalovirus [54] and Epstein Barr Virus [60]. A recent 2018 study found that, in addition to HSV-1, herpesvirus types HHV-6 [46,47] and HHV-7 were highly present in AD patients [61]. In this study, by Readhead et al., HHV-6 and HHV-7 were also observed to be involved in regulatory processes critical to characteristic features of the disease [61].

Bacterial infection has likewise been associated with the progression of AD. The presence of bacteria in the brain has been determined in previous studies, suggesting the presence of a brain microbiome [62–64]. Even though bacterial presence has been seen in the brains of healthy individuals, tissue samples from AD brains have greater levels of bacterial species [64], indicating a greater level of infiltration. *Chlamydia pneumoniae* is the most widely-studied bacteria regarding association to AD [25,49]. A clinical investigation, made up of a healthy control group and an AD group, detected *C. pneumoniae* in 90% of the AD patients, whereas the control group were all negative [53]. *Escherichia coli*, likewise identified in AD brains [65], has been found to be capable of synthesizing extracellular amyloid [66]. Stains such as *Borrelia burgdorferi* [67], spirochetes [48], and *Porphyromonas gingivalis*, a pathogen commonly linked to chronic periodontitis, have also been identified in AD brain samples [68].

Interestingly, fungal infection from species primarily associated with periodontal disease has recently been suggested to be involved in AD progression. Researchers in a 2014 study detected multiple fungal species in AD brain samples, including *Saccharomyces cerevisiae*, *Malassezia globosa*, *Malassezia restricta*, *Penicillium* and *Phoma* [69]. Pisa et al. have since followed up on this initial discovery by analyzing the presence of these species between brain regions [70]. With any of these studies that have been conducted, however, it is important to recognize the technical limitations that arise when studying microorganisms and neurodegenerative disease. Many of these studies are limited to the use of post-mortem brain samples, and thus present the risk of contamination due to death or the passage of microbes from other areas of the body, such as the gut to the brain, due to the lack of a functioning BBB to prevent this leakage.

3.2. *Invasion of the CNS and Role in Aβ Generation of AD-Associated Microorganisms and Viruses*

Depending on the organism, there are several ways that pathogens can infiltrate the CNS and potentially further the progression of AD. The first is through a compromised BBB. Whereas a healthy and functional BBB normally provides a selective barrier to the passage of cells and molecules into the brain, a compromised BBB can allow direct entry into the cerebral spinal fluid via the bloodstream [21]. This places aging populations and those with weakened immune responses especially at risk, as some viruses, such as herpesvirus, can remain latent after initial infection and then reactivate in aging individuals long after, to introduce delayed adverse complications [71]. Even with a healthy BBB, however, bacteria and viruses are still able to be introduced into the brain through various mechanisms. HIV, for example, is carried from the immune system to the brain by infected leukocytes that are able to cross the BBB. P. gingivalis and other oral spirochetes have also been suggested to be capable of invading the CNS via the oral cavity, through the trigeminal nerves and ganglia [42]. Additionally, pathogens such as bacteria and viruses can bypass the BBB altogether by entering through the olfactory system, as the nasal cavity connects the peripheral environment to brain regions such as the olfactory bulb [72], the entorhinal cortex and the hippocampus, which traditionally receive smell sensory signals. *C. pneumoniae*, a respiratory pathogen, has specifically been suggested to enter the brain through the olfactory system, with its presence detected in the entorhinal cortex and hippocampal formation of AD patients [73]. A recent 2020 study model demonstrated that exposure to *C. pneumoniae* via the olfactory system was sufficient to induce Aβ plaque and NFT formation in the olfactory cortex and hippocampus of immunocompromised individuals [74]. This is further evidenced by Little et al., who found that intranasal inoculation of *C. pneumoniae* was sufficient to induce AD-like traits in mice [75]. Once in the brain, there are several ways in which these pathogens contribute to the Aβ production that is characteristic to AD. One mechanism is through the alteration of gene expression. The study previously mentioned by Readhead et al. found that HHV-6 and HHV-7 interact with known regulatory genes responsible for amyloid processing, such as the amyloid beta A4 precursor protein-binding family (*APBB2*), clusterin (*CLU*), and gamma-secretase subunit presenilin-1 (*PSEN1*) [61]. Similarly, infection of *C. pneumoniae* within human neuronal cell cultures possibly alters calcium-related gene expression such that they express patterns similar to those reported in AD brain samples [76]. Another way viruses can influence Aβ production is through protein misfolding. Specifically, viruses such as HHV, cytomegalovirus and Epstein Barr Virus have been shown to contain prion-like domains that may trigger the misfolding of proteins like Aβ [77]. Through these varying mechanisms, many microorganisms and viruses have been found to initiate Aβ plaque formation.

In-vivo studies have noted a correlation between viral and bacterial infections and the accumulation of Aβ peptides. Mice infected with HSV-1 [78], pseudorabies virus [79], *C. pneumoniae* [75] and *P. gingivalis* [68,80] were found to have a significantly increased level of $A\beta_{1\text{–}42}$ in the brain. In addition, Aβ expression was found to be upregulated in rats exposed to bacterial pathogens. HSV-1 has also been found to infect the hippocampus region at a greater rate; the same area found to have greater amounts of Aβ plaques in AD [81]. In vitro studies observed cells co-cultured with either HSV-1, HSV-2, *P. gingivalis* or *B. burgdoferi* to have increased intracellular concentrations of Aβ [80,82–89]. Furthermore, HSV-1 has been associated with the inhibition of the non-amyloidogenic pathway of APP metabolism, and the increased expression of β-secretase. This is evidenced by a 2011 study indicating the direct and frequent interaction between HSV-1 and AβPP [90]. Aβ plaques have also been identified in the brains of HIV-1-infected individuals. Autopsies performed on 162 HIV positive individuals found roughly half of them to contain Aβ plaques [91]. Cell culture studies observed an increase in Aβ production and secretion following exposure to mRNA and proteins from the HIV Nef gene [92].

It must be noted, however, that the association between infection and Aβ plaque formation is not consistent across all populations. For example, in the study previously mentioned by Sundar et al., younger and healthier individuals exposed to *C. pneumoniae* did not exhibit the same AB peptide and NFT formation as older and immunocompromised subjects [74]. Moreover, it has been observed that genetic discrepancies, especially in the APOE gene, influence one's susceptibility to HSV-1 infection and subsequent AD development. Specifically, the APOEε4 allele places individuals at a greater risk of developing HSV-1-associated AD, with a combination of APOEε4 and HSV-1 comprising 60% of all AD cases [93,94]. This finding has been recapitulated in animal studies, where mice with the APOEε4 allele display a greater viral load than mice with other allele types after HSV-1 infection [95].

3.3. *Aβ as an Antimicrobial Peptide*

It is clear that a relation exists between bacterial and viral infections and Aβ production rate, as described in the previous section. Aβ peptides have long been thought to lack any physiological function; however, this notion has been challenged in recent years. Clinical studies have observed the depletion of Aβ peptides, through anti-Aβ therapies, to increase the rate of infections in some participants. Furthermore, Aβ plaques have been found to contain microbial and viral DNA, such as HSV-1. One study identified HSV-1 virus DNA in roughly 90% of Aβ plaques [96]. In addition, AD brains have been associated with 5- to 10-fold increases in bacterial read compared to control brains [64]. In the presence of bacterial lipopolysaccharides, microglial cells have also been shown to upregulate Aβ production [97]. From these and similar findings, it has been suggested that the pathogenesis of AD could be triggered by viral and/or microbial infections. These observations led to the recent development of the antimicrobial protection hypothesis for AD, which explores the notion of Aβ peptides having a role in innate immunity as an AMP that aids in the entrapment and degradation of invading bacteria and viruses.

The innate immune system utilizes AMPs to target invading microorganisms, such as bacteria, viruses, fungi, and in some instances cancerous cells. Mammalian AMPs exist in three main families: defensins, histatins and cathelicidins [98]. Similar to how Aβ peptides are generated through the two-step cleavage of APP, AMPs are also formed from the breakdown of larger precursor proteins. Examples of amyloid AMPs that have a role in immunity are present in the human body. Amyloidogenic major basic protein-1 (MBP-1) is implemented in eosinophils against pathogens [99]. Like Aβ peptides, MBP-1 also forms aggregates, specifically at the surface of the bacteria to limit its spread. Further support of Aβ peptides being AMP stems from their similarity to AMP LL-37, the only cathelicidin identified thus far in humans. Both compounds exhibit tendencies to form cytotoxic soluble oligomers and insoluble fibrils, characteristic features of tinctorial amyloid [98]. Additionally, deficiency in the latter can result in Kostmann syndrome, an immunodeficiency disorder that, if left untreated, can result in death due to infection within the first year of life [81]. High levels of LL-37 are likewise dangerous, as they has been associated with the development of plaques in atherosclerosis and other non-infectious diseases [98]. Protein analyses comparing known AMPs and Aβ peptides demonstrate structural similarities between these peptides, as well sequential similarities pointing to a shared homology between Aβ peptides and a specific family of bacteriocins [100]. This is particularly notable as bacteriocins are traditionally synthesized by bacteria as part of an antimicrobial response to contact with closely-related strains [101]. If verified as an AMP, Aβ would not be the only AMP suggested to be involved in AD. Often expressed in epithelial cells, β-defensin-1 is significantly elevated in astrocytes of the hippocampus, the choroid plexus, and the granulovacuolar degeneration structures of AD brain samples [102].

In vitro studies suggest the ability of Aβ peptides to be an AMP, and inhibit growth of a number of bacteria and viruses. In respect to the latter, Aβ has been shown to have antiviral activity against both HSV-1 and the influenza virus A by inhibiting the infectivity of HSV-1 [82], influenza virus A [103], H3N2 [103] and H1N1 [103]. Researchers found that in mouse and human neural cell cultures, Aβ peptide deposition was accelerated in response to HSV-1 and HHV6 infection, with the oligomers binding to the viruses as part of a protective entrapment mechanism [104]. Aβ peptides have also been shown to have antimicrobial properties against both Gram-positive and Gram-negative bacteria, including *Enterococcus faecalis*, *Escherichia coli*, *Listeria monocytogenes*, *Salmonella typhimurium*, *Staphylococcus aureus*, *Staphylococcus epidermidis*, *Streptococcus agalactiae* and *Streptococcus pneumoniae* [30]. In a study comparing Aβ peptides and LL-37, Soscia et al. found that, against eight different bacteria and viruses, Aβ peptides demonstrated an antimicrobial activity equivalent to, and sometimes even greater than, the known AMP LL-37. The same study also found that, when comparing the brain homogenates of Aβ-enriched areas from both AD and non-AD brains, the AD brain samples had elevated antimicrobial and antiviral activity. These discrepancies were eliminated once the AD brain tissues were immunodepleted using anti-Aβ antibodies. It is important to consider the sequence length of the Aβ peptide when examining its antimicrobial capabilities, as $A\beta_{1-42}$ was shown to be capable of binding to the surface of bacteria and aggregating into clusters, whereas other peptide lengths were not [105].

Animal studies have further evidenced the important potential role that Aβ peptides have in protecting against infections. In a 2016 study, researchers tested Aβ's functionality as an AMP in mice and nematode models [106]. Kumar et al. found that transgenic 5XFAD mice, which constitutively express human Aβ peptides, survived significantly longer than wild-type mice after the injection of *Salmonella typhimurium* into their brains. The 5XFAD mice were observed to have accelerated Aβ deposition that closely co-localized with the bacteria, reducing their cerebral viral load compared to wild-type mice. These findings were recapitulated again in worm models, as nematodes expressing Aβ were found to have increased survival following fungal infection of *C. albicans*, compared to nematodes that did not produce Aβ peptides. Additionally, it has been observed that impaired mice that lacked the ability to generate Aβ peptides have been shown to have increased postpartum mortality, which was only reversed by maintaining a sterile environment [107]. Furthermore, APP knockout mice were also observed to have increased rates of mortality. Altogether, these studies support the notion that Aβ peptides are an AMP, and an integral part of the brain's innate immune response against invading pathogens.

The mechanism by which Aβ peptides have been suggested to exert their antimicrobial and antiviral effect has been based on entrapment and lytic activity, as illustrated in Figure 2. Being a self-complimentary peptide containing two distinct hydrophobic and hydrophilic surfaces, Aβ peptides can self-assemble into oligomers. As oligomerization continues, a fibril network is created which targets, captures and agglutinates microbes, limiting their proliferation and impact on their environment. Aβ peptide affinity towards microbes has been suggested to be due to its positive charge and the microbe's negatively charged membrane [30,108,109]. Furthermore, the ability of Aβ peptides to agglutinate microbes stems from its heparin-binding activity, which is able to target carbohydrates present on the surface of microbes. Once entrapped, it is suggested that Aβ peptides induce cell membrane disruption by forming cation channels. These channels cause ion dyshomeostasis and subsequent cell death. This mechanism is similar to the activity observed in AMPs such as LL-37, which assert protection through microbial agglutination and entrapment. The oligomerization activity observed with Aβ peptides is a common trait of AMPs, which mediates their ability to entrap and lyse pathogens while maintaining their resistance to protease activity. The entrapment of microbes and viruses can also enhance their uptake by neutrophils and macrophages. In respect to HSV-1, Aβ peptides have been proposed to interfere with its ability to fuse with the plasma membrane of cells, hindering its infective ability [110].

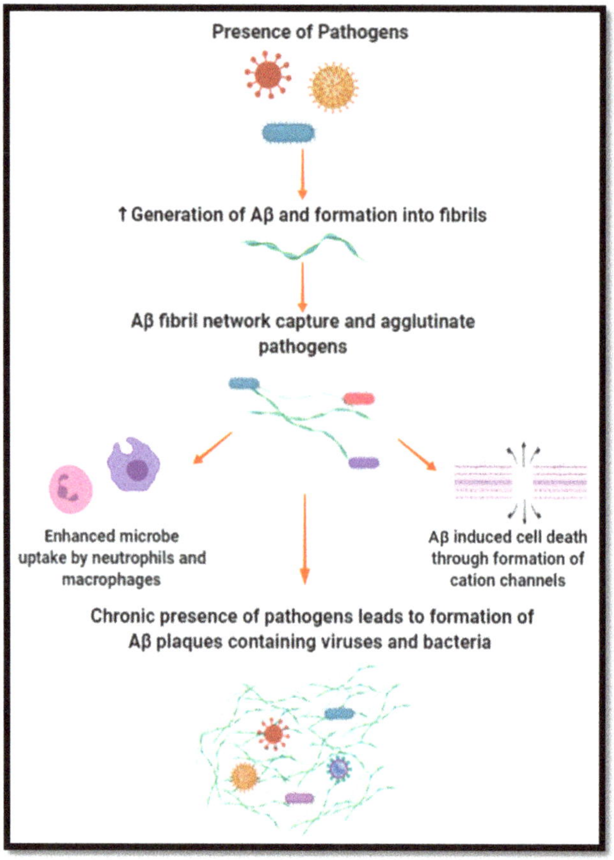

Figure 2. Aβ as an antimicrobial peptide.

4. Potential Novel Therapeutics

In aligning with the antimicrobial hypothesis of AD, the use of antimicrobial and antiviral therapeutics could prove to be effective in targeting the root cause of AD. It should be noted, however, that the chronic over-production of Aβ peptides, which form numerous insoluble plaques, would also need to be addressed. As such, the primary aim of these drugs would be to target bacteria and viruses, but a secondary aim would be to reduce the already-present Aβ burden that the brain is under.

4.1. Antiviral Drugs

4.1.1. Acyclovir

Acyclovir is an antiviral drug that is used for HSV-1 infections, which has been found to be well tolerated and safe [111]. By inhibiting the virus's replication, this antiviral agent is able to reduce the viral load exhibited by HSV-1. Administration of acyclovir in HSV-1-infected cells has been found to cause a significant reduction of HSV-1 proteins, along with a reduction of roughly 28% in Aβ accumulation, compared to untreated cells. The study associated this with reduced levels of β-secretase and a component of γ-secretase which metabolizes APP into Aβ [112]. Acyclovir administration has also shown to prevent HSV-1-related neuronal death [113]. In relation to the cognitive impairment observed in AD, a study by Hui et al. investigated the effects of acyclovir on Aβ oligomer-induced spatial cognitive impairments. The study found that the co-administration of acyclovir with dexamethasone attenuated impairments in spatial cognition. Furthermore, this combination reduced the levels of neuroinflammation markers such as TNF-α and IL-6, along with microglia activation. Interestingly, the study found these effects to only occur when acyclovir and dexamethasone were administered together [114].

4.1.2. Penciclovir

Penciclovir is another antiviral drug that targets HSV-1 DNA replication by blocking chain elongation. Cell cultures infected with HSV-1 displayed a reduction of virus and Aβ accumulation when penciclovir was administered. This was paralleled with a reduction in β-secretase and a component of γ-secretase [112].

4.1.3. Foscarnet

Foscarnet has been tested for its ability to reduce HSV-1 levels in vitro. A study found it was able to reduce Aβ accumulation, although only at higher doses. It was also unable to significantly reduce virus levels. Furthermore, foscarnet was not as effective as acyclovir or penciclovir, and hence currently is not seen as the optimal antiviral drug available for AD [112].

4.1.4. Valacyclovir

Valacyclovir, an antiviral medication used in HSV-1 and HSV-2 infections, has been determined to positively impact cognition by improving visual object learning, verbal memory and working memory in patients with schizophrenia [115]. Due to its effects on working memory, its effectiveness against HSV-1 and HSV-2, and its generally safe consumption, valacyclovir has been suggested as a potential therapeutic for AD. A clinical study is currently underway in which patients that both have mild AD and tested positive for HSV-1 or HSV-2 will receive valacyclovir. The aim of the study is to determine the impact of this treatment on cognition and the accumulation of amyloid and tau [116].

4.1.5. Bay 57-1293

Numerous studies have determined the antiviral agent Bay 57-1293 to be effective in combating HSV-1 [117–121]. By targeting the helicase–primase complex, Bay 57-1293 can inhibit viral DNA replication, and has been found to be more potent than acyclovir. The severity and frequency of recurring HSV was also found to be reduced by use of this drug [121]. Furthermore, it was able to decrease levels of Aβ and reduce P-tau production in Vero cells infected with HSV-1 [117].

4.1.6. Biflavonoids

Our lab has investigated the use of bioflavonoids, including ginkgetin, isoginkgetin and ginkgolic acid, derived from the leaves of Ginkgo biloba. The antiviral capabilities of these compounds has been well established in previous studies [122–126]. Hayashi et al. determined ginkgetin to successfully inhibit the viral replication of HSV-1, HSV-2 and the human cytomegalovirus, while also suppressing viral protein synthesis [122]. Additionally, a study by Miki et al. found ginkgetin to have anti-influenza virus activity [123]. Ginkgetin has been studied for use in AD by Zeng et al., who administered the drug to APP/PS1 transgenic mice. They observed a significant reduction in Aβ plaques and an improvement in inflammation [127]. Borenstein et al. have demonstrated the ability of ginkgolic acid to limit virus infectivity by inhibiting its fusion. The study found ginkgolic acid to be successful in inhibiting HSV-1, human cytomegalovirus and zika virus. Furthermore, it was effective in inhibiting viral protein synthesis and genome replication, in HSV-1 and human cytomegalovirus, respectively [124]. Ginkgolic acid has also demonstrated antimicrobial properties, specifically against *E. coli* and *Staphlylococcus aureus* [125]. Isoginkgetin has been shown to provide neuroprotection against the cytotoxic effects of excessive Aβ accumulation [128,129], while also having anti-microbial and anti-fungal activity [126]. Our lab's preliminary work in testing these three compounds in AD determined their effectiveness in reducing Aβ load in vitro, further supporting their therapeutic potential in AD.

4.2. Antimicrobial Drugs

4.2.1. Doxycycline

Doxycycline is a tetracycline antibiotic that has been studied for its therapeutic efficacy in AD models. Contrary to other tetracyclines, doxycycline has been determined to be safe and is able to penetrate the BBB [130], allowing it to exert its effect directly in the CNS. In vivo models, in which doxycycline was administered to mice, observed its accumulation in amyloid deposits, including Aβ plaques [131]. With respect to the production and formation of Aβ oligomers, it was observed that although doxycycline administration in transgenic mice did not cause a shift in Aβ monomers, there was a significant reduction in Aβ 18-mer levels when compared to control [132]. The same study also observed a significant memory recovery in animals that received treatment; however, there was no reduction in Aβ plaque size [132]. The paper suggested this was possibly due to the short two-month period of the study, as a previous three-month study found plaque size to be significantly reduced [132]. In respect to neuroinflammation, a reduction in microglia activation has also been associated with doxycycline administration [132]. A drosophila model, which administered doxycycline to $A\beta_{1-42}$-expressing flies, observed that the treated group's locomotor deficits developed slower than the control group. The same study also observed doxycycline administration to be associated with reduced Aβ fibrilization, suggesting the production of smaller amyloid structures [133]. Another study associated doxycycline with the destabilization of Aβ fibrils [134]. Clinical trials, however, were not as successful. One study, which administered doxycycline and rifampicin, observed improvements in cognitive function, as assessed by the Standardized Alzheimer's Disease Assessment Scale–Cognitive Subscale (SADAScog) score [135]. However, a second study did not find any improvements in the cognition or function of patients with mild to moderate AD with doxycycline/rifampicin administration [136]. Further investigations would be needed to understand why the benefits seen in murine models do not translate into clinical trials.

4.2.2. Propranolol

Propranolol hydrochloride, an antihypertensive drug shown to have antimicrobial properties [137], has also been found to impact Aβ production. Cortico-hippocampal neuronal cultures treated with this drug manifested reduced levels of Aβ production. Furthermore, the one-month treatment of Tg2576 mice resulted in roughly a 40% reduction of $A\beta_{1-40}$ and $A\beta_{1-42}$ levels in the brain. When administered over a period of 6 months, Aβ peptide levels were still reduced in the brain; however, no improvement in spatial memory function was observed [138].

4.2.3. Rifampicin

Rifampicin is an antibiotic derived from *Nocardia mediterranei*, which has been investigated for use in neurodegenerative diseases such as Parkinson's and AD [139,140]. Rifampicin has been found to provide neuroprotection through its anti-oxidant and anti-inflammatory properties [139,141]. Furthermore, in vitro studies found that its administration improved neuronal survival and reduced microglial activation [141]. Studies by Tomiyama et al. found rifampicin to protect neurons from cytotoxicity by scavenging free radicals [142,143]. In relation to the antimicrobial hypothesis, rifampicin has been previously studied for use in bacterial cerebral infections [144]. As rifampicin is able to cross the BBB [144], it can exert its antimicrobial effect directly in the brain. In the presence of rifampicin, a reduction of Aβ fibril formation [142] has been observed in addition to augmented Aβ clearance [145]. A study by Umeda et al., in which rifampicin was administered to APPOSK mice, found the treatment to reduce Aβ accumulation, provide synaptic protection, and reduce microglial activation [146].

Clinical studies exploring the impact of rifampicin on cognitive function have also been investigated, as mentioned in previous sections. Even with its many benefits, the oral intake of rifampicin has also been associated with liver injury in humans. To circumvent this limitation, administering rifampicin intranasally or subcutaneously has been suggested [147]. These routes of rifampicin administration have been shown to be more effective in improving memory than oral administration [147].

4.2.4. Gingipain Inhibitors

The use of gingipain inhibitors in AD is another approach that has been taken to alleviate the negative impact of the disease. Gingipains are virulence factors that are produced by *P. gingivalis* [148]. They are made up of a group of cysteine proteinases, specifically arginin–gingipain A, arginine–gingipain B, and lysign–gingipain [148,149]. Given the key role gingipains play in host colonization [148] and the inactivation of host defenses [150–152], they are essential for the survival and pathogenicity of *P. gingivalis*. Regarding $A\beta_{1-42}$ peptide production, *P. gingivalis* infection was found to increase $A\beta_{1-42}$ levels. Furthermore, incubating *P. gingivalis* with $A\beta_{1-42}$ peptides led to a significant increase in *P. gingivalis* death. These two findings further support the antimicrobial hypothesis for Aβ peptides [153]. Gingipain inhibitors, such as COR286, COR271 and COR388, have been found to be effective in inducing *P. gingivalis* death and reducing the bacterial load in the brain, more so than other antibiotics, such as moxifloxacin [153,154]. In addition, COR271 was found to provide some level of neuroprotection as well [153]. The administration of gingipain inhibitors has also been associated with a decrease in host $A\beta_{1-42}$ response to *P. gingivalis* infection [68].

4.3. *Limitations*

Even with the benefits associated with the antimicrobial and antiviral drugs listed above, insights into their mechanism of action and their impact on Aβ peptide levels are needed. A greater understanding as to whether their administration indirectly reduces the presence of Aβ peptides by reducing the viral/bacterial load on the brain, or if they act directly in reducing Aβ peptides level, is needed. If it is the latter, and the antimicrobial hypothesis for Aβ peptides holds true, their efficacy might not be as positive as hoped. In addition, it is important that the chosen antimicrobial or antiviral drug does not have any adverse effects that could take away from its benefits. For example, cefepime is an antibiotic that has shown to be able to cross the blood-brain barrier and cause neurotoxic symptoms [154].

5. Conclusions and Future Directions

The findings of the numerous studies highlighted in this review present a clear indication of the role bacteria and viruses can have in AD development. Even with this conclusion, it is clear that a specific bacteria or virus alone is not responsible for AD development, as no specific bacteria or virus has been identified to be universally present in all AD brains. Rather, a number of viruses and bacteria could exacerbate the progression of neurodegenerative diseases, either independently or along with other pathogens. By exploring the presence of multiple viruses/bacteria in AD brains, future investigations can give insights into which microorganisms are most present, and whether all AD brains have both a detected and increased level of selected bacteria/virus. The use of antiviral and antimicrobial drugs early on, while the patient is still in the presymptomatic phase of AD, could have potential effectiveness in targeting the root cause of AD pathogenesis and alleviating the viral/microbial load on the brain. Further investigations into their use in AD would give greater insight regarding their efficacy and limitations.

Author Contributions: U.H.I. and G.M.P. conceptualized the manuscript. U.H.I. and E.Z. wrote the manuscript. U.H.I., E.Z., and G.M.P. reviewed and edited the manuscript. All authors have read and agree to the published version of the manuscript.

Funding: This study was supported by the Altschul Foundation and in part by grant number P50 AT008661-01 from the NCCIH and ODS. We acknowledge that the contents of this review do not represent the views of the NCCIH, the ODS, the NIH, or the United States Government.

Acknowledgments: Figures were made using BioRender.

Conflicts of Interest: The authors declare no conflict of interest.

Abbreviations

AD	Alzheimer's Disease
Aβ	amyloid-β
AMP	antimicrobial peptide
APP	amyloid precursor proteins
BACE1	B-site ABPP cleaving enzyme
BBB	blood brain barrier
CNS	central nervous system
HSV-1	herpes simplex virus-1
MBP-1	major basic protein-1
NFTS	neurofibrillary tangles

References

1. Alzheimer's Association. 2019 Alzheimer's Disease Facts and Figures. *Alzheimer's Dement.* **2019**, *15*, 321–387. [CrossRef]
2. Soria Lopez, J.A.; Gonzalez, H.M.; Leger, G.C. Alzheimer's disease. *Handb. Clin. Neurol.* **2019**, *167*, 231–255. [CrossRef] [PubMed]
3. Hardy, J.; Allsop, D. Amyloid deposition as the central event in the aetiology of Alzheimer's disease. *Trends Pharmacol. Sci.* **1991**, *12*, 383–388. [CrossRef]
4. Hardy, J.A.; Higgins, G.A. Alzheimer's disease: The amyloid cascade hypothesis. *Science* **1992**, *256*, 184–185. [CrossRef]
5. Surgucheva, I.; Newell, K.L.; Burns, J.; Surguchov, A. New α- and γ-synuclein immunopathological lesions in human brain. *Acta Neuropathol. Commun.* **2014**, *2*, 132. [CrossRef] [PubMed]
6. Gomez-Isla, T.; Hollister, R.; West, H.; Mui, S.; Growdon, J.H.; Petersen, R.C.; Parisi, J.E.; Hyman, B.T. Neuronal loss correlates with but exceeds neurofibrillary tangles in Alzheimer's disease. *Ann. Neurol.* **1997**, *41*, 17–24. [CrossRef]
7. Tetz, G.; Pinho, M.; Pritzkow, S.; Mendez, N.; Soto, C.; Tetz, V. Bacterial DNA promotes Tau aggregation. *Sci. Rep.* **2020**, *10*, 2369. [CrossRef]
8. Nelson, P.T.; Alafuzoff, I.; Bigio, E.H.; Bouras, C.; Braak, H.; Cairns, N.J.; Castellani, R.J.; Crain, B.J.; Davies, P.; Del Tredici, K.; et al. Correlation of Alzheimer disease neuropathologic changes with cognitive status: A review of the literature. *J. Neuropathol. Exp. Neurol.* **2012**, *71*, 362–381. [CrossRef]
9. Giacobini, E.; Gold, G. Alzheimer disease therapy–moving from amyloid-β to tau. *Nat. Rev. Neurol.* **2013**, *9*, 677–686. [CrossRef]
10. Sato, C.; Barthelemy, N.R.; Mawuenyega, K.G.; Patterson, B.W.; Gordon, B.A.; Jockel-Balsarotti, J.; Sullivan, M.; Crisp, M.J.; Kasten, T.; Kirmess, K.M.; et al. Tau Kinetics in Neurons and the Human Central Nervous System. *Neuron* **2018**, *97*, 1284–1298. [CrossRef]
11. Hardy, J.; Selkoe, D.J. The amyloid hypothesis of Alzheimer's disease: Progress and problems on the road to therapeutics. *Science* **2002**, *297*, 353–356. [CrossRef] [PubMed]
12. Cai, Z.; Hussain, M.D.; Yan, L.J. Microglia, neuroinflammation, and beta-amyloid protein in Alzheimer's disease. *Int. J. Neurosci.* **2014**, *124*, 307–321. [CrossRef] [PubMed]
13. Panza, F.; Lozupone, M.; Logroscino, G.; Imbimbo, B.P. A critical appraisal of amyloid-beta-targeting therapies for Alzheimer disease. *Nat. Rev. Neurol.* **2019**, *15*, 73–88. [CrossRef]
14. Arbor, S.C.; LaFontaine, M.; Cumbay, M. Amyloid-beta Alzheimer targets-protein processing, lipid rafts, and amyloid-beta pores. *Yale J. Biol. Med.* **2016**, *89*, 5–21. [PubMed]
15. Murphy, M.P.; LeVine, H., 3rd. Alzheimer's disease and the amyloid-beta peptide. *J. Alzheimer's Dis.* **2010**, *19*, 311–323. [CrossRef] [PubMed]

16. Nunan, J.; Small, D.H. Regulation of APP cleavage by alpha-, beta- and gamma-secretases. *FEBS Lett.* **2000**, *483*, 6–10. [CrossRef]
17. Chen, G.F.; Xu, T.H.; Yan, Y.; Zhou, Y.R.; Jiang, Y.; Melcher, K.; Xu, H.E. Amyloid beta: Structure, biology and structure-based therapeutic development. *Acta Pharmacol. Sin.* **2017**, *38*, 1205–1235. [CrossRef]
18. Gravina, S.A.; Ho, L.; Eckman, C.B.; Long, K.E.; Otvos, L., Jr.; Younkin, L.H.; Suzuki, N.; Younkin, S.G. Amyloid beta protein (A beta) in Alzheimer's disease brain. Biochemical and immunocytochemical analysis with antibodies specific for forms ending at A beta 40 or A beta 42(43). *J. Biol. Chem.* **1995**, *270*, 7013–7016. [CrossRef]
19. McGowan, E.; Pickford, F.; Kim, J.; Onstead, L.; Eriksen, J.; Yu, C.; Skipper, L.; Murphy, M.P.; Beard, J.; Das, P.; et al. Abeta42 is essential for parenchymal and vascular amyloid deposition in mice. *Neuron* **2005**, *47*, 191–199. [CrossRef]
20. Kim, J.; Onstead, L.; Randle, S.; Price, R.; Smithson, L.; Zwizinski, C.; Dickson, D.W.; Golde, T.; McGowan, E. Abeta40 inhibits amyloid deposition in vivo. *J. Neurosci.* **2007**, *27*, 627–633. [CrossRef]
21. Orgogozo, J.M.; Gilman, S.; Dartigues, J.F.; Laurent, B.; Puel, M.; Kirby, L.C.; Jouanny, P.; Dubois, B.; Eisner, L.; Flitman, S.; et al. Subacute meningoencephalitis in a subset of patients with AD after Abeta42 immunization. *Neurology* **2003**, *61*, 46–54. [CrossRef] [PubMed]
22. Gilman, S.; Koller, M.; Black, R.S.; Jenkins, L.; Griffith, S.G.; Fox, N.C.; Eisner, L.; Kirby, L.; Rovira, M.B.; Forette, F.; et al. Clinical effects of Abeta immunization (AN1792) in patients with AD in an interrupted trial. *Neurology* **2005**, *64*, 1553–1562. [CrossRef] [PubMed]
23. Weggen, S.; Eriksen, J.L.; Das, P.; Sagi, S.A.; Wang, R.; Pietrzik, C.U.; Findlay, K.A.; Smith, T.E.; Murphy, M.P.; Bulter, T.; et al. A subset of NSAIDs lower amyloidogenic Abeta42 independently of cyclooxygenase activity. *Nature* **2001**, *414*, 212–216. [CrossRef] [PubMed]
24. Kukar, T.L.; Ladd, T.B.; Bann, M.A.; Fraering, P.C.; Narlawar, R.; Maharvi, G.M.; Healy, B.; Chapman, R.; Welzel, A.T.; Price, R.W.; et al. Substrate-targeting gamma-secretase modulators. *Nature* **2008**, *453*, 925–929. [CrossRef] [PubMed]
25. Green, R.C.; Schneider, L.S.; Amato, D.A.; Beelen, A.P.; Wilcock, G.; Swabb, E.A.; Zavitz, K.H. Effect of tarenflurbil on cognitive decline and activities of daily living in patients with mild Alzheimer disease: A randomized controlled trial. *JAMA* **2009**, *302*, 2557–2564. [CrossRef]
26. Doody, R.S.; Raman, R.; Farlow, M.; Iwatsubo, T.; Vellas, B.; Joffe, S.; Kieburtz, K.; He, F.; Sun, X.; Thomas, R.G.; et al. A phase 3 trial of semagacestat for treatment of Alzheimer's disease. *N. Engl. J. Med.* **2013**, *369*, 341–350. [CrossRef]
27. McLaurin, J.; Golomb, R.; Jurewicz, A.; Antel, J.P.; Fraser, P.E. Inositol stereoisomers stabilize an oligomeric aggregate of Alzheimer amyloid beta peptide and inhibit abeta -induced toxicity. *J. Biol. Chem.* **2000**, *275*, 18495–18502. [CrossRef]
28. Salloway, S.; Sperling, R.; Keren, R.; Porsteinsson, A.P.; van Dyck, C.H.; Tariot, P.N.; Gilman, S.; Arnold, D.; Abushakra, S.; Hernandez, C.; et al. A phase 2 randomized trial of ELND005, scyllo-inositol, in mild to moderate Alzheimer disease. *Neurology* **2011**, *77*, 1253–1262. [CrossRef]
29. Therapeutics-Elenbecestat. Available online: https://www.alzforum.org/therapeutics/elenbecestat (accessed on 7 May 2020).
30. Gosztyla, M.L.; Brothers, H.M.; Robinson, S.R. Alzheimer's Amyloid-β is an Antimicrobial Peptide: A Review of the Evidence. *J. Alzheimer's Dis.* **2018**, *62*, 1495–1506. [CrossRef]
31. Caselli, R.J.; Reiman, E.M. Characterizing the preclinical stages of Alzheimer's disease and the prospect of presymptomatic intervention. *J. Alzheimer's Dis.* **2013**, *33*, S405–S416. [CrossRef]
32. Villemagne, V.L.; Burnham, S.; Bourgeat, P.; Brown, B.; Ellis, K.A.; Salvado, O.; Szoeke, C.; Macaulay, S.L.; Martins, R.; Maruff, P.; et al. Amyloid β deposition, neurodegeneration, and cognitive decline in sporadic Alzheimer's disease: A prospective cohort study. *Lancet Neurol.* **2013**, *12*, 357–367. [CrossRef]
33. Hampel, H.; O'Bryant, S.E.; Molinuevo, J.L.; Zetterberg, H.; Masters, C.L.; Lista, S.; Kiddle, S.J.; Batrla, R.; Blennow, K. Blood-based biomarkers for Alzheimer disease: Mapping the road to the clinic. *Nat. Rev. Neurol.* **2018**, *14*, 639–652. [CrossRef] [PubMed]
34. Bousiges, O.; Cretin, B.; Lavaux, T.; Philippi, N.; Jung, B.; Hezard, S.; Heitz, C.; Demuynck, C.; Gabel, A.; Martin-Hunyadi, C.; et al. Diagnostic Value of Cerebrospinal Fluid Biomarkers (Phospho-Tau181, total-Tau, Aβ42, and Aβ40) in Prodromal Stage of Alzheimer's Disease and Dementia with Lewy Bodies. *J. Alzheimer's Dis.* **2016**, *51*, 1069–1083. [CrossRef] [PubMed]

35. Janelidze, S.; Stomrud, E.; Palmqvist, S.; Zetterberg, H.; van Westen, D.; Jeromin, A.; Song, L.; Hanlon, D.; Tan Hehir, C.A.; Baker, D.; et al. Plasma β-amyloid in Alzheimer's disease and vascular disease. *Sci. Rep.* **2016**, *6*, 26801. [CrossRef]
36. Weston, P.S.J.; Poole, T.; Ryan, N.S.; Nair, A.; Liang, Y.; Macpherson, K.; Druyeh, R.; Malone, I.B.; Ahsan, R.L.; Pemberton, H.; et al. Serum neurofilament light in familial Alzheimer disease: A marker of early neurodegeneration. *Neurology* **2017**, *89*, 2167–2175. [CrossRef] [PubMed]
37. Salvadores, N.; Shahnawaz, M.; Scarpini, E.; Tagliavini, F.; Soto, C. Detection of misfolded Aβ oligomers for sensitive biochemical diagnosis of Alzheimer's disease. *Cell Rep.* **2014**, *7*, 261–268. [CrossRef] [PubMed]
38. Kraus, A.; Saijo, E.; Metrick, M.A., 2nd; Newell, K.; Sigurdson, C.J.; Zanusso, G.; Ghetti, B.; Caughey, B. Seeding selectivity and ultrasensitive detection of tau aggregate conformers of Alzheimer disease. *Acta Neuropathol.* **2019**, *137*, 585–598. [CrossRef]
39. Kinney, J.W.; Bemiller, S.M.; Murtishaw, A.S.; Leisgang, A.M.; Salazar, A.M.; Lamb, B.T. Inflammation as a central mechanism in Alzheimer's disease. *Alzheimer's Dement.* **2018**, *4*, 575–590. [CrossRef]
40. DiSabato, D.J.; Quan, N.; Godbout, J.P. Neuroinflammation: The devil is in the details. *J. Neurochem.* **2016**, *139*, 136–153. [CrossRef] [PubMed]
41. Skaper, S.D.; Facci, L.; Zusso, M.; Giusti, P. An Inflammation-Centric View of Neurological Disease: Beyond the Neuron. *Front. Cell Neurosci.* **2018**, *12*, 72. [CrossRef]
42. Vinayagam, R.; Bhuvaneswarri, J.; Paddmanabhan, P.; Manisundar, N. Alziemer's Disease and Periodontal Disease Bidirectional Interrelationships. *Biosci. Biotechnol. Res. Asia* **2014**, *11*, 259–261. [CrossRef]
43. Klein, R.S.; Garber, C.; Howard, N. Infectious immunity in the central nervous system and brain function. *Nat. Immunol.* **2017**, *18*, 132–141. [CrossRef]
44. Hensley, K. Neuroinflammation in Alzheimer's disease: Mechanisms, pathologic consequences, and potential for therapeutic manipulation. *J. Alzheimer's Dis.* **2010**, *21*, 1–14. [CrossRef] [PubMed]
45. De Chiara, G.; Marcocci, M.E.; Sgarbanti, R.; Civitelli, L.; Ripoli, C.; Piacentini, R.; Garaci, E.; Grassi, C.; Palamara, A.T. Infectious agents and neurodegeneration. *Mol. Neurobiol.* **2012**, *46*, 614–638. [CrossRef] [PubMed]
46. Fulop, T.; Itzhaki, R.F.; Balin, B.J.; Miklossy, J.; Barron, A.E. Role of Microbes in the Development of Alzheimer's Disease: State of the Art—An International Symposium Presented at the 2017 IAGG Congress in San Francisco. *Front. Genet.* **2018**, *9*, 362. [CrossRef] [PubMed]
47. Itzhaki, R.F.; Lathe, R.; Balin, B.J.; Ball, M.J.; Bearer, E.L.; Braak, H.; Bullido, M.J.; Carter, C.; Clerici, M.; Cosby, S.L.; et al. Microbes and Alzheimer's Disease. *J. Alzheimer's Dis.* **2016**, *51*, 979–984. [CrossRef] [PubMed]
48. Miklossy, J. Historic evidence to support a causal relationship between spirochetal infections and Alzheimer's disease. *Front. Aging Neurosci.* **2015**, *7*, 46. [CrossRef]
49. Maheshwari, P.; Eslick, G.D. Bacterial infection and Alzheimer's disease: A meta-analysis. *J. Alzheimer's Dis.* **2015**, *43*, 957–966. [CrossRef]
50. Ashraf, G.M.; Tarasov, V.V.; Makhmutova, A.; Chubarev, V.N.; Avila-Rodriguez, M.; Bachurin, S.O.; Aliev, G. The Possibility of an Infectious Etiology of Alzheimer Disease. *Mol. Neurobiol.* **2019**, *56*, 4479–4491. [CrossRef]
51. Murphy, J.V.; Yunis, E.J. Encephalopathy following measles infection in children with chronic illness. *J. Pediatrics* **1976**, *88*, 937–942. [CrossRef]
52. Dobson, C.B.; Itzhaki, R.F. Herpes simplex virus type 1 and Alzheimer's disease. *Neurobiol. Aging* **1999**, *20*, 457–465. [CrossRef]
53. Balin, B.J.; Gerard, H.C.; Arking, E.J.; Appelt, D.M.; Branigan, P.J.; Abrams, J.T.; Whittum-Hudson, J.A.; Hudson, A.P. Identification and localization of Chlamydia pneumoniae in the Alzheimer's brain. *Med. Microbiol. Immunol.* **1998**, *187*, 23–42. [CrossRef] [PubMed]
54. Lurain, N.S.; Hanson, B.A.; Martinson, J.; Leurgans, S.E.; Landay, A.L.; Bennett, D.A.; Schneider, J.A. Virological and immunological characteristics of human cytomegalovirus infection associated with Alzheimer disease. *J. Infect. Dis.* **2013**, *208*, 564–572. [CrossRef] [PubMed]
55. Lovheim, H.; Olsson, J.; Weidung, B.; Johansson, A.; Eriksson, S.; Hallmans, G.; Elgh, F. Interaction between Cytomegalovirus and Herpes Simplex Virus Type 1 Associated with the Risk of Alzheimer's Disease Development. *J. Alzheimer's Dis.* **2018**, *61*, 939–945. [CrossRef] [PubMed]

56. Linard, M.; Letenneur, L.; Garrigue, I.; Doize, A.; Dartigues, J.F.; Helmer, C. Interaction between APOE4 and herpes simplex virus type 1 in Alzheimer's disease. *Alzheimer's Dement. J. Alzheimer's Assoc.* **2020**, *16*, 200–208. [CrossRef]
57. De Chiara, G.; Piacentini, R.; Fabiani, M.; Mastrodonato, A.; Marcocci, M.E.; Limongi, D.; Napoletani, G.; Protto, V.; Coluccio, P.; Celestino, I.; et al. Recurrent herpes simplex virus-1 infection induces hallmarks of neurodegeneration and cognitive deficits in mice. *PLoS Pathog.* **2019**, *15*, e1007617. [CrossRef]
58. Wozniak, M.A.; Shipley, S.J.; Combrinck, M.; Wilcock, G.K.; Itzhaki, R.F. Productive herpes simplex virus in brain of elderly normal subjects and Alzheimer's disease patients. *J. Med. Virol.* **2005**, *75*, 300–306. [CrossRef]
59. Jamieson, G.A.; Maitland, N.J.; Wilcock, G.K.; Craske, J.; Itzhaki, R.F. Latent herpes simplex virus type 1 in normal and Alzheimer's disease brains. *J. Med. Virol.* **1991**, *33*, 224–227. [CrossRef]
60. Carbone, I.; Lazzarotto, T.; Ianni, M.; Porcellini, E.; Forti, P.; Masliah, E.; Gabrielli, L.; Licastro, F. Herpes virus in Alzheimer's disease: Relation to progression of the disease. *Neurobiol. Aging* **2014**, *35*, 122–129. [CrossRef]
61. Readhead, B.; Haure-Mirande, J.V.; Funk, C.C.; Richards, M.A.; Shannon, P.; Haroutunian, V.; Sano, M.; Liang, W.S.; Beckmann, N.D.; Price, N.D.; et al. Multiscale Analysis of Independent Alzheimer's Cohorts Finds Disruption of Molecular, Genetic, and Clinical Networks by Human Herpesvirus. *Neuron* **2018**, *99*, 64–82. [CrossRef]
62. Westfall, S.; Dinh, D.M.; Pasinetti, G.M. Investigation of Potential Brain Microbiome in Alzheimer's Disease: Implications of Study Bias. *J. Alzheimer's Dis.* **2020**, *75*, 559–570. [CrossRef]
63. Branton, W.G.; Lu, J.Q.; Surette, M.G.; Holt, R.A.; Lind, J.; Laman, J.D.; Power, C. Brain microbiota disruption within inflammatory demyelinating lesions in multiple sclerosis. *Sci. Rep.* **2016**, *6*, 37344. [CrossRef] [PubMed]
64. Emery, D.C.; Shoemark, D.K.; Batstone, T.E.; Waterfall, C.M.; Coghill, J.A.; Cerajewska, T.L.; Davies, M.; West, N.X.; Allen, S.J. 16S rRNA Next Generation Sequencing Analysis Shows Bacteria in Alzheimer's Post-Mortem Brain. *Front. Aging Neurosci.* **2017**, *9*, 195. [CrossRef]
65. Zhan, X.; Stamova, B.; Jin, L.W.; DeCarli, C.; Phinney, B.; Sharp, F.R. Gram-negative bacterial molecules associate with Alzheimer disease pathology. *Neurology* **2016**, *87*, 2324–2332. [CrossRef] [PubMed]
66. Zhan, X.; Stamova, B.; Sharp, F.R. Lipopolysaccharide Associates with Amyloid Plaques, Neurons and Oligodendrocytes in Alzheimer's Disease Brain: A Review. *Front. Aging Neurosci.* **2018**, *10*, 42. [CrossRef] [PubMed]
67. Miklossy, J.; Khalili, K.; Gern, L.; Ericson, R.L.; Darekar, P.; Bolle, L.; Hurlimann, J.; Paster, B.J. Borrelia burgdorferi persists in the brain in chronic lyme neuroborreliosis and may be associated with Alzheimer disease. *J. Alzheimer's Dis.* **2004**, *6*, 639–649. [CrossRef]
68. Dominy, S.S.; Lynch, C.; Ermini, F.; Benedyk, M.; Marczyk, A.; Konradi, A.; Nguyen, M.; Haditsch, U.; Raha, D.; Griffin, C.; et al. Porphyromonas gingivalis in Alzheimer's disease brains: Evidence for disease causation and treatment with small-molecule inhibitors. *Sci. Adv.* **2019**, *5*. [CrossRef]
69. Alonso, R.; Pisa, D.; Marina, A.I.; Morato, E.; Rábano, A.; Carrasco, L. Fungal infection in patients with Alzheimer's disease. *J. Alzheimer's Dis.* **2014**, *41*, 301–311. [CrossRef]
70. Pisa, D.; Alonso, R.; Rábano, A.; Rodal, I.; Carrasco, L. Different Brain Regions are Infected with Fungi in Alzheimer's Disease. *Sci. Rep.* **2015**, *5*, 15015. [CrossRef]
71. Romagnoli, M.; Porcellini, E.; Carbone, I.; Veerhuis, R.; Licastro, F. Impaired Innate Immunity Mechanisms in the Brain of Alzheimer's Disease. *Int. J. Mol. Sci.* **2020**, *21*, 1126. [CrossRef]
72. Rey, N.L.; Wesson, D.W.; Brundin, P. The olfactory bulb as the entry site for prion-like propagation in neurodegenerative diseases. *Neurobiol. Dis.* **2018**, *109*, 226–248. [CrossRef]
73. Hammond, C.J.; Hallock, L.R.; Howanski, R.J.; Appelt, D.M.; Little, C.S.; Balin, B.J. Immunohistological detection of Chlamydia pneumoniae in the Alzheimer's disease brain. *BMC Neurosci.* **2010**, *11*, 121. [CrossRef] [PubMed]
74. Sundar, S.; Battistoni, C.; McNulty, R.; Morales, F.; Gorky, J.; Foley, H.; Dhurjati, P. An agent-based model to investigate microbial initiation of Alzheimer's via the olfactory system. *Theor. Biol. Med. Model.* **2020**, *17*, 5. [CrossRef] [PubMed]
75. Little, C.S.; Hammond, C.J.; MacIntyre, A.; Balin, B.J.; Appelt, D.M. Chlamydia pneumoniae induces Alzheimer-like amyloid plaques in brains of BALB/c mice. *Neurobiol. Aging* **2004**, *25*, 419–429. [CrossRef]

76. Cappellini, C.A.C.; Ahmad, B.; Williams, K.G.; Zoga, J.; Hingley, S.T. Chlamydia Pneumoniae Infection of Neuronal Cells Induces Changes in Calcium-Associated Gene Expression Consistent with Alzheimer's Disease. Available online: https://digitalcommons.pcom.edu/posters/6 (accessed on 22 May 2020).
77. Tetz, G.; Tetz, V. Prion-like Domains in Eukaryotic Viruses. *Sci. Rep.* **2018**, *8*, 8931. [CrossRef]
78. Wozniak, M.A.; Itzhaki, R.F.; Shipley, S.J.; Dobson, C.B. Herpes simplex virus infection causes cellular beta-amyloid accumulation and secretase upregulation. *Neurosci. Lett.* **2007**, *429*, 95–100. [CrossRef]
79. Tanaka, S.; Nagashima, H. Establishment of an Alzheimer's disease model with latent herpesvirus infection using PS2 and Tg2576 double transgenic mice. *Exp. Anim.* **2018**, *67*, 185–192. [CrossRef]
80. Ishida, N.; Ishihara, Y.; Ishida, K.; Tada, H.; Funaki-Kato, Y.; Hagiwara, M.; Ferdous, T.; Abdullah, M.; Mitani, A.; Michikawa, M.; et al. Periodontitis induced by bacterial infection exacerbates features of Alzheimer's disease in transgenic mice. *NPJ Aging Mech. Dis.* **2017**, *3*, 15. [CrossRef]
81. Moir, R.D.; Lathe, R.; Tanzi, R.E. The antimicrobial protection hypothesis of Alzheimer's disease. *Alzheimer's Dement. J. Alzheimer's Assoc.* **2018**, *14*, 1602–1614. [CrossRef]
82. Bourgade, K.; Le Page, A.; Bocti, C.; Witkowski, J.M.; Dupuis, G.; Frost, E.H.; Fulop, T., Jr. Protective Effect of Amyloid-beta Peptides Against Herpes Simplex Virus-1 Infection in a Neuronal Cell Culture Model. *J. Alzheimer's Dis.* **2016**, *50*, 1227–1241. [CrossRef]
83. De Chiara, G.; Marcocci, M.E.; Civitelli, L.; Argnani, R.; Piacentini, R.; Ripoli, C.; Manservigi, R.; Grassi, C.; Garaci, E.; Palamara, A.T. APP processing induced by herpes simplex virus type 1 (HSV-1) yields several APP fragments in human and rat neuronal cells. *PLoS ONE* **2010**, *5*, e13989. [CrossRef]
84. Santana, S.; Recuero, M.; Bullido, M.J.; Valdivieso, F.; Aldudo, J. Herpes simplex virus type I induces the accumulation of intracellular beta-amyloid in autophagic compartments and the inhibition of the non-amyloidogenic pathway in human neuroblastoma cells. *Neurobiol. Aging* **2012**, *33*, 430.e19–430.e33. [CrossRef] [PubMed]
85. Li Puma, D.D.; Piacentini, R.; Leone, L.; Gironi, K.; Marcocci, M.E.; De Chiara, G.; Palamara, A.T.; Grassi, C. Herpes Simplex Virus Type-1 Infection Impairs Adult Hippocampal Neurogenesis via Amyloid-beta Protein Accumulation. *Stem Cells* **2019**, *37*, 1467–1480. [CrossRef] [PubMed]
86. Piacentini, R.; Li Puma, D.D.; Ripoli, C.; Marcocci, M.E.; De Chiara, G.; Garaci, E.; Palamara, A.T.; Grassi, C. Herpes Simplex Virus type-1 infection induces synaptic dysfunction in cultured cortical neurons via GSK-3 activation and intraneuronal amyloid-beta protein accumulation. *Sci. Rep.* **2015**, *5*, 15444. [CrossRef] [PubMed]
87. Kristen, H.; Santana, S.; Sastre, I.; Recuero, M.; Bullido, M.J.; Aldudo, J. Herpes simplex virus type 2 infection induces AD-like neurodegeneration markers in human neuroblastoma cells. *Neurobiol. Aging* **2015**, *36*, 2737–2747. [CrossRef] [PubMed]
88. Miklossy, J.; Kis, A.; Radenovic, A.; Miller, L.; Forro, L.; Martins, R.; Reiss, K.; Darbinian, N.; Darekar, P.; Mihaly, L.; et al. Beta-amyloid deposition and Alzheimer's type changes induced by Borrelia spirochetes. *Neurobiol. Aging* **2006**, *27*, 228–236. [CrossRef] [PubMed]
89. Nie, R.; Wu, Z.; Ni, J.; Zeng, F.; Yu, W.; Zhang, Y.; Kadowaki, T.; Kashiwazaki, H.; Teeling, J.L.; Zhou, Y. Porphyromonas gingivalis Infection Induces Amyloid-beta Accumulation in Monocytes/Macrophages. *J. Alzheimer's Dis.* **2019**, *72*, 479–494. [CrossRef] [PubMed]
90. Cheng, S.B.; Ferland, P.; Webster, P.; Bearer, E.L. Herpes simplex virus dances with amyloid precursor protein while exiting the cell. *PLoS ONE* **2011**, *6*, e17966. [CrossRef] [PubMed]
91. Green, D.A.; Masliah, E.; Vinters, H.V.; Beizai, P.; Moore, D.J.; Achim, C.L. Brain deposition of beta-amyloid is a common pathologic feature in HIV positive patients. *AIDS* **2005**, *19*, 407–411. [CrossRef]
92. Khan, M.B.; Lang, M.J.; Huang, M.B.; Raymond, A.; Bond, V.C.; Shiramizu, B.; Powell, M.D. Nef exosomes isolated from the plasma of individuals with HIV-associated dementia (HAD) can induce Aβ(1-42) secretion in SH-SY5Y neural cells. *J. Neurovirol.* **2016**, *22*, 179–190. [CrossRef]
93. Lopatko Lindman, K.; Weidung, B.; Olsson, J.; Josefsson, M.; Kok, E.; Johansson, A.; Eriksson, S.; Hallmans, G.; Elgh, F.; Lövheim, H. A genetic signature including apolipoprotein Eε4 potentiates the risk of herpes simplex-associated Alzheimer's disease. *Alzheimers Dement* **2019**, *5*, 697–704. [CrossRef]
94. Itzhaki, R.F.; Lin, W.R.; Shang, D.; Wilcock, G.K.; Faragher, B.; Jamieson, G.A. Herpes simplex virus type 1 in brain and risk of Alzheimer's disease. *Lancet* **1997**, *349*, 241–244. [CrossRef]
95. Burgos, J.S.; Ramirez, C.; Sastre, I.; Bullido, M.J.; Valdivieso, F. ApoE4 is more efficient than E3 in brain access by herpes simplex virus type 1. *Neuroreport* **2003**, *14*, 1825–1827. [CrossRef] [PubMed]

96. Wozniak, M.A.; Mee, A.P.; Itzhaki, R.F. Herpes simplex virus type 1 DNA is located within Alzheimer's disease amyloid plaques. *J. Pathol.* **2009**, *217*, 131–138. [CrossRef] [PubMed]
97. Bitting, L.; Naidu, A.; Cordell, B.; Murphy, G.M., Jr. Beta-amyloid peptide secretion by a microglial cell line is induced by beta-amyloid-(25-35) and lipopolysaccharide. *J. Biol. Chem.* **1996**, *271*, 16084–16089. [CrossRef]
98. Soscia, S.J.; Kirby, J.E.; Washicosky, K.J.; Tucker, S.M.; Ingelsson, M.; Hyman, B.; Burton, M.A.; Goldstein, L.E.; Duong, S.; Tanzi, R.E.; et al. The Alzheimer's disease-associated amyloid beta-protein is an antimicrobial peptide. *PLoS ONE* **2010**, *5*, e9505. [CrossRef]
99. Soragni, A.; Yousefi, S.; Stoeckle, C.; Soriaga, A.B.; Sawaya, M.R.; Kozlowski, E.; Schmid, I.; Radonjic-Hoesli, S.; Boutet, S.; Williams, G.J.; et al. Toxicity of eosinophil MBP is repressed by intracellular crystallization and promoted by extracellular aggregation. *Mol. Cell* **2015**, *57*, 1011–1021. [CrossRef]
100. Pastore, A.; Raimondi, F.; Rajendran, L.; Temussi, P.A. Why does the Aβ peptide of Alzheimer share structural similarity with antimicrobial peptides? *Commun. Biol.* **2020**, *3*, 135. [CrossRef] [PubMed]
101. Dicks, L.M.T.; Dreyer, L.; Smith, C.; van Staden, A.D. A Review: The Fate of Bacteriocins in the Human Gastro-Intestinal Tract: Do They Cross the Gut-Blood Barrier? *Front. Microbiol.* **2018**, *9*, 2297. [CrossRef]
102. Williams, W.M.; Torres, S.; Siedlak, S.L.; Castellani, R.J.; Perry, G.; Smith, M.A.; Zhu, X. Antimicrobial peptide β-defensin-1 expression is upregulated in Alzheimer's brain. *J. Neuroinflamm.* **2013**, *10*, 127. [CrossRef]
103. White, M.R.; Kandel, R.; Tripathi, S.; Condon, D.; Qi, L.; Taubenberger, J.; Hartshorn, K.L. Alzheimer's associated β-amyloid protein inhibits influenza A virus and modulates viral interactions with phagocytes. *PLoS ONE* **2014**, *9*, e101364. [CrossRef]
104. Eimer, W.A.; Vijaya Kumar, D.K.; Navalpur Shanmugam, N.K.; Rodriguez, A.S.; Mitchell, T.; Washicosky, K.J.; György, B.; Breakefield, X.O.; Tanzi, R.E.; Moir, R.D. Alzheimer's Disease-Associated β-Amyloid Is Rapidly Seeded by Herpesviridae to Protect against Brain Infection. *Neuron* **2018**, *99*, 56–63. [CrossRef] [PubMed]
105. Spitzer, P.; Condic, M.; Herrmann, M.; Oberstein, T.J.; Scharin-Mehlmann, M.; Gilbert, D.F.; Friedrich, O.; Grömer, T.; Kornhuber, J.; Lang, R.; et al. Amyloidogenic amyloid-β-peptide variants induce microbial agglutination and exert antimicrobial activity. *Sci. Rep.* **2016**, *6*, 32228. [CrossRef] [PubMed]
106. Kumar, D.K.V.; Choi, S.H.; Washicosky, K.J.; Eimer, W.A.; Tucker, S.; Ghofrani, J.; Lefkowitz, A.; McColl, G.; Goldstein, L.E.; Tanzi, R.E.; et al. Amyloid-β peptide protects against microbial infection in mouse and worm models of Alzheimer's disease. *Sci. Transl. Med.* **2016**, *8*. [CrossRef]
107. Dominguez, D.; Tournoy, J.; Hartmann, D.; Huth, T.; Cryns, K.; Deforce, S.; Serneels, L.; Camacho, I.E.; Marjaux, E.; Craessaerts, K.; et al. Phenotypic and Biochemical Analyses of BACE1- and BACE2-deficient Mice. *J. Biol. Chem.* **2005**, *280*, 30797–30806. [CrossRef] [PubMed]
108. Bishop, G.M.; Robinson, S.R. The amyloid hypothesis: Let sleeping dogmas lie? *Neurobiol. Aging* **2002**, *23*, 1101–1105. [CrossRef]
109. Robinson, S.R.; Bishop, G.M. Abeta as a bioflocculant: Implications for the amyloid hypothesis of Alzheimer's disease. *Neurobiol. Aging* **2002**, *23*, 1051–1072. [CrossRef]
110. Bourgade, K.; Dupuis, G.; Frost, E.H.; Fülöp, T., Jr. Anti-Viral Properties of Amyloid-β Peptides. *J. Alzheimer's Dis.* **2016**, *54*, 859–878. [CrossRef]
111. Gnann, J.W., Jr.; Barton, N.H.; Whitley, R.J. Acyclovir: Mechanism of action, pharmacokinetics, safety and clinical applications. *Pharmacotherapy* **1983**, *3*, 275–283. [CrossRef] [PubMed]
112. Wozniak, M.A.; Frost, A.L.; Preston, C.M.; Itzhaki, R.F. Antivirals reduce the formation of key Alzheimer's disease molecules in cell cultures acutely infected with herpes simplex virus type 1. *PLoS ONE* **2011**, *6*, e25152. [CrossRef]
113. Zambrano, A.; Solis, L.; Salvadores, N.; Cortes, M.; Lerchundi, R.; Otth, C. Neuronal cytoskeletal dynamic modification and neurodegeneration induced by infection with herpes simplex virus type 1. *J. Alzheimer's Dis.* **2008**, *14*, 259–269. [CrossRef]
114. Hui, Z.; Zhijun, Y.; Yushan, Y.; Liping, C.; Yiying, Z.; Difan, Z.; Chunglit, C.T.; Wei, C. The combination of acyclovir and dexamethasone protects against Alzheimer's disease-related cognitive impairments in mice. *Psychopharmacology* **2020**. [CrossRef] [PubMed]
115. Prasad, K.M.; Eack, S.M.; Keshavan, M.S.; Yolken, R.H.; Iyengar, S.; Nimgaonkar, V.L. Antiherpes virus-specific treatment and cognition in schizophrenia: A test-of-concept randomized double-blind placebo-controlled trial. *Schizophr. Bull.* **2013**, *39*, 857–866. [CrossRef]

116. Devanand, D.P.; Andrews, H.; Kreisl, W.C.; Razlighi, Q.; Gershon, A.; Stern, Y.; Mintz, A.; Wisniewski, T.; Acosta, E.; Pollina, J.; et al. Antiviral therapy: Valacyclovir Treatment of Alzheimer's Disease (VALAD) Trial: Protocol for a randomised, double-blind, placebo-controlled, treatment trial. *BMJ Open* **2020**, *10*, e032112. [CrossRef] [PubMed]
117. Wozniak, M.A.; Frost, A.L.; Itzhaki, R.F. The helicase-primase inhibitor BAY 57-1293 reduces the Alzheimer's disease-related molecules induced by herpes simplex virus type 1. *Antivir. Res.* **2013**, *99*, 401–404. [CrossRef] [PubMed]
118. Baumeister, J.; Fischer, R.; Eckenberg, P.; Henninger, K.; Ruebsamen-Waigmann, H.; Kleymann, G. Superior efficacy of helicase-primase inhibitor BAY 57-1293 for herpes infection and latency in the guinea pig model of human genital herpes disease. *Antivir. Chem. Chemother.* **2007**, *18*, 35–48. [CrossRef] [PubMed]
119. Betz, U.A.; Fischer, R.; Kleymann, G.; Hendrix, M.; Rubsamen-Waigmann, H. Potent in vivo antiviral activity of the herpes simplex virus primase-helicase inhibitor BAY 57-1293. *Antimicrob. Agents Chemother.* **2002**, *46*, 1766–1772. [CrossRef]
120. Kleymann, G. Helicase primase: Targeting the Achilles heel of herpes simplex viruses. *Antivir. Chem. Chemother.* **2004**, *15*, 135–140. [CrossRef]
121. Kleymann, G.; Fischer, R.; Betz, U.A.; Hendrix, M.; Bender, W.; Schneider, U.; Handke, G.; Eckenberg, P.; Hewlett, G.; Pevzner, V.; et al. New helicase-primase inhibitors as drug candidates for the treatment of herpes simplex disease. *Nat. Med.* **2002**, *8*, 392–398. [CrossRef]
122. Hayashi, K.; Hayashi, T.; Morita, N. Mechanism of action of the antiherpesvirus biflavone ginkgetin. *Antimicrob. Agents Chemother.* **1992**, *36*, 1890–1893. [CrossRef]
123. Miki, K.; Nagai, T.; Suzuki, K.; Tsujimura, R.; Koyama, K.; Kinoshita, K.; Furuhata, K.; Yamada, H.; Takahashi, K. Anti-influenza virus activity of biflavonoids. *Bioorganic Med. Chem. Lett.* **2007**, *17*, 772–775. [CrossRef]
124. Borenstein, R.; Hanson, B.A.; Markosyan, R.M.; Gallo, E.S.; Narasipura, S.D.; Bhutta, M.; Shechter, O.; Lurain, N.S.; Cohen, F.S.; Al-Harthi, L.; et al. Ginkgolic acid inhibits fusion of enveloped viruses. *Sci. Rep.* **2020**, *10*, 4746. [CrossRef] [PubMed]
125. Lee, J.H.; Kim, Y.G.; Ryu, S.Y.; Cho, M.H.; Lee, J. Ginkgolic acids and Ginkgo biloba extract inhibit Escherichia coli O157:H7 and Staphylococcus aureus biofilm formation. *Int. J. Food Microbiol.* **2014**, *174*, 47–55. [CrossRef]
126. Bagla, V.P.; McGaw, L.J.; Elgorashi, E.E.; Eloff, J.N. Antimicrobial activity, toxicity and selectivity index of two biflavonoids and a flavone isolated from Podocarpus henkelii (Podocarpaceae) leaves. *BMC Complementary Altern. Med.* **2014**, *14*, 383. [CrossRef]
127. Zeng, Y.Q.; Wang, Y.J.; Zhou, X.F. Ginkgetin Ameliorates Neuropathological Changes in APP/PS1 Transgenical Mice Model. *J. Prev. Alzheimer's Dis.* **2016**, *3*, 24–29. [CrossRef]
128. Kang, S.S.; Lee, J.Y.; Choi, Y.K.; Song, S.S.; Kim, J.S.; Jeon, S.J.; Han, Y.N.; Son, K.H.; Han, B.H. Neuroprotective effects of naturally occurring biflavonoids. *Bioorganic Med. Chem. Lett.* **2005**, *15*, 3588–3591. [CrossRef] [PubMed]
129. Sasaki, H.; Kitoh, Y.; Tsukada, M.; Miki, K.; Koyama, K.; Juliawaty, L.D.; Hakim, E.H.; Takahashi, K.; Kinoshita, K. Inhibitory activities of biflavonoids against amyloid-beta peptide 42 cytotoxicity in PC-12 cells. *Bioorganic Med. Chem. Lett.* **2015**, *25*, 2831–2833. [CrossRef]
130. Balducci, C.; Forloni, G. Doxycycline for Alzheimer's Disease: Fighting beta-Amyloid Oligomers and Neuroinflammation. *Front. Pharmacol.* **2019**, *10*, 738. [CrossRef]
131. Merlini, G.; Ascari, E.; Amboldi, N.; Bellotti, V.; Arbustini, E.; Perfetti, V.; Ferrari, M.; Zorzoli, I.; Marinone, M.G.; Garini, P.; et al. Interaction of the anthracycline 4′-iodo-4′-deoxydoxorubicin with amyloid fibrils: Inhibition of amyloidogenesis. *Proc. Natl. Acad. Sci. USA* **1995**, *92*, 2959–2963. [CrossRef]
132. Balducci, C.; Santamaria, G.; La Vitola, P.; Brandi, E.; Grandi, F.; Viscomi, A.R.; Beeg, M.; Gobbi, M.; Salmona, M.; Ottonello, S.; et al. Doxycycline counteracts neuroinflammation restoring memory in Alzheimer's disease mouse models. *Neurobiol. Aging* **2018**, *70*, 128–139. [CrossRef]
133. Costa, R.; Speretta, E.; Crowther, D.C.; Cardoso, I. Testing the therapeutic potential of doxycycline in a Drosophila melanogaster model of Alzheimer disease. *J. Biol. Chem.* **2011**, *286*, 41647–41655. [CrossRef]
134. Gautieri, A.; Beeg, M.; Gobbi, M.; Rigoldi, F.; Colombo, L.; Salmona, M. The Anti-Amyloidogenic Action of Doxycycline: A Molecular Dynamics Study on the Interaction with Abeta42. *Int. J. Mol. Sci.* **2019**, *20*, 4641. [CrossRef] [PubMed]
135. Loeb, M.B.; Molloy, D.W.; Smieja, M.; Standish, T.; Goldsmith, C.H.; Mahony, J.; Smith, S.; Borrie, M.; Decoteau, E.; Davidson, W.; et al. A randomized, controlled trial of doxycycline and rifampin for patients with Alzheimer's disease. *J. Am. Geriatr. Soc.* **2004**, *52*, 381–387. [CrossRef] [PubMed]

136. Molloy, D.W.; Standish, T.I.; Zhou, Q.; Guyatt, G. A multicenter, blinded, randomized, factorial controlled trial of doxycycline and rifampin for treatment of Alzheimer's disease: The DARAD trial. *Int. J. Geriatr. Psychiatry* **2013**, *28*, 463–470. [CrossRef] [PubMed]
137. Mandal, A.; Sinha, C.; Kumar Jena, A.; Ghosh, S.; Samanta, A. An Investigation on in vitro and in vivo Antimicrobial Properties of the Antidepressant: Amitriptyline Hydrochloride. *Braz. J. Microbiol.* **2010**, *41*, 635–645. [CrossRef]
138. Wang, J.; Zhao, Z.; Lin, E.; Zhao, W.; Qian, X.; Freire, D.; Bilski, A.E.; Cheng, A.; Vempati, P.; Ho, L.; et al. Unintended effects of cardiovascular drugs on the pathogenesis of Alzheimer's disease. *PLoS ONE* **2013**, *8*, e65232. [CrossRef]
139. Yulug, B.; Hanoglu, L.; Kilic, E.; Schabitz, W.R. RIFAMPICIN: An antibiotic with brain protective function. *Brain Res. Bull.* **2014**, *107*, 37–42. [CrossRef]
140. Yulug, B.; Hanoglu, L.; Ozansoy, M.; Isık, D.; Kilic, U.; Kilic, E.; Schabitz, W.R. Therapeutic role of rifampicin in Alzheimer's disease. *Psychiatry Clin. Neurosci.* **2018**, *72*, 152–159. [CrossRef]
141. Bi, W.; Zhu, L.; Jing, X.; Liang, Y.; Tao, E. Rifampicin and Parkinson's disease. *Neurol. Sci.* **2013**, *34*, 137–141. [CrossRef]
142. Tomiyama, T.; Asano, S.; Suwa, Y.; Morita, T.; Kataoka, K.; Mori, H.; Endo, N. Rifampicin prevents the aggregation and neurotoxicity of amyloid beta protein in vitro. *Biochem. Biophys. Res. Commun.* **1994**, *204*, 76–83. [CrossRef]
143. Tomiyama, T.; Shoji, A.; Kataoka, K.; Suwa, Y.; Asano, S.; Kaneko, H.; Endo, N. Inhibition of amyloid beta protein aggregation and neurotoxicity by rifampicin. Its possible function as a hydroxyl radical scavenger. *J. Biol. Chem.* **1996**, *271*, 6839–6844. [CrossRef]
144. Mindermann, T.; Landolt, H.; Zimmerli, W.; Rajacic, Z.; Gratzl, O. Penetration of rifampicin into the brain tissue and cerebral extracellular space of rats. *J. Antimicrob. Chemother.* **1993**, *31*, 731–737. [CrossRef] [PubMed]
145. Qosa, H.; Abuznait, A.H.; Hill, R.A.; Kaddoumi, A. Enhanced brain amyloid-β clearance by rifampicin and caffeine as a possible protective mechanism against Alzheimer's disease. *J. Alzheimer's Dis.* **2012**, *31*, 151–165. [CrossRef]
146. Umeda, T.; Ono, K.; Sakai, A.; Yamashita, M.; Mizuguchi, M.; Klein, W.L.; Yamada, M.; Mori, H.; Tomiyama, T. Rifampicin is a candidate preventive medicine against amyloid-β and tau oligomers. *Brain* **2016**, *139*, 1568–1586. [CrossRef]
147. Umeda, T.; Tanaka, A.; Sakai, A.; Yamamoto, A.; Sakane, T.; Tomiyama, T. Intranasal rifampicin for Alzheimer's disease prevention. *Alzheimer's Dement.* **2018**, *4*, 304–313. [CrossRef] [PubMed]
148. Li, N.; Collyer, C.A. Gingipains from Porphyromonas gingivalis—Complex domain structures confer diverse functions. *Eur. J. Microbiol. Immunol.* **2011**, *1*, 41–58. [CrossRef] [PubMed]
149. Curtis, M.A.; Kuramitsu, H.K.; Lantz, M.; Macrina, F.L.; Nakayama, K.; Potempa, J.; Reynolds, E.C.; Aduse-Opoku, J. Molecular genetics and nomenclature of proteases of Porphyromonas gingivalis. *J. Periodontal Res.* **1999**, *34*, 464–472. [CrossRef] [PubMed]
150. Guo, Y.; Nguyen, K.A.; Potempa, J. Dichotomy of gingipains action as virulence factors: From cleaving substrates with the precision of a surgeon's knife to a meat chopper-like brutal degradation of proteins. *Periodontology 2000* **2010**, *54*, 15–44. [CrossRef] [PubMed]
151. Potempa, J.; Banbula, A.; Travis, J. Role of bacterial proteinases in matrix destruction and modulation of host responses. *Periodontology 2000* **2000**, *24*, 153–192. [CrossRef]
152. Jagels, M.A.; Ember, J.A.; Travis, J.; Potempa, J.; Pike, R.; Hugli, T.E. Cleavage of the human C5A receptor by proteinases derived from Porphyromonas gingivalis: Cleavage of leukocyte C5a receptor. *Adv. Exp. Med. Biol.* **1996**, *389*, 155–164. [CrossRef]
153. Arastu-Kapur, S.; Nguyen, M.; Raha, D.; Ermini, F.; Haditsch, U.; Araujo, J.; De Lannoy, I.A.M.; Ryder, M.I.; Dominy, S.S.; Lynch, C.; et al. Treatment of Porphyromonas gulae infection and downstream pathology in the aged dog by lysine-gingipain inhibitor COR388. *Pharmcol. Res. Perspect.* **2020**, *8*, e00562. [CrossRef]
154. Payne, L.E.; Gagnon, D.J.; Riker, R.R.; Seder, D.B.; Glisic, E.K.; Morris, J.G.; Fraser, G.L. Cefepime-induced neurotoxicity: A systematic review. *Crit. Care* **2017**, *21*, 276. [CrossRef] [PubMed]

International Journal of
Molecular Sciences

MDPI

Review

Natural Medicines and Their Underlying Mechanisms of Prevention and Recovery from Amyloid Β-Induced Axonal Degeneration in Alzheimer's Disease

Tomoharu Kuboyama [1,2], Ximeng Yang [1] and Chihiro Tohda [1,*]

1 Section of Neuromedical Science, Institute of Natural Medicine, University of Toyama, Sugitani 2630, Toyama 930-0194, Japan; t-kuboyama@daiichi-cps.ac.jp (T.K.); ximeng@inm.u-toyama.ac.jp (X.Y.)
2 Laboratory of Pharmacognosy, Daiichi University of Pharmacy, 22-1 Tamagawa-cho, Minami-ku, Fukuoka 815-8511, Japan
* Correspondence: chihiro@inm.u-toyama.ac.jp; Tel.: +81-76-434-7646

Received: 17 June 2020; Accepted: 28 June 2020; Published: 30 June 2020

Abstract: In Alzheimer's disease (AD), amyloid β (Aβ) induces axonal degeneration, neuronal network disruption, and memory impairment. Although many candidate drugs to reduce Aβ have been clinically investigated, they failed to recover the memory function in AD patients. Reportedly, Aβ deposition occurred before the onset of AD. Once neuronal networks were disrupted by Aβ, they could hardly be recovered. Therefore, we speculated that only removal of Aβ was not enough for AD therapy, and prevention and recovery from neuronal network disruption were also needed. This review describes the challenges related to the condition of axons for AD therapy. We established novel in vitro models of Aβ-induced axonal degeneration. Using these models, we found that several traditional medicines and their constituents prevented or helped recover from Aβ-induced axonal degeneration. These drugs also prevented or helped recover from memory impairment in in vivo models of AD. One of these drugs ameliorated memory decline in AD patients in a clinical study. These results indicate that prevention and recovery from axonal degeneration are possible strategies for AD therapy.

Keywords: amyloid β; axon; traditional medicines; Polygalae Radix; diosgenin; naringenin; kihito

1. Introduction

Dementia is one of serious refractory diseases in the world, especially in aging societies such as Japan, Europe, and United States. Alzheimer's disease (AD) is the most frequent cause of dementia. Estimated 35 million people were suffered from AD in 2017 [1]. AD is a progressive and irreversible neurodegenerative disease. Not only is the quality of life in AD patients extremely low, but oftentimes, the family bears a heavy burden of care for the patient. The mean simulated survival of AD patients is 19.0 years [2]. Amyloid β (Aβ) is a critical cause of AD [3–6]. Aβ is cleaved from amyloid precursor protein (APP) by processing enzymes β-secretase and γ-secretase [7]. There are two types of AD, namely familial and sporadic. In both cases, Aβ induces neurodegeneration and memory deficits [5,7]. In familial AD, mutations in *APP* or presenilin (PS), a component of γ-secretase, result in Aβ overproduction [7]. In sporadic AD, other gene mutations such as mutations in *APOE* and *TREM2*, are possibly related to Aβ accumulation [7].

Recently, many candidates targeting Aβ reduction such as β-secretase and γ-secretase inhibitors and anti-Aβ antibodies have been clinically investigated for AD therapy. Unfortunately, these candidates have not been clinically approved, as most of them failed to achieve positive effects in primary outcomes [8,9]. Reportedly, a considerable amount of Aβ had already been accumulated in the brain at the onset of AD [10]. Therefore, prevention of Aβ production and removal of Aβ might not be adequate to treat AD patients. Currently, only four medicines are available for AD therapy, namely donepezil, galantamine, rivastigmine, and memantine, which regulate the synaptic function to improve memory. Although these drugs are effective for AD, their effects are limited. These drugs only delay deterioration of AD symptoms, but do not help patients recover from AD. Moreover, they show limited benefits when compared with their adverse effects [11–14]. Therefore, the French minister of health delisted these medicines for AD therapy in 2018 [15]. In AD, Aβ induces irreversible and fundamental neural atrophy in the brain [16–18], leading to severe memory deficits. Therefore, drugs that only modulate the synaptic function might not be adequate to cure AD. Structural degeneration of neuronal networks are proceeded and promoted by Aβ.

At the onset of AD, neurodegeneration and memory deficits are not severe [10]. Therefore, we hypothesized that if Aβ-induced neural atrophy is prevented or recovered, memory function might be maintained or restored, respectively. Aβ induces axonal degeneration and neuronal death in vitro and in the brain of AD patients [16,17,19]. We established new in vitro models of Aβ-induced irreversible axonal degeneration [20–22]. The present review summarizes our challenges in proving our hypothesis using these models. We focused especially on natural medicines and investigated their effects using the in vitro AD models, in vivo AD models, and human subjects. Our study proposes new strategies for the prevention and treatment of AD by protecting against and recovering from Aβ-induced axonal degeneration.

2. Prevention of Aβ-Induced Axonal Degeneration

Aβ induces neuritic degeneration in in vitro models, in vivo mouse models, and AD patients, leading to disruption of neuronal networks and memory deficits [17,21,23–28]. Although several studies have suggested a variety of mechanisms of Aβ-induced neuritic degeneration, such as abnormality of autophagy and activation of calcineurin [29,30], its critical cause has not been identified. We observed Aβ-induced acute collapse of axonal growth cone within 1 h in cultured neurons [20,31]. This culture system made it possible to visualize and analyze early events in the Aβ-induced axonal degeneration process. Various inhibitors and mutated gene transfection revealed that Aβ induced Ca^{2+} signaling via N-methyl-D-aspartate receptor and transient receptor potential channel, activated calcineurin and calpain, facilitated clathrin-mediated endocytosis, and induced growth cone collapse (Figure 1). In case of repulsive axon guidance, asymmetric clathrin-mediated endocytosis removes the plasma membrane and β1-integrin from the surface of the growth cone, leading to turning the growth cone away from the reduced membrane [32]. Therefore, Aβ might remove some functional molecules from the surface of the growth cone, leading to the collapse of the growth cone and then causing atrophy of axons.

Intracerebroventricular (i.c.v.) injection of Aβ to adult mice induced axonal degeneration in the brain and memory deficits [33–35]. When specific inhibitors of clathrin-mediated endocytosis, namely pitstop 2 and myristoylated dynamin inhibitory peptide, were simultaneously i.c.v. injected with Aβ, axonal degeneration and memory deficits were prevented [31]. Simultaneous treatments with Aβ and the inhibitors prevented growth cone collapse and axonal atrophy in cultured neurons. These results show that inhibition of clathrin-mediated endocytosis prevents not only the early events elicited by Aβ such as collapse and degeneration of axons, but also the late events elicited by Aβ such as memory deficits. Clathrin-mediated endocytosis is a novel and critical target for AD prevention.

5XFAD mouse is a transgenic model of AD, which overexpresses familial AD mutations, namely K670N/M671L (Swedish), I716V (Florida), and V717I (London) mutations in APP and M146L/L286V mutations in PS1, and shows Aβ deposition, phosphorylated tau deposition, neuronal loss, axonal degeneration, and memory deficits [36–39]. In the brain of 4-month-old 5XFAD mice, Aβ plaques and degenerated axons were few in number and novel object recognition memory was normal [40]. However, in 5-month-old mice, Aβ plaques and degenerated axons were significantly increased and novel object recognition memory was impaired. We found that water extract of Polygalae Radix (PR) (roots of *Polygala tenuifolia*, a traditional herbal medicine) inhibited Aβ-induced endocytosis in growth cones in cultured neurons [40]. When PR was orally administered to 4-month-old 5XFAD mice for 56 days, degenerated axons were significantly decreased in number when compared with that in vehicle-administered mice. Object recognition memory was not impaired, but there was no change in the increase in Aβ plaques after PR administration. In cultured neurons, PR prevented Aβ-induced growth cone collapse even in the presence of Aβ. These findings indicate that PR prevented Aβ-induced toxicity even in the presence of Aβ possibly via inhibition of endocytosis. Another group also showed that administration of PR extract to 4-month-old 5XFAD mice for 2 months prevented memory deficits [41]. The authors showed neuroprotective effects of the PR extract against Aβ. Tenuifolin, a constituent in PR, counteracted Aβ toxicity in in vitro and in vivo [42]. PR is traditionally used for memory loss in East Asia [43,44]. PR has been shown to enhance cognitive functions in healthy elderly individuals [45]. Therefore, PR is a promising preventive candidate drug for AD.

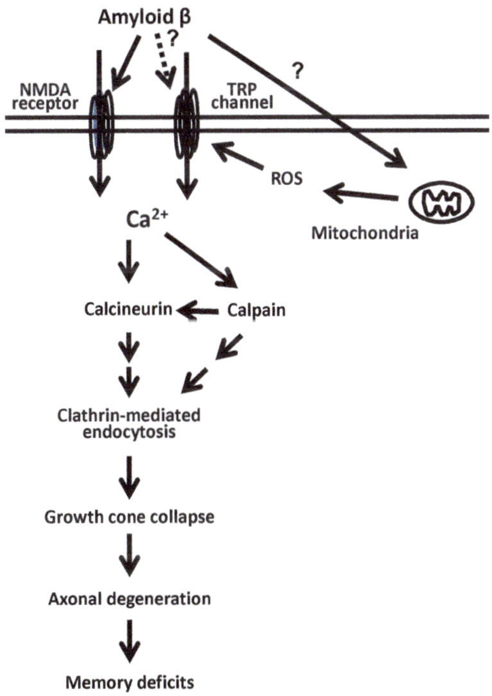

Figure 1. Aβ-induced growth cone collapse via clathrin-mediated endocytosis [31].

3. Recovery from Aβ-Induced Axonal Degeneration

3.1. Ashwagandha

Ashwagandha (a root of *Withania somnifera* Dunal), also called Indian Ginseng, is one of the most important herbal drugs in Indian traditional Ayurvedic medicine; namely, as a rasayana drug that is used for longevity and increasing vital energy and intellectual power. Ashwagandha is clinically applied for dyspepsia, anxiety, depression, insomnia, and loss of memory. Various pharmacological studies of Ashwagandha and its constituents have been investigated, such as anti-inflammation [46,47], anti-stress [48,49], and neuroprotection [50,51]. These facts stimulated our curiosity about effects of Ashwagandha and its constituents on AD. We found that a methanol extract of Ashwagandha induced neurite outgrowth in cultured human neuroblastoma SK-N-SH cells [52]. Then, six compounds were isolated from the methanol extract as active constituents showing neurite outgrowth activities [53,54].

Effects of the active constituents of Ashwagandha in Aβ-induced degenerated condition were investigated. Aβ induces axonal atrophy, and the atrophied axons never recover, even after removal of

Aβ in cultured cortical neurons [21,22], namely, Aβ induced irreversible axonal atrophy. Among the active constituents, withanolide A, withanoside IV, and withanoside VI (Figure 2) were administered to cultured cortical neurons after Aβ induced axonal atrophy. As a result, these three constituents induced axonal growth in the Aβ-treated neuron culture [21,55,56]. These three constituents were orally administered to the Aβ-i.c.v.-injected mice after memory deficits occurred. They ameliorated degenerated axons in the brain and recovered memory. Among these three constituents, withanoside IV was shown to be metabolized into sominone (Figure 2) after oral administration [55]. Sominone induced axonal growth in Aβ-treated cultured neurons and recovered degenerated axons in the brain and memory in 5XFAD mice at 6–8 months of age when the mice already showed axonal degeneration and memory deficits. These results indicate that sominone is an active principle of withanoside IV after oral administration. Sominone induced axonal growth in normal cultured neurons and in the brain of normal adult mice [57]. Sominone also enhanced spatial memory in normal adult mice. In those cases, sominone increased RET phosphorylation, which meant activation of RET. RET is a receptor of the glial cell line-derived neurotrophic factor (GDNF). In cultured normal neurons, GDNF induced axonal growth, knocking down RET diminished sominone-induced axonal growth, while sominone enhanced the secretion of GDNF. These results indicate that sominone induces axonal growth via GDNF-RET signaling and enhances memory in normal adult mice. Considering that transplantation of GDNF-overexpressing cells ameliorated memory in 5XFAD mice [58], withanoside IV and sominone might improve memory via GDNF-RET signaling in 5XFAD mice. We synthesized a novel compound, denosomin (Figure 2), which is a derivative of sominone. Denosomin induced axonal growth in Aβ-treated neuron culture as much as sominone did [59]. Denosomin showed stronger neuroprotective effects against Aβ than sominone at low concentrations. Therefore, denosomin is expected to be a novel anti-AD drug.

Withanoside IV and denosomin recovered motor function in spinal cord injured mice [60,61]. Axonal growth activities of these compounds might contribute to the recovery from spinal cord injury. Ashwagandha and its constituents were reportedly effective for various neurological disorder models besides AD and spinal cord injury, such as Parkinson's disease, addiction, anxiety, schizophrenia, dyslexia, depression, and so on [62]. Ashwagandha can be a source of novel drugs for not only AD, but also other various refractory neurological disorders.

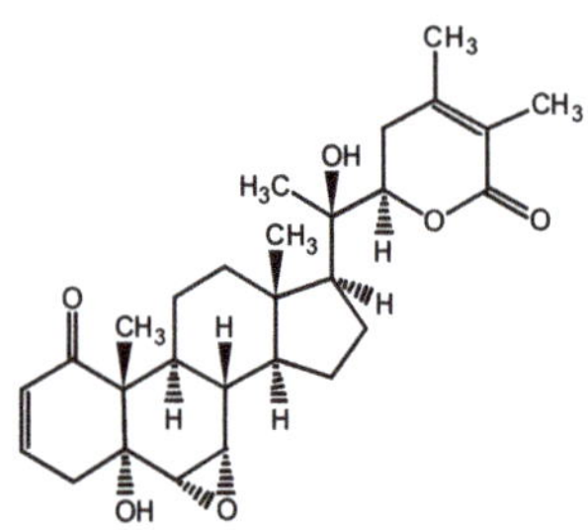

Withanolide A

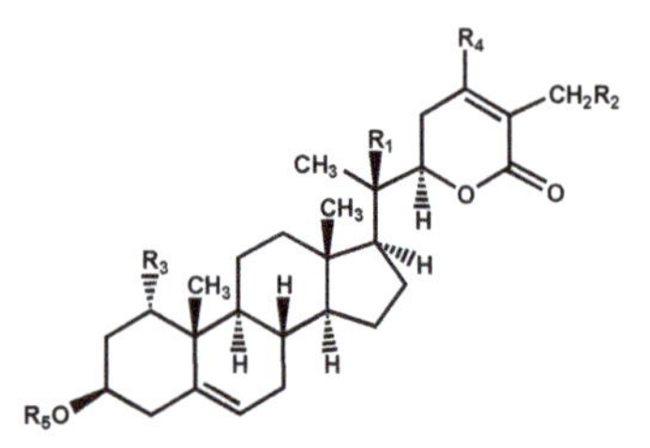

	R_1	R_2	R_3	R_4	R_5
Withanoside IV	H	OH	OH	CH_3	-^{1}Glc^{6}-^{1}Glc
Withanoside VI	OH	H	OH	CH_3	-^{1}Glc^{6}-^{1}Glc
Sominone	H	OH	OH	CH_3	H
Denosomin	H	H	H	H	H

Figure 2. Structures of Ashwagandha-related compounds.

3.2. *Diosgenin*

Diosgenin is a steroid sapogenin derived from Dioscorea wild yam. Rhizome of *Dioscorea japonica* Thunberg is used as a tonic agent for aging people in Japanese traditional Kampo medicine. We screened traditional medicine-derived compounds and found diosgenin to be a potent stimulator of axonal growth. Therefore, we postulated that diosgenin had anti-AD activities.

Treatment with diosgenin significantly induced axonal growth after axons were already atrophied by Aβ in cultured neurons [63]. Diosgenin was administered for 20 days to 6- to 8-month-old 5XFAD mice when axonal degeneration and memory impairment had already occurred. Diosgenin significantly improved the degenerated axons and memory function. These results indicated that diosgenin normalized degenerated axons, induced axonal growth, helped in reconstruction of neuronal networks, and improved memory in 5XFAD mice.

To clarify the mechanisms of diosgenin, a protein capable of direct binding with diosgenin was identified using the drug affinity responsive target stability (DARTS) method [64]. Briefly, diosgenin was mixed with neuron lysate and was treated with a protease. During this reaction, diosgenin binds to a target protein, leading to conformational modification of the protein. Vulnerability of the protein against the protease is possibly changed. After poly-acrylamide gel electrophoresis, band intensity of the target protein should be different between drug treatment and control treatment. The changed band is considered a candidate for direct binding with diosgenin. Using the DARTS method, various proteins that can directly bind with compounds or molecules have been identified [65,66]. In case of diosgenin, 1,25D_3-membrane-associated rapid response, steroid-binding protein (1,25D_3-MARRS) was identified, which is a membrane receptor for 1α,25-dyhydroxyvitamin D_3 [67]. Direct binding of diosgenin with 1,25D_3-MARRS was indicated by several other experiments, such as docking simulation of diosgenin to 1,25D_3-MARRS. By knocking down 1,25D_3-MARRS in cultured neurons, axonal growth activity of diosgenin was completely blocked. Treatment with a functional blocking antibody for 1,25D_3-MARRS also inhibited diosgenin-induced axonal growth. By using kinase inhibitors, it was suggested that diosgenin induced axonal growth via phosphoinositide 3-kinase, extracellular signal-regulated kinase, protein kinase C, and protein kinase A. Since these kinases are reported to be involved in 1,25D_3-MARRS signaling [67–70], diosgenin probably induced axonal growth via these kinases under 1,25D_3-MARRS signaling (Figure 3).

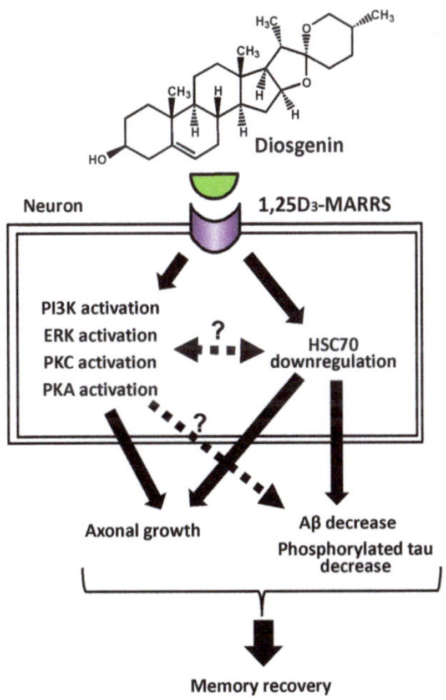

Figure 3. Signaling mechanism of diosgenin-induced memory recovery in Alzheimer's disease models [63,71,72].

In normal cultured neurons, diosgenin induced axonal growth [73]. In normal young mice, diosgenin increased axonal density in the brain cortex and enhanced memory. These in vitro and in vivo effects of diosgenin were completely blocked by the functional blocking antibody for 1,25D_3-MARRS. These results indicate that 1,25D_3-MARRS is a receptor for diosgenin with positive effects on axons and memory in AD-like conditions as well as in normal conditions.

To further investigate the downstream signaling of diosgenin, gene expression in the brain cortex of diosgenin-administered 5XFAD mice was comprehensively analyzed [71]. Expression of heat shock cognate 70 (HSC70) was decreased by diosgenin administration. On the other hand, HSC70 was increased in the brain cortex of 5XFAD mice when compared with HSC70 expression in wild-type mice. In cultured neurons, Aβ treatment increased HSC70 expression, while diosgenin-treatment decreased it. This in vitro effect of diosgenin was canceled by the functional blocking antibody for 1,25D_3-MARRS. Knocking down of HSC70 alone induced axonal growth in cultured neurons. VER-155008, an inhibitor of HSC70, also induced axonal growth in the Aβ-treated neuron culture [72]. VER-155008 attenuated degenerated axons and improved memory in 5XFAD mice. Diosgenin decreased Aβ plaques and phosphorylated tau in the brain of 5XFAD mice [63]. Phosphorylated tau is another cause of AD, which is accumulated by Aβ-induced activation of glycogen synthase kinase 3β, Cyclin dependent kinase-5, and so on, leading to microtubule disassembly and neuritic degeneration [74–76]. VER-155008 also decreased Aβ plaques and phosphorylated tau in the brain of 5XFAD mice [72]. These results indicate that diosgenin binds to 1,25D_3-MARRS and downregulates HSC70, leading to attenuation of degenerated axons, decrease in Aβ plaques and phosphorylated tau, and recovery from memory deficits in 5XFAD mice (Figure 2). We speculated that HSC70 promoted α-tubulin degradation in axons and diosgenin attenuated the degenerated axons via downregulating HSC70 and retaining α-tubulin expression in axons [71]. Further investigations are needed to clarify how 1,25D_3-MARRS signaling downregulates HSC70 and how HSC70 decreases Aβ plaques, decreases phosphorylated tau, and attenuates degenerated axons.

We also investigated the effects of diosgenin in humans [77]. In the investigation, diosgenin-rich yam extract was used. Healthy men and women (mean age was 46.50 years) were randomly divided into two groups. One group took the extract, and the other group took placebo for 12 weeks; then, memory function was assessed by Repeatable Battery for the Assessment of Neuropsychological Status (RBANS). After a 6-week interval period, the group previously taken placebo took the extract, and the other group took placebo for 12 weeks; then, memory function was assessed in the same way. This was a placebo-controlled, randomized, double-blind, and crossover study. As a result, the total score of RBANS was significantly increased by taking the extract. The diosgenin-rich yam extract enhanced cognitive function in healthy adult humans. This result as well as the results obtained in mice, in which diosgenin enhanced memory in both normal adult mice and 5XFAD mice [63,73], suggest that the diosgenin-rich yam extract might be effective for AD patients. We are planning to perform a clinical study for AD patients using this yam extract.

3.3. *Drynariae Rhizoma*

Drynariae Rhizoma (DR) is a rhizome of *Drynaria fortunei* (Kunze ex Mett.) J. Sm. (Polypodiaceae), which is a Chinese herbal drug used for bone fracture and for tonifying the kidneys [78]. We found that water extract of DR induced axonal growth in the Aβ-treated neuron culture [79]. Oral administration of DR extract attenuated degenerated axons and improved memory in 5XFAD mice [80]. In vitro bioassay-guided isolation is usually performed to identify the active compounds in herbal medicines [81]. However, this method may sometimes lead to the misidentification of the real active compounds, as the metabolization process in the body and permeability of the blood-brain barrier are ignored. Therefore, we aimed to identify the real active compounds by detecting compounds in the brain of 5XFAD mice after oral administration of DR extract (in vivo screening) [80]. Naringenin was identified as a compound that crossed the blood-brain barrier. Naringenin showed axonal growth activity in the Aβ-treated neuron culture, ameliorated degenerated axons, and improved memory in 5XFAD mice. The DR extract contained a low amount of Naringenin itself. However, it contained an abundant amount of naringin, a glycoside of naringenin [80]. After oral administration of the DR extract, naringin is metabolized into naringenin and absorbed [82], and then likely transferred into the brain. The DARTS method showed that collapsin response mediator protein 2 (CRMP2) is a candidate of direct binding protein with naringenin. After treatment with naringenin to cultured neurons, the neuron lysate was immunoprecipitated with an anti-CRMP2 antibody. As a result, naringenin was detected from the immunoprecipitation. These results mean that naringenin directly binds to CRMP2. Knocking down of CRMP2 diminished axonal growth activity of naringenin. In Alzheimer's disease, Aβ induced CRMP2 phosphorylation and axonal degeneration [83–85]. Our culture experiments also showed that Aβ increased phosphorylated CRMP2, while naringenin decreased it. These results indicate that naringin in the DR extract is metabolized into naringenin after oral administration of DR. Subsequently, naringenin is transferred into the brain. It binds to CRMP2, dephosphorylates CRMP2, ameliorates degenerated axons, and recovers memory (Figure 4).

Oral administration of DR extract and naringenin reduced brain Aβ plaques in 5XFAD mice [80]. We focused on microglia that might contribute to the effect of naringenin against Aβ plaques. Microglia are divided into at least two phenotypes: M1 and M2. Aβ increases M1 microglia that secrete pro-inflammatory cytokines, such as interferon-γ, interleukin (IL)-1β, and tumor necrosis factor-α, and induces inflammation that leads to neural damage [86]. On the other hand, M2 microglia secrete anti-inflammatory cytokines such as IL-4, IL-10, and transforming growth factor-β [87]. M2 microglia also express Aβ degradation enzymes, insulin degradation enzyme (IDE), and neprilysin [88] and reduce Aβ plaques [89]. We established a method to distinguish M1 and M2 phenotypes in cultured microglia [90]. Using this protocol, the effects of naringenin on microglia were investigated [91]. Aβ increased M1 predominance and decreased M2 predominance. In contrast, treatment with naringenin decreased M1 and increased M2. Naringenin also increased expression of IDE and neprilysin. These findings indicate that M2 polarization by naringenin contributes to the reduction of Aβ plaques in 5XFAD mice (Figure 4).

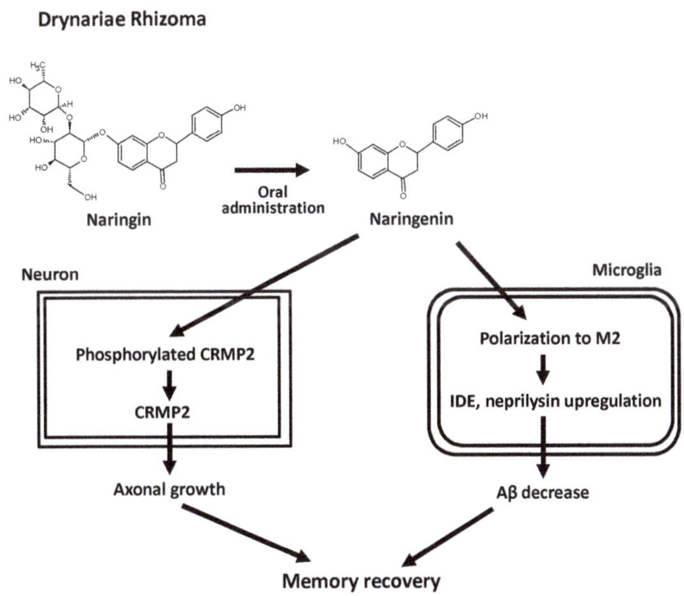

Figure 4. Mechanisms of memory recovery induced by Drynariae Rhizoma extract in Alzheimer's disease models [80,91].

Rhizome of *Eleutherococcus senticosus* (Rupr. & Maxim.) Maxim., also known as Siberian ginseng, is used in Japanese traditional Kampo medicine as a tonic drug. Leafs of *E. senticosus* are not used for medicine, but consumed as tea, wine, and so on. We previously found that *E. senticosus* leaf extract and its constituents enhanced object recognition memory in healthy young mice [92]. The combination of DR and *E. senticosus* leaf extract has been investigated in healthy humans [93]. Healthy men and women were randomly divided into two groups, where one group (mean age was 63.0 years) took the combination, and the other group (mean age was 64.3 years) took a placebo for 12 weeks. After that, memory function was assessed by RBANS and stress responses were measured by the public health research foundation stress checklist short form. This was a placebo-controlled, randomized, and double-blind study. As a result, the combination group significantly increased the figure recall subscore of RBANS, and improved the anxiety/uncertainly score in the stress checklist. The combination of DR and *E. senticosus* leaf extract showed memory enhancing and anti-stress effects in healthy humans. It may be worthwhile to clinically investigate this combination in AD patients.

3.4. Kihito and Kamikihito

Kihito is a Japanese traditional Kampo formula used for insomnia, anemia, amnesia, depression, and neurosis. Kihito is composed of 12 herbal medicines, namely Ginseng Radix (roots of *Panax ginseng* C.A. Meyer), PR (roots of *Polygala tenuifolia* Willd.), Astragali Radix (roots of *Astragalus membranaceus* Bunge), Zizyphi Fructus (fruits of *Zizyphus jujube* Mill. var. *inermis* [Bunge] Rehd.), Zizyphi Spinosi Semen (seeds of *Z. jujube* Mill. var. *spinosa* [Bunge] Hu ex H.F. Chou), Angelicae Radix (roots of *Angelica acutiloba* Kitagawa), Glycyrrhizae Radix (roots of *Glycyrrhiza uralensis* Fisch. ex DC.), Atractylodis Rhizoma (rhizomes of *Atractylodes ovata* DC.), Zingiberis Rhizoma (rhizomes of *Zingiber officinale* Roscoe), Poria (sclerotium of *Poria cocos* Wolf), Saussureae Radix (roots of *Saussurea lappa* Clarke), and Longanae Arillus (pulp of *Euphoria longana* Lam.). Among these components, we reported that metabolite M1 derived from Ginseng Radix and water extract of Astragali Radix induced axonal growth in the Aβ-treated neuron culture and improved memory in AD model mice [35,94]. We have discussed previously that PR prevented axonal degeneration and memory deficits in AD model mice [40]. We also clarified that PR induced axonal growth in the Aβ-treated neuron culture [95]. In Japanese traditional Kampo medicine, clinical application of a single herbal medicine is rare, but formulae containing a combination of several herbal medicines are frequently used for treatment. Therefore, we focused on kihito as a candidate for an anti-AD formula and investigated its effects in in vitro and in vivo AD models before initiation of a clinical study. As expected, kihito induced axonal growth after Aβ-induced axonal atrophy in cultured neurons [96]. Kihito attenuated degenerated axons and memory in AD model mice i.c.v. injected with Aβ. It also increased expression of myelin basic protein (myelin marker) and synaptophysin (synaptic marker) in the brain of AD model mice. Thus, kihito probably reconstructed neuronal networks in the brain and improved memory in AD model mice.

The effects of kihito were clinically investigated in an open-label crossover study [97]. All AD patients in this investigation were recruited from the Toyama University hospital, Japan. Duration of kihito administration was 16 weeks. The Japanese version of the Mini-Mental State Examination (MMSE-J) was used for the evaluation. The mean of the MMSE-J score at the start of the study was 20.5, and mean age of the participants was 71.8 years. All groups took acetylcholinesterase inhibitors as usual. The MMSE-J scores significantly increased during the kihito administration period when compared with the period with no kihito administration. The temporal orientation subtest score in MMSE-J was significantly improved by the kihito administration. These findings showed that kihito improved cognitive function in AD patients.

Kamikihito is a modified formula of kihito, containing kihito, Bupleuri Radix (roots of *Bupleurum falcatum* L.), and Gardeniae Fructus (fruits of *Gardenia jasminoides* Ellis). We found that kamikihito also induced axonal growth in the Aβ-treated neuron culture and induced improvement in degenerated axons and memory in 5XFAD mice [39,98]. Using the DARTS method, cytosolic aspartate aminotransferase (cAST) was identified as the direct binding protein candidate of kamikihito [99].

Knock-down and pharmacological inhibition of cAST diminished the axonal growth activity of kamikihito. Treatment with recombinant cAST induced axonal growth in the normal neuron culture and in the brains of normal adult mice and enhanced memory function in normal adult mice [100]. Thus, kamikihito induces axonal growth via cAST activation, enhances memory function, and possibly improves memory in 5XFAD mice.

4. Conclusions

We developed two in vitro models of AD, namely the Aβ-induced axonal growth cone collapse model and the Aβ-induced axonal atrophy model. Using these models, we found several drugs derived from natural medicines, which prevented growth cone collapse or enhanced axonal growth. These agents prevented or induced recovery from memory deficits in AD model mice, and some of them were observed to be effective in humans. These findings indicate that axonal growth is possibly an important target for AD therapy and our models have high predictive validity for anti-AD activity.

Clarifying molecular mechanisms of natural medicines is complicated. Natural medicines contain thousands of constituents, and each of them shows various pharmacological effects. It is very complicated to identify active compounds and molecular targets of their compounds. These have made it difficult to analyze natural medicines scientifically. We advocated to resolve these problems by a combination of in vivo screening and the DARTS method. Our strategies could comprehensively identify active compounds in natural medicines and their target molecules, allowing to gain a foothold for usages of the natural medicines based on scientific evidence.

Several target molecules of anti-AD agents were identified in our studies. Most of these molecules had not been reported as AD-related molecules. Further detailed analysis of the signaling cascade of these molecules might reveal novel and effective pathways for AD prevention and cure. Natural medicines are clinically used for thousands of years. Safety of natural medicines has already been established. Studies of natural medicine might be a shortcut for the development of novel anti-AD drugs.

Author Contributions: All authors wrote the manuscript and drew the figures. All authors have read and agreed to the published version of the manuscript.

Funding: This work was partially funded by JSPS KAKENHI, Grant Numbers 18K07389 (T.K.) and 19K16288 (X.Y.), and research grants from the Takeda Science Foundation, Japan.

Acknowledgments: We would like to express our gratitude to all members in Division of Neuromedical Science, Institute of Natural Medicine, University of Toyama.

Conflicts of Interest: T.K. and C.T. declare that they received research fund from Kobayashi Pharmaceutical Co., Ltd., Japan. The funders had no role in the design of the study; in the collection, analyses, or interpretation of data; in the writing of the manuscript, or in the decision to publish the results.

Abbreviations

AD	Alzheimer's disease
APP	Amyloid precursor protein
Aβ	Amyloid β
PS	Presenilin
i.c.v.	Intracerebroventricular
PR	Polygalae Radix
DARTS	Drug affinity responsive target stability
HSC70	Heat shock cognate 70
RBANS	Repeatable Battery for the Assessment of Neuropsychological Status
DR	Drynariae Rhizoma
CRMP2	Collapsin response mediator protein 2
IL	Interleukin
IDE	Insulin degradation enzyme
MMSE-J	The Japanese version of the Mini-Mental State Examination
cAST	Cytosolic aspartate aminotransferase

References

1. Kumar, N. *Global Alzheimer's Disease Prevalent Cases Forecast, 2017–2027*; DRG Epidemiology, Decision Resources Group: Burlington, MA, USA, 2016.
2. Wimo, A.; Handels, R.; Winblad, B.; Black, C.M.; Johansson, G.; Salomonsson, S.; Eriksdotter, M.; Khandker, R.K. Quantifying and Describing the Natural History and Costs of Alzheimer's Disease and Effects of Hypothetical Interventions. *J. Alzheimers Dis.* **2020**, *75*, 891–902. [CrossRef] [PubMed]
3. Hardy, J.; Higgins, G. Alzheimer's disease: The amyloid cascade hypothesis. *Science* **1992**, *256*, 184–185. [CrossRef] [PubMed]
4. Insel, P.S.; Ossenkoppele, R.; Gessert, D.; Jagust, W.; Landau, S.; Hansson, O.; Weiner, M.W.; Mattsson, N. Time to Amyloid Positivity and Preclinical Changes in Brain Metabolism, Atrophy, and Cognition: Evidence for Emerging Amyloid Pathology in Alzheimer's Disease. *Front. Neurosci.* **2017**, *11*, 281. [CrossRef]
5. Reiss, A.B.; Arain, H.A.; Stecker, M.M.; Siegart, N.M.; Kasselman, L.J. Amyloid toxicity in Alzheimer's disease. *Rev. Neurosci.* **2018**, *29*, 613–627. [CrossRef] [PubMed]
6. Kondo, T.; Asai, M.; Tsukita, K.; Kutoku, Y.; Ohsawa, Y.; Sunada, Y.; Imamura, K.; Egawa, N.; Yahata, N.; Okita, K.; et al. Modeling Alzheimer's disease with iPSCs reveals stress phenotypes associated with intracellular Aβ and differential drug responsiveness. *Cell Stem Cell* **2013**, *12*, 487–496. [CrossRef]
7. Dorszewska, J.; Prendecki, M.; Oczkowska, A.; Dezor, M.; Kozubski, W. Molecular Basis of Familial and Sporadic Alzheimer's Disease. *Curr. Alzheimer Res.* **2016**, *13*, 952–963. [CrossRef]
8. Graham, W.V.; Bonito-Oliva, A.; Sakmar, T.P. Update on Alzheimer's Disease Therapy and Prevention Strategies. *Annu. Rev. Med.* **2017**, *68*, 413–430. [CrossRef]
9. Weller, J.; Budson, A. Current understanding of Alzheimer's disease diagnosis and treatment. *F1000Research* **2018**, 7. [CrossRef]
10. Jack, C.R., Jr.; Knopman, D.S.; Jagust, W.J.; Shaw, L.M.; Aisen, P.S.; Weiner, M.W.; Petersen, R.C.; Trojanowski, J.Q. Hypothetical model of dynamic biomarkers of the Alzheimer's pathological cascade. *Lancet Neurol.* **2010**, *9*, 119–128. [CrossRef]
11. Birks, J.S.; Harvey, R.J. Donepezil for dementia due to Alzheimer's disease. *Cochrane Database Syst. Rev.* **2018**, *6*, Cd001190. [CrossRef]
12. Loy, C.; Schneider, L. Galantamine for Alzheimer's disease and mild cognitive impairment. *Cochrane Database Syst. Rev.* **2006**, *1*, Cd001747. [CrossRef] [PubMed]
13. Birks, J.S.; Chong, L.Y.; Grimley Evans, J. Rivastigmine for Alzheimer's disease. *Cochrane Database Syst. Rev.* **2015**, *9*, Cd001191. [PubMed]
14. McShane, R.; Westby, M.J.; Roberts, E.; Minakaran, N.; Schneider, L.; Farrimond, L.E.; Maayan, N.; Ware, J.; Debarros, J. Memantine for dementia. *Cochrane Database Syst. Rev.* **2019**, *3*, Cd003154. [CrossRef] [PubMed]
15. Walsh, S.; King, E.; Brayne, C. France removes state funding for dementia drugs. *BMJ* **2019**, *367*, l6930. [CrossRef] [PubMed]
16. Benes, F.M.; Farol, P.A.; Majocha, R.E.; Marotta, C.A.; Bird, E.D. Evidence for axonal loss in regions occupied by senile plaques in Alzheimer cortex. *Neuroscience* **1991**, *42*, 651–660. [CrossRef]
17. Dickson, T.C.; Vickers, J.C. The morphological phenotype of beta-amyloid plaques and associated neuritic changes in Alzheimer's disease. *Neuroscience* **2001**, *105*, 99–107. [CrossRef]
18. Mattson, M.P. Pathways towards and away from Alzheimer's disease. *Nature* **2004**, *430*, 631–639. [CrossRef]
19. Alobuia, W.M.; Xia, W.; Vohra, B.P. Axon degeneration is key component of neuronal death in amyloid-beta toxicity. *Neurochem. Int.* **2013**, *63*, 782–789. [CrossRef]
20. Kuboyama, T. Visualizing Axonal Growth Cone Collapse and Early Amyloid β Effects in Cultured Mouse Neurons. *JoVE* **2018**, *140*, e58229. [CrossRef]
21. Kuboyama, T.; Tohda, C.; Komatsu, K. Neuritic regeneration and synaptic reconstruction induced by withanolide A. *Br. J. Pharmacol.* **2005**, *144*, 961–971. [CrossRef]
22. Kuboyama, T. Development of New Therapies for Neurodegenerative Diseases via Axonal Growth. *Yakugaku Zasshi* **2019**, *139*, 1385–1390. [CrossRef] [PubMed]
23. Yankner, B.A.; Dawes, L.R.; Fisher, S.; Villa-Komaroff, L.; Oster-Granite, M.L.; Neve, R.L. Neurotoxicity of a fragment of the amyloid precursor associated with Alzheimer's disease. *Science* **1989**, *245*, 417–420. [CrossRef] [PubMed]

24. Postuma, R.B.; He, W.; Nunan, J.; Beyreuther, K.; Masters, C.L.; Barrow, C.J.; Small, D.H. Substrate-bound beta-amyloid peptides inhibit cell adhesion and neurite outgrowth in primary neuronal cultures. *J. Neurochem.* **2000**, *74*, 1122–1130. [CrossRef] [PubMed]
25. Jawhar, S.; Trawicka, A.; Jenneckens, C.; Bayer, T.A.; Wirths, O. Motor deficits, neuron loss, and reduced anxiety coinciding with axonal degeneration and intraneuronal Abeta aggregation in the 5XFAD mouse model of Alzheimer's disease. *Neurobiol. Aging* **2012**, *33*, 196-e29. [CrossRef]
26. Tsai, J.; Grutzendler, J.; Duff, K.; Gan, W.B. Fibrillar amyloid deposition leads to local synaptic abnormalities and breakage of neuronal branches. *Nat. Neurosci.* **2004**, *7*, 1181–1183. [CrossRef]
27. Wirths, O.; Weis, J.; Kayed, R.; Saido, T.C.; Bayer, T.A. Age-dependent axonal degeneration in an Alzheimer mouse model. *Neurobiol. Aging* **2007**, *28*, 1689–1699. [CrossRef]
28. Perl, D.P. Neuropathology of Alzheimer's disease. *Mt. Sinai J. Med.* **2010**, *77*, 32–42. [CrossRef]
29. Liu, J.; Li, L. Targeting Autophagy for the Treatment of Alzheimer's Disease: Challenges and Opportunities. *Front. Mol. Neurosci.* **2019**, *12*, 203. [CrossRef]
30. Wu, H.Y.; Hudry, E.; Hashimoto, T.; Kuchibhotla, K.; Rozkalne, A.; Fan, Z.; Spires-Jones, T.; Xie, H.; Arbel-Ornath, M.; Grosskreutz, C.L.; et al. Amyloid beta induces the morphological neurodegenerative triad of spine loss, dendritic simplification, and neuritic dystrophies through calcineurin activation. *J. Neurosci.* **2010**, *30*, 2636–2649. [CrossRef]
31. Kuboyama, T.; Lee, Y.A.; Nishiko, H.; Tohda, C. Inhibition of clathrin-mediated endocytosis prevents amyloid beta-induced axonal damage. *Neurobiol. Aging* **2015**, *36*, 1808–1819. [CrossRef]
32. Tojima, T.; Hines, J.H.; Henley, J.R.; Kamiguchi, H. Second messengers and membrane trafficking direct and organize growth cone steering. *Nat. Rev. Neurosci.* **2011**, *12*, 191–203. [CrossRef] [PubMed]
33. Maurice, T.; Lockhart, B.P.; Privat, A. Amnesia induced in mice by centrally administered beta-amyloid peptides involves cholinergic dysfunction. *Brain Res.* **1996**, *706*, 181–193. [CrossRef]
34. Tohda, C.; Tamura, T.; Komatsu, K. Repair of amyloid beta(25-35)-induced memory impairment and synaptic loss by a Kampo formula, Zokumei-to. *Brain Res.* **2003**, *990*, 141–147. [CrossRef]
35. Tohda, C.; Matsumoto, N.; Zou, K.; Meselhy, M.R.; Komatsu, K. Abeta(25-35)-induced memory impairment, axonal atrophy, and synaptic loss are ameliorated by M1, A metabolite of protopanaxadiol-type saponins. *Neuropsychopharmacology* **2004**, *29*, 860–868. [CrossRef] [PubMed]
36. Oakley, H.; Cole, S.L.; Logan, S.; Maus, E.; Shao, P.; Craft, J.; Guillozet-Bongaarts, A.; Ohno, M.; Disterhoft, J.; Van Eldik, L.; et al. Intraneuronal beta-amyloid aggregates, neurodegeneration, and neuron loss in transgenic mice with five familial Alzheimer's disease mutations: Potential factors in amyloid plaque formation. *J. Neurosci.* **2006**, *26*, 10129–10140. [CrossRef] [PubMed]
37. Urano, T.; Tohda, C. Icariin improves memory impairment in Alzheimer's disease model mice (5xFAD) and attenuates amyloid beta-induced neurite atrophy. *Phytother. Res.* **2010**, *24*, 1658–1663. [CrossRef]
38. Joyashiki, E.; Matsuya, Y.; Tohda, C. Sominone improves memory impairments and increases axonal density in Alzheimer's disease model mice, 5XFAD. *Int. J. Neurosci.* **2011**, *121*, 181–190. [CrossRef]
39. Tohda, C.; Nakada, R.; Urano, T.; Okonogi, A.; Kuboyama, T. Kamikihi-to (KKT) rescues axonal and synaptic degeneration associated with memory impairment in a mouse model of Alzheimer's disease, 5XFAD. *Int. J. Neurosci.* **2011**, *121*, 641–648. [CrossRef]
40. Kuboyama, T.; Hirotsu, K.; Arai, T.; Yamasaki, H.; Tohda, C. Polygalae Radix Extract Prevents Axonal Degeneration and Memory Deficits in a Transgenic Mouse Model of Alzheimer's Disease. *Front. Pharmacol.* **2017**, *8*, 805. [CrossRef]
41. Park, H.; Kang, S.; Nam, E.; Suh, Y.H.; Chang, K.A. The Protective Effects of PSM-04 Against Beta Amyloid-Induced Neurotoxicity in Primary Cortical Neurons and an Animal Model of Alzheimer's Disease. *Front. Pharmacol.* **2019**, *10*, 2. [CrossRef]
42. Liu, Y.M.; Li, Z.Y.; Hu, H.; Xu, S.P.; Chang, Q.; Liao, Y.H.; Pan, R.L.; Liu, X.M. Tenuifolin, a secondary saponin from hydrolysates of polygalasaponins, counteracts the neurotoxicity induced by Abeta25-35 peptides in vitro and in vivo. *Pharmacol. Biochem. Behav.* **2015**, *128*, 14–22. [CrossRef] [PubMed]
43. Wu, J.G.; Wang, Y.Y.; Zhang, Z.L.; Yu, B. Herbal medicine in the treatment of Alzheimer's disease. *Chin. J. Integr. Med.* **2015**, *21*, 102–107. [CrossRef]
44. May, B.H.; Lu, C.; Lu, Y.; Zhang, A.L.; Xue, C.C. Chinese herbs for memory disorders: A review and systematic analysis of classical herbal literature. *J. Acupunct. Meridian Stud.* **2013**, *6*, 2–11. [CrossRef] [PubMed]

45. Shin, K.Y.; Lee, J.-Y.; Won, B.Y.; Jung, H.Y.; Chang, K.-A.; Koppula, S.; Suh, Y.-H. BT-11 is effective for enhancing cognitive functions in the elderly humans. *Neurosci. Lett.* **2009**, *465*, 157–159. [CrossRef]
46. Bhattacharya, S.K.; Satyan, K.S.; Ghosal, S. Antioxidant activity of glycowithanolides from Withania somnifera. *Indian J. Exp. Biol.* **1997**, *35*, 236–239.
47. Anbalagan, K.; Sadique, J. Influence of an Indian medicine (Ashwagandha) on acute-phase reactants in inflammation. *Indian J. Exp. Biol.* **1981**, *19*, 245–249. [PubMed]
48. Bhattacharya, S.K.; Goel, R.K.; Kaur, R.; Ghosal, S. Anti-stress activity of sitoindosides VII and VIII, new acylsterylglucosides from Withania somnifera. *Phytother. Res.* **1987**, *1*, 32–37. [CrossRef]
49. Singh, N.; Nath, R.; Lata, A.; Singh, S.P.; Kohli, R.P.; Bhargava, K.P. *Withania Somnifera* (Ashwagandha), a rejuvenating herbal drug which enhances survival during stress (an adaptogen). *Int. J. Crude Drug Res.* **1982**, *20*, 29–35. [CrossRef]
50. Ahmad, M.; Saleem, S.; Ahmad, A.S.; Ansari, M.A.; Yousuf, S.; Hoda, M.N.; Islam, F. Neuroprotective effects of *Withania somnifera* on 6-hydroxydopamine induced Parkinsonism in rats. *Hum. Exp. Toxicol.* **2005**, *24*, 137–147. [CrossRef]
51. Jain, S.; Shukla, S.D.; Sharma, K.; Bhatnagar, M. Neuroprotective effects of *Withania somnifera* Dunn. in hippocampal sub-regions of female albino rat. *Phytother. Res.* **2001**, *15*, 544–548. [CrossRef]
52. Tohda, C.; Kuboyama, T.; Komatsu, K. Dendrite extension by methanol extract of Ashwagandha (roots of *Withania somnifera*) in SK-N-SH cells. *Neuroreport* **2000**, *11*, 1981–1985. [CrossRef]
53. Zhao, J.; Nakamura, N.; Hattori, M.; Kuboyama, T.; Tohda, C.; Komatsu, K. Withanolide derivatives from the roots of *Withania somnifera* and their neurite outgrowth activities. *Chem. Pharm. Bull. (Tokyo)* **2002**, *50*, 760–765. [CrossRef] [PubMed]
54. Kuboyama, T.; Tohda, C.; Zhao, J.; Nakamura, N.; Hattori, M.; Komatsu, K. Axon- or dendrite-predominant outgrowth induced by constituents from Ashwagandha. *Neuroreport* **2002**, *13*, 1715–1720. [CrossRef] [PubMed]
55. Kuboyama, T.; Tohda, C.; Komatsu, K. Withanoside IV and its active metabolite, sominone, attenuate A beta(25-35)-induced neurodegeneration. *Eur. J. Neurosci.* **2006**, *23*, 1417–1426. [CrossRef] [PubMed]
56. Tohda, C.; Komatsu, K.; Kuboyama, T. Scientific basis for the anti-dementia drugs of constituents from Ashwagandha (*Withania somnifera*). *J. Trad. Med.* **2005**, *22*, 176–182.
57. Tohda, C.; Joyashiki, E. Sominone enhances neurite outgrowth and spatial memory mediated by the neurotrophic factor receptor, RET. *Br. J. Pharmacol.* **2009**, *157*, 1427–1440. [CrossRef] [PubMed]
58. Petukhova, E.O.; Mukhamedshina, Y.O.; Salafutdinov, II; Garanina, E.E.; Kaligin, M.S.; Leushina, A.V.; Rizvanov, A.A.; Reis, H.J.; Palotás, A.; Zefirov, A.L.; et al. Effects of Transplanted Umbilical Cord Blood Mononuclear Cells Overexpressing GDNF on Spatial Memory and Hippocampal Synaptic Proteins in a Mouse Model of Alzheimer's Disease. *J. Alzheimers Dis.* **2019**, *69*, 443–453.
59. Matsuya, Y.; Yamakawa, Y.; Tohda, C.; Teshigawara, K.; Yamada, M.; Nemoto, H. Synthesis of sominone and its derivatives based on an RCM strategy: Discovery of a novel anti-Alzheimer's disease medicine candidate "denosomin". *Org. Lett.* **2009**, *11*, 3970–3973. [CrossRef]
60. Nakayama, N.; Tohda, C. Withanoside IV improves hindlimb function by facilitating axonal growth and increase in peripheral nervous system myelin level after spinal cord injury. *Neurosci. Res.* **2007**, *58*, 176–182. [CrossRef]
61. Teshigawara, K.; Kuboyama, T.; Shigyo, M.; Nagata, A.; Sugimoto, K.; Matsuya, Y.; Tohda, C. A novel compound, denosomin, ameliorates spinal cord injury via axonal growth associated with astrocyte-secreted vimentin. *Br. J. Pharmacol.* **2013**, *168*, 903–919. [CrossRef]
62. Zahiruddin, S.; Basist, P.; Parveen, A.; Parveen, R.; Khan, W.; Gaurav; Ahmad, S. Ashwagandha in brain disorders: A review of recent developments. *J. Ethnopharmacol.* **2020**, *257*, 112876. [CrossRef] [PubMed]
63. Tohda, C.; Urano, T.; Umezaki, M.; Nemere, I.; Kuboyama, T. Diosgenin is an exogenous activator of 1,25D$_3$-MARRS/Pdia3/ERp57 and improves Alzheimer's disease pathologies in 5XFAD mice. *Sci. Rep.* **2012**, *2*, 535. [CrossRef] [PubMed]
64. Lomenick, B.; Jung, G.; Wohlschlegel, J.A.; Huang, J. Target identification using drug affinity responsive target stability (DARTS). *Curr. Protoc. Chem. Biol.* **2011**, *3*, 163–180. [CrossRef] [PubMed]
65. Lomenick, B.; Hao, R.; Jonai, N.; Chin, R.M.; Aghajan, M.; Warburton, S.; Wang, J.; Wu, R.P.; Gomez, F.; Loo, J.A.; et al. Target identification using drug affinity responsive target stability (DARTS). *Proc. Natl. Acad. Sci. USA* **2009**, *106*, 21984–21989. [CrossRef]

66. Lomenick, B.; Olsen, R.W.; Huang, J. Identification of direct protein targets of small molecules. *ACS Chem. Biol.* **2011**, *6*, 34–46. [CrossRef]
67. Larsson, B.; Nemere, I. Effect of growth and maturation on membrane-initiated actions of 1,25-dihydroxyvitamin D3-II: Calcium transport, receptor kinetics, and signal transduction in intestine of female chickens. *J. Cell. Biochem.* **2003**, *90*, 901–913. [CrossRef]
68. Nemere, I.; Farach-Carson, M.C.; Rohe, B.; Sterling, T.M.; Norman, A.W.; Boyan, B.D.; Safford, S.E. Ribozyme knockdown functionally links a 1,25(OH)2D3 membrane binding protein (1,25D3-MARRS) and phosphate uptake in intestinal cells. *Proc. Natl. Acad. Sci. USA* **2004**, *101*, 7392–7397. [CrossRef]
69. Rosso, A.; Pansera, M.; Zamoner, A.; Zanatta, L.; Bouraïma-Lelong, H.; Carreau, S.; Silva, F.R. 1α,25(OH)2-Vitamin D3 stimulates rapid plasma membrane calcium influx via MAPK activation in immature rat Sertoli cells. *Biochimie* **2012**, *94*, 146–154. [CrossRef]
70. Buitrago, C.G.; Arango, N.S.; Boland, R.L. 1α,25(OH)2D3-dependent modulation of Akt in proliferating and differentiating C2C12 skeletal muscle cells. *J. Cell. Biochem.* **2012**, *113*, 1170–1181. [CrossRef]
71. Yang, X.; Tohda, C. Diosgenin restores Aβ-induced axonal degeneration by reducing the expression of heat shock cognate 70 (HSC70). *Sci. Rep.* **2018**, *8*, 11707. [CrossRef]
72. Yang, X.; Tohda, C. Heat Shock Cognate 70 Inhibitor, VER-155008, Reduces Memory Deficits and Axonal Degeneration in a Mouse Model of Alzheimer's Disease. *Front. Pharmacol.* **2018**, *9*, 48. [CrossRef] [PubMed]
73. Tohda, C.; Lee, Y.A.; Goto, Y.; Nemere, I. Diosgenin-induced cognitive enhancement in normal mice is mediated by 1,25D3-MARRS. *Sci. Rep.* **2013**, *3*, 3395. [CrossRef]
74. Jin, M.; Shepardson, N.; Yang, T.; Chen, G.; Walsh, D.; Selkoe, D.J. Soluble amyloid beta-protein dimers isolated from Alzheimer cortex directly induce Tau hyperphosphorylation and neuritic degeneration. *Proc. Natl. Acad. Sci. USA* **2011**, *108*, 5819–5824. [CrossRef]
75. Hanger, D.P.; Anderton, B.H.; Noble, W. Tau phosphorylation: The therapeutic challenge for neurodegenerative disease. *Trends Mol. Med.* **2009**, *15*, 112–119. [CrossRef]
76. Mandelkow, E.M.; Biernat, J.; Drewes, G.; Gustke, N.; Trinczek, B.; Mandelkow, E. Tau domains, phosphorylation, and interactions with microtubules. *Neurobiol. Aging* **1995**, *16*, 355–362. [CrossRef]
77. Tohda, C.; Yang, X.; Matsui, M.; Inada, Y.; Kadomoto, E.; Nakada, S.; Watari, H.; Shibahara, N. Diosgenin-Rich Yam Extract Enhances Cognitive Function: A Placebo-Controlled, Randomized, Double-Blind, Crossover Study of Healthy Adults. *Nutrients* **2017**, *9*, 1160. [CrossRef] [PubMed]
78. Chinese Pharmacopoeia Commission. *Pharmacopeia of the People's Republic of China*; People's Medicine Publishing House Press: Beijing, China, 2010; Volume I.
79. Yang, Z.Y.; Kuboyama, T.; Kazuma, K.; Konno, K.; Tohda, C. Active Constituents from Drynaria fortunei Rhizomes on the Attenuation of Abeta25-35-Induced Axonal Atrophy. *J. Nat. Prod.* **2015**, *78*, 2297–2300. [CrossRef] [PubMed]
80. Yang, Z.; Kuboyama, T.; Tohda, C. A Systematic Strategy for Discovering a Therapeutic Drug for Alzheimer's Disease and Its Target Molecule. *Front. Pharmacol.* **2017**, *8*, 340. [CrossRef]
81. Koehn, F.E.; Carter, G.T. The evolving role of natural products in drug discovery. *Nat. Rev. Drug Discov.* **2005**, *4*, 206–220. [CrossRef] [PubMed]
82. Fang, T.; Wang, Y.; Ma, Y.; Su, W.; Bai, Y.; Zhao, P. A rapid LC/MS/MS quantitation assay for naringin and its two metabolites in rats plasma. *J. Pharm. Biomed. Anal.* **2006**, *40*, 454–459. [CrossRef] [PubMed]
83. Hensley, K.; Kursula, P. Collapsin Response Mediator Protein-2 (CRMP2) is a Plausible Etiological Factor and Potential Therapeutic Target in Alzheimer's Disease: Comparison and Contrast with Microtubule-Associated Protein Tau. *J. Alzheimers Dis.* **2016**, *53*, 1–14. [CrossRef] [PubMed]
84. Cole, A.R.; Knebel, A.; Morrice, N.A.; Robertson, L.A.; Irving, A.J.; Connolly, C.N.; Sutherland, C. GSK-3 phosphorylation of the Alzheimer epitope within collapsin response mediator proteins regulates axon elongation in primary neurons. *J. Biol. Chem.* **2004**, *279*, 50176–50180. [CrossRef] [PubMed]
85. Cole, A.R.; Noble, W.; van Aalten, L.; Plattner, F.; Meimaridou, R.; Hogan, D.; Taylor, M.; LaFrancois, J.; Gunn-Moore, F.; Verkhratsky, A.; et al. Collapsin response mediator protein-2 hyperphosphorylation is an early event in Alzheimer's disease progression. *J. Neurochem.* **2007**, *103*, 1132–1144. [CrossRef] [PubMed]
86. Tang, Y.; Le, W. Differential Roles of M1 and M2 Microglia in Neurodegenerative Diseases. *Mol. Neurobiol.* **2016**, *53*, 1181–1194. [CrossRef]
87. Chen, Z.; Trapp, B.D. Microglia and neuroprotection. *J. Neurochem.* **2016**, *136*, 10–17. [CrossRef] [PubMed]

88. Shimizu, E.; Kawahara, K.; Kajizono, M.; Sawada, M.; Nakayama, H. IL-4-induced selective clearance of oligomeric beta-amyloid peptide(1-42) by rat primary type 2 microglia. *J. Immunol.* **2008**, *181*, 6503–6513. [CrossRef]
89. Cherry, J.D.; Olschowka, J.A.; O'Banion, M.K. Arginase 1+ microglia reduce Abeta plaque deposition during IL-1beta-dependent neuroinflammation. *J. Neuroinflamm.* **2015**, *12*, 203. [CrossRef]
90. Kuboyama, T.; Wahane, S.; Huang, Y.; Zhou, X.; Wong, J.K.; Koemeter-Cox, A.; Martini, M.; Friedel, R.H.; Zou, H. HDAC3 inhibition ameliorates spinal cord injury by immunomodulation. *Sci. Rep.* **2017**, *7*, 8641. [CrossRef]
91. Yang, Z.; Kuboyama, T.; Tohda, C. Naringenin promotes microglial M2 polarization and Abeta degradation enzyme expression. *Phytother. Res.* **2019**, *33*, 1114–1121. [CrossRef]
92. Yamauchi, Y.; Ge, Y.W.; Yoshimatsu, K.; Komastu, K.; Kuboyama, T.; Yang, X.; Tohda, C. Memory Enhancement by Oral Administration of Extract of Eleutherococcus senticosus Leaves and Active Compounds Transferred in the Brain. *Nutrients* **2019**, *11*, 1142. [CrossRef]
93. Tohda, C.; Matsui, M.; Inada, Y.; Yang, X.; Kuboyama, T.; Kimbara, Y.; Watari, H. Combined Treatment with Two Water Extracts of Eleutherococcus senticosus Leaf and Rhizome of Drynaria fortunei Enhances Cognitive Function: A Placebo-Controlled, Randomized, Double-Blind Study in Healthy Adults. *Nutrients* **2020**, *12*, 303. [CrossRef]
94. Tohda, C.; Tamura, T.; Matsuyama, S.; Komatsu, K. Promotion of axonal maturation and prevention of memory loss in mice by extracts of Astragalus mongholicus. *Br. J. Pharmacol.* **2006**, *149*, 532–541. [CrossRef] [PubMed]
95. Naito, R.; Tohda, C. Characterization of anti-neurodegenerative effects of Polygala tenuifolia in Abeta(25-35)-treated cortical neurons. *Biol. Pharm. Bull.* **2006**, *29*, 1892–1896. [CrossRef] [PubMed]
96. Tohda, C.; Naito, R.; Joyashiki, E. Kihi-to, a herbal traditional medicine, improves Abeta(25-35)-induced memory impairment and losses of neurites and synapses. *BMC Complement. Altern. Med.* **2008**, *8*, 49. [CrossRef] [PubMed]
97. Watari, H.; Shimada, Y.; Matsui, M.; Tohda, C. Kihito, a Traditional Japanese Kampo Medicine, Improves Cognitive Function in Alzheimer's Disease Patients. *Evid. Based Complement. Alternat. Med.* **2019**, *2019*, 4086749. [CrossRef]
98. Watari, H.; Shimada, Y.; Tohda, C. New Treatment for Alzheimer's Disease, Kamikihito, Reverses Amyloid-β Induced Progression of Tau Phosphorylation and Axonal Atrophy. *Evid. Based Complement. Alternat. Med.* **2014**, *2014*, 706487. [CrossRef]
99. Watari, H.; Shimada, Y.; Tohda, C. Cytosolic aspartate aminotransferase, a direct binding protein of kamikihito, regulates axon growth. *Tradit. Kampo Med.* **2016**, *3*, 41–49.
100. Kobayashi, R.; Tohda, C. Extracellular Cytosolic Aspartate Aminotransferase Promotes Axonal Growth and Object Recognition Memory. *Neurochem. Res.* **2017**, *42*, 3465–3473. [CrossRef]

International Journal of
Molecular Sciences

MDPI

Review

High-Speed Atomic Force Microscopy Reveals the Structural Dynamics of the Amyloid-β and Amylin Aggregation Pathways

Takahiro Watanabe-Nakayama [1,*], Bikash R. Sahoo [2], Ayyalusamy Ramamoorthy [3] and Kenjiro Ono [4]

1 WPI Nano Life Science Institute, Kanazawa University, Kakuma-machi, Kanazawa 920-1192, Japan
2 Biophysics Program, Department of Chemistry, Macromolecular Science and Engineering, and Biomedical Engineering, The University of Michigan, Ann Arbor, MI 48109-1055, USA; bsahoo@umich.edu
3 Biophysics and Department of Chemistry, University of Michigan, Ann Arbor, MI 48109-1055, USA; ramamoor@umich.edu
4 Division of Neurology, Department of Internal Medicine, School of Medicine, Showa University, Hatanodai, Shinagawa district, Tokyo 142-8666, Japan; onoken@med.showa-u.ac.jp
* Correspondence: tnakawata@se.kanazawa-u.ac.jp; Tel.: +81-76-234-4573

Received: 13 May 2020; Accepted: 14 June 2020; Published: 16 June 2020

Abstract: Individual Alzheimer's disease (AD) patients have been shown to have structurally distinct amyloid-β (Aβ) aggregates, including fibrils, in their brain. These findings suggest the possibility of a relationship between AD progression and Aβ fibril structures. Thus, the characterization of the structural dynamics of Aβ could aid the development of novel therapeutic strategies and diagnosis. Protein structure and dynamics have typically been studied separately. Most of the commonly used biophysical approaches are limited in providing substantial details regarding the combination of both structure and dynamics. On the other hand, high-speed atomic force microscopy (HS-AFM), which simultaneously visualizes an individual protein structure and its dynamics in liquid in real time, can uniquely link the structure and the kinetic details, and it can also unveil novel insights. Although amyloidogenic proteins generate heterogeneously aggregated species, including transient unstable states during the aggregation process, HS-AFM elucidated the structural dynamics of individual aggregates in real time in liquid without purification and isolation. Here, we review and discuss the HS-AFM imaging of amyloid aggregation and strategies to optimize the experiments showing findings from Aβ and amylin, which is associated with type II diabetes, shares some common biological features with Aβ, and is reported to be involved in AD.

Keywords: molecular imaging; amyloid β-peptides; islet amyloid polypeptide; high-speed atomic force microscopy

1. Introduction

The relationship between neurodegenerative disease symptoms and the structural dynamics of associated amyloidogenic proteins has been revealed. For amyloid-β (Aβ), different fibril structures and morphologies have been observed in Alzheimer's disease (AD) patients [1,2]. Therefore, the Aβ polymorphism was correlated with pathological phenotypes. Similarly, fibril polymorphism and the associated toxicity/infectivity was identified in other amyloidogenic proteins, such as human tau, α-synuclein, and yeast prion-derived Sup35. Tau protein amyloid fibrils demonstrated different isoforms and structures between AD [3], Pick's disease [4], progressive supranuclear palsy (PSP) [5], corticobasal degeneration (CBD) [5], and chronic traumatic encephalopathy (CTE) in tauopathy [6]. α-synuclein fibrils with different structures exhibited different toxicities [7] and caused different symptoms in α-synucleinopathy [8,9]. One of the best studied prion model proteins, yeast Sup35, formed fibrils with different structures and mechanical stiffness depending on the temperature during aggregation, which was reflected in prion activities, such as fragmentation and propagation [10–12]. Elucidating the structural dynamics of the amyloid protein is indispensable to uncovering the mechanism underlying disease onset and the action of drug candidates. Examination of the structure could lead to the diagnosis and treatment methods for the progression of the disease in each individual [13].

Amyloidogenic proteins commonly have intrinsically disordered regions in some or all of their monomeric structure, and they change these to form aggregation core structures along the aggregation pathway. The structural dynamics during the aggregation process depend on the surrounding physicochemical conditions in vivo, including the free-in-solution or membrane-bound forms, pH, and electrolytes. Aβ40, a 40 residue Aβ variant cleaved from the amyloid precursor protein (APP), is unstructured [14,15] and can also form a lowly populated 3_{10} helical structure in solution [16]; disordered oligomers were also reported [17]. This variant formed disordered helical structures upon interaction with the lipid membrane [18], and could disrupt the membrane structure via a two-step mechanism consisting of fiber-independent pore formation and fiber-dependent 'detergent-like' membrane fragmentation [19]. Aβ42—a 42 amino acid chain of Aβ variants—showed different structural dynamics from Aβ40 through the interaction with a membrane. Aβ42 oligomerization was accelerated by the lipid membrane [20]. Aβ42 oligomers assembled into the pore-forming oligomers with three distinct pore sizes that functioned as ion channels, while Aβ40 did not form such pores [21]. Upon interaction with a reconstituted membrane, Aβ42 assembled into a β-barrel structure, while Aβ40 formed fibrils [22]. Tau [23], α-synuclein, and other amyloidogenic proteins, including amylin, also formed ion-channel oligomers in the membrane [24]. Cytosolic acidification caused by oxidative stress promoted AD [25] and Parkinson's disease (PD) [26]. A change in pH altered the structural dynamics and aggregation pathways of Aβ [27–30], tau [31], α-synuclein [32–35], and amylin [36]. Some amyloid oligomeric conformers were more toxic than amyloid fibrils [37–39]. Different oligomers showed different structures and toxicities [40,41]. Thus, amyloid fibrils with different structures may reflect variable toxicity, aggregation pathways, and surrounding microenvironments in different patients and symptoms [1,42].

To date, investigations of the structural dynamics of amyloidogenic protein aggregation examined the structure and dynamics separately. X-ray crystallography, nuclear magnetic resonance (NMR), and cryo-electron microscopy (cryo-EM) revealed the spatial coordinates of constituent atoms in the protein structure. X-ray crystallography demonstrated the fibril structures of amyloidogenic protein fragments [43,44] and oligomers of Aβ peptides [45–48]. NMR studies reported the high-resolution structural details of membrane-bound amyloid oligomers [41,49,50]. Solution- and solid-state NMR was used to investigate the monomer and transient intermediate structures in Aβ assemblies [16,51–58]. In addition, solid-state NMR studies revealed the in-register parallel β-sheets in Aβ fibril structures [1, 42,59–64]. Cryo-EM studies discovered the Aβ42 fibril structure [65] and individual Aβ40 strain structures from AD patients [2]. However, the structural images were static and average across the applied protein samples. Thus, those methods require structural homogeneity in the analyzed protein samples.

Single molecule observation using fluorescence dyes under optical microscopy was used to visualize the structural dynamics of amyloidogenic protein aggregations. Thioflavin T (ThT), specific for the cross-β structure, [66,67] was used to visualize the fibril elongation of Aβ40 [68], α-synuclein [69], and prion protein (PrP) [70]. Other fluorophores co-assembled or cross-linked to proteins were used to visualize the aggregation dynamics of α-synuclein [71–76] and the fibril growth of Aβ [77] and Sup35 [78]. However, the captured images did not show the protein structure but the spatial distribution of fluorescence spots.

Atomic force microscopy (AFM) visualizes individual molecules at the nanometer spatial resolution in solution, although the temporal resolution is low in conventional AFM [79–84].

AFM was used to capture structural images and measure the nanomechanical properties of individual amyloid aggregates [85–87] in the ongoing heterologous aggregation processes of Aβ [20,88–98], synuclein [90,99–106], and amylin [107,108]. High-speed AFM (HS-AFM) enabled the kinetic measurement of the structural dynamics of biological molecular processes [79–84] including amyloid aggregation [93,108–116]. Here, we show that HS-AFM links structural and dynamics studies, reviewing recent HS-AFM studies and including our findings for Aβ42 [93] and amylin [116], which is associated with not only type II diabetes but also AD [117–123].

2. HS-AFM Observation of Aβ42 Fibril Growth

2.1. HS-AFM Observation of Self-Replicative Aβ42 Fibril Growth

Aβ fibril formation has been thought to be similar to the self-replication mechanism for prion proteins. In this mechanism, the soluble conformer assembles and changes its structure to an abnormal form, and some of the abnormal conformers grow into fibrous structures with the incorporation of the soluble conformers [60]. The time-lapse AFM with a lower scanning rate observed Aβ40 protofibrils and mature fibrils growing in a self-replicative manner [89,107]. To assess Aβ42, characterized by faster aggregation, we used HS-AFM.

The sample preparation procedure and the imaging conditions were critical for the HS-AFM observation, as described in Section 8.1. We prepared low molecular weight (LMW) and high molecular weight (HMW) Aβ42 fractions in 10 mM sodium phosphate, with pH 7.4 [93,124]. Aβ42 fibril formation and elongation was observed in LMW Aβ42 within approximately 1 h after the addition of 0.1 M NaCl [93] (Figure 1). The LMW Aβ42 was introduced to the HS-AFM sample chamber (Figure 1a). Sodium chloride was added just before or after the peptide introduction; this was immediately followed by HS-AFM observation. Species in solution were adsorbed to the surface and became detectable by HS-AFM (Figure 1a,b) [93]. The bound aggregates interacted with species in the solution and were able to elongate as shown in Figure 1b.

The HS-AFM images clearly distinguished the three structurally distinct types of fibrils in this condition: (1) the spiral fibrils with a ≈100 nm periodicity in height, (2) the straight fibrils without any structural periodicity, and (3) the hybrid fibrils in which the spiral and the straight parts were mixed (see Figure 2) [93]. This result indicates that the manner of Aβ42 fibril growth followed the prion-like self-replication in the spiral and the straight parts [93]. In addition, HS-AFM also uncovered the structural switch in the hybrid fibril elongation and the presence of spherical oligomers in the sample mixture [93]. A recent solution NMR study identified fibril elongation and the formation of a heterogeneous mixture (fibrils and oligomers) when Aβ monomers were added to sonicated preformed seeds in real time [125]. However, a solid-state NMR investigation is needed to obtain additional high-resolution structural information; whereas, HS-AFM experiments eased the retrieval of the structural switch and fibril polymorphism in real time.

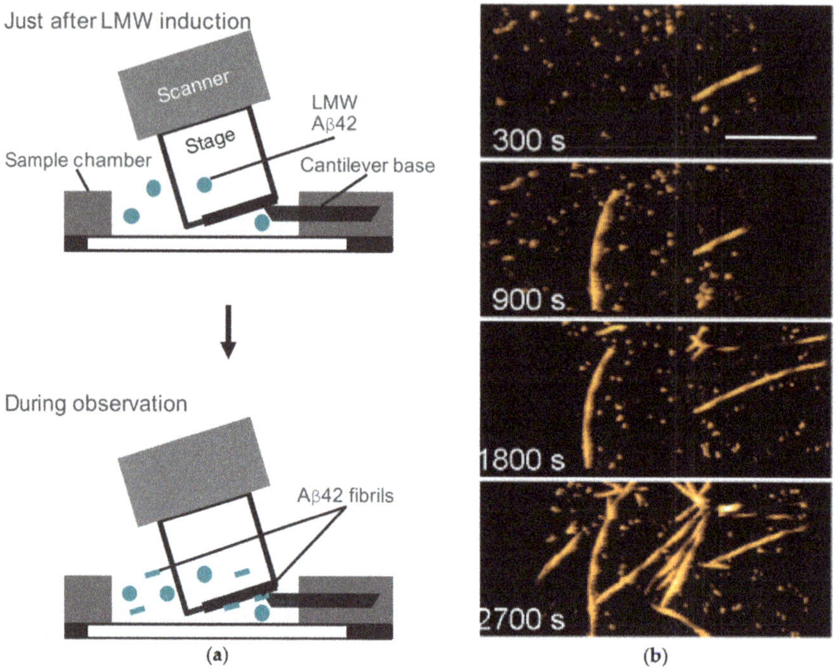

Figure 1. High-speed atomic force microscopy (HS-AFM) observation of amyloid-β (Aβ)42 fibril elongation. (**a**) Schematic view of HS-AFM observation of low molecular weight (LMW) Aβ42 incubation. LMW Aβ42 in 10 mM sodium phosphate, at pH 7.4, was introduced with 0.1 M NaCl for aggregation acceleration. Some aggregates in the solution bound to the mica surface and the fibrous aggregates in them were elongated by the incorporation of the LMW Aβ42 in free solution. (**b**) Representative HS-AFM images of LMW Aβ42 incubation at the indicated time after addition of 0.1 M NaCl. The scale bar is 300 nm. Reproduced from [93].

The molecular process of Aβ aggregation was characterized by HS-AFM observation under various physicochemical conditions. The structural switch of the growth mode was also characterized to be an inherent process in the Aβ42 self-replication reaction. To characterize the structural switch, Aβ42 fibril formation and elongation was observed after the addition of 0.1 M sodium chloride or potassium chloride [93]. The interaction between the HS-AFM stage surface and the observed Aβ42 molecules was different in the presence of sodium and potassium ions, as described in Section 8.3. As shown in Figure 3a, the fibril type distribution was different under the sodium and potassium conditions [93]. The spiral and the hybrid types were dominant in the sodium and potassium ion buffers, respectively, with no significant difference in the fibril length between the two conditions (Figure 3a,b,d) [93].

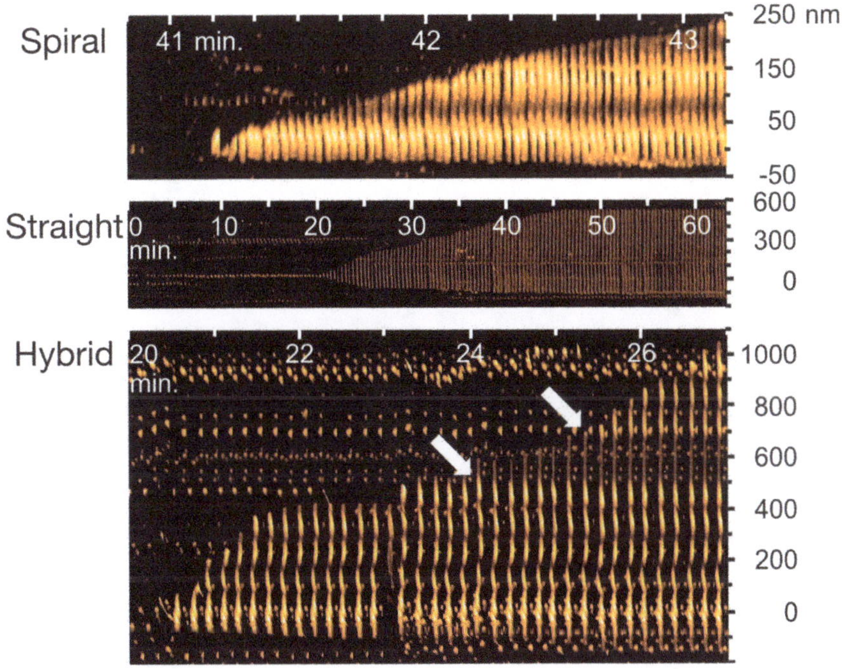

Figure 2. Kymographs of spiral, straight, and hybrid types of Aβ42 fibrils from HS-AFM images of the indicated time ranges after addition of 0.1 M NaCl. Arrows indicate the positions at which the growth mode switched from the spiral to the straight at 24 min and from the straight to the spiral at around 25 min. Reproduced from [93].

The appearance frequency of the spiral and the straight growth modes decreased and increased in the potassium buffer, respectively, when compared with the sodium buffer (Figure 3c) [93]. The lengths of the spiral and straight sections in the potassium condition were shorter and longer than in the sodium condition, respectively [93]. These results indicated that the structural switch process could be modulated by changes in the physicochemical conditions, and that the sodium buffer constrained fibril growth to the spiral mode, while the potassium buffer decreased the activation energy and the free energy difference between the spiral and the straight states of the structure switch process [93]. These findings revise the conventional model of the self-templating replication of amyloid fibrils, suggesting that the fibril structure strain can be altered by the external physicochemical environment even after fibril seed formation (nucleation) [93].

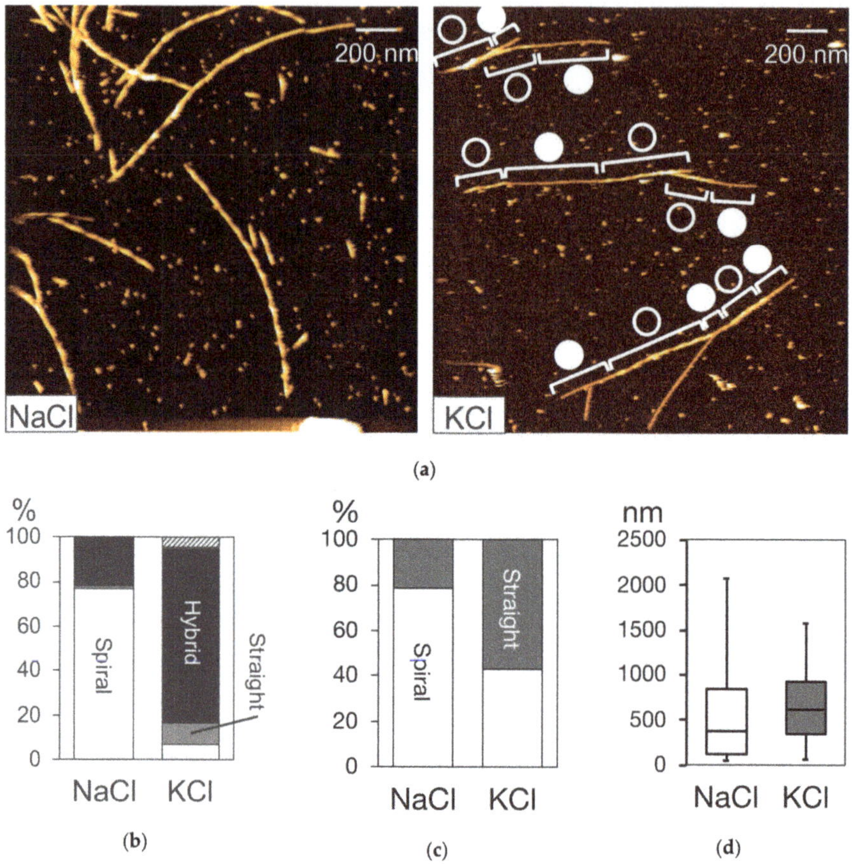

Figure 3. HS-AFM observation of fibril elongation from LMW Aβ42 incubation on mica surface in 10 mM sodium phosphate, pH 7.4 titrated with 0.1 M NaCl, or KCl. (**a**) HS-AFM images of Aβ 42 fibrils approximately 1 h after the addition of NaCl or KCl as indicated. The spiral and straight parts in the hybrid-type fibrils are indicated by open and closed circles. (**b–d**) Distributions of the fibril type (**b**), the growth mode (**c**), and the fibril length (**d**) under NaCl and KCl conditions. The top hatched portion for KCl in (**b**) indicates the fibrils whose structure was not determined due to their length being shorter than the spiral pitch. Reproduced from [93] except for the image in NaCl shown in (**a**).

2.2. Kinetic Analysis of Aβ42 Fibril Growth

The high spatiotemporal resolution of HS-AFM enables kinetic analysis of fibril elongation according to the fibril structure. Aβ42 fibrils, especially at the fast ends, repeated the pause (dwell) time and the growth phase during their elongation in both the spiral and the straight growth modes (Figure 4a) [93]. The dwell time, step time, and step size showed the single exponential distributions (Figure 4b), suggesting that the transition between the dwell phase and growth phase proceeds with first-order kinetics [93]. The kinetic parameters for this transition were different between the spiral and straight growth modes (Figure 4c) [93]. The step sizes (62 nm and 36 nm of mean values for the spiral and the straight growth modes, respectively) were much longer than the width of the single β-strand (0.47 nm) in amyloid fibrils, which suggests that a number of peptides were taken into the fibrils during the growth phase, and that the transition between the dwell and growth phases did not correspond to the incorporation or dissociation of peptides (peptide concentration independence of the dwell phase needs to be examined) (Figure 4d).

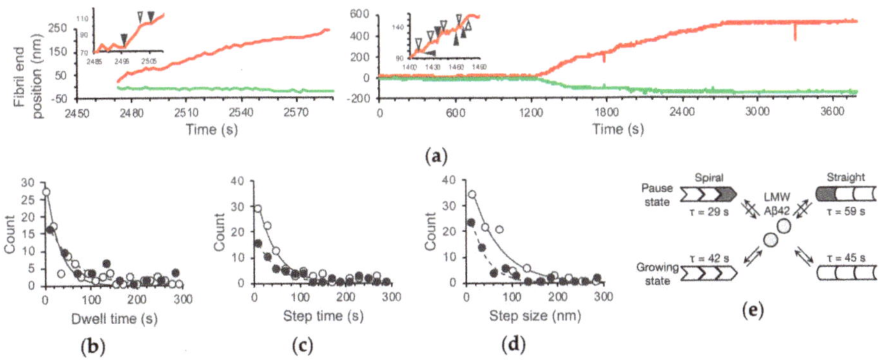

Figure 4. Stepwise growth of Aβ42 fibrils. (**a**) Time courses of the fast (red) and the slow (green) ends of the spiral and straight fibrils in Figure 2. The insets are the enlarged time courses of the fast ends. The open and closed triangles indicate the start and end of pause states (dwell time). (**b–d**) Distribution of the dwell time (**b**), time for step (**c**), and step size (**d**) with single exponential fits giving the mean life times of the pause and growing states in (**e**), and mean step sizes, for the spiral (open circles with solid lines) and straight (closed circles with dashed lines) type fibrils from the LMW Aβ42 incubation. (**e**) Mechanistic model of the stepwise growth of the spiral and straight Aβ42 fibrils with the kinetic parameters from (**b**) and (**c**). Reproduced from [93].

The similar stepwise growth and its kinetics were characterized in Aβ25–35 fibril growth by AFM [126]. The transitions between the pause and growing phases were also identified as first-order kinetic processes with the structure conversion at the fibril ends [126]. The fibril ends can neither incorporate nor release peptides at the pause phase (the blocked state), while they consecutively grow with 7 nm or its integer multiples during the growing phase [126]. The differences in the kinetic parameters between the spiral and the straight growth modes of Aβ42 indicate that the energy landscape for the transition between the dwell and growth modes was different between the two types of fibrils. The fibril end structure at the dwell phase may also be different between the spiral and straight fibrils (Figure 4e).

3. Mechanical Force Modulates the Transition from Oligomeric to Filamentous States of Aβ

HS-AFM can be used to investigate the force-dependent molecular processes. The physical environment is also one of essential factors that determine biochemical responses [127]. In the amyloid research field, the application of external mechanical forces, such as shaking, agitation, and sheer stress, has been used to accelerate aggregation and fibrillation [128–132]. A recent study by Wang et al. showed the effects of mechanical sample rotation in the commonly used magic angle spinning NMR experiments to study amyloid-β aggregation and the effect of (−)-epigallocatechin gallate (EGCG), which is a small molecule polyphenolic compound found in green tea extract [133]. However, the underlying mechanism of the physical effects has remained unclear. Although the exerted force to the sample by the AFM tip is usually kept as small as possible (see Section 8.4), methods for controlling and actively utilizing the applied force to characterize the molecular processes have been developed [134,135].

Using the external force applied by HS-AFM, Tashiro et al. found the linearly organized globular Aβ42 oligomers that showed the force-dependent thin filament extrusion [112]. They first observed the aggregate species in 1-day Aβ42 incubation bound to the stage surface, and then temporarily increased the tapping force of the cantilever tip, which is immediately followed by the observation of the same area with the original tapping force [112]. As shown in Figure 5, the breakage of the connected oligomers induced the extension of the thin straight filaments [112], which suggests that the stability of the oligomers may suppress their structural conversion into the filaments and that its removal may be promoted by the applied force. This result provides insights into the importance of the mechanistic stability of amyloid oligomers.

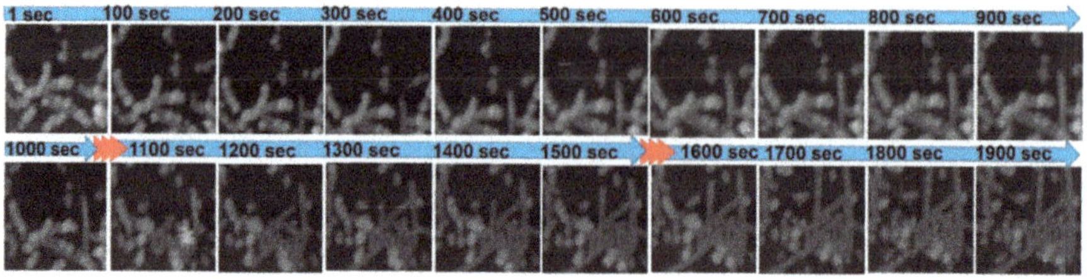

Figure 5. HS-AFM imaging of tip-induced Aβ42 filament growth. The HS-AFM tapping amplitude was transiently increased at the time indicated by the red arrows. Reprint with permission [112]; Copyright (2019), John Wiley and Sons.

4. HS-AFM Observation of Aβ42 Aggregates on the Off-Pathway

4.1. HS-AFM Observation of Dissociating Aβ42 Aggregates

HS-AFM visualizes the structural dynamics of non-fibrous Aβ42 aggregates. The HMW fraction, which was obtained by gel filtration, mainly contained globular aggregates [93]. They were classified into three different types of structural dynamics after the addition of 0.1 M sodium chloride: (1) gradually dissociating aggregates, (2) aggregates with unchanging sizes, and (3) stepwise dissociating aggregates (Figure 6a–f) [93]. The type I aggregates exponentially decreased in size (measured as the height from the HS-AFM stage) and then reached certain non-zero sizes, which suggests that the type I aggregates were composed of at least two parts with different dissociation constants (Figure 6d) [93]. The time course of the type I size, h(t), can be expressed as Equation (1):

$$h(t) = h_{a0} \cdot \exp(-k_{a0} \cdot t) + h_{b0} \cdot \exp(-k_{b0} \cdot t) \quad (1)$$

where h_{a0} and k_{a0} are the initial height and dissociation constant of part a, and h_{b0} and k_{b0} are the initial height and dissociation constant of part b. When the dissociation rate is extremely low ($k_{b0} \fallingdotseq 0$), Equation (1) can be rewritten as Equation (2):

$$h(t) = h_{a0} \cdot \exp(-k_{a0} \cdot t) + h_{b0}. \quad (2)$$

Equation (2) fitted the time courses of the type I aggregates well (Figure 6d) [93]. These analyses indicated that the type I aggregates formed during the sample preparation in 10 mM sodium phosphate, pH 7.4, and were then dissociated by the equilibrium shift after the addition of 0.1 M sodium chloride. For the relationship between the type I, II, and III aggregates, there were at least two possibilities: (1) they were distinct assemblies or (2) type II and III correspond to the slow dissociating part (the part b) of the type I aggregates [93]. The fast dissociating parts were completely dissociated before the aggregates appeared in the observation area, which may correspond to the type II aggregates (Figure 6b,d) [93]. We should also note that the tapping AFM probe may accidentally break the type II aggregates, resulting in a stepwise decrease in the aggregate size and thus generate type III aggregates (Figure 6f) [93].

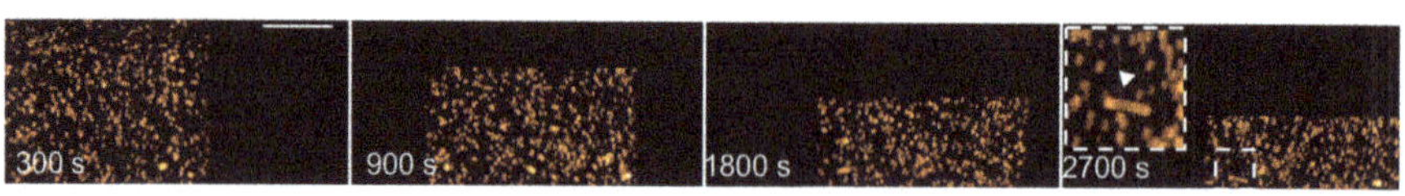

(a)

Figure 6. *Cont.*

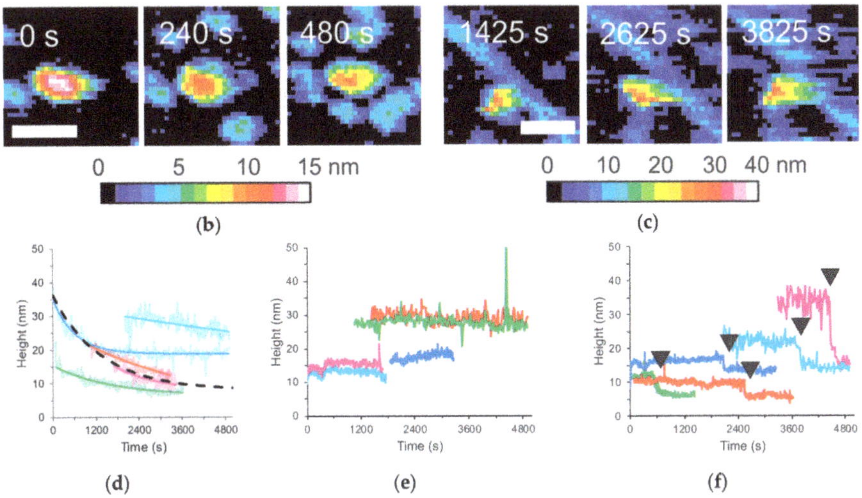

Figure 6. HS-AFM observation of non-fibrous aggregates in high molecular weight (HMW) Aβ42 incubation. (**a**) HS-AFM images of HMW Aβ42 incubation at the indicated time after the addition of 0.1 M NaCl. A closed triangle in the highlighted dashed box indicates a representative short fibril in HMW incubation. Bar, 300 nm. (**b**,**c**) HS-AFM image of two types of spherical aggregates in HMW incubation: type I gradually decreased in height (**b**); type II maintained its height (**c**). Bars, 50 nm. (**d**–**f**) Five representative time courses of height of type I (**d**), type II (**e**), and type III (**f**) aggregates in HMW incubation after an addition of 0.1 M NaCl. Each of the trajectories for type I is shown with the fitting curve of Equation (2). The dashed line corresponds to Equation (2), with the median values obtained from the best-fit values for individual type I aggregates. The type III aggregates show the sudden decrease in their heights at the time indicated by closed triangles. Different colors correspond to different single aggregates. Reproduced from [93].

4.2. Characterization of Aggregation Pathway by Statistical Analysis

HS-AFM analysis of the frequency of appearance of specific amyloid aggregates identified the position of individual aggregates during the molecular process. Figure 7a shows the time course of the cumulative number of fibrils appearing in the observation area after the addition of 0.1 M sodium chloride to the LMW or the HMW Aβ42 fractions [93]. The graph shows an upward trend in which fibrils appeared and accumulated. Although the LMW appeared throughout the time period, the HMW began to increase in the number of fibrils 20 to 30 min after the incubation. This tendency can be confirmed remarkably on the distribution of the time when the fibrils appeared, and there was a significant difference in the variance of the two distributions (Figure 7b) [93]. The bulk assay with gel filtration also indicated that the HMW incubation temporarily produced the LMW fraction [93]. There was no difference in the distribution of the fibril types between the incubation of the LMW and the HMW Aβ42 (Figure 7c) [93]. No significant difference was observed in the size of the fibril seeds at the time of appearance (Figure 7d) [93]. These results suggest that the on-pathway for fibril formation was common to both the LMW and HMW Aβ42 incubations, and that the HMW took time to form fibril nuclei [93]. As shown in Section 4.1, the HMW contained aggregates that dissociated according to the equilibrium shift after the addition of 0.1 M sodium chloride. Therefore, the HMW was located in the off-pathway and dissociated to generate the LMW; then, the fibrils were formed along the on-pathway [93]. This HS-AFM observation correlated to a recent kinetic study reported by Knowles et al., which calculated a fast Aβ42 oligomer formation (approximately 8×10^{-7} S^{-1}) and slow oligomer dissociation (approximately 9×10^{-5} S^{-1}) [136]. Thus, the delay in HMW dissociation to form a seed competent LMW is connected to the delay in fibril elongation.

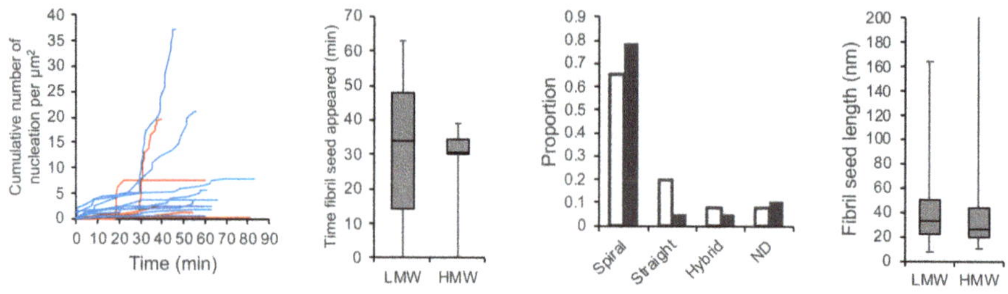

(a) (b) (c) (d)

Figure 7. Timing of the fibril appearance in the incubation of LMW and HWW Aβ42. (**a**) Time courses of cumulative number of fibrils in the observation area from LMW (blue lines) and HMW (red lines) Aβ42 incubation. Different lines correspond to different experiments. (**b–d**) The distribution of times at which individual fibril seeds appeared (**b**), the fibril type (**c**), and the length of fibril seeds when they first appeared in the observation area (**d**). The open and closed bars in (**c**) correspond to LMW and HMW incubation, respectively. Reproduced from [93].

5. HS-AFM Observation of Early Aggregation Stages of Aβ

Although highly challenging, there is considerable interest in probing the early events of amyloid aggregation to trace the formation of toxic oligomeric intermediates. While several biophysical approaches, including a combination of NMR techniques, have been employed to monitor these events at high resolution [48], they are limited by many factors, including the sample size, sensitivity, and timescale for measurements. On the other hand, HF-AFM is well suited for this purpose and can complement well with other studies by providing high-throughput measurements in real time.

Banerjee et al. characterized the structural features of low-order Aβ42 aggregates, measuring the spatial size of monomers through decamers and observing intramolecular structural dynamics of trimers, pentamers, and heptamers [109]. They prepared stable oligomers up to decamers by photochemical cross-linking and measured the sizes [109]. They found that the oligomeric order/size (volume) relationship showed two different proportionalities with the boundary at the tetramer, suggesting that at least two types of monomer packing patterns exist in their assembly [109]. Using HS-AFM, they observed intramolecular structural dynamics of the representative oligomers (trimer, pentamer, and heptamer) at subsecond temporal resolution [109]. The trimers sustained single blobs with almost constant width and length, while the pentamers and heptamers showed transition between a single compact globular shape and the extended multi-lobe structure (Figure 8). The pentamer had the two blobs with different sizes to each other in the extended state (Figure 8b). The heptamer can be extended to three lobes with different size, in addition to the double-lobe state (Figure 8c). These results suggest that Aβ42 oligomers are the lobe-linked structures, in which each lobe is composed of two or three peptides and that each oligomeric state can transiently extend and shrink between a compact globular shape and the multi-lobe structure [109].

Feng et al. identified that Aβ42 at the early aggregation stage can be classified into four structural types and characterized the kinetic interactions between those aggregates using HS-AFM [113]. They observed Aβ42 aggregates shortly after the dissolution of 1,1,1,3,3,3-hexafluoro-2-propanol (HFIP)-treated Aβ42 in phosphate-buffered saline. Their statistical analyses based on the three dimensional size measurements indicated four classes of Aβ42 aggregates: $A\beta_{15\text{-}20\,nm}$ with 15–20 nm (length and width) and height of 2.8 nm; $A\beta_{36\,nm}$ of a bilobed structure with 34 nm length, 17 nm width, and 2.8 nm height; and $A\beta_{Agg}$ with 34–36 nm (length and width) and height of 9.2 nm; disordered chain-like structure [113]. They also observed binding/dissociating interactions between the different type aggregates. The on/off interactions of $A\beta_{15\text{–}20\,nm}$–$A\beta_{36\,nm}$ and $A\beta_{15\text{–}20\,nm}$–$A\beta_{15\text{–}20\,nm}$ were respectively characterized to be a single step process [113] (Figure 9b,c,e,f,h,i). The $A\beta_{15\text{–}20\,nm}$–$A\beta_{36\,nm}$ binding showed slightly higher affinity compared with the $A\beta_{15\text{–}20\,nm}$–$A\beta_{15\text{–}20\,nm}$ interactions [113]

(Figure 9h,i). In contrast, the binding time for $A\beta_{15\text{–}20\,nm}$–$A\beta_{Agg}$ was distributed randomly (Figure 9a,d,g), which suggests that the $A\beta_{15\text{–}20\,nm}$–$A\beta_{Agg}$ interaction was not a single step but accepted various binding patterns [113]. Inevitably, this "permissive" interaction may produce various types of oligomers with different toxicities [113].

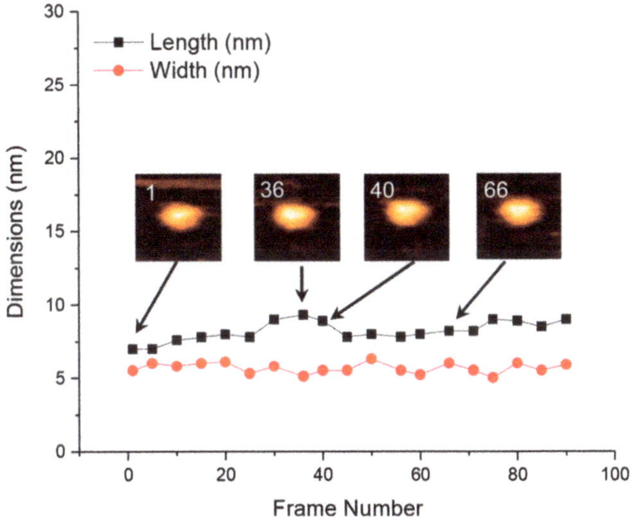

(a)

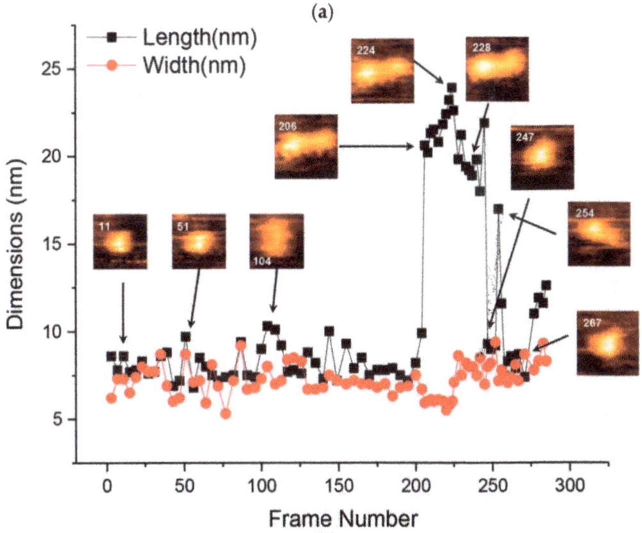

(b)

Figure 8. *Cont.*

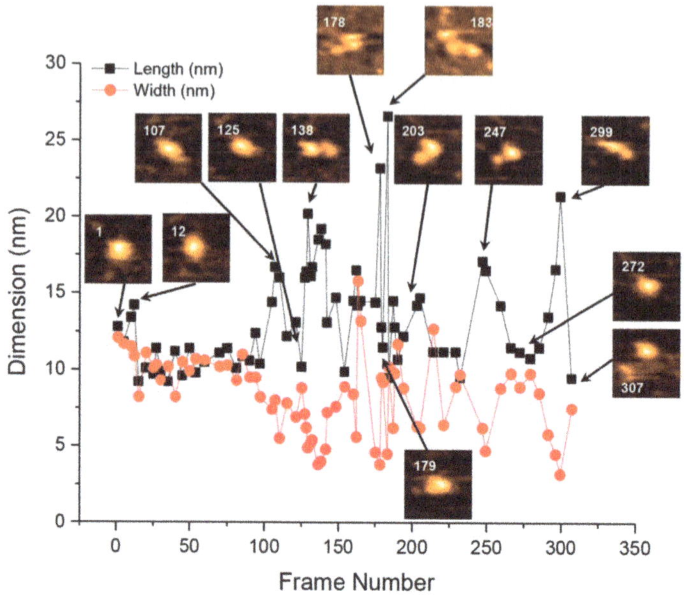

(c)

Figure 8. HS-AFM images of Aβ42 oligomers. Time courses of width and length obtained from HS-AFM images at the indicated times for trimers (**a**), pentamers (**b**), and heptamers (**c**). Reprinted with permission from [109]. Copyright (2017) American Chemical Society.

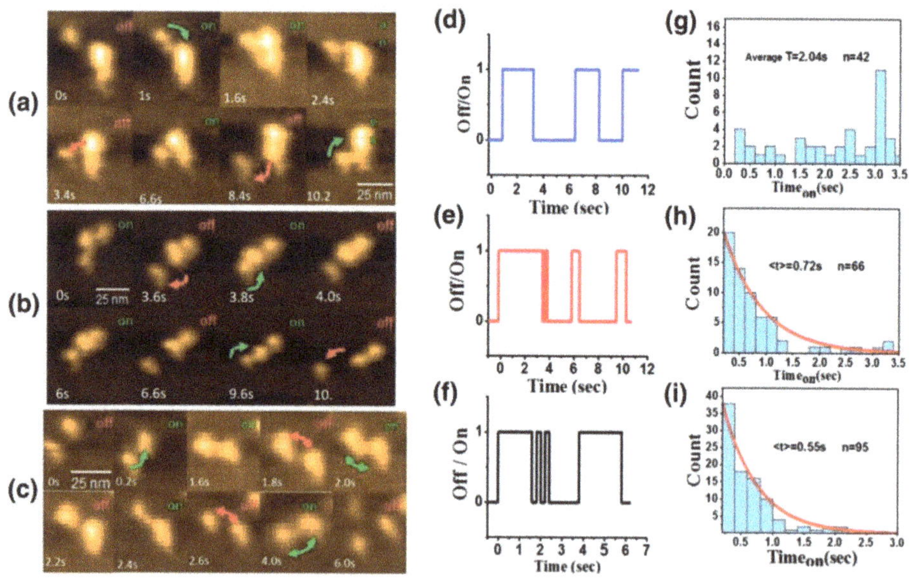

Figure 9. HS-AFM imaging and kinetic analyses of interaction between Aβ oligomers. Successive HS-AFM images (**a–c**), representative time courses of binding/dissociation (**d–f**), and distributions of the bound state (**g–i**) for the interaction of $A\beta_{15\text{–}20\,nm}$–$A\beta_{Agg}$ (**a**,**d**,**g**), $A\beta_{15\text{–}20\,nm}$–$A\beta_{36\,nm}$ (**b**,**e**,**h**), and $A\beta_{15\text{–}20\,nm}$–$A\beta_{15\text{–}20\,nm}$. Reprint with permission [113]: Copyright (2019), Elsevier.

6. HS-AFM Observation of Interaction between Amyloidogenic Proteins and Other Chemical Compounds

HS-AFM can also identify the interactions between amyloidogenic proteins and other chemical compounds, such as lipids and the potential anti-amyloidogenic inhibitor target(s) in the amyloid aggregation pathway by comparing observations in the presence and absence of the compounds.

6.1. Interaction between a Toxic Aβ Oligomer and Lipid Bilayer

Membrane–amyloidogenic protein interactions have been thought to lead specific aggregate conformers and play an important role in neurotoxicity as described in the Introduction. Aβ aggregation on a membrane has been known to depend on the physicochemical properties of the membrane (lipid composition, gel/liquid phases, phase separation, charge on head group, and oxidation level), which influences the toxicity of Aβ [137–143]. HS-AFM can visualize the effect of the lipid–amyloid interaction on the lipid membrane and characterize the interaction in real time.

Ewald et al. demonstrated the importance of the lipid composition on the interaction with a toxic Aβ oligomer, showing HS-AFM movies of lipid composition-dependent membrane disruption [111]. They prepared the toxic Aβ42 oG37C oligomers in which the 37th residue was changed from glycine to cysteine and added them to the planar lipid bilayer on the stage, changing the lipid composition in the mixture of sphingomyelin (SM)/1-palmitoyl-2-oleoylphosphatidylcholine (POPC)/cholesterol (Chol)/GM1-ganglioside (GM1). They visualized the GM1-dependent oligomer anchoring and the requirement of GM1/Chol coexistence at an appropriate ratio for membrane solubilization (Figure 10) [111].

6.2. Aggregation Inhibition by Natural Phenolic Compounds

Some natural phenolic compounds in foods have been known to inhibit amyloid aggregation [144–153]. The binding of the grape extract polyphenol myricetin to monomeric Aβ42 following the inhibition of amyloid aggregation was identified [152]. We investigated how myricetin altered the structural dynamics of Aβ42 amyloid fibril formation using HS-AFM. The observation was initiated by administering 2.5 μM Aβ42 (19:1 = LMW:seeds) and 10 μM myricetin to the HS-AFM sample chamber [93]. Some seeds appeared on the stage and they barely extended (Figure 11a) [93]. Then, the solution in the sample chamber was replaced with fresh 2.5 μM Aβ42 containing neither seed nor myricetin [93]. Immediately after the exchange, the fibrils started to grow, and the fibril amount also increased (Figure 11b) [93]. This result indicated that myricetin reversibly inhibited the fibril elongation reaction [93]. Further, the HS-AFM results suggested that the myricetin–monomer Aβ42 complex perturbed the dynamic equilibrium between the monomer and LMW that restricted the recruitment of LMW into the fibril ends. This could be by either the unavailability of seed competent Aβ42 species to recruit into the fibril ends or the reversible binding of myricetin to the fibril ends or both.

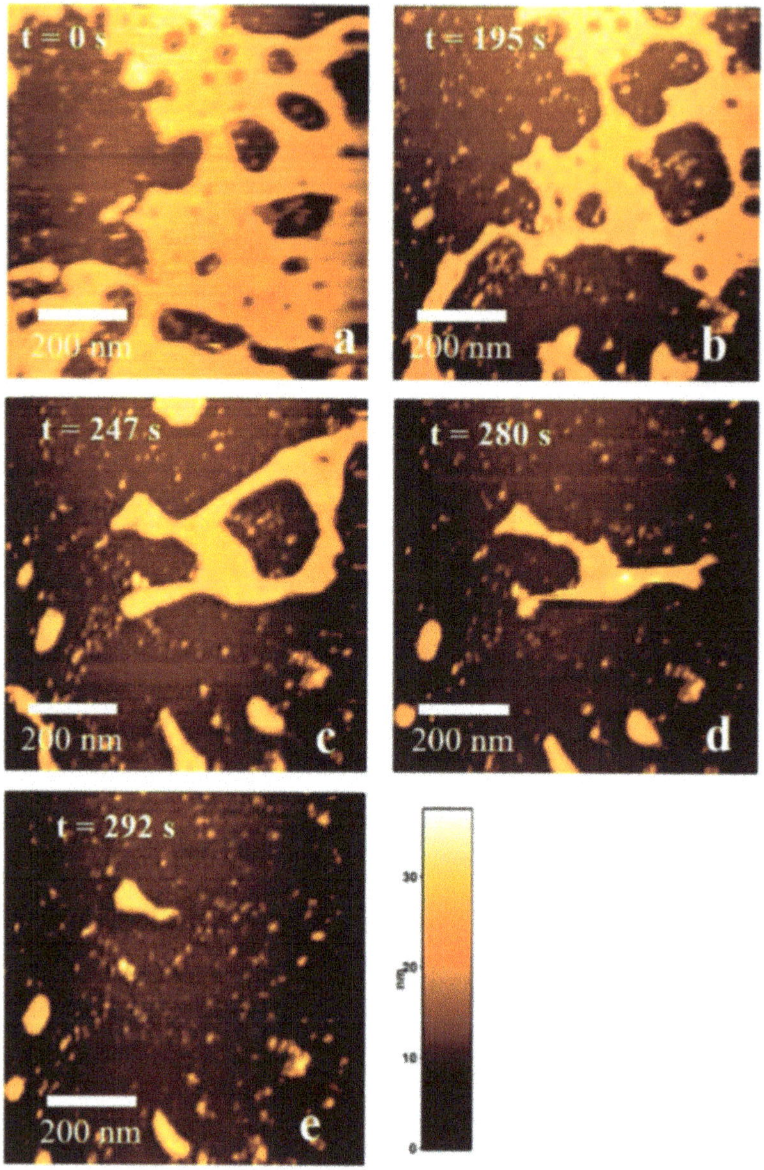

Figure 10. HS-AFM images of sphingomyelin (SM)/1-palmitoyl-2-oleoylphosphatidylcholine (POPC)/cholesterol (Chol)/GM1-ganglioside (GM1) membrane disruption after the addition of Aβ42 oG37C oligomers (**a**–**e**). The membrane with holes was removed from its edge. Reproduced from [111] with permission from The Royal Society of Chemistry.

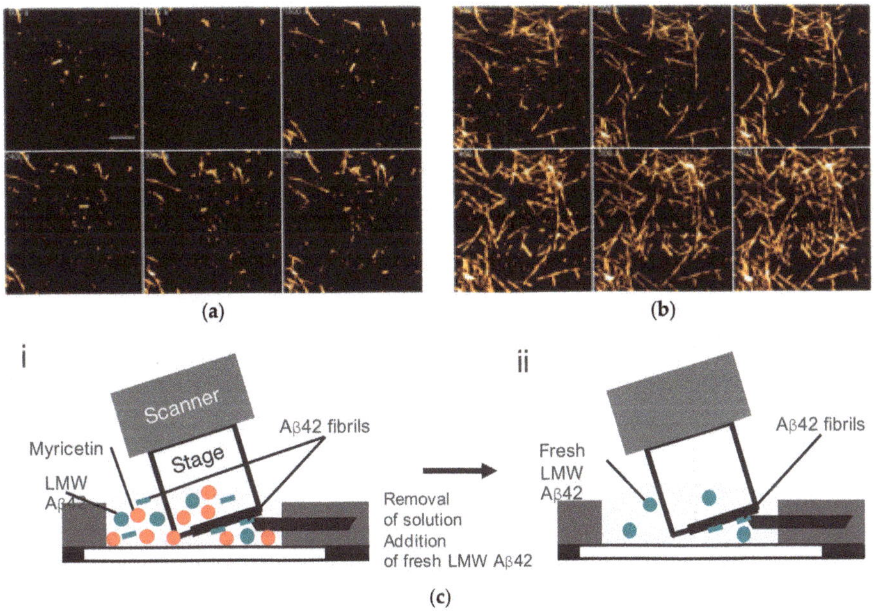

Figure 11. HS-AFM observation of Aβ42 fibril growth in the presence/absence of myricetin. (**a**) HS-AFM images of LMW Aβ42 incubation with fibril seeds and myricetin at the indicated time after the addition of 0.1 M NaCl. Scale bar, 300 nm. (**b**) HS-AFM images of the same observation area after the replacement of the solution to fresh LMW Aβ42. (**c**) Schematic views of HS-AFM observation condition (i) and (ii) correspond to (**a**) and (**b**) Reproduced from [93].

6.3. Aggregation Inhibition by Synthetic Polymers

Amylin (also known as islet amyloid polypeptide protein (IAPP)) is a 37-residue peptide that is produced and co-secreted with insulin from pancreatic β cells. The amylin amyloid aggregates deposit on the β cells in the type II diabetes. Amylin shares some common characteristics with Aβ, such as folding into similar β-sheet structures [154], binding to amylin-3 receptor [155], and being digested by insulin-degrading enzyme [156]. Amylin crosses the blood–brain barrier (BBB) [157–159], and its aggregate deposition is found in the brains of type II diabetes patients with AD [121]. The mechanisms underlying the pathological [122,123] and suppressive [117–120] effects of amylin to AD remain controversial [160].

We observed amylin aggregation and fibril formation in the presence or absence of a polymethacrylate-derived copolymer (PMAQA) that has been used in various biological research fields, including lipid–nanodisc formation, Aβ–nanodisc interaction, the improvement of drug delivery, and the bioavailability on microencapsulation [41,161–163]. In this observation, amylin fibril seeds were immobilized on the stage beforehand, and then amylin monomers were added to the sample chamber alone or together with PMAQA [116]. In the absence of PMAQA, the original fibril seeds grew as observed in Aβ42, and some of the newly created fibrils in the chamber bound to and extended on the stage (Figure 12a,b) [116]. In the presence of PMAQA, the original fibril did not elongate, and de novo fibrils did not appear (Figure 12c,d) [116]. Unlike the HS-AFM observations for Aβ42 that showed fiber polymorphism (Figure 2), amylin showed a majority of straight fibers (Figure 12). NMR analysis of the PMAQA–amylin complex indicated that PMAQA bound to the amyloid core domain (NFGAIL) of amylin [116]. These results suggested that the binding of PMAQA to amylin monomer inhibited nucleation and self-replicative fibril elongation [116]. In this way, the consecutive imaging of fibril seeds adhered on the stage in advance and in the same observation area after the addition of monomers could not only observe fibril elongation immediately, but also distinguished between the original fibrils and the de novo fibrils by identifying the appearance time and place of the fibrils.

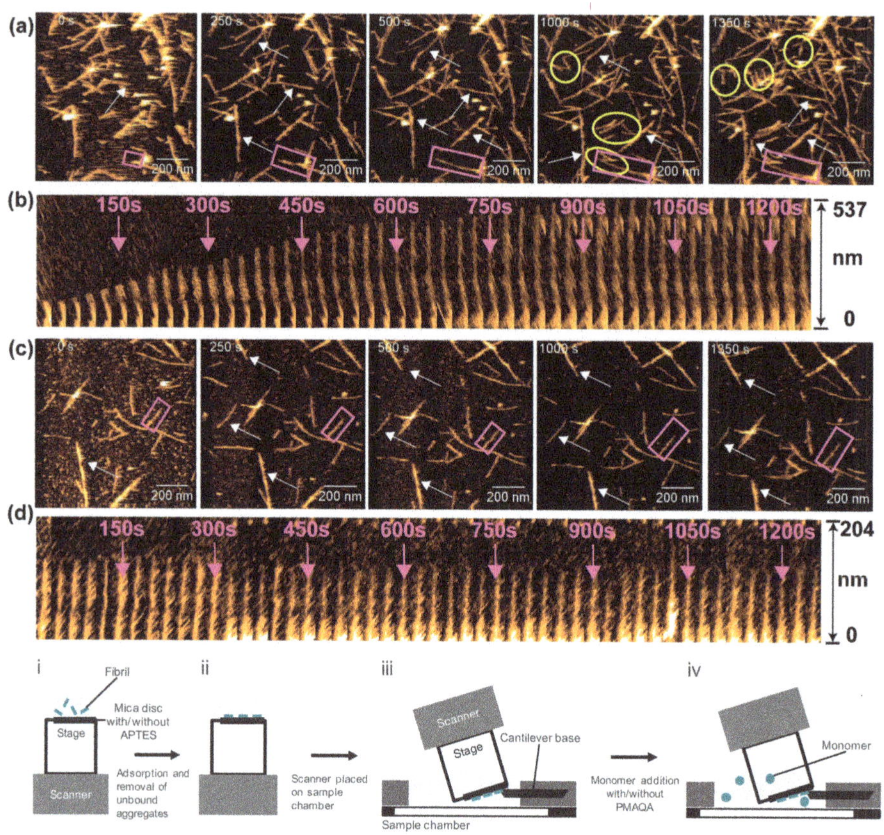

Figure 12. HS-AFM observation of amylin fibril growth. (**a**,**c**) HS-AFM images of the amylin seeding reaction in the absence (**a**) or presence (**c**) of polymethacrylate-derived copolymer (PMAQA). The de novo nucleated fibrils are highlighted by yellow circles in (**a**). The growth extents of fibrils are indicated by the white arrows. (**b**,**d**) Kymographs of individual amylin fibrils highlighted by purple boxes in (**a**) and (**c**) in the absence (**b**) or presence (**d**) of PMAQA. The experimental setups are shown in (**e**). Reprint with permission [116]: Copyright (2019), the Royal Society of Chemistry.

The effects of styrene–maleic acid copolymers varying with charge (SMAEA/SMAQA) on amylin were investigated using HS-AFM [164]. The results from this study identified morphologically distinct amylin species. HS-AFM showed the cationic SMAQA polymer [165] interaction with amylin generates de novo spherical globulomers that are incompetent to grow in size or recruit to the fibril ends to proceed with the seeding reaction. In contrast, the anionic SMAEA polymer [166] accelerated amylin fibrillation and generated de novo spherical globulomers that grew in size and proceeded with the seeding reaction. These observations indicate that SMAQA and SMAEA acted as an inhibitor and promotor for amylin aggregation, respectively.

6.4. Aggregation Inhibition by Heterologous Aggregation

Although the heterologous aggregation of amyloidogenic proteins has been found in vivo, the molecular mechanism has been still unclear. In addition, designing peptides that lead to coaggregation with amyloidogenic proteins can potentially become the candidates for the therapeutic drugs.

Kakinen et al. observed the structural dynamics of homologous fibril growth of full-length amylin and the heterologous assembly of full-length amylin/its shorter component, 8–20 or 19–29 S20G, using HS-AFM [108]. The 8–20 and 19–29 S20G are the peptides from the 8th to the 20th and from the 19th to the 29th with replacement at the 20th residue of serine with glycine. Both of the two regions build up a cross β structure in amylin fibrils [108]. Using HS-AFM, they found and characterized the inhibition of amylin fibril growth in coaggregation with the peptides [108]. The fibril morphology of pure full-length amylin differed from that of coaggregation (Figure 13a–g) [108]. The fibril thickness of homologous full-length amylin was widely distributed from 9 to 20 nm, while that of the coaggregation fibrils showed a narrower distribution (Figure 13d) [108]. The authors interpreted that the mature fibrils formed at the pure full-length amylin on the HS-AFM stage, which was reflected in the wide distribution of fibril thickness (Figure 13d) [108].

Compared with the homologous aggregation, the coaggregation increased the surface roughness, which resulted from the higher production of small aggregates (Figure 13a–c,e–g). Similar to our Aβ42 study [93], kymographs showed that both the homologous and heterologous fibril growth of amylin followed the stepwise and polarized manner at the fast and slow ends (Figure 13h–j). The kymograph analysis also indicated that the coaggregation reduced the apparent growth speed (Figure 13k) due to decreases in the step speed (Figure 13l) and increases in the pause time (Figure 13m) [108]. These results were interpreted as the incorrect docking of the shorter peptides that should be removed or converted to the correct docking, which reduced the fibril elongation [108].

In addition, the authors focused on the relationship between step size and step time. This proportionality was much higher in the self-elongation of full-length amylin than in the coaggregation (Figure 13n) [108]. This analysis suggested that the step speed was kept at a constant value in the self-assembly of full-length amylin, while, for the coaggregation, the step speed was widely distributed [108]. They also showed significant recovery effects of coaggregation with 19–29 S20G on the survival rate, hatching rate, and phenotypic normality, using an in vivo model with zebrafish embryos [108].

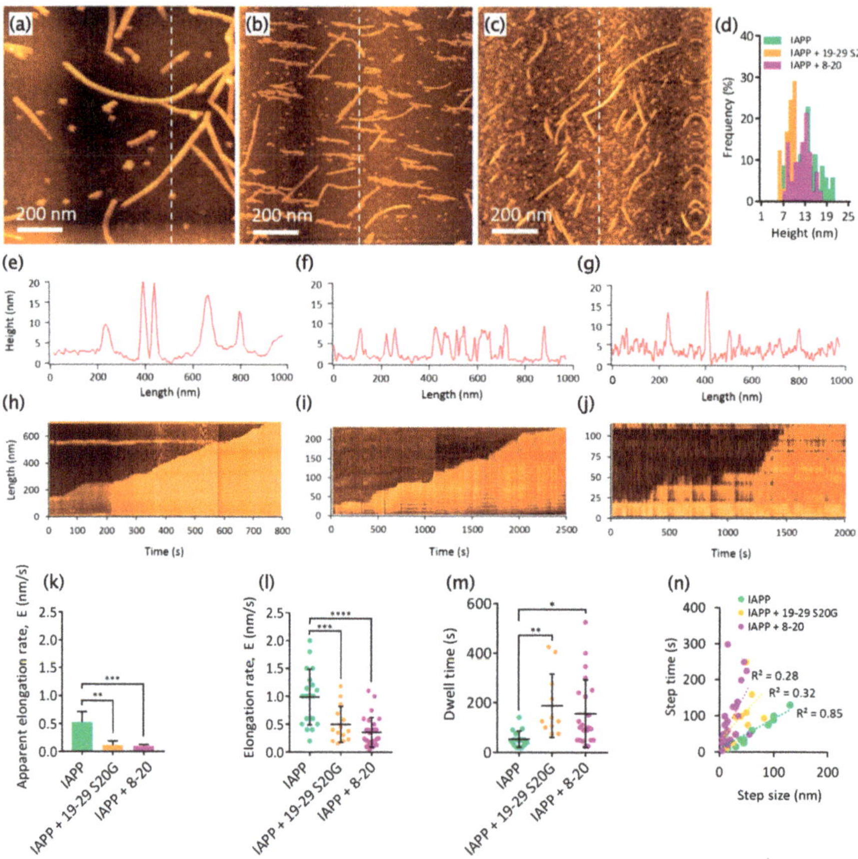

Figure 13. HS-AFM observation of homologous and heterologous amylin fibril growth. (**a**–**c**) HS-AFM images of self-assembly of full-length amylin (**a**), coaggregations with 19–29 S20G (**b**) and 8–20 (**c**). (**d**) Distribution of fibril thickness. (**e**–**g**) Cross-sections at the dashed lines in (**a**–**c**). (**h**–**j**) Representative kymographs of homologous full-length amylin fibrils (**h**), heterologous fibrils with full-length amylin/19–29 S20G (**i**), and with full-length amylin/8–20 (**j**). (**k**) Apparent fibril growth speed. (**l**) Pause-free elongation speed. (**m**) Distribution of pause time. (**n**) Relationship between the step size and step time. Error bars correspond to the mean ± S.D. Asterisks represent statistically significant differences between the sample mean and control mean (ANOVA; * $p \leq 0.05$; ** $p \leq 0.01$; *** $p \leq 0.005$; **** $p \leq 0.001$). Reprinted with permission from [108]. Copyright (2019) American Chemical Society.

7. HS-AFM Observation of Other Amyloidogenic Proteins

High-speed AFM reveals the structural dynamics of not only Aβ and amylin, but also other amyloid protein aggregations. Milhiet et al. showed protofilament elongtation and its stacking into polymorphic mature fibrils of lithostathine, which is overexpressed in the pre-clinical stage of AD [115]. Zhang et al. studied the structural dynamics of α-synuclein monomers and dimers using HS-AFM [110]. The monomer showed a transition between a spherical structure and a protruding tail-like structure and a structure completely extended, similar to a string [110]. The dimers were less flexible and basically maintained a dumbbell structure in which two spherical structures were connected [110]. Konno and Watanabe-Nakayama et al. observed yeast prion Sup35 monomer, oligomer, and fiber elongation [114]. HS-AFM revealed the structural dynamics of the intrinsically disordered (IDR) and partially folded regions of the Sup35 monomer, differences in the core structure and in the IDR between Sup35 oligomers and fibrils, the stepwise growth of oligomers with distinct core size, and the continuous unidirectional elongation of Sup35 fibrils [114].

8. Optimization of HS-AFM Observation of Amyloid Aggregation

HS-AFM observation depended on the sample preparation, surrounding buffer composition, and imaging parameters. These conditions should be optimized so that the structure and dynamics on the HS-AFM stage are consistent with the conventional structural and dynamic analyses. In this section, we discuss the conditions for HS-AFM observation of amyloid aggregation.

8.1. Sample Preparation and Control of Aggregation Initiation

The sample preparation procedure should be optimized according to the aggregation process to be observed by HS-AFM. As shown in Sections 2 and 4, structural dynamics differ depending on the preparation of Aβ42 (LMW and HMW). In addition, the operational efficiency at the start of HS-AFM observation and the reproducibility of the results should be considered. The sample should be stored under conditions in which the aggregation does not progress, and the aggregation reaction needs to proceed during HS-AFM observation. In our studies, the size fractionated Aβ42 samples were stored in low ionic strength 10 mM sodium phosphate, at pH 7.4 [93,150], and then the aggregation was initiated by the addition of 0.1 M sodium chloride or potassium chloride [93].

8.2. Sample Density

Since the size of the HS-AFM observation field is limited as described in Section 8.4, a low density of sample molecules makes it difficult to find the molecules to be observed, while a high density means that the molecules are in contact with other molecules and have constrained structural dynamics; both cases should be avoided. For example, when the growing end of an observed fibril encounters another fibril, the growth stops there, and no further elongation occurs (Figure 1b). For the statistical analysis described in Sections 2.1 and 4.2, a sufficient number of aggregates are required within the observation area. Considering these conditions, the optimization of sample concentration is important. In our study, we set the concentration of Aβ42 to 2.5 μM in terms of monomers as in [93].

8.3. Stage Materials and Sample Solution

HS-AFM can only visualize the molecules at the interface between the solid surface of the stage and the liquid phase of the sample solution. HS-AFM cannot image molecules that are weakly bound to the stage and that are free in the sample solution. In addition, conditions in which molecules bind strongly to the stage could affect the structural dynamics and thus should be avoided. Therefore, we optimized a sample molecule binding condition in which the molecules stay at their positions during the time required to acquire their images in one frame, and where the structural dynamics are not affected (or the effect on the structural dynamics can be estimated). The binding force to the stage was determined by the chemical properties of the sample molecules (isoelectric point (pI), hydrophobicity, etc.), stage material, and composition of the sample solution. Thus, depending on the properties of the target molecule under investigation, one can optimize the chemical property of the mica surface (e.g., bare mica or 3-aminopropyltriethoxysilane (APTES) modified mica) and/or solution pH and ionic strength to obtain the desirable structural and dynamic information.

An atomically flat surface of freshly cleaved mica ($[KAl_2(OH)_2AlSi_3O_{10}]$) was used as a first candidate for the HS-AFM stage. The surface was negatively charged. The pH and ionic strength of the solution and the pI of the sample molecules determined the electrostatic interaction with the mica surface. As the pI of Aβ42 was 5.31 (estimated by the ProtParam [167] in Expasy), the net charge was negative in a neutral pH solution, and electrostatic repulsion was expected to act between the peptide surface and mica. In fact, we observed Aβ42 to be less adsorbed to mica in a low ionic strength buffer of 10 mM sodium phosphate, pH 7.4.

Electrostatic interactions between mica and sample molecules can be controlled by the addition of salts, as observed in ion exchange chromatography. Aβ42 aggregates bound to the mica surface immediately after the addition of 0.1 M sodium chloride to its low ionic strength buffer solution (10 mM sodium phosphate, pH 7.4), suggesting that the electrostatic repulsion acting between the Aβ42 and mica surface was canceled by the addition of salt and the hydrophobic interaction bound Aβ42 to the mica surface. The effect of salt on the electrostatic interaction between sample molecules and the mica surface varied depending on the type of cation, even at the same concentration. Potassium ions had a greater canceling effect of this electrostatic interaction than did sodium ions [168,169]. We used this difference in potassium and sodium ions to change the interaction between Aβ42 and the mica surface to characterize the structure switch of Aβ42 fibril elongation, as described in Section 2.1.

Chemical modification of the mica surface or other materials can be used to modulate the interaction between sample molecules and the stage surface. Modification of the mica surface with 3-aminopropyltriethoxysilane (APTES) made the surface positively charged [170]. In addition, the amino group of APTES can be used for the covalent immobilization of sample molecules with glutaraldehyde [170]. Highly ordered pyrolytic graphite (HOPG) can be used for the immobilization of sample molecules with hydrophobic interactions. HOPG was used for the immobilization of Aβ25-35 fibrils [126].

The effect that the sample-stage interaction had on the structural dynamics of sample molecules needed to be verified. To examine this effect, we confirmed the consistency between the time courses of the ThT assay in the in vitro aggregation reaction and the total amount of aggregate in the HS-AFM observation area. The trends of the total aggregate amount and the ThT fluorescence intensity were consistent with each other (Figure 14a,b) [93]. They rapidly increased in the HMW Aβ42 incubation and gradually increased at a low level in the LMW Aβ42 incubation (Figure 14a,b) [94]. The aggregate structures were also consistent between the HS-AFM image and the transmission electron microscope (TEM) image observed at the same time points after the initiation of aggregation (Figures 1b, 6a and 14c,d). For our HS-AFM observation of the amylin aggregation, we used a low ionic strength solution and confirmed no significant difference in observation between bare and APTES mica, indicating that the electrostatic interaction between the sample and the surface did not affect the structural dynamics of amylin in our observation [116].

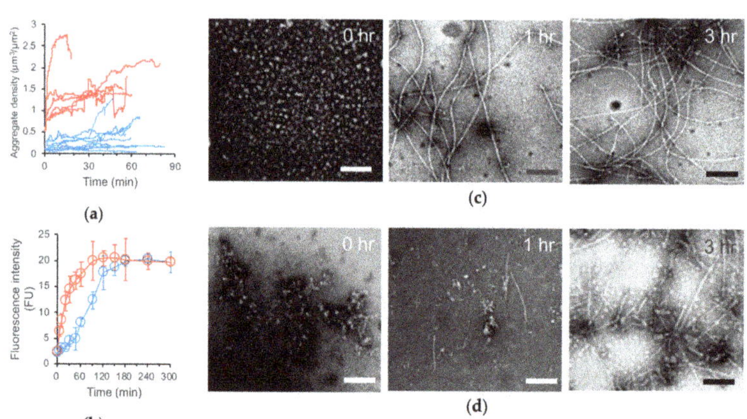

Figure 14. Relationship between HS-AFM observation, thioflavin T (ThT) assay, and transmission electron microscopy (TEM) of LMW and HMW Aβ42 incubation. The time evolution of the total Aβ42 aggregate density in HS-AFM observation (**a**); time course of ThT fluorescence intensity (**b**); TEM images of LMW (**c**; blue lines in (**a**,**b**)) and HMW ((**d**); red lines in (**a**,**b**)) Aβ42 incubations. Bars, 100 nm. Reproduced from [93].

8.4. Size of Scanning Area and Time for Image Acquisition

The observation area and imaging speed are in an inversely proportional relationship. For the statistical analysis of structural dynamics of amyloid aggregates, the size of the observation area is needed prior to the imaging speed. In this case, the typical scale of the observation field and the typical imaging speed are micrometers and several seconds to ten seconds per frame, respectively. For the imaging of faster structural dynamics of individual aggregates, the typical imaging speed is subseconds per frame, which requires a reduction in the observation area. The time 'T', also referred as the scanning rate, for the acquisition of one frame in the observation area (W nm in width × N lines in the y-direction) is expressed as follows [170]:

$$T = \pi WN/(2\lambda f_B \theta_m) \tag{3}$$

where λ is the periodicity of the sample surface with the sinusoidal shape, f_B is the feedback bandwidth, and θ_m is the maximum allowable phase delay of feedback control. λ is usually several nm, f_B is about 100 kHz at maximum, and θ_m is $\approx 20°$ or less for fragile samples and $\approx 45°$ or more for stiff samples. In addition to Equation (3), the number of frames captured in the same observation field should be considered. The tapping HS-AFM probe applies a mechanical force to the sample molecules. Thus, the larger the number of frames, the higher the probability that the sample will be broken. In addition to the imaging speed and applied force, the number of frames should be reduced for the analysis of increasing numbers of amyloid fibrils because the fragmented amyloid fibrils individually serve as seeds and then grow.

9. Conclusions

Different Aβ fibril structures have been found in different AD patients, which evokes the relationship between the fibril formation and AD progression. The structural dynamics of amyloid aggregation need to be elucidated for diagnosis and drug discovery. In previous studies, the structural dynamics were characterized by separately studying the structure and dynamics. However, the conventional structure and dynamics analyses lack information from the other missing component. Thus, the information that could not be gathered from those studies has remained unknown. The development of HS-AFM allowed video recording with nanometer spatial resolution, which has enabled the simultaneous analysis of structure and dynamics. We were able to analyze the structural dynamics of individual amyloid aggregates, including fibrils, even when different types of aggregates coexisted.

Recent studies have shown that amyloid proteins formed aggregates with diverse structures under physiological conditions that differed from those in vitro [171]. Under physiological conditions, amyloidogenic proteins underwent interactions with biological membranes, metal ions, other amyloidogenic proteins, and variants with different amino acid sequences causing heterologous aggregation. When reproducing the various physiological conditions by immobilizing a planar membrane on the stage and/or adding metal ions or variants of amyloidogenic proteins to the sample chamber, HS-AFM was used to visualize the structural dynamics in the aggregation processes. In addition, comparison with the observations in the presence of designed inhibitors may contribute to therapeutic development to characterize the desired drug targeting specific toxic aggregates.

HS-AFM observation has some unique characteristics: the visualization was limited to the structural dynamics that occurred in the solid–liquid interface of the stage and the sample solution; the imaging speed and size of the observation area were restricted by their inverse proportion relationship; and the statistical analysis and the macroscopic trend of the HS-AFM observations must be confirmed to be consistent with the conventional structure and dynamics analyses. In this way, HS-AFM did not only link the previous structure and dynamics analyses but did identify the structural dynamics that could not be elucidated using conventional methods.

Funding: This work was supported by Kanazawa University CHOZEN project. Resources for the amylin study were supported by NIH (AG048934 to A.R.).

Acknowledgments: We thank Hiroki Konno for his help and Noriyuki Kodera and Toshio Ando for their development of HS-AFM.

Conflicts of Interest: The authors declare no conflict of interest.

Abbreviations

AD	Alzheimer's disease
Aβ	Amyloid β
HS-AFM	High-speed atomic force microscopy
HMW	High molecular weight
LMW	Low molecular weight
ThT	Thioflavin T

References

1. Lu, J.X.; Qiang, W.; Yau, W.M.; Schwieters, C.D.; Meredith, S.C.; Tycko, R. Molecular structure of β-amyloid fibrils in alzheimer's disease brain tissue. *Cell* **2013**, *154*, 1257–1268. [CrossRef] [PubMed]
2. Kollmer, M.; Close, W.; Funk, L.; Rasmussen, J.; Bsoul, A.; Schierhorn, A.; Schmidt, M.; Sigurdson, C.J.; Jucker, M.; Fändrich, M. Cryo-EM structure and polymorphism of Aβ amyloid fibrils purified from Alzheimer's brain tissue. *Nat. Commun.* **2019**, *10*, 4760. [CrossRef] [PubMed]
3. Fitzpatrick, A.W.P.; Falcon, B.; He, S.; Murzin, A.G.; Murshudov, G.; Garringer, H.J.; Crowther, R.A.; Ghetti, B.; Goedert, M.; Scheres, S.H.W. Cryo-EM structures of tau filaments from Alzheimer's disease. *Nature* **2017**, *547*, 185–190. [CrossRef]
4. Falcon, B.; Zhang, W.; Murzin, A.G.; Murshudov, G.; Garringer, H.J.; Vidal, R.; Crowther, R.A.; Ghetti, B.; Scheres, S.H.W.; Goedert, M. Structures of filaments from Pick's disease reveal a novel tau protein fold. *Nature* **2018**, *561*, 137–140. [CrossRef] [PubMed]
5. Narasimhan, S.; Guo, J.L.; Changolkar, L.; Stieber, A.; McBride, J.D.; Silva, L.V.; He, Z.; Zhang, B.; Gathagan, R.J.; Trojanowski, J.Q.; et al. Pathological Tau Strains from Human Brains Recapitulate the Diversity of Tauopathies in Nontransgenic Mouse Brain. *J. Neurosci.* **2017**, *37*, 11406–11423. [CrossRef]
6. Falcon, B.; Zivanov, J.; Zhang, W.; Murzin, A.G.; Garringer, H.J.; Vidal, R.; Crowther, R.A.; Newell, K.L.; Ghetti, B.; Goedert, M.; et al. Novel tau filament fold in chronic traumatic encephalopathy encloses hydrophobic molecules. *Nature* **2019**, *568*, 420–423. [CrossRef]
7. Bousset, L.; Pieri, L.; Ruiz-Arlandis, G.; Gath, J.; Jensen, P.H.; Habenstein, B.; Madiona, K.; Olieric, V.; Böckmann, A.; Meier, B.H.; et al. Structural and functional characterization of two alpha-synuclein strains. *Nat. Commun.* **2013**, *4*, 2575. [CrossRef]
8. Peng, C.; Gathagan, R.J.; Covell, D.J.; Medellin, C.; Stieber, A.; Robinson, J.L.; Zhang, B.; Pitkin, R.M.; Olufemi, M.F.; Luk, K.C.; et al. Cellular milieu imparts distinct pathological α-synuclein strains in α-synucleinopathies. *Nature* **2018**, *557*, 558–563. [CrossRef]
9. Peelaerts, W.; Bousset, L.; Van der Perren, A.; Moskalyuk, A.; Pulizzi, R.; Giugliano, M.; Van den Haute, C.; Melki, R.; Baekelandt, V. α-Synuclein strains cause distinct synucleinopathies after local and systemic administration. *Nature* **2015**, *522*, 340–344. [CrossRef]
10. Tanaka, M.; Chien, P.; Naber, N.; Cooke, R.; Weissman, J.S. Conformational variations in an infectious protein determine prion strain differences. *Nature* **2004**, *428*, 323–328. [CrossRef]
11. Tanaka, M.; Collins, S.R.; Toyama, B.H.; Weissman, J.S. The physical basis of how prion conformations determine strain phenotypes. *Nature* **2006**, *442*, 585–589. [CrossRef]
12. Toyama, B.H.; Kelly, M.J.S.; Gross, J.D.; Weissman, J.S. The structural basis of yeast prion strain variants. *Nature* **2007**, *449*, 233–237. [CrossRef] [PubMed]
13. Atilio Gerez, J.; Riek, R. Neurodegenerative diseases distinguished through protein-structure analysis. *Nature* **2020**, *578*, 223–224. [CrossRef] [PubMed]
14. Barnes, C.A.; Robertson, A.J.; Louis, J.M.; Anfinrud, P.; Bax, A. Observation of β-Amyloid peptide oligomerization by pressure-jump NMR spectroscopy. *J. Am. Chem. Soc.* **2019**, *141*, 13762–13766. [CrossRef]

15. Roche, J.; Shen, Y.; Lee, J.H.; Ying, J.; Bax, A. Monomeric Aβ $^{1-40}$ and Aβ $^{1-42}$ Peptides in solution adopt very similar ramachandran map distributions that closely resemble random coil. *Biochemistry* **2016**, *55*, 762–775. [CrossRef] [PubMed]
16. Vivekanandan, S.; Brender, J.R.; Lee, S.Y.; Ramamoorthy, A. A partially folded structure of amyloid-beta(1-40) in an aqueous environment. *Biochem. Biophys. Res. Commun.* **2011**, *411*, 312–316. [CrossRef] [PubMed]
17. Kotler, S.A.; Brender, J.R.; Vivekanandan, S.; Suzuki, Y.; Yamamoto, K.; Monette, M.; Krishnamoorthy, J.; Walsh, P.; Cauble, M.; Holl, M.M.B.; et al. High-resolution NMR characterization of low abundance oligomers of amyloid-β without purification. *Sci. Rep.* **2015**, *5*, 11811. [CrossRef] [PubMed]
18. Korshavn, K.J.; Bhunia, A.; Lim, M.H.; Ramamoorthy, A. Amyloid-β adopts a conserved, partially folded structure upon binding to zwitterionic lipid bilayers prior to amyloid formation. *Chem. Commun.* **2016**, *52*, 882–885. [CrossRef]
19. Sciacca, M.F.M.; Kotler, S.A.; Brender, J.R.; Chen, J.; Lee, D.K.; Ramamoorthy, A. Two-step mechanism of membrane disruption by Aβ through membrane fragmentation and pore formation. *Biophys. J.* **2012**, *103*, 702–710. [CrossRef]
20. Banerjee, S.; Hashemi, M.; Zagorski, K.; Lyubchenko, Y.L. Interaction of Aβ42 with membranes triggers the self-assembly into oligomers. *Int. J. Mol. Sci.* **2020**, *21*, 1129. [CrossRef]
21. Bode, D.C.; Baker, M.D.; Viles, J.H. Ion Channel Formation by Amyloid-β42 oligomers but not Amyloid-β40 in cellular membranes. *J. Biol. Chem.* **2017**, *292*, 1404–1413. [CrossRef] [PubMed]
22. Serra-Batiste, M.; Ninot-Pedrosa, M.; Bayoumi, M.; Gairí, M.; Maglia, G.; Carulla, N. Aβ42 assembles into specific β-barrel pore-forming oligomers in membrane-mimicking environments. *Proc. Natl. Acad. Sci. USA* **2016**, *113*, 10866–10871. [CrossRef] [PubMed]
23. Patel, N.; Ramachandran, S.; Azimov, R.; Kagan, B.L.; Lal, R. Ion Channel formation by tau protein: Implications for Alzheimer's Disease and tauopathies. *Biochemistry* **2015**, *54*, 7320–7325. [CrossRef] [PubMed]
24. Quist, A.; Doudevski, I.; Lin, H.; Azimova, R.; Ng, D.; Frangione, B.; Kagan, B.; Ghiso, J.; Lal, R. Amyloid ion channels: A common structural link for protein-misfolding disease. *Proc. Natl. Acad. Sci. USA* **2005**, *102*, 10427–10432. [CrossRef] [PubMed]
25. Demetrius, L.A.; Simon, D.K. An inverse-Warburg effect and the origin of Alzheimer's disease. *Biogerontology* **2012**, *13*, 583–594. [CrossRef] [PubMed]
26. Mattson, M.P.; Pedersen, W.A.; Duan, W.; Culmsee, C.; Camandola, S. Cellular and molecular mechanisms underlying perturbed energy metabolism and neuronal degeneration in Alzheimer's and Parkinson's diseases. *Ann. N. Y. Acad. Sci.* **1999**, *893*, 154–175. [CrossRef]
27. Su, Y.; Chang, P.-T. Acidic pH promotes the formation of toxic fibrils from β-amyloid peptide. *Brain Res.* **2001**, *893*, 287–291. [CrossRef]
28. Xu, W.; Zhang, C.; Derreumaux, P.; Gräslund, A.; Morozova-Roche, L.; Mu, Y. Intrinsic determinants of Aβ12–24 pH-Dependent self-assembly revealed by combined computational and experimental studies. *PLoS ONE* **2011**, *6*, e24329. [CrossRef]
29. Shi, J.M.; Lv, J.M.; Gao, B.X.; Zhang, L.; Ji, S.R. Endosomal pH favors shedding of membrane-inserted amyloid-β peptide. *Protein Sci.* **2019**, *28*, 889–899. [CrossRef]
30. Gregori, M.; Cassina, V.; Brogioli, D.; Salerno, D.; De Kimpe, L.; Scheper, W.; Masserini, M.; Mantegazza, F. Stability of Aβ (1-42) peptide fibrils as consequence of environmental modifications. *Eur. Biophys. J.* **2010**, *39*, 1613–1623. [CrossRef]
31. Jebarupa, B.; Muralidharan, M.; Arun, A.; Mandal, A.K.; Mitra, G. Conformational heterogeneity of tau: Implication on intrinsic disorder, acid stability and fibrillation in Alzheimer's disease. *Biophys. Chem.* **2018**, *241*, 27–37. [CrossRef] [PubMed]
32. Cole, N.B.; DiEuliis, D.; Leo, P.; Mitchell, D.C.; Nussbaum, R.L. Mitochondrial translocation of α-synuclein is promoted by intracellular acidification. *Exp. Cell Res.* **2008**, *314*, 2076–2089. [CrossRef] [PubMed]
33. Ranjan, P.; Kumar, A. Perturbation in long-range contacts modulates the kinetics of amyloid formation in α-synuclein familial mutants. *ACS Chem. Neurosci.* **2017**, *8*, 2235–2246. [CrossRef]
34. Buell, A.K.; Galvagnion, C.; Gaspar, R.; Sparr, E.; Vendruscolo, M.; Knowles, T.P.J.; Linse, S.; Dobson, C.M. Solution conditions determine the relative importance of nucleation and growth processes in α-synuclein aggregation. *Proc. Natl. Acad. Sci. USA* **2014**, *111*, 7671–7676. [CrossRef] [PubMed]

35. Flynn, J.D.; McGlinchey, R.P.; Walker, R.L.; Lee, J.C. Structural features of -synuclein amyloid fibrils revealed by raman spectroscopy. *J. Biol. Chem.* **2018**, *293*, 767–776. [CrossRef]
36. Zhang, J.; Tan, J.; Pei, R.; Ye, S. Acidic environment significantly alters aggregation pathway of human islet amyloid polypeptide at negative Lipid Membrane. *Langmuir* **2020**, *36*, 1530–1537. [CrossRef]
37. Verma, M.; Vats, A.; Taneja, V. Toxic species in amyloid disorders: Oligomers or mature fibrils. *Ann. Indian Acad. Neurol.* **2015**, *18*, 138–145.
38. Chang, Y.J.; Chen, Y.R. The coexistence of an equal amount of Alzheimer's amyloid-β 40 and 42 forms structurally stable and toxic oligomers through a distinct pathway. *FEBS J.* **2014**, *281*, 2674–2687. [CrossRef]
39. Sahoo, B.R.; Bekier, M.E.; Liu, Z.; Kocman, V.; Stoddard, A.K.; Anantharamaiah, G.M.; Nowick, J.; Fierke, C.A.; Wang, Y.; Ramamoorthy, A. Structural Interaction of Apolipoprotein A-I Mimetic Peptide with Amyloid-β Generates Toxic Hetero-oligomers. *J. Mol. Biol.* **2020**, *432*, 1020–1034. [CrossRef]
40. Strodel, B.; Lee, J.W.L.; Whittleston, C.S.; Wales, D.J. Transmembrane Structures for Alzheimer's Aβ_{1-42} Oligomers. *J. Am. Chem. Soc.* **2010**, *132*, 13300–13312. [CrossRef]
41. Sahoo, B.R.; Genjo, T.; Bekier, M.; Cox, S.J.; Stoddard, A.K.; Ivanova, M.; Yasuhara, K.; Fierke, C.A.; Wang, Y.; Ramamoorthy, A. Alzheimer's amyloid-beta intermediates generated using polymer-nanodiscs. *Chem. Commun.* **2018**, *54*, 12883–12886. [CrossRef] [PubMed]
42. Qiang, W.; Yau, W.-M.; Lu, J.-X.; Collinge, J.; Tycko, R. Structural variation in amyloid-β fibrils from Alzheimer's disease clinical subtypes. *Nature* **2017**, *541*, 217–221. [CrossRef] [PubMed]
43. Sawaya, M.R.; Sambashivan, S.; Nelson, R.; Ivanova, M.I.; Sievers, S.A.; Apostol, M.I.; Thompson, M.J.; Balbirnie, M.; Wiltzius, J.J.W.; McFarlane, H.T.; et al. Atomic structures of amyloid cross-β spines reveal varied steric zippers. *Nature* **2007**, *447*, 453–457. [CrossRef]
44. Colletier, J.P.; Laganowsky, A.; Landau, M.; Zhao, M.; Soriaga, A.B.; Goldschmidt, L.; Flot, D.; Cascio, D.; Sawaya, M.R.; Eisenberg, D. Molecular basis for amyloid-β polymorphism. *Proc. Natl. Acad. Sci. USA* **2011**, *108*, 16938–16943. [CrossRef] [PubMed]
45. Kreutzer, A.G.; Hamza, I.L.; Spencer, R.K.; Nowick, J.S. X-ray Crystallographic Structures of a Trimer, Dodecamer, and Annular Pore Formed by an Aβ17-36 β-Hairpin. *J. Am. Chem. Soc.* **2016**, *138*, 4634–4642. [CrossRef]
46. Kreutzer, A.G.; Spencer, R.K.; McKnelly, K.J.; Yoo, S.; Hamza, I.L.; Salveson, P.J.; Nowick, J.S. A hexamer of a peptide derived from Aβ16-36. *Biochemistry* **2017**, *56*, 6061–6071. [CrossRef]
47. Salveson, P.J.; Spencer, R.K.; Kreutzer, A.G.; Nowick, J.S. X-ray crystallographic structure of a compact dodecamer from a peptide derived from Aβ_{16-36}. *Org. Lett.* **2017**, *19*, 3462–3465. [CrossRef] [PubMed]
48. Kreutzer, A.G.; Yoo, S.; Spencer, R.K.; Nowick, J.S. Stabilization, assembly, and toxicity of trimers derived from Aβ. *J. Am. Chem. Soc.* **2017**, *139*, 966–975. [CrossRef] [PubMed]
49. Rodriguez Camargo, D.C.; Korshavn, K.J.; Jussupow, A.; Raltchev, K.; Goricanec, D.; Fleisch, M.; Sarkar, R.; Xue, K.; Aichler, M.; Mettenleiter, G.; et al. Stabilization and structural analysis of a membrane-associated hIAPP aggregation intermediate. *Elife* **2017**, *6*, e31226. [CrossRef] [PubMed]
50. Sahoo, B.R.; Genjo, T.; Cox, S.J.; Stoddard, A.K.; Anantharamaiah, G.M.; Fierke, C.; Ramamoorthy, A. Nanodisc-Forming scaffold protein promoted retardation of amyloid-beta aggregation. *J. Mol. Biol.* **2018**, *430*, 4230–4244. [CrossRef]
51. Sahoo, B.R.; Cox, S.J.; Ramamoorthy, A. High-resolution probing of early events in amyloid-β aggregation related to Alzheimer's disease. *Chem. Commun.* **2020**, *56*, 4627–4639. [CrossRef] [PubMed]
52. Yu, L.; Edalji, R.; Harlan, J.E.; Holzman, T.F.; Lopez, A.P.; Labkovsky, B.; Hillen, H.; Barghorn, S.; Ebert, U.; Richardson, P.L.; et al. Structural characterization of a soluble amyloid β-Peptide oligomer. *Biochemistry* **2009**, *48*, 1870–1877. [CrossRef] [PubMed]
53. Scheidt, H.A.; Morgado, I.; Huster, D. Solid-state NMR reveals a close structural relationship between amyloid-?? protofibrils and oligomers. *J. Biol. Chem.* **2012**, *287*, 22822–22826. [CrossRef] [PubMed]
54. Doi, T.; Masuda, Y.; Irie, K.; Akagi, K.; Monobe, Y.; Imazawa, T.; Takegoshi, K. Solid-state NMR analysis of the β-strand orientation of the protofibrils of amyloid β-protein. *Biochem. Biophys. Res. Commun.* **2012**, *428*, 458–462. [CrossRef] [PubMed]
55. Gu, L.; Liu, C.; Guo, Z. Structural insights into A??42 oligomers using site-directed spin labeling. *J. Biol. Chem.* **2013**, *288*, 18673–18683. [CrossRef] [PubMed]

56. Tay, W.M.; Huang, D.; Rosenberry, T.L.; Paravastu, A.K. The Alzheimer's amyloid-β(1-42) peptide forms off-pathway oligomers and fibrils that are distinguished structurally by intermolecular organization. *J. Mol. Biol.* **2013**, *425*, 2494–2508. [CrossRef]
57. Potapov, A.; Yau, W.-M.; Ghirlando, R.; Thurber, K.R.; Tycko, R. Successive stages of amyloid-β self-assembly characterized by solid-state nuclear magnetic resonance with dynamic nuclear polarization. *J. Am. Chem. Soc.* **2015**, *137*, 8294–8307. [CrossRef]
58. Ahmed, R.; Akcan, M.; Khondker, A.; Rheinstädter, M.C.; Bozelli, J.C., Jr.; Epand, R.M.; Huynh, V.; Wylie, R.G.; Boulton, S.; Huang, J.; et al. Atomic resolution map of the soluble amyloid beta assembly toxic surfaces. *Chem. Sci.* **2019**, *10*, 6072–6082. [CrossRef]
59. Benzinger, T.L.S.; Gregory, D.M.; Burkoth, T.S.; Miller-Auer, H.; Lynn, D.G.; Botto, R.E.; Meredith, S.C. Propagating structure of Alzheimer's -amyloid(10-35) is parallel -sheet with residues in exact register. *Proc. Natl. Acad. Sci. USA* **1998**, *95*, 13407–13412. [CrossRef]
60. Petkova, A.T.; Leapman, R.D.; Guo, Z.; Yau, W.-M.; Mattson, M.P.; Tycko, R. Self-propagating, molecular-level polymorphism in Alzheimer's β-amyloid fibrils. *Science* **2005**, *307*, 262–265. [CrossRef]
61. Luhrs, T.; Ritter, C.; Adrian, M.; Riek-Loher, D.; Bohrmann, B.; Dobeli, H.; Schubert, D.; Riek, R. 3D structure of Alzheimer's amyloid- (1-42) fibrils. *Proc. Natl. Acad. Sci. USA* **2005**, *102*, 17342–17347. [CrossRef] [PubMed]
62. Petkova, A.T.; Yau, W.-M.; Tycko, R. Experimental Constraints on Quaternary Structure in Alzheimer's β-Amyloid Fibrils. *Biochemistry* **2006**, *45*, 498–512. [CrossRef] [PubMed]
63. Paravastu, A.K.; Petkova, A.T.; Tycko, R. Polymorphic fibril formation by residues 10–40 of the Alzheimer's β-Amyloid peptide. *Biophys. J.* **2006**, *90*, 4618–4629. [CrossRef] [PubMed]
64. Xiao, Y.; Ma, B.; McElheny, D.; Parthasarathy, S.; Long, F.; Hoshi, M.; Nussinov, R.; Ishii, Y. Aβ(1–42) fibril structure illuminates self-recognition and replication of amyloid in Alzheimer's disease. *Nat. Struct. Mol. Biol.* **2015**, *22*, 499–505. [CrossRef] [PubMed]
65. Gremer, L.; Schölzel, D.; Schenk, C.; Reinartz, E.; Labahn, J.; Ravelli, R.B.G.; Tusche, M.; Lopez-Iglesias, C.; Hoyer, W.; Heise, H.; et al. Fibril structure of amyloid-β(1–42) by cryo–electron microscopy. *Science (80-)* **2017**, *358*, 116–119. [CrossRef] [PubMed]
66. Khurana, R.; Coleman, C.; Ionescu-Zanetti, C.; Carter, S.A.; Krishna, V.; Grover, R.K.; Roy, R.; Singh, S. Mechanism of thioflavin T binding to amyloid fibrils. *J. Struct. Biol.* **2005**, *151*, 229–238. [CrossRef]
67. Biancalana, M.; Koide, S. Molecular mechanism of Thioflavin-T binding to amyloid fibrils. *Biochim. Biophys. Acta Proteins Proteom.* **2010**, *1804*, 1405–1412. [CrossRef]
68. Ban, T.; Hoshino, M.; Takahashi, S.; Hamada, D.; Hasegawa, K.; Naiki, H.; Goto, Y. Direct observation of Aβ amyloid fibril growth and inhibition. *J. Mol. Biol.* **2004**, *344*, 757–767. [CrossRef]
69. Wördehoff, M.M.; Bannach, O.; Shaykhalishahi, H.; Kulawik, A.; Schiefer, S.; Willbold, D.; Hoyer, W.; Birkmann, E. Single fibril growth kinetics of α-synuclein. *J. Mol. Biol.* **2015**, *427*, 1428–1435. [CrossRef]
70. Sang, J.C.; Meisl, G.; Thackray, A.M.; Hong, L.; Ponjavic, A.; Knowles, T.P.J.; Bujdoso, R.; Klenerman, D. Direct observation of murine prion protein replication in Vitro. *J. Am. Chem. Soc.* **2018**, *140*, 14789–14798. [CrossRef]
71. Cremades, N.; Cohen, S.I.; Deas, E.; Abramov, A.Y.; Chen, A.Y.; Orte, A.; Sandal, M.; Clarke, R.W.; Dunne, P.; Aprile, F.A.; et al. Direct observation of the interconversion of normal and toxic forms of α-synuclein. *Cell* **2012**, *149*, 1048–1059. [CrossRef] [PubMed]
72. Pinotsi, D.; Buell, A.K.; Galvagnion, C.; Dobson, C.M.; Kaminski Schierle, G.S.; Kaminski, C.F. Direct observation of heterogeneous amyloid fibril growth kinetics via two-color super-resolution microscopy. *Nano Lett.* **2014**, *14*, 339–345. [CrossRef] [PubMed]
73. Lv, Z.; Krasnoslobodtsev, A.V.V.; Zhang, Y.; Ysselstein, D.; Rochet, J.-C.; Blanchard, S.C.C.; Lyubchenko, Y.L.L. Direct Detection of α-Synuclein dimerization dynamics: Single-Molecule fluorescence analysis. *Biophys. J.* **2015**, *108*, 2038–2047. [CrossRef] [PubMed]
74. Fauerbach, J.A.; Jovin, T.M. Pre-aggregation kinetics and intermediates of α-synuclein monitored by the ESIPT probe 7MFE. *Eur. Biophys. J.* **2018**, *47*, 345–362. [CrossRef] [PubMed]
75. Tosatto, L.; Horrocks, M.H.; Dear, A.J.; Knowles, T.P.J.; Dalla Serra, M.; Cremades, N.; Dobson, C.M.; Klenerman, D.; González-Hernández, T.; Cruz-Muros, I.; et al. Single-molecule FRET studies on alpha-synuclein oligomerization of Parkinson's disease genetically related mutants. *Sci. Rep.* **2015**, *5*, 16696. [CrossRef]

76. Nath, S.; Meuvis, J.; Hendrix, J.; Carl, S.A.; Engelborghs, Y. Early aggregation steps in alpha-synuclein as measured by FCS and FRET: Evidence for a contagious conformational change. *Biophys. J.* **2010**, *98*, 1302–1311. [CrossRef]
77. Liang, Y.; Lynn, D.G.; Berland, K.M. Direct observation of nucleation and growth in amyloid self-assembly. *J. Am. Chem. Soc.* **2010**, *132*, 6306–6308. [CrossRef]
78. Inoue, Y.; Kishimoto, A.; Hirao, J.; Yoshida, M.; Taguchi, H. Strong growth polarity of yeast prion fiber revealed by single fiber imaging. *J. Biol. Chem.* **2001**, *276*, 35227–35230. [CrossRef]
79. Ando, T.; Kodera, N.; Takai, E.; Maruyama, D.; Saito, K.; Toda, A. A high-speed atomic force microscope for studying biological macromolecules. *Proc. Natl. Acad. Sci. USA* **2001**, *98*, 12468–12472. [CrossRef]
80. Eghiaian, F.; Rico, F.; Colom, A.; Casuso, I.; Scheuring, S. High-speed atomic force microscopy: Imaging and force spectroscopy. *FEBS Lett.* **2014**, *588*, 3631–3638. [CrossRef]
81. Uchihashi, T.; Scheuring, S. Applications of high-speed atomic force microscopy to real-time visualization of dynamic biomolecular processes. *Biochim. Biophys. Acta Gen. Subj.* **2018**, *1862*, 229–240. [CrossRef] [PubMed]
82. Ando, T.; Uchihashi, T.; Scheuring, S. Filming biomolecular processes by high-speed atomic force microscopy. *Chem. Rev.* **2014**, *114*, 3120–3188. [CrossRef] [PubMed]
83. Ando, T. High-speed atomic force microscopy. *Curr. Opin. Chem. Biol.* **2019**, *51*, 105–112. [CrossRef] [PubMed]
84. Uchihashi, T.; Ganser, C. Recent advances in bioimaging with high-speed atomic force microscopy. *Biophys. Rev.* **2020**, *12*, 363–369. [CrossRef]
85. Ruggeri, F.S.; Šneideris, T.; Vendruscolo, M.; Knowles, T.P.J. Atomic force microscopy for single molecule characterisation of protein aggregation. *Arch. Biochem. Biophys.* **2019**, *664*, 134–148. [CrossRef]
86. Adamcik, J.; Mezzenga, R. Study of amyloid fibrils via atomic force microscopy. *Curr. Opin. Colloid Interface Sci.* **2012**, *17*, 369–376. [CrossRef]
87. Adamcik, J.; Jung, J.-M.; Flakowski, J.; De Los Rios, P.; Dietler, G.; Mezzenga, R. Understanding amyloid aggregation by statistical analysis of atomic force microscopy images. *Nat. Nanotechnol.* **2010**, *5*, 423–428. [CrossRef]
88. Blackley, H.K.; Sanders, G.H.; Davies, M.C.; Roberts, C.J.; Tendler, S.J.; Wilkinson, M.J. In-Situ atomic force microscopy study of beta-amyloid fibrillization. *J. Mol. Biol.* **2000**, *298*, 833–840. [CrossRef]
89. Goldsbury, C.; Frey, P.; Olivieri, V.; Aebi, U.; Müller, S.A. Multiple assembly pathways underlie Amyloid-β Fibril polymorphisms. *J. Mol. Biol.* **2005**, *352*, 282–298. [CrossRef]
90. Ruggeri, F.S.; Charmet, J.; Kartanas, T.; Peter, Q.; Chia, S.; Habchi, J.; Dobson, C.M.; Vendruscolo, M.; Knowles, T.P.J. Microfluidic deposition for resolving single-molecule protein architecture and heterogeneity. *Nat. Commun.* **2018**, *9*, 3890. [CrossRef]
91. Ladiwala, A.R.A.; Litt, J.; Kane, R.S.; Aucoin, D.S.; Smith, S.O.; Ranjan, S.; Davis, J.; Van Nostrand, W.E.; Tessier, P.M.; Nostrand, W.E.V.; et al. Conformational differences between two amyloid β oligomers of similar size and dissimilar toxicity. *J. Biol. Chem.* **2012**, *287*, 24765–24773. [CrossRef] [PubMed]
92. Mastrangelo, I.A.; Ahmed, M.; Sato, T.; Liu, W.; Wang, C.; Hough, P.; Smith, S.O. High-resolution atomic force microscopy of soluble Aβ42 oligomers. *J. Mol. Biol.* **2006**, *358*, 106–119. [CrossRef]
93. Watanabe-Nakayama, T.; Ono, K.; Itami, M.; Takahashi, R.; Teplow, D.B.D.B.; Yamada, M. High-speed atomic force microscopy reveals structural dynamics of amyloid β1-42 aggregates. *Proc. Natl. Acad. Sci. USA* **2016**, *113*, 5835–5840. [CrossRef]
94. Qiang, W.; Kelley, K.; Tycko, R. Polymorph-specific kinetics and thermodynamics of β-amyloid fibril growth. *J. Am. Chem. Soc.* **2013**, *135*, 6860–6871. [CrossRef] [PubMed]
95. Hasecke, F.; Miti, T.; Perez, C.; Barton, J.; Schölzel, D.; Gremer, L.; Grüning, C.S.R.; Matthews, G.; Meisl, G.; Knowles, T.P.J.; et al. Origin of metastable oligomers and their effects on amyloid fibril self-assembly. *Chem. Sci.* **2018**, *9*, 5937–5948. [CrossRef] [PubMed]
96. Lin, Y.-C.; Li, C.; Fakhraai, Z. Kinetics of surface-mediated fibrillization of amyloid-β (12–28) peptides. *Langmuir* **2018**, *34*, 4665–4672. [CrossRef]
97. Jeong, J.S.; Ansaloni, A.; Mezzenga, R.; Lashuel, H.A.; Dietler, G. Novel mechanistic insight into the molecular basis of amyloid polymorphism and secondary nucleation during amyloid formation. *J. Mol. Biol.* **2013**, *425*, 1765–1781. [CrossRef]
98. Sahoo, B.R.; Liang, W.; Tang, W.; Ramamoorthy, A. Degradation of Alzheimer' s Amyloid- β by a catalytically inactive insulin degrading enzyme. *BioRxiv* **2020**. [CrossRef]

99. Sweers, K.K.M.; Stöckl, M.; Bennink, M.L.; Subramaniam, V. Characterizing nanoscale morphologic and mechanical properties of α-synuclein amyloid fibrils with atomic force microscopy. *Bio-Nanoimaging* **2014**, 309–322. [CrossRef]
100. Ruggeri, F.S.; Benedetti, F.; Knowles, T.P.J.; Lashuel, H.A.; Sekatskii, S.; Dietler, G. Identification and nanomechanical characterization of the fundamental single-strand protofilaments of amyloid α-synuclein fibrils. *Proc. Natl. Acad. Sci. USA* **2018**, *115*, 7230–7235. [CrossRef]
101. Khalaf, O.; Fauvet, B.; Oueslati, A.; Dikiy, I.; Mahul-Mellier, A.-L.; Ruggeri, F.S.; Mbefo, M.K.; Vercruysse, F.; Dietler, G.; Lee, S.-J.; et al. The H50Q Mutation enhances α-synuclein aggregation, secretion, and toxicity. *J. Biol. Chem.* **2014**, *289*, 21856–21876. [CrossRef] [PubMed]
102. Kim, H.J.; Chatani, E.; Goto, Y.; Paik, S.R. Seed-dependent accelerated fibrillation of alpha-synuclein induced by periodic ultrasonication treatment. *J. Microbiol. Biotechnol.* **2007**, *17*, 2027–2032. [PubMed]
103. Sidhu, A.; Segers-Nolten, I.; Raussens, V.; Claessens, M.M.A.E.; Subramaniam, V. Distinct mechanisms determine α-synuclein fibril morphology during growth and maturation. *ACS Chem. Neurosci.* **2017**, *8*, 538–547. [CrossRef] [PubMed]
104. Kim, H.-Y.; Cho, M.-K.; Kumar, A.; Maier, E.; Siebenhaar, C.; Becker, S.; Fernandez, C.O.; Lashuel, H.A.; Benz, R.; Lange, A.; et al. Structural properties of pore-forming oligomers of α-Synuclein. *J. Am. Chem. Soc.* **2009**, *131*, 17482–17489. [CrossRef] [PubMed]
105. Hoyer, W.; Cherny, D.; Subramaniam, V.; Jovin, T.M. Rapid Self-assembly of α-Synuclein observed by In Situ atomic force microscopy. *J. Mol. Biol.* **2004**, *340*, 127–139. [CrossRef]
106. Hong, D.-P.; Han, S.; Fink, A.L.; Uversky, V.N. Characterization of the Non-Fibrillar α-Synuclein Oligomers. *Protein Pept. Lett.* **2011**, *18*, 230–240. [CrossRef]
107. Goldsbury, C.; Kistler, J.; Aebi, U.; Arvinte, T.; Cooper, G.J. Watching amyloid fibrils grow by time-lapse atomic force microscopy. *J. Mol. Biol.* **1999**, *285*, 33–39. [CrossRef]
108. Kakinen, A.; Xing, Y.; Hegoda Arachchi, N.; Javed, I.; Feng, L.; Faridi, A.; Douek, A.M.; Sun, Y.; Kaslin, J.; Davis, T.P.; et al. Single-Molecular heteroamyloidosis of human islet amyloid polypeptide. *Nano Lett.* **2019**, *19*, 6535–6546. [CrossRef]
109. Banerjee, S.; Sun, Z.; Hayden, E.Y.; Teplow, D.B.; Lyubchenko, Y.L. Nanoscale dynamics of amyloid β-42 oligomers as revealed by high-speed atomic force microscopy. *ACS Nano* **2017**, *11*, 12202–12209. [CrossRef]
110. Zhang, Y.; Hashemi, M.; Lv, Z.; Williams, B.; Popov, K.I.; Dokholyan, N.V.; Lyubchenko, Y.L. High-speed atomic force microscopy reveals structural dynamics of α-synuclein monomers and dimers. *J. Chem. Phys.* **2018**, *148*, 123322. [CrossRef]
111. Ewald, M.; Henry, S.; Lambert, E.; Feuillie, C.; Bobo, C.; Cullin, C.; Lecomte, S.; Molinari, M. High speed atomic force microscopy to investigate the interactions between toxic Aβ 1-42 peptides and model membranes in real time: Impact of the membrane composition. *Nanoscale* **2019**, *11*, 7229–7238. [CrossRef] [PubMed]
112. Tashiro, R.; Taguchi, H.; Hidaka, K.; Endo, M.; Sugiyama, H. Effects of physical damage in the intermediate phase on the progression of Amyloid β fibrillization. *Chem. Asian J.* **2019**, *14*, 4140–4145. [CrossRef] [PubMed]
113. Feng, L.; Watanabe, H.; Molino, P.; Wallace, G.G.; Phung, S.L.; Uchihashi, T.; Higgins, M.J. Dynamics of inter-molecular interactions between single Aβ42 oligomeric and aggregate species by high-speed atomic force microscopy. *J. Mol. Biol.* **2019**, *431*, 2687–2699. [CrossRef] [PubMed]
114. Konno, H.; Watanabe-Nakayama, T.; Uchihashi, T.; Okuda, M.; Zhu, L.; Kodera, N.; Kikuchi, Y.; Ando, T.; Taguchi, H. Dynamics of oligomer and amyloid fibril formation by yeast prion Sup35 observed by high-speed atomic force microscopy. *Proc. Natl. Acad. Sci. USA* **2020**, *117*, 7831–7836. [CrossRef]
115. Milhiet, P.-E.; Yamamoto, D.; Berthoumieu, O.; Dosset, P.; Le Grimellec, C.; Verdier, J.-M.; Marchal, S.; Ando, T. Deciphering the structure, growth and assembly of amyloid-like fibrils using high-speed atomic force microscopy. *PLoS ONE* **2010**, *5*, e13240. [CrossRef] [PubMed]
116. Sahoo, B.R.; Genjo, T.; Nakayama, T.W.; Stoddard, A.K.; Ando, T.; Yasuhara, K.; Fierke, C.A.; Ramamoorthy, A. A cationic polymethacrylate-copolymer acts as an agonist for β-amyloid and an antagonist for amylin fibrillation. *Chem. Sci.* **2019**, *10*, 3976–3986. [CrossRef]
117. Zhu, H.; Wang, X.; Wallack, M.; Li, H.; Carreras, I.; Dedeoglu, A.; Hur, J.-Y.; Zheng, H.; Li, H.; Fine, R.; et al. Intraperitoneal injection of the pancreatic peptide amylin potently reduces behavioral impairment and brain amyloid pathology in murine models of Alzheimer's disease. *Mol. Psychiatry* **2015**, *20*, 252–262. [CrossRef]

118. Zhu, H.; Xue, X.; Wang, E.; Wallack, M.; Na, H.; Hooker, J.M.; Kowall, N.; Tao, Q.; Stein, T.D.; Wolozin, B.; et al. Amylin receptor ligands reduce the pathological cascade of Alzheimer's disease. *Neuropharmacology* **2017**, *119*, 170–181. [CrossRef]
119. Wang, E.; Zhu, H.; Wang, X.; Gower, A.C.; Wallack, M.; Blusztajn, J.K.; Kowall, N.; Qiu, W.Q. Amylin treatment reduces neuroinflammation and ameliorates abnormal patterns of gene expression in the cerebral cortex of an Alzheimer's Disease mouse model. *J. Alzheimer's Dis.* **2017**, *56*, 47–61. [CrossRef]
120. Mohamed, L.A.; Zhu, H.; Mousa, Y.M.; Wang, E.; Qiu, W.Q.; Kaddoumi, A. Amylin enhances Amyloid-β peptide brain to blood efflux across the blood-brain barrier. *J. Alzheimer's Dis.* **2017**, *56*, 1087–1099. [CrossRef]
121. Jackson, K.; Barisone, G.A.; Diaz, E.; Jin, L.; DeCarli, C.; Despa, F. Amylin deposition in the brain: A second amyloid in Alzheimer disease? *Ann. Neurol.* **2013**, *74*, 517–526. [CrossRef] [PubMed]
122. Oskarsson, M.E.; Paulsson, J.F.; Schultz, S.W.; Ingelsson, M.; Westermark, P.; Westermark, G.T. In Vivo seeding and cross-seeding of localized amyloidosis: A molecular link between type 2 diabetes and alzheimer disease. *Am. J. Pathol.* **2015**, *185*, 834–846. [CrossRef] [PubMed]
123. Mousa, Y.M.; Abdallah, I.M.; Hwang, M.; Martin, D.R.; Kaddoumi, A. Amylin and pramlintide modulate γ-secretase level and APP processing in lipid rafts. *Sci. Rep.* **2020**, *10*, 3751. [CrossRef] [PubMed]
124. Watanabe-Nakayama, T.; Ono, K. High-Speed atomic force microscopy of individual amyloidogenic protein assemblies. In *Methods in Molecular Biology*; Humana Press: Clifton, NJ, USA, 2018; Volume 1814, pp. 201–212.
125. Brender, J.R.; Ghosh, A.; Kotler, S.A.; Krishnamoorthy, J.; Bera, S.; Morris, V.; Sil, T.B.; Garai, K.; Reif, B.; Bhunia, A.; et al. Probing transient non-native states in amyloid beta fiber elongation by NMR. *Chem. Commun.* **2019**, *55*, 4483–4486. [CrossRef] [PubMed]
126. Kellermayer, M.S.Z.; Karsai, A.; Benke, M.; Soós, K.; Penke, B. Stepwise dynamics of epitaxially growing single amyloid fibrils. *Proc. Natl. Acad. Sci. USA* **2008**, *105*, 141–144. [CrossRef] [PubMed]
127. Howard, J. *Mechanics of Motor Proteins and the Cytoskeleton*; Sinauer Associates: Sunderland, MA, USA, 2001; ISBN 9780878933334.
128. Dunstan, D.E.; Hamilton-Brown, P.; Asimakis, P.; Ducker, W.; Bertolini, J. Shear flow promotes amyloid-fibrilization. *Protein Eng. Des. Sel.* **2009**, *22*, 741–746. [CrossRef] [PubMed]
129. Collins, S.R.; Douglass, A.; Vale, R.D.; Weissman, J.S. Mechanism of prion propagation: Amyloid growth occurs by monomer addition. *PLoS Biol.* **2004**, *2*, e321. [CrossRef]
130. Liu, J.-J.; Lindquist, S. Oligopeptide-repeat expansions modulate 'protein-only' inheritance in yeast. *Nature* **1999**, *400*, 573–576. [CrossRef] [PubMed]
131. Serpell, L.C.; Berriman, J.; Jakes, R.; Goedert, M.; Crowther, R.A. Fiber diffraction of synthetic α-synuclein filaments shows amyloid-like cross-β conformation. *Proc. Natl. Acad. Sci. USA* **2000**, *97*, 4897–4902. [CrossRef] [PubMed]
132. Hill, E.K.; Krebs, B.; Goodall, D.G.; Howlett, G.J.; Dunstan, D.E. Shear flow induces amyloid fibril formation. *Biomacromolecules* **2006**, *7*, 10–13. [CrossRef] [PubMed]
133. Wang, J.; Yamamoto, T.; Bai, J.; Cox, S.J.; Korshavn, K.J.; Monette, M.; Ramamoorthy, A. Real-time monitoring of the aggregation of Alzheimer's amyloid-β via ^{1}H magic angle spinning NMR spectroscopy. *Chem. Commun.* **2018**, *54*, 2000–2003. [CrossRef]
134. Owa, M.; Uchihashi, T.; Yanagisawa, H.; Yamano, T.; Iguchi, H.; Fukuzawa, H.; Wakabayashi, K.; Ando, T.; Kikkawa, M. Inner lumen proteins stabilize doublet microtubules in cilia and flagella. *Nat. Commun.* **2019**, *10*, 1143. [CrossRef]
135. Fukui, T.; Uchihashi, T.; Sasaki, N.; Watanabe, H.; Takeuchi, M.; Sugiyasu, K. Direct observation and manipulation of supramolecular polymerization by high-speed atomic force microscopy. *Angew. Chem.* **2018**, *57*, 15465–15470. [CrossRef] [PubMed]
136. Michaels, T.C.T.; Šarić, A.; Curk, S.; Bernfur, K.; Arosio, P.; Meisl, G.; Dear, A.J.; Cohen, S.I.A.; Dobson, C.M.; Vendruscolo, M.; et al. Dynamics of oligomer populations formed during the aggregation of Alzheimer's Aβ42 peptide. *Nat. Chem.* **2020**, *12*, 445–451. [CrossRef]
137. Relini, A.; Cavalleri, O.; Rolandi, R.; Gliozzi, A. The two-fold aspect of the interplay of amyloidogenic proteins with lipid membranes. *Chem. Phys. Lipids* **2009**, *158*, 1–9. [CrossRef]
138. Sasahara, K.; Morigaki, K.; Shinya, K. Effects of membrane interaction and aggregation of amyloid β-peptide on lipid mobility and membrane domain structure. *Phys. Chem. Chem. Phys.* **2013**, *15*, 8929–8939. [CrossRef]

139. Choucair, A.; Chakrapani, M.; Chakravarthy, B.; Katsaras, J.; Johnston, L.J. Preferential accumulation of Aβ(1−42) on gel phase domains of lipid bilayers: An AFM and fluorescence study. *Biochim. Biophys. Acta (BBA) Biomembr.* **2007**, *1768*, 146–154. [CrossRef] [PubMed]
140. Koppaka, V.; Axelsen, P.H. Accelerated accumulation of amyloid β proteins on oxidatively damaged lipid membranes. *Biochemistry* **2000**, *39*, 10011–10016. [CrossRef] [PubMed]
141. Drolle, E.; Hane, F.; Lee, B.; Leonenko, Z. Atomic force microscopy to study molecular mechanisms of amyloid fibril formation and toxicity in Alzheimer's disease. *Drug Metab. Rev.* **2014**, *46*, 207–223. [CrossRef] [PubMed]
142. Drolle, E.; Negoda, A.; Hammond, K.; Pavlov, E.; Leonenko, Z. Changes in lipid membranes may trigger amyloid toxicity in Alzheimer's disease. *PLoS ONE* **2017**, *12*, e0182194. [CrossRef]
143. Hane, F.; Drolle, E.; Gaikwad, R.; Faught, E.; Leonenko, Z. Amyloid-β aggregation on model lipid membranes: An atomic force microscopy study. *J. Alzheimer's Dis.* **2011**, *26*, 485–494. [CrossRef] [PubMed]
144. Pithadia, A.; Brender, J.R.; Fierke, C.A.; Ramamoorthy, A. Inhibition of IAPP aggregation and toxicity by natural products and derivatives. *J. Diabetes Res.* **2016**, *2016*. [CrossRef]
145. Ono, K.; Yoshiike, Y.; Takashima, A.; Hasegawa, K.; Naiki, H.; Yamada, M. Potent anti-amyloidogenic and fibril-destabilizing effects of polyphenols in vitro: Implications for the prevention and therapeutics of Alzheimer's disease. *J. Neurochem.* **2003**, *87*, 172–181. [CrossRef] [PubMed]
146. Ono, K.; Hasegawa, K.; Naiki, H.; Yamada, M. Anti-amyloidogenic activity of tannic acid and its activity to destabilize Alzheimer's β-amyloid fibrils in vitro. *Biochim. Biophys. Acta Mol. Basis Dis.* **2004**, *1690*, 193–202. [CrossRef] [PubMed]
147. Ono, K.; Hirohata, M.; Yamada, M. Ferulic acid destabilizes preformed beta-amyloid fibrils in vitro. *Biochem. Biophys. Res. Commun.* **2005**, *336*, 444–449. [CrossRef] [PubMed]
148. Ono, K.; Yamada, M. Antioxidant compounds have potent anti-fibrillogenic and fibril-destabilizing effects for alpha-synuclein fibrils in vitro. *J. Neurochem.* **2006**, *97*, 105–115. [CrossRef]
149. Hirohata, M.; Hasegawa, K.; Tsutsumi-Yasuhara, S.; Ohhashi, Y.; Ookoshi, T.; Ono, K.; Yamada, M.; Naiki, H. The anti-amyloidogenic effect is exerted against Alzheimer's beta-amyloid fibrils in vitro by preferential and reversible binding of flavonoids to the amyloid fibril structure. *Biochemistry* **2007**, *46*, 1888–1899. [CrossRef]
150. Ono, K.; Condron, M.M.; Ho, L.; Wang, J.; Zhao, W.; Pasinetti, G.M.; Teplow, D.B. Effects of grape seed-derived polyphenols on amyloid beta-protein self-assembly and cytotoxicity. *J. Biol. Chem.* **2008**, *283*, 32176–32187. [CrossRef]
151. Hamaguchi, T.; Ono, K.; Murase, A.; Yamada, M. Phenolic compounds prevent Alzheimer's pathology through different effects on the Amyloid-β aggregation pathway. *Am. J. Pathol.* **2009**, *175*, 2557–2565. [CrossRef]
152. Ono, K.; Li, L.; Takamura, Y.; Yoshiike, Y.; Zhu, L.; Han, F.; Mao, X.; Ikeda, T.; Takasaki, J.; Nishijo, H.; et al. Phenolic compounds prevent amyloid β-protein oligomerization and synaptic dysfunction by site-specific binding. *J. Biol. Chem.* **2012**, *287*, 14631–14643. [CrossRef] [PubMed]
153. Takahashi, R.; Ono, K.; Takamura, Y.; Mizuguchi, M.; Ikeda, T.; Nishijo, H.; Yamada, M. Phenolic compounds prevent the oligomerization of α-synuclein and reduce synaptic toxicity. *J. Neurochem.* **2015**, *134*, 943–955. [CrossRef]
154. Lim, Y.-A.; Ittner, L.M.; Lim, Y.L.; Götz, J. Human but not rat amylin shares neurotoxic properties with Aβ42 in long-term hippocampal and cortical cultures. *FEBS Lett.* **2008**, *582*, 2188–2194. [CrossRef] [PubMed]
155. Fu, W.; Ruangkittisakul, A.; MacTavish, D.; Shi, J.Y.; Ballanyi, K.; Jhamandas, J.H. Amyloid β (Aβ) peptide directly activates amylin-3 receptor subtype by triggering multiple intracellular signaling pathways. *J. Biol. Chem.* **2012**, *287*, 18820–18830. [CrossRef]
156. Qiu, W.Q.; Walsh, D.M.; Ye, Z.; Vekrellis, K.; Zhang, J.; Podlisny, M.B.; Rosner, M.R.; Safavi, A.; Hersh, L.B.; Selkoe, D.J. Insulin-degrading enzyme regulates extracellular levels of amyloid beta-protein by degradation. *J. Biol. Chem.* **1998**, *273*, 32730–32738. [CrossRef] [PubMed]
157. Olsson, M.; Herrington, M.K.; Reidelberger, R.D.; Permert, J.; Arnelo, U. Comparison of the effects of chronic central administration and chronic peripheral administration of islet amyloid polypeptide on food intake and meal pattern in the rat. *Peptides* **2007**, *28*, 1416–1423. [CrossRef]
158. Banks, W.A.; Kastin, A.J. Differential permeability of the blood–brain barrier to two pancreatic peptides: Insulin and amylin. *Peptides* **1998**, *19*, 883–889. [CrossRef]

159. Banks, W.A.; Kastin, A.J.; Maness, L.M.; Huang, W.; Jaspan, J.B. Permeability of the blood-brain barrier to amylin. *Life Sci.* **1995**, *57*, 1993–2001. [CrossRef]
160. Mietlicki-Baase, E.G. Amylin in Alzheimer's disease: Pathological peptide or potential treatment? *Neuropharmacology* **2018**, *136*, 287–297. [CrossRef]
161. Paolino, D.; Vero, A.; Cosco, D.; Pecora, T.M.G.; Cianciolo, S.; Fresta, M.; Pignatello, R. Improvement of oral bioavailability of curcumin upon microencapsulation with methacrylic copolymers. *Front. Pharmacol.* **2016**, *7*. [CrossRef]
162. Wang, W.; Shao, A.; Zhang, N.; Fang, J.; Ruan, J.J.; Ruan, B.H. Cationic polymethacrylate-modified liposomes significantly enhanced doxorubicin delivery and antitumor activity. *Sci. Rep.* **2017**, *7*, 43036. [CrossRef]
163. Yasuhara, K.; Arakida, J.; Ravula, T.; Ramadugu, S.K.; Sahoo, B.; Kikuchi, J.; Ramamoorthy, A. Spontaneous lipid nanodisc fomation by amphiphilic polymethacrylate copolymers. *J. Am. Chem. Soc.* **2017**, *139*, 18657–18663. [CrossRef] [PubMed]
164. Sahoo, B.R.; Ii, L.S.; Ivanova, M.; Deng, Z.; Nakayama, T.W.; Suladze, S.; Reif, B.; Ando, T.; Martyniuk, C.J. Conformational tuning of amylin by charged SMA copolymers. *BioRxiv* **2020**. [CrossRef]
165. Ravula, T.; Hardin, N.Z.; Ramadugu, S.K.; Cox, S.J.; Ramamoorthy, A. Formation of pH-Resistant monodispersed polymer–lipid nanodiscs. *Angew. Chem. Int. Ed.* **2018**, *57*, 1342–1345. [CrossRef] [PubMed]
166. Ravula, T.; Ramadugu, S.K.; Di Mauro, G.; Ramamoorthy, A. Bioinspired, size-tunable self-assembly of polymer–lipid bilayer nanodiscs. *Angew. Chem. Int. Ed.* **2017**, *56*, 11466–11470. [CrossRef]
167. Gasteiger, E.; Hoogland, C.; Gattiker, A.; Duvaud, S.; Wilkins, M.R.; Appel, R.D.; Bairoch, A. Protein identification and analysis tools on the ExPASy server. In *The Proteomics Protocols Handbook*; Humana Press: Clifton, NJ, USA, 2005.
168. Czajkowsky, D.M.; Shao, Z. Inhibition of protein adsorption to muscovite mica by monovalent cations. *J. Microsc.* **2003**, *211*, 1–7. [CrossRef]
169. Watanabe-Nakayama, T.; Itami, M.; Kodera, N.; Ando, T.; Konno, H. High-speed atomic force microscopy reveals strongly polarized movement of clostridial collagenase along collagen fibrils. *Sci. Rep.* **2016**, *6*, 28975. [CrossRef]
170. Uchihashi, T.; Kodera, N.; Ando, T. Guide to video recording of structure dynamics and dynamic processes of proteins by high-speed atomic force microscopy. *Nat. Protoc.* **2012**, *7*, 1193–1206. [CrossRef]
171. Cendrowska, U.; Silva, P.J.; Ait-Bouziad, N.; Müller, M.; Guven, Z.P.; Vieweg, S.; Chiki, A.; Radamaker, L.; Kumar, S.T.; Fändrich, M.; et al. Unraveling the complexity of amyloid polymorphism using gold nanoparticles and cryo-EM. *Proc. Natl. Acad. Sci. USA* **2020**, *117*, 6866–6874. [CrossRef]

International Journal of
Molecular Sciences

MDPI

Review

Polyphenols with Anti-Amyloid β Aggregation Show Potential Risk of Toxicity Via Pro-Oxidant Propertie

Hatasu Kobayashi [1], Mariko Murata [1], Shosuke Kawanishi [2] and Shinji Oikawa [1,*]

1 Department of Environmental and Molecular Medicine, Mie University Graduate School of Medicine, Tsu, Mie 514-8507, Japan; hatasuk@doc.medic.mie-u.ac.jp (H.K.); mmurata@doc.medic.mie-u.ac.jp (M.M.)

2 Faculty of Pharmaceutical Sciences, Suzuka University of Medical Science, Suzuka, Mie 513-8670, Japan; kawanisi@suzuka-u.ac.jp

* Correspondence: s-oikawa@doc.medic.mie-u.ac.jp; Tel.: +81-59-231-5011

Received: 8 April 2020; Accepted: 13 May 2020; Published: 18 May 2020

Abstract: Alzheimer's disease (AD) is the most common form of dementia among older people. Amyloid β (Aβ) aggregation has been the focus for a therapeutic target for the treatment of AD. Naturally occurring polyphenols have an inhibitory effect on Aβ aggregation and have attracted a lot of attention for the development of treatment strategies which could mitigate the symptoms of AD. However, considerable evidence has shown that the pro-oxidant mechanisms of polyphenols could have a deleterious effect. Our group has established an assay system to evaluate the pro-oxidant characteristics of chemical compounds, based on their reactivity with DNA. In this review, we have summarized the anti-Aβ aggregation and pro-oxidant properties of polyphenols. These findings could contribute to understanding the mechanism underlying the potential risk of polyphenols. We would like to emphasize the importance of assessing the pro-oxidant properties of polyphenols from a safety point of view.

Keywords: Amyloid β; Polyphenol; Pro-oxidant; Alzheimer's Disease

---

1. Amyloid β Aggregation in Alzheimer's Disease

Alzheimer's disease (AD) is the leading cause of dementia, and disease prevalence has been increasing dramatically with a worldwide increase in the aging population [1]. Numerous studies have suggested that accumulation of the Amyloid β (Aβ) peptide in the brain is the initial pathological event for AD [2]. The Aβ peptide is a soluble, extracellular fragment generated from the sequential cleavage of the amyloid precursor protein (APP) by β- and γ-secretases [3]. Aβ accumulation promotes conformational changes in the peptide, resulting in the formation of oligomers and fibrils; ultimately, resulting in plaque deposition—one of the hallmarks of AD pathology [2,4]. The nucleation-dependent polymerization mechanism, which separates the amyloid fibrillization process into a nucleation phase and an elongation phase [5], is currently proposed as an aggregation mechanism for the Aβ peptide (Figure 1). During the nucleation phase, soluble Aβ monomers undergo conformational changes and self-associate to form oligomeric nuclei that are rich in β-sheets. During the elongation phase, these oligomeric nuclei act as a template and associate with monomers to initiate polymerization [6]. There are currently four approved medications for AD (three cholinesterase inhibitors and one uncompetitive NMDA receptor modulator), but they have a small effect size and show no effect on long-term disease progression [7]. Therefore, new drugs directed against various identified targets of AD, such as Aβ, tau, ApoE, and neuroinflammation are urgently needed [7]. Among these therapeutic targets, researchers have largely focused on Aβ aggregation for the prevention and treatment of AD, based on the "amyloid cascade hypothesis".

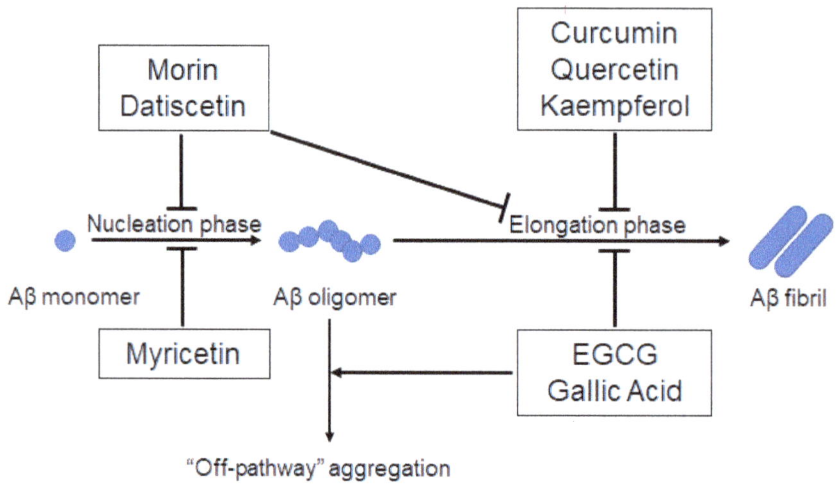

Figure 1. A schematic model showing the inhibitory effects of polyphenols on Aβ aggregation, based on the "amyloid cascade hypothesis." Myricetin inhibits nucleation [8]. Morin and datiscetin inhibit nucleation and elongation [9]. Curcumin [10], quercetin [11], and kaempferol [9] inhibit elongation. EGCG [12] and gallic acid [13] inhibit elongation and redirect Aβ oligomers to "off-pathway" aggregation. Aβ: amyloid β, EGCG: epigallocatechin gallate.

2. Beneficial Anti-Aβ Aggregation and Adverse Pro-Oxidant Effects of Polyphenols

Researchers have investigated the inhibitory effects of various chemical and biological molecules on Aβ aggregation to develop a strategy for mitigating AD. These compounds include small organic molecules, peptide derivatives, chemical and molecular chaperones, and antibodies, to name a few [4]. Polyphenols are naturally occurring secondary metabolites found in large quantities in fruits, vegetables, seeds, and plant-derived oils; thus exhibit easy availability [14,15]. In vitro studies have shown that several polyphenols reduce Aβ aggregation by inhibiting the nucleation phase or elongation phase, or both, and redirecting the Aβ oligomers to the less-toxic "off-pathway" aggregation (Figure 1). Details of the anti-Aβ aggregation activity of each polyphenol are described below (Section 2.1–2.4). The anti-Aβ aggregation activities of some compounds have been confirmed in animal studies, and clinical studies have either been performed or are being performed to test these selected polyphenols [16,17]. However, considerable evidence has raised the concern that polyphenols could exert deleterious effects through their pro-oxidant mechanisms [18–20]. Many polyphenols involved in anti-Aβ aggregation have been reported to display pro-oxidant activities, which are potentially linked with toxic effects (Table 1, Figure 2).

A common feature of polyphenols, especially those harboring hydroxyl groups in the phenol ring, is that they can readily participate in redox reactions [21], which is associated with both their antioxidant and pro-oxidant properties. Our group has established an assay to evaluate the pro-oxidant characteristics of chemical compounds on the basis of their ability to induce oxidative DNA damage, and investigated the mechanisms of the reactive oxygen species (ROS) generation [22]. Based on our results and that of others in the literature, in this review we have focused on polyphenols whose mechanisms of inhibiting Aβ aggregation have been well-studied, and have summarized their pro-oxidant properties. In addition, recent studies have suggested that biological activities of polyphenols were attributed to not only the polyphenols themselves but also their metabolites generated in vivo [23,24]. Therefore, we also have described some cases showing that metabolites are involved in pro-oxidant properties.

Table 1. Toxic effects associated with pro-oxidant properties of naturally occurring polyphenols harboring anti-Aβ aggregation activity.

Anti-Aβ Aggregation Effect	Polyphenol	Toxic Effects Associated with Pro-Oxidant Properties	Concentration or Dose Showing Toxic Effects of Polyphenols
Inhibiting nucleation	Myricetin	**Cytotoxicity**	
		Cytotoxicity linked with ROS generation	Cell: 20 μM [25], 50 μM [26,27]
		Genotoxicity	
		Oxidative DNA damage	Cell: 20 μM [28], 50 μM [29] DNA: 5 μM [30], 200 μM [31]
		Mutagenic activity	Bacteria: 0.628 μmol/plate [32] Cell: 42 μM [33]
Inhibiting nucleation and elongation	Morin	**Genotoxicity**	
		Oxidative DNA damage	Cell: 100 μM [34] DNA: 5 μM [30],10 μM [35], 20 μM [36], 100 μM [37]
		Mutagenic activity	Bacteria: 0.149 μmol/plate [38]
	Datiscetin	No report	
Inhibiting elongation	Curcumin	**Cytotoxicity**	
		Cytotoxicity linked with ROS generation	Cell: 5 μM [39], 50 μM [40]
		Genotoxicity	
		DNA damage in cultured cell	Cell: 50 μM [41]
		Curcumin metabolite-mediated oxidative damage in isolated DNA	DNA: 2 μM [42]
		Tumorigenicity	
		Colon mucosal hyperplasia and hepatocellular adenoma in rats and mice treated with turmeric oleoresin containing curcumin (79%-85%), respectively	Colon hyperplasia: 2000 mg/kg/day (male rats) [43] Hepatocellular adenoma: 520 mg/kg/day (male mice) [43], 1620 mg/kg/day (female mice) [43]

Table 1. *Cont.*

Anti-Aβ Aggregation Effect	Polyphenol	Toxic Effects Associated with Pro-Oxidant Properties	Concentration or Dose Showing Toxic Effects of Polyphenols
Inhibiting elongation (continued)	Quercetin	**Cytotoxicity**	
		Cytotoxicity linked with ROS generation	Cell: 50 μM [44]
		Genotoxicity	
		Oxidative DNA damage	Cell: 30 μM [45], 50 μM [29], 100 μM [34] DNA: 10 μM [46]
		Mutagenic activity	Bacteria: 0.121 μmol/plate [47] Cell: 2.2 μM [48], 32.5 μM [49]
		Carcinogenesis	
		Renal tubule adenocarcinomas and intestinal and bladder cancer in rats	Renal tubule adenocarcinomas: 1900 mg/kg/day (male rats) [48] Intestinal and bladder cancer: 27.8 mM/rat (male, cumulative dose) [50], 25.3 mM/rat (female, cumulative dose) [50]
	Kaempferol	**Genotoxicity**	
		Oxidative DNA damage	Cell: 50 μM [29]
		Mutagenic activity	Bacteria: 0.143 μmol/plate [47]

Table 1. *Cont.*

Anti-Aβ Aggregation Effect	Polyphenol	Toxic Effects Associated with Pro-Oxidant Properties	Concentration or Dose Showing Toxic Effects of Polyphenols
Inhibiting elongation and redirecting to "off-pathway" aggregation	EGCG	**Cytotoxicity**	
		Cytotoxicity linked with ROS generation	Cell: 2 μM [51], 12.5 μM [52]
		Genotoxicity	
		Oxidative DNA damage	Cell: 100 μM [53], 200 μM [54] DNA: 5 μM [54]
		Hepatotoxicity and gastrointestinal toxicity	
		Gastrointestinal tract and liver lesion in rats and mice treated with green tea extract containing EGCG (48.4%)	Gastrointestinal tract lesion: 1000 mg/kg/day (male and female rats) [55] Liver lesion: 1000 mg/kg/day (male and female rats) [55], 300 mg/kg/day (male mice) [55]
		High dose intake-associated liver damage in humans	Human: 704 mg/day [56]
	Gallic acid	**Cytotoxicity**	
		Cytotoxicity linked with ROS generation	Cell: 74 μM [57], 294 μM [58,59]
		Genotoxicity	
		Oxidative DNA damage	DNA: 5 μM [60], 200 μM [61]
		Hepatotoxicity and nephrotoxicity	
		Liver damage in mice and rats, and renal injury in rats	Liver damage: 200 mg/kg/day (male mice) [62], 100 mg/kg/day (male rats) [63] Renal injury: 100 mg/kg/day (male rats) [63]

Aβ: amyloid β, ROS: reactive oxygen species, EGCG: epigallocatechin gallate.

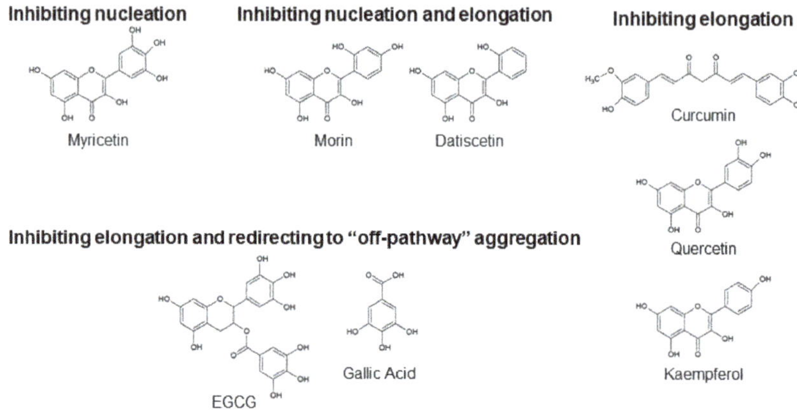

Figure 2. Chemical structures of polyphenols shown in Table 1. EGCG: epigallocatechin gallate

2.1. Polyphenols Involved in Inhibiting Nucleation

Myricetin

Myricetin is one of the most common naturally occurring compounds found in a large variety of plants and has been reported to show good biological activity as an antioxidant, anti-inflammatory, and anti-tumorigenic agent [64,65]. Studies using fluorescence spectroscopy with thioflavin T and electron microscopy have shown that myricetin inhibits the formation of Aβ fibrils [66]. Ono et al. demonstrated that myricetin blocked Aβ oligomer formation and bound to monomeric Aβ by an assay using a photoinduced cross-linking agent and nuclear magnetic resonance (NMR) [8]. These findings suggested that myricetin could prevent nucleation via direct binding to the Aβ monomer. Myricetin was also shown to reduce the number of high molecular weight oligomers and prevent the development of AD pathology in an AD mouse model [67].

Despite these encouraging results, myricetin has been reported to have mutagenic activity [32,33]. A recent study showed that myricetin tested positive in a bacterial mutagenicity assay and in vitro micronuclei formation assay [32]. Metal-mediated DNA damage induced by myricetin has been demonstrated in studies using plasmid DNA, isolated nuclei, and cultured cells [28–31]. The inhibitory effects of several ROS scavengers on DNA damage [28,31] and the generation of 8-oxo-7,8-dihydro-2′-deoxyguanosine (8-oxodG), an indicator of oxidative DNA damage [31], have indicated pro-oxidant mechanisms of myricetin-induced DNA damage.

2.2. Polyphenols Involved in Inhibiting Nucleation and Elongation

Morin and Datiscetin

Morin a member of the flavonoid family, was originally isolated from the members of the *Moraceae* family and is found in a wide variety of fruits, vegetables, and herbs [68,69]. Morin, which has antioxidant and anti-inflammatory activities, has been reported to show pharmacological effects in several diseases [68–70]. The inhibitory effect of morin towards Aβ aggregation has been reported in several in vitro studies that tested naturally occurring compounds [66,71]. Furthermore, sustained treatment with morin could reduce the production of insoluble Aβ and the formation of amyloid plaques [72] and rescue cognitive impairment [72,73] in AD and dementia animal models. NMR analysis has shown that morin could prevent both the nucleation and the elongation phases during Aβ42 aggregation by interacting with His13, His14, and Gln15, which are close to the intermolecular β-sheet region of Aβ42 [9]. This anti-Aβ aggregation activity has been attributed to the C-1 oxygen of the C-ring and the 2′-hydroxyl group of the B-ring (Figure 3), which stabilize the flatness between the A-, B-, and C-rings of morin and enable it to interact with the intermolecular β-sheet region [74]. Datiscetin, which has the same structure as morin except for the 4′-hydroxyl group of the B-ring, also prevents Aβ aggregation by the same mechanism [9,74].

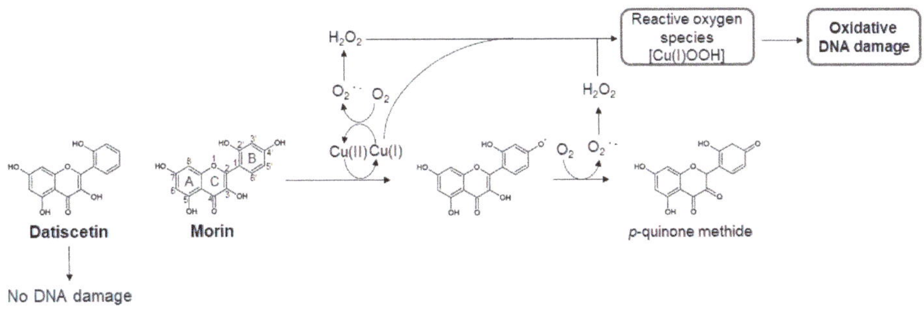

Figure 3. Possible mechanism of oxidative DNA damage induced by morin in the presence of Cu(II). The 4′-hydroxyl group of the B-ring of morin is responsible for the generation of Cu(I)-hydroperoxide (Cu(I)OOH) and the resultant oxidative DNA damage. Datiscetin, an analog of morin, without the 4′-hydroxyl group, does not damage DNA.

Previously, morin has been shown to promote ROS generation. Morin could induce metal-mediated lipid peroxidation of the nuclear membrane and DNA strand break in isolated nuclei [35]. The morin-Cu(II) complex could cleave plasmid DNA via an oxidative pathway [30,37]. Cell model studies have suggested that morin can cause DNA strand breaks though ROS production [34]. Morin was shown to have a mutagenic activity with the *Salmonella*/microsomal activation system [38]. Recently, we have shown that in the presence of Cu(II), morin induces not only DNA strand breaks but also base modification, including 8-oxodG formation, in isolated DNA [36]. By testing the effects of various ROS scavengers and Cu(I) chelators on DNA damage, we proposed that morin undergoes autoxidation via the Cu(I)/Cu(II) redox cycle, resulting in H_2O_2 generation to produce Cu(I)-hydroperoxide, which causes oxidative DNA damage (Figure 3) [36]. However, datiscetin, which lacks the 4′-hydroxyl group of the B-ring, did not induce DNA damage under our experimental condition (unpublished data). These results indicated that the 4′-hydroxyl group of the B-ring plays an important role in the pro-oxidant activity of morin.

2.3. *Polyphenols Involved in Inhibiting Elongation*

2.3.1. Curcumin

Curcumin is the main naturally occurring polyphenol found in turmeric, which is isolated from the rhizome of *Curcuma longa* and is extensively used as a spice in curries and mustards [75]. Turmeric has also been traditionally used as a medicinal herb for the treatment of various diseases in Ayurvedic and traditional Chinese medicine [76]. Research on curcumin has shown that it possesses several protective and therapeutic properties, including anti-inflammatory, antioxidant, anti-microbial, and anti-cancer activity [77,78]. Recently, in the context of therapies for AD and other neurodegenerative diseases, Schubert et al. proposed a novel drug screening platform which finds candidates with multiple neuroprotective activities, and identified curcumin as a lead compound from the screening of natural product libraries [79]. As one of the neuroprotective properties, several in vitro [80,81] and in vivo [81,82] studies have demonstrated that curcumin can inhibit Aβ aggregation. Curcumin has been shown to prevent the formation and extension of Aβ fibrils and destabilize preformed Aβ fibrils in vitro [80]. Curcumin inhibits Aβ aggregation by directly binding Aβ to block its self-assembly in an in vitro aggregation assay [81]. NMR analysis has indicated that curcumin interacts with residue number 12 and 17–21, included in the β-sheet structure of the Aβ42 fibrils [10], suggesting that it has an inhibitory effect on fibril elongation. On the other hand, some researchers caution that these wide-range bioactivities of curcumin are characteristics of a pan-assay interference compound (PAINS) [83,84]. PAINS are compounds displaying activities which do not depend on a specific and drug-like interaction between molecule and protein, leading to artifact in multiple types of assays [83,84]. Thus, further research is needed to explore the therapeutic value of curcumin.

Curcumin has also been reported to have pro-oxidant properties under some conditions [85]. Several studies have reported ROS generation and DNA damage in cultured cells exposed to curcumin [39–41]. A National Toxicology Program study has revealed that the dietary intake of turmeric oleoresin, which contains a high curcumin content (79–85%), induced hyperplasia of the colon mucosa in rats and increased hepatocellular adenoma in mice [43]. These findings raise the possibility that curcumin-induced oxidative DNA damage may promote tumorigenesis [78]. However, we have previously shown that curcumin does not cause damage to isolated DNA by itself, even in the presence of Cu(II) [42]. Curcumin/Cu(II)-mediated oxidative DNA damage has occurred only when curcumin was pre-treated with cytochrome P450 (CYP) enzymes, suggesting that metabolites of curcumin act as DNA-damaging agents [42]. Mass spectral analysis indicated CYP2D6-mediated *o*-demethyl curcumin formation, which was considered to generate Cu(I)-hydroperoxide during the autoxidation of *o*-demethyl curcumin, resulting in DNA damage [42]. CYP enzymes are known to be related to bioactivation of several chemical carcinogens [86], suggesting that curcumin-mediated hepatocellular adenoma in mice [43] might be explained by particular metabolisms of hepatocytes. *o*-demethyl curcumin has a catechol moiety (phenol with two hydroxy groups in the ortho-position), resulting from CYP-mediated demethylation of curcumin, that is likely to play a critical role in oxidative damage (Figure 4A). Thus, some antioxidants could be converted to pro-oxidants in particular metabolic conditions. These findings suggest that it is important to evaluate not only target polyphenols but also their metabolites.

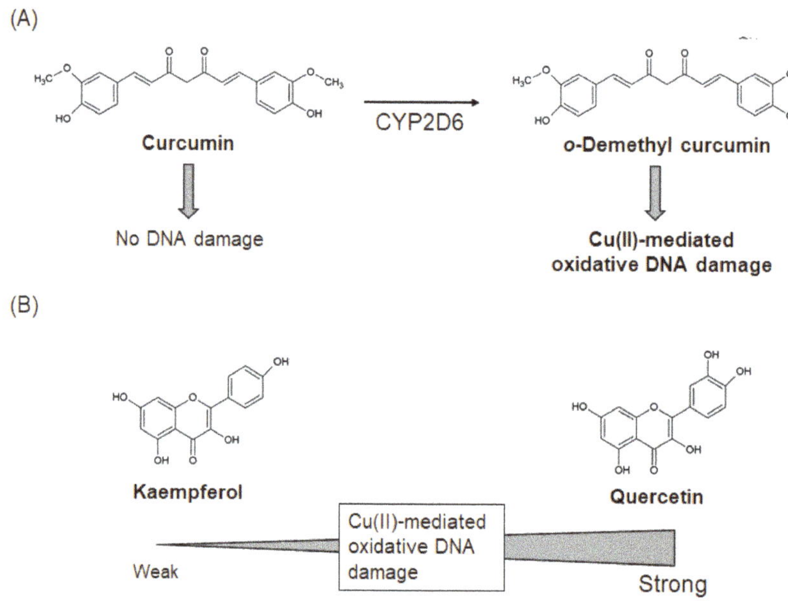

Figure 4. The role of catechol moieties in *o*-demethyl curcumin- and quercetin-mediated oxidative DNA damage in the presence of Cu(II). (**A**) *o*-Demethyl curcumin (with a catechol moiety) induced Cu(II)-mediated oxidative damage, while curcumin, its parent compound (without catechol moieties), did not. (**B**) Quercetin (with a catechol moiety) induced stronger oxidative damage than kaempferol, its analog (without catechol moieties) in the presence of Cu(II). CYP: cytochrome P450

2.3.2. Quercetin and Kaempferol

Quercetin, a readily available naturally occurring polyphenol, is found abundantly in vegetables and fruits, such as onions and apples [87]. Quercetin is well known to have antioxidant and anti-inflammatory properties, and is expected to play protective roles in a wide range of diseases [88–90]. In vitro aggregation studies show that quercetin inhibits Aβ fibril formation by strengthening the hydrophobic interactions between the Aβ β-sheet structure and the aromatic ring by hydrogen bonding [91]. Quercetin exerts an anti-amyloidogenic effect in vitro by preferentially binding to Aβ fibrils at the growth edge, rather than to Aβ monomers, resulting in inhibition of fibril elongation [11]. Quercetin also reduces Aβ-induced neurotoxicity in a cell system overexpressing mutant APP, which is associated with early-onset familial AD [91]. Treatment with quercetin reduced the number and size of Aβ plaques and improves cognitive function in an AD mouse model [92]. Kaempferol, an analog of quercetin, also showed anti-Aβ aggregation activity [66]

Although the beneficial effects of quercetin are widely accepted, there is a concern about its potential pro-oxidant and cytotoxic activities when used therapeutically [93]. Results from several in vitro and in vivo studies suggest that quercetin has a pro-oxidant effect in addition to its antioxidant effect [34,94]. Quercetin has been reported to be mutagenic [47–49], and induce renal tubule adenocarcinomas [48] and intestinal and bladder cancer [50] in rats. We have previously shown that quercetin induced oxidative DNA damage both in isolated and cellular DNA [45,46]. Quercetin caused 8-oxodG formation in HL-60 cells, but not in their H_2O_2-resistant clones, HP100 cells, indicating that H_2O_2 is the main mediator of DNA damage and cytotoxicity in this context [45]. The pro-oxidant activity of quercetin is likely due to the presence of the catechol moiety and the resultant susceptibility to autoxidation, leading to conversion into ortho-semiquinone and ortho-quinone [95,96]. This finding is supported by the observation that kaempferol, a quercetin analog without catechol moieties, induces markedly weaker oxidative DNA damage than quercetin (Figure 4B) [46]. Furthermore, quercetin exhibits both mutagenicity and carcinogenicity [48–50], whereas kaempferol exhibits only mutagenicity (Table 1) [47], which might reflect the different extent of oxidative DNA damage caused by quercetin and kaempferol.

2.4. Polyphenols Involved in Inhibiting Elongation and Redirecting Aβ Monomers to "Off-Pathway" Aggregation

2.4.1. Epigallocatechin Gallate (EGCG) and Other Green Tea Catechins

Numerous epidemiological studies have demonstrated that consumption of green tea has many health benefits [97]. Among green tea catechins, EGCG is most abundant (65% of the total catechin content in green tea) and most biologically active [98]. EGCG is a powerful antioxidant, anti-inflammatory, and anti-infective agent, and is suggested to have protective effects in fighting many diseases [99–101]. EGCG inhibits Aβ fibrillogenesis by directly binding to natively unfolded polypeptides and promoting the formation of unstructured and nontoxic oligomers (so-called "off-pathway" aggregation) instead of toxic β-sheet–rich fibril [12,102]. EGCG oxidation products, such as quinones, may be involved in redirecting "off-pathway" aggregation by covalently binding to lysine of Aβ through a Schiff base formation [103]. In vitro studies have also demonstrated the ability of EGCG to convert mature Aβ fibrils into "off-pathway" aggregation by directly binding to the β-sheet-rich fibril and mediating conformational change [104]. In addition, Rezai-Zadeh et al. have reported that EGCG treatment decreases the Aβ plaque burden in the brain and improves working memory, using an AD mouse model [105,106].

However, many reports have suggested links between intake of high dose of EGCG and damage in several organs, especially the liver, in humans [56,107,108]. In 2018, the European Food Safety Authority concluded that intake of doses equal or above 800 mg EGCG/day, taken as a food supplement, can induce a significant increase of serum transaminases, which is indicative of liver injury [107]. A National Toxicology Program study reported that oral administration of green tea extracts containing EGCG (48.4% by weight) induced lesions in the gastrointestinal tract and liver in rats and mice [55]. A few cell model studies have shown that EGCG induces cellular DNA damage [53,109]. These potential harmful effects of EGCG have been attributed to its pro-oxidant activity [110–112]. Our previous report has indicated that EGCG significantly increases the content of 8-oxodG of DNA in cultured cells, but not in its H_2O_2-resistant clone cell [54], which is consistent with studies demonstrating intracellular ROS generation in cultured cells treated with EGCG [113,114]. Furthermore, EGCG caused oxidative damage to isolated DNA in the presence of Fe(III) and Cu(II) [54]. This was likely due to the generation of different ROS: hydroxy radical from the reaction of Fe(II) with H_2O_2 and Cu(I)-hydroperoxide from the reaction of Cu(I) with H_2O_2 [54]. To investigate the association between the chemical structure of green tea polyphenols and metal-mediated ROS generation, we compared EGCG-induced oxidative DNA damage in the presence of Fe(III) and Cu(II) with epicatechin gallate [115], epigallocatechin [116] and catechin [116], which are the other main green tea polyphenols that exert anti-Aβ aggregation activities. The results showed that EGCG, epicatechin gallate and epigallocatechin induced oxidative DNA damage in the presence of Fe(III) and Cu(II), whereas catechin did so in the presence of Cu(II) alone [54], suggesting that the pyrogallol moiety (phenolic three hydroxyl group) may be critical for Fe(III)-mediated ROS generation in green tea catechins (Figure 5).

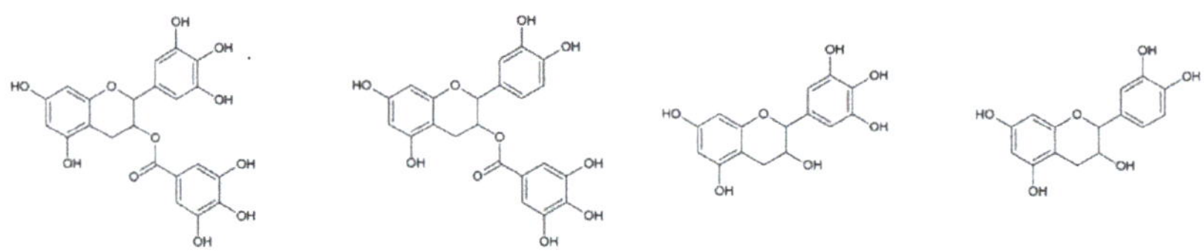

	EGCG	Epicatechin gallate	Epigallocatechin	Catechin
Fe(III)-mediated oxidative DNA damage	(++)	(+)	(++)	(-)
Cu(II)-mediated oxidative DNA damage	(+)	(+)	(+++)	(++)

Figure 5. Fe(III)- and Cu(II)-mediated DNA damage caused by green tea polyphenols. Three tea polyphenols with pyrogallol moieties (EGCG, epicatechin gallate, and epigallocatechin) can induce Fe(III)- and Cu(II)-mediated oxidative DNA damage although, catechins, which harbor no pyrogallol moieties, only cause Cu(II)-mediated oxidative DNA damage. (+++), (++), (+), and (-) represent the extent of DNA damage. EGCG: epigallocatechin gallate

2.4.2. Propyl Gallate and Gallic Acid

Propyl gallate and gallic acid have been reported to inhibit Aβ aggregation [117]. The anti-Aβ aggregation activities of gallic acid have been well-studied; the mechanism by which propyl gallate inhibits Aβ aggregation remains unknown. Gallic acid is an abundantly found polyphenol in the plant kingdom and is present in tea, wine, and fruits, such as grape and berries [118]. Gallic acid has been reported to have a beneficial effect on health and is pharmacologically effective in many diseases [119]. In relation to AD, several in vitro studies have demonstrated that gallic acid can reduce Aβ aggregation [13,117,120,121]. Molecular docking studies have shown that gallic acid interacts with Aβ aggregates and inhibits Aβ fibril formation by disrupting the Lys28-Ala42 salt bridge of Aβ [13]. Alternatively, gallic acid may convert toxic Aβ aggregates into "off-pathway" aggregation [122], similar to previously reported properties of EGCG [12,102]. Recently, Yu et al. have reported that gallic acid treatment alleviates cognitive decline in an AD mouse model at both early and late stages [13].

In contrast, potential harmful effects of propyl gallate and gallic acid, associated with their pro-oxidant properties, have also been reported [59–63]. Propyl gallate, but not gallic acid, is carcinogenic in mice and rats [123]. While propyl gallate led to 8-oxodG formation in cultured cells, it did not induce damage in isolated DNA [60]. Propyl gallate has been known to convert to gallic acid by an esterase (Figure 6) [124]. Therefore, to clarify its mechanism of carcinogenicity, we studied isolated DNA damage caused by gallic acid. Gallic acid and esterase-treated propyl gallate could induce Fe(III)- and Cu(II)-dependent oxidative DNA damage in isolated DNA through metal-mediated autoxidation [60]. These results suggest that gallic acid converted from propyl gallate plays an important role in propyl gallate-mediated carcinogenicity. To understand why gallic acid, but not propyl gallate, induces oxidative DNA damage, highest occupied molecular orbital (HOMO) energy estimation [125] was performed. The HOMO energy of the anionic form of gallic acid is smaller than that of propyl gallate, suggesting that gallic acid can readily undergo autoxidation compared to propyl gallate (Figure 6) [60]. Furthermore, gallic acid has been reported to display toxic effects other than carcinogenesis [59,62,63]. Administration of gallic acid induces liver injury [62,63] in mice and rats, and renal damage [63] in rats. ROS-associated cytotoxicity of gallic acid against noncancerous cell has been demonstrated using rat primary cultured hepatocytes [62] and vascular smooth muscle cells [59].

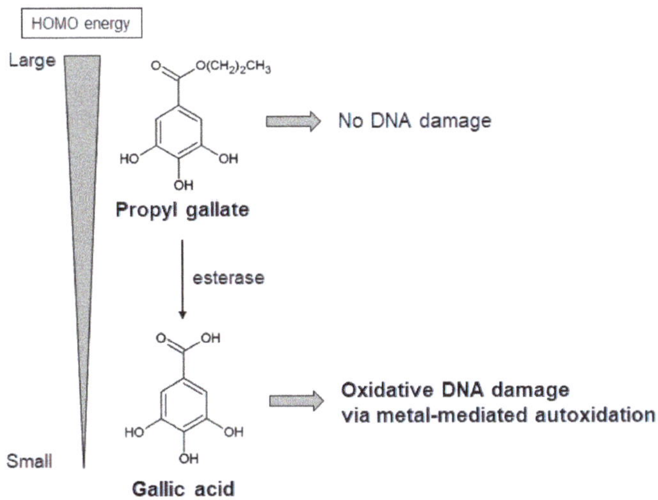

Figure 6. Estimation of HOMO energy to assess the pro-oxidant reactivity of gallic acid and propyl gallate. HOMO energies of gallic acid and propyl gallate were estimated from ab initio molecular orbital calculations at Hartree–Fock 6-31G* level. Calculations were performed using Spartan 02 for Windows (Wavefunction Inc., CA) [125]. HOMO energy: highest occupied molecular orbital energy

3. The Role of Phenolic Hydroxyl Groups in Anti-Aβ Aggregation and Pro-Oxidant Activities of Polyphenols

The phenolic hydroxyl groups of polyphenols are considered to be essential for its anti-Aβ aggregation activity. Quinones generated from phenolic hydroxyl groups can react with the lysine side chains of proteins [126]. Lys28 of Aβ has been reported to be critical for Aβ42 aggregation [127]. Therefore, quinones, especially catechol-type quinones, may contribute to the inhibition of Aβ aggregation. This is supported by the finding that the interactions between quinones from several polyphenols and lysine of Aβ play an important role in the inhibition of Aβ aggregation [103,128]. In contrast, our studies have shown ROS generation by several polyphenols through their autoxidation and quinone formation in the presence of metal ions such as Cu(II) [36,42,45,46,54,60,129]. In addition, some metabolites of target polyphenols also display pro-oxidant activities via quinone formation, even though target polyphenols themselves are not pro-oxidant [42,60].

Some studies have reported binding of the phenolic hydroxyl groups with histidine in anti-amyloid aggregation activity [9,130]. Histidine residues of Aβ impact Aβ aggregation by affecting the oligomeric equilibria [131] and interacting with metal ions [132]. Morin interacts with His13, His14, and Gln15 of Aβ42, corresponding to the intermolecular regions of β-sheets, and prevents Aβ assembly likely via its aromatic rings [9]. In the case of islet amyloid polypeptide, curcumin was shown to prevent inter-peptide interaction between Phe15 and His18, which is important for the aggregation of amyloids [130]. However, we have suggested that phenolic hydroxyl groups of morin and a metabolite of curcumin react with Cu(II), which leads to ROS generation and oxidative DNA damage [36,42].

Interestingly, copper is also thought to be associated with the enhancement of Aβ aggregation. The level of copper is elevated in the blood of AD patients [133] and Aβ plaques in an AD mouse model [134]. Cu(II) interacts with Aβ and enables the formation of β-sheets via its binding to His13 and His14, thereby forming a brace between Aβ strands [135]. Several polyphenols enable the chelating of various metal ions [136,137]. A recent report has shown that EGCG inhibits Cu(II)-associated amyloid aggregation of α-synuclein [138]. These findings suggest that polyphenols may inhibit Aβ aggregation via a Cu(II) chelating mechanism. However, as mentioned above, the interaction of polyphenols with Cu(II) leads to concomitant oxidative DNA damage [36,42,45,46,54,60,129].

These findings suggest that polyphenols can block Aβ aggregation and cause oxidative damage under certain circumstances, such as when they are in proximity to DNA.

4. Conclusions

Naturally occurring polyphenols are generally regarded as safe, based on their long history of use in the diet. However, when used at pharmacological concentrations, they have potential risks [18–20]. In this review, the pro-oxidant properties and the associated toxic effects of several naturally occurring polyphenols with anti-Aβ aggregation activity have been summarized. The pro-oxidant and anti-Aβ aggregation effects can be attributed to the structural features of polyphenols, suggesting a potential risk of oxidative damage. Therefore, we would like to emphasize the importance of assessing pro-oxidant properties of polyphenols from the point of view of safety.

Author Contributions: H.K. and S.O. wrote the first draft of the manuscript; M.M. and S.K. read the draft and made substantial and critical revisions. All authors have read and agreed to the published version of the manuscript.

Conflicts of Interest: The authors declare no conflict of interest.

Abbreviations

AD	Alzheimer's disease
Aβ	amyloid β
APP	amyloid precursor protein
ROS	reactive oxygen species
EGCG	epigallocatechin gallate
NMR	nuclear magnetic resonance
8-oxodG	8-oxo-7,8-dihydro-2′-deoxyguanosine
PAINS	pan-assay interference compound
CYP	cytochrome P450
HOMO energy	highest occupied molecular orbital energy

References

1. Winblad, B.; Amouyel, P.; Andrieu, S.; Ballard, C.; Brayne, C.; Brodaty, H.; Cedazo-Minguez, A.; Dubois, B.; Edvardsson, D.; Feldman, H.; et al. Defeating Alzheimer's disease and other dementias: A priority for European science and society. *Lancet. Neurol.* **2016**, *15*, 455–532. [CrossRef]
2. Panza, F.; Lozupone, M.; Logroscino, G.; Imbimbo, B.P. A critical appraisal of amyloid-beta-targeting therapies for Alzheimer disease. *Nat. Rev. Neurol.* **2019**, *15*, 73–88. [CrossRef] [PubMed]

3. LaFerla, F.M.; Green, K.N.; Oddo, S. Intracellular amyloid-β in Alzheimer's disease. *Nat. Rev. Neurosci.* **2007**, *8*, 499–509. [CrossRef] [PubMed]
4. Ayala, S.; Genevaux, P.; Hureau, C.; Faller, P. (Bio)chemical Strategies To Modulate Amyloid-beta Self-Assembly. *Acs Chem. Neurosci.* **2019**, *10*, 3366–3374. [CrossRef]
5. Arosio, P.; Knowles, T.P.; Linse, S. On the lag phase in amyloid fibril formation. *Phys. Chem. Chem. Phys. Pccp* **2015**, *17*, 7606–7618. [CrossRef]
6. Kumar, S.; Walter, J. Phosphorylation of amyloid beta (Aβ) peptides–A trigger for formation of toxic aggregates in Alzheimer's disease. *Aging (Albany Ny)* **2011**, *3*, 803. [CrossRef]
7. Long, J.M.; Holtzman, D.M. Alzheimer Disease: An Update on Pathobiology and Treatment Strategies. *Cell* **2019**, *179*, 312–339. [CrossRef]
8. Ono, K.; Li, L.; Takamura, Y.; Yoshiike, Y.; Zhu, L.; Han, F.; Mao, X.; Ikeda, T.; Takasaki, J.; Nishijo, H.; et al. Phenolic compounds prevent amyloid beta-protein oligomerization and synaptic dysfunction by site-specific binding. *J. Biol. Chem.* **2012**, *287*, 14631–14643. [CrossRef]
9. Hanaki, M.; Murakami, K.; Akagi, K.; Irie, K. Structural insights into mechanisms for inhibiting amyloid beta42 aggregation by non-catechol-type flavonoids. *Bioorganic Med. Chem.* **2016**, *24*, 304–313. [CrossRef]
10. Masuda, Y.; Fukuchi, M.; Yatagawa, T.; Tada, M.; Takeda, K.; Irie, K.; Akagi, K.-i.; Monobe, Y.; Imazawa, T.; Takegoshi, K. Solid-state NMR analysis of interaction sites of curcumin and 42-residue amyloid β-protein fibrils. *Bioorganic Med. Chem.* **2011**, *19*, 5967–5974. [CrossRef]
11. Hirohata, M.; Hasegawa, K.; Tsutsumi-Yasuhara, S.; Ohhashi, Y.; Ookoshi, T.; Ono, K.; Yamada, M.; Naiki, H. The anti-amyloidogenic effect is exerted against Alzheimer's beta-amyloid fibrils in vitro by preferential and reversible binding of flavonoids to the amyloid fibril structure. *Biochemistry* **2007**, *46*, 1888–1899. [CrossRef] [PubMed]
12. Ehrnhoefer, D.E.; Bieschke, J.; Boeddrich, A.; Herbst, M.; Masino, L.; Lurz, R.; Engemann, S.; Pastore, A.; Wanker, E.E. EGCG redirects amyloidogenic polypeptides into unstructured, off-pathway oligomers. *Nat. Struct. Mol. Biol.* **2008**, *15*, 558–566. [CrossRef] [PubMed]
13. Yu, M.; Chen, X.; Liu, J.; Ma, Q.; Zhuo, Z.; Chen, H.; Zhou, L.; Yang, S.; Zheng, L.; Ning, C.; et al. Gallic acid disruption of Abeta1-42 aggregation rescues cognitive decline of APP/PS1 double transgenic mouse. *Neurobiol. Dis.* **2019**, *124*, 67–80. [CrossRef] [PubMed]
14. Dhouafli, Z.; Cuanalo-Contreras, K.; Hayouni, E.A.; Mays, C.E.; Soto, C.; Moreno-Gonzalez, I. Inhibition of protein misfolding and aggregation by natural phenolic compounds. *Cell. Mol. Life Sci. Cmls* **2018**, *75*, 3521–3538. [CrossRef] [PubMed]
15. Perez-Jimenez, J.; Neveu, V.; Vos, F.; Scalbert, A. Systematic analysis of the content of 502 polyphenols in 452 foods and beverages: An application of the phenol-explorer database. *J. Agric. Food Chem.* **2010**, *58*, 4959–4969. [CrossRef] [PubMed]
16. Velander, P.; Wu, L.; Henderson, F.; Zhang, S.; Bevan, D.R.; Xu, B. Natural product-based amyloid inhibitors. *Biochem. Pharmacol.* **2017**, *139*, 40–55. [CrossRef]
17. Yamada, M.; Ono, K.; Hamaguchi, T.; Noguchi-Shinohara, M. Natural Phenolic Compounds as Therapeutic and Preventive Agents for Cerebral Amyloidosis. *Adv. Exp. Med. Biol.* **2015**, *863*, 79–94.
18. Sen, S.; Chakraborty, R. Antioxidant Supplements: Friend or Foe? *In Free Radicals in Human Health and Disease* **2015**, 293–322. [CrossRef]
19. Martin, K.R.; Appel, C.L. Polyphenols as dietary supplements: A double-edged sword. *Nutr. Diet. Suppl.* **2009**, *2*, 1–12. [CrossRef]
20. Lambert, J.D.; Sang, S.; Yang, C.S. Possible controversy over dietary polyphenols: Benefits vs risks. *Chem. Res. Toxicol.* **2007**, *20*, 583–585. [CrossRef]
21. Cieślik, E.; Gręda, A.; Adamus, W. Contents of polyphenols in fruit and vegetables. *Food Chem.* **2006**, *94*, 135–142. [CrossRef]
22. Murata, M.; Kawanishi, S. Mechanisms of oxidative DNA damage induced by carcinogenic arylamines. *Front. Biosci. (Landmark Ed.)* **2011**, *16*, 1132–1143. [CrossRef] [PubMed]
23. Almeida, A.F.; Dos Santos, C.N.; Ventura, M.R. Polyphenols, their Metabolites and Derivatives as Drug Leads. *Curr. Pharm. Des.* **2018**, *24*, 2188–2207. [CrossRef] [PubMed]
24. Carregosa, D.; Carecho, R.; Figueira, I.; Santos, C.N. Low-Molecular Weight Metabolites from Polyphenols as Effectors for Attenuating Neuroinflammation. *J. Agric. Food Chem.* **2020**, *68*, 1790–1807. [CrossRef]

25. Sahu, S.C.; Flynn, T.J.; Bradlaw, J.A.; Roth, W.L.; Barton, C.N.; Yates, J.G. Pro-oxidant effects of the flavonoid myricetin on rat hepatocytes in culture. *Toxicol. Methods* **2001**, *11*, 277–283. [CrossRef]
26. Park, H.; Park, S.; Bazer, F.W.; Lim, W.; Song, G. Myricetin treatment induces apoptosis in canine osteosarcoma cells by inducing DNA fragmentation, disrupting redox homeostasis, and mediating loss of mitochondrial membrane potential. *J. Cell. Physiol.* **2018**, *233*, 7457–7466. [CrossRef]
27. Knickle, A.; Fernando, W.; Greenshields, A.L.; Rupasinghe, H.P.V.; Hoskin, D.W. Myricetin-induced apoptosis of triple-negative breast cancer cells is mediated by the iron-dependent generation of reactive oxygen species from hydrogen peroxide. *Food Chem. Toxicol. Int. J. Publ. Br. Ind. Biol. Res. Assoc.* **2018**, *118*, 154–167. [CrossRef]
28. Sahu, S.C.; Gray, G.C. Interactions of flavonoids, trace metals, and oxygen: Nuclear DNA damage and lipid peroxidation induced by myricetin. *Cancer Lett.* **1993**, *70*, 73–79. [CrossRef]
29. Arif, H.; Sohail, A.; Farhan, M.; Rehman, A.A.; Ahmad, A.; Hadi, S.M. Flavonoids-induced redox cycling of copper ions leads to generation of reactive oxygen species: A potential role in cancer chemoprevention. *Int. J. Biol. Macromol.* **2018**, *106*, 569–578. [CrossRef]
30. Jomova, K.; Hudecova, L.; Lauro, P.; Simunkova, M.; Alwasel, S.H.; Alhazza, I.M.; Valko, M. A Switch between Antioxidant and Prooxidant Properties of the Phenolic Compounds Myricetin, Morin, 3′,4′-Dihydroxyflavone, Taxifolin and 4-Hydroxy-Coumarin in the Presence of Copper(II) Ions: A Spectroscopic, Absorption Titration and DNA Damage Study. *Molecules (Basel, Switz.)* **2019**, *24*, 4335. [CrossRef]
31. Yoshino, M.; Haneda, M.; Naruse, M.; Murakami, K. Prooxidant activity of flavonoids: Copper-dependent strand breaks and the formation of 8-hydroxy-2′-deoxyguanosine in DNA. *Mol. Genet. Metab.* **1999**, *68*, 468–472. [CrossRef] [PubMed]
32. Hobbs, C.A.; Swartz, C.; Maronpot, R.; Davis, J.; Recio, L.; Koyanagi, M.; Hayashi, S.M. Genotoxicity evaluation of the flavonoid, myricitrin, and its aglycone, myricetin. *Food Chem. Toxicol. Int. J. Publ. Br. Ind. Biol. Res. Assoc.* **2015**, *83*, 283–292. [CrossRef] [PubMed]
33. Silva, I.D.; Gaspar, J.; Rodrigues, A.; da Costa, G.G.; Laires, A.; Rueff, J. Mechanisms of myricetin mutagenicity in V79 cells: Involvement of radicalar species. *Teratog. Carcinog. Mutagenesis* **1996**, *16*, 253–268. [CrossRef]
34. Yen, G.C.; Duh, P.D.; Tsai, H.L.; Huang, S.L. Pro-oxidative properties of flavonoids in human lymphocytes. *Biosci. Biotechnol. Biochem.* **2003**, *67*, 1215–1222. [CrossRef]
35. Sahu, S.C.; Gray, G.C. Lipid peroxidation and DNA damage induced by morin and naringenin in isolated rat liver nuclei. *Food Chem. Toxicol. Int. J. Publ. Br. Ind. Biol. Res. Assoc.* **1997**, *35*, 443–447. [CrossRef]
36. Mori, Y.; Kato, S.; Fujisawa, Y.; Ohnishi, S.; Hiraku, Y.; Kawanishi, S.; Murata, M.; Oikawa, S. Mechanisms of DNA damage induced by morin, an inhibitor of amyloid beta-peptide aggregation. *Free Radic. Res.* **2019**, *53*, 115–123. [CrossRef]
37. Roy, A.S.; Samanta, S.K.; Ghosh, P.; Tripathy, D.R.; Ghosh, S.K.; Dasgupta, S. Cell cytotoxicity and serum albumin binding capacity of the morin-Cu(ii) complex and its effect on deoxyribonucleic acid. *Mol. Biosyst.* **2016**, *12*, 2818–2833. [CrossRef]
38. Jurado, J.; Alejandre-Duran, E.; Alonso-Moraga, A.; Pueyo, C. Study on the mutagenic activity of 13 bioflavonoids with the Salmonella Ara test. *Mutagenesis* **1991**, *6*, 289–295. [CrossRef]
39. Yoshino, M.; Haneda, M.; Naruse, M.; Htay, H.H.; Tsubouchi, R.; Qiao, S.L.; Li, W.H.; Murakami, K.; Yokochi, T. Prooxidant activity of curcumin: Copper-dependent formation of 8-hydroxy-2′-deoxyguanosine in DNA and induction of apoptotic cell death. *Toxicol. Vitr. Int. J. Publ. Assoc. Bibra* **2004**, *18*, 783–789. [CrossRef]
40. Woo, J.H.; Kim, Y.H.; Choi, Y.J.; Kim, D.G.; Lee, K.S.; Bae, J.H.; Min, D.S.; Chang, J.S.; Jeong, Y.J.; Lee, Y.H.; et al. Molecular mechanisms of curcumin-induced cytotoxicity: Induction of apoptosis through generation of reactive oxygen species, down-regulation of Bcl-XL and IAP, the release of cytochrome c and inhibition of Akt. *Carcinogenesis* **2003**, *24*, 1199–1208. [CrossRef]
41. Urbina-Cano, P.; Bobadilla-Morales, L.; Ramírez-Herrera, M.A.; Corona-Rivera, J.R.; Mendoza-Magaña, M.L.; Troyo-Sanromán, R.; Corona-Rivera, A. DNA damage in mouse lymphocytes exposed to curcumin and copper. *J. Appl. Genet.* **2006**, *47*, 377–382. [CrossRef] [PubMed]
42. Sakano, K.; Kawanishi, S. Metal-mediated DNA damage induced by curcumin in the presence of human cytochrome P450 isozymes. *Arch. Biochem. Biophys.* **2002**, *405*, 223–230. [CrossRef]

43. National Toxicology Program. NTP Toxicology and Carcinogenesis Studies of Turmeric Oleoresin (CAS No. 8024-37-1) (Major Component 79%–85% Curcumin, CAS No. 458-37-7) in F344/N Rats and B6C3F1 Mice (Feed Studies). *Natl. Toxicol. Program Tech. Rep. Ser.* **1993**, *427*, 1–275.
44. Lapidot, T.; Walker, M.D.; Kanner, J. Antioxidant and prooxidant effects of phenolics on pancreatic β-cells in vitro. *J. Agric. Food Chem.* **2002**, *50*, 7220–7225. [CrossRef] [PubMed]
45. Yamashita, N.; Kawanishi, S. Distinct mechanisms of DNA damage in apoptosis induced by quercetin and luteolin. *Free Radic. Res.* **2000**, *33*, 623–633. [CrossRef]
46. Yamashita, N.; Tanemura, H.; Kawanishi, S. Mechanism of oxidative DNA damage induced by quercetin in the presence of Cu(II). *Mutat. Res.* **1999**, *425*, 107–115. [CrossRef]
47. Resende, F.A.; Vilegas, W.; Dos Santos, L.C.; Varanda, E.A. Mutagenicity of flavonoids assayed by bacterial reverse mutation (Ames) test. *Molecules (Basel, Switz.)* **2012**, *17*, 5255–5268. [CrossRef]
48. National Toxicology Program. Toxicology and Carcinogenesis Studies of Quercetin (CAS No. 117-39-5) in F344 Rats (Feed Studies). *Natl. Toxicol. Program Tech. Rep. Ser.* **1992**, *409*, 1–171.
49. Gaspar, J.; Rodrigues, A.; Laires, A.; Silva, F.; Costa, S.; Monteiro, M.J.; Monteiro, C.; Rueff, J. On the mechanisms of genotoxicity and metabolism of quercetin. *Mutagenesis* **1994**, *9*, 445–449. [CrossRef]
50. Pamukcu, A.M.; Yalciner, S.; Hatcher, J.F.; Bryan, G.T. Quercetin, a rat intestinal and bladder carcinogen present in bracken fern (Pteridium aquilinum). *Cancer Res.* **1980**, *40*, 3468–3472.
51. Chan, M.M.; Soprano, K.J.; Weinstein, K.; Fong, D. Epigallocatechin-3-gallate delivers hydrogen peroxide to induce death of ovarian cancer cells and enhances their cisplatin susceptibility. *J. Cell. Physiol.* **2006**, *207*, 389–396. [CrossRef] [PubMed]
52. Nakagawa, H.; Hasumi, K.; Woo, J.-T.; Nagai, K.; Wachi, M. Generation of hydrogen peroxide primarily contributes to the induction of Fe (II)-dependent apoptosis in Jurkat cells by (−)-epigallocatechin gallate. *Carcinogenesis* **2004**, *25*, 1567–1574. [CrossRef] [PubMed]
53. Johnson, M.K.; Loo, G. Effects of epigallocatechin gallate and quercetin on oxidative damage to cellular DNA. *Mutat. Res.* **2000**, *459*, 211–218. [CrossRef]
54. Furukawa, A.; Oikawa, S.; Murata, M.; Hiraku, Y.; Kawanishi, S. (-)-Epigallocatechin gallate causes oxidative damage to isolated and cellular DNA. *Biochem. Pharmacol.* **2003**, *66*, 1769–1778. [CrossRef]
55. National Toxicology Program. Toxicology Studies of green tea extract in F344/NTac Rats and B6C3F1/N mice and toxicology and carcinogenesis studies of green tea extract in Wistar Han [Crl:WI(Han)] rats and B6C3F1/N Mice (gavage studies). *Natl. Toxicol. Program Tech. Rep. Ser.* **2016**, *585*, 1–192.
56. Hu, J.; Webster, D.; Cao, J.; Shao, A. The safety of green tea and green tea extract consumption in adults—Results of a systematic review. *Regul. Toxicol. Pharmacol. Rtp* **2018**, *95*, 412–433. [CrossRef]
57. Lee, K.W.; Hur, H.J.; Lee, H.J.; Lee, C.Y. Antiproliferative effects of dietary phenolic substances and hydrogen peroxide. *J. Agric. Food Chem.* **2005**, *53*, 1990–1995. [CrossRef]
58. Isuzugawa, K.; Inoue, M.; Ogihara, Y. Catalase contents in cells determine sensitivity to the apoptosis inducer gallic acid. *Biol. Pharm. Bull.* **2001**, *24*, 1022–1026. [CrossRef]
59. Qiu, X.; Takemura, G.; Koshiji, M.; Hayakawa, Y.; Kanoh, M.; Maruyama, R.; Ohno, Y.; Minatoguchi, S.; Akao, S.; Fukuda, K.; et al. Gallic acid induces vascular smooth muscle cell death via hydroxyl radical production. *Heart Vessel.* **2000**, *15*, 90–99. [CrossRef]
60. Kobayashi, H.; Oikawa, S.; Hirakawa, K.; Kawanishi, S. Metal-mediated oxidative damage to cellular and isolated DNA by gallic acid, a metabolite of antioxidant propyl gallate. *Mutat. Res.* **2004**, *558*, 111–120. [CrossRef]
61. Yoshino, M.; Haneda, M.; Naruse, M.; Htay, H.; Iwata, S.; Tsubouchi, R.; Murakami, K. Prooxidant action of gallic acid compounds: Copper-dependent strand breaks and the formation of 8-hydroxy-2′-deoxyguanosine in DNA. *Toxicol. Vitr.* **2002**, *16*, 705–709. [CrossRef]
62. Galati, G.; Lin, A.; Sultan, A.M.; O'Brien, P.J. Cellular and in vivo hepatotoxicity caused by green tea phenolic acids and catechins. *Free Radic. Biol. Med.* **2006**, *40*, 570–580. [CrossRef]
63. Abarikwu, S.O.; Durojaiye, M.; Alabi, A.; Asonye, B.; Akiri, O. Curcumin protects against gallic acid-induced oxidative stress, suppression of glutathione antioxidant defenses, hepatic and renal damage in rats. *Ren. Fail.* **2016**, *38*, 321–329. [CrossRef] [PubMed]
64. Jiang, M.; Zhu, M.; Wang, L.; Yu, S. Anti-tumor effects and associated molecular mechanisms of myricetin. *Biomed. Pharmacother. Biomed. Pharmacother.* **2019**, *120*, 109506. [CrossRef] [PubMed]

65. Semwal, D.K.; Semwal, R.B.; Combrinck, S.; Viljoen, A. Myricetin: A Dietary Molecule with Diverse Biological Activities. *Nutrients* **2016**, *8*, 90. [CrossRef] [PubMed]
66. Ono, K.; Yoshiike, Y.; Takashima, A.; Hasegawa, K.; Naiki, H.; Yamada, M. Potent anti-amyloidogenic and fibril-destabilizing effects of polyphenols in vitro: Implications for the prevention and therapeutics of Alzheimer's disease. *J. Neurochem.* **2003**, *87*, 172–181. [CrossRef] [PubMed]
67. Hamaguchi, T.; Ono, K.; Murase, A.; Yamada, M. Phenolic compounds prevent Alzheimer's pathology through different effects on the amyloid-beta aggregation pathway. *Am. J. Pathol.* **2009**, *175*, 2557–2565. [CrossRef]
68. Kataria, R.; Sobarzo-Sanchez, E.; Khatkar, A. Role of Morin in Neurodegenerative Diseases: A Review. *Curr. Top. Med. Chem.* **2018**, *18*, 901–907. [CrossRef]
69. Caselli, A.; Cirri, P.; Santi, A.; Paoli, P. Morin: A Promising Natural Drug. *Curr. Med. Chem.* **2016**, *23*, 774–791. [CrossRef]
70. Sinha, K.; Ghosh, J.; Sil, P.C. Morin and Its Role in Chronic Diseases. *Adv. Exp. Med. Biol.* **2016**, *928*, 453–471.
71. Kim, H.; Park, B.S.; Lee, K.G.; Choi, C.Y.; Jang, S.S.; Kim, Y.H.; Lee, S.E. Effects of naturally occurring compounds on fibril formation and oxidative stress of beta-amyloid. *J. Agric. Food Chem.* **2005**, *53*, 8537–8541. [CrossRef] [PubMed]
72. Du, Y.; Qu, J.; Zhang, W.; Bai, M.; Zhou, Q.; Zhang, Z.; Li, Z.; Miao, J. Morin reverses neuropathological and cognitive impairments in APPswe/PS1dE9 mice by targeting multiple pathogenic mechanisms. *Neuropharmacology* **2016**, *108*, 1–13. [CrossRef] [PubMed]
73. Sharma, D.; Singh, M.; Kumar, P.; Vikram, V.; Mishra, N. Development and characterization of morin hydrate loaded microemulsion for the management of Alzheimer's disease. *Artif. Cells Nanomed. Biotechnol.* **2017**, *45*, 1620–1630. [CrossRef] [PubMed]
74. Murakami, K.; Irie, K. Three Structural Features of Functional Food Components and Herbal Medicine with Amyloid beta42 Anti-Aggregation Properties. *Molecules (Basel,Switz.)* **2019**, *24*.
75. Gupta, S.C.; Sung, B.; Kim, J.H.; Prasad, S.; Li, S.; Aggarwal, B.B. Multitargeting by turmeric, the golden spice: From kitchen to clinic. *Mol. Nutr. Food Res.* **2013**, *57*, 1510–1528. [CrossRef] [PubMed]
76. Kocaadam, B.; Sanlier, N. Curcumin, an active component of turmeric (Curcuma longa), and its effects on health. *Crit. Rev. Food Sci. Nutr.* **2017**, *57*, 2889–2895. [CrossRef]
77. Jin, T.R. Curcumin and dietary polyphenol research: Beyond drug discovery. *Acta Pharmacol. Sin.* **2018**, *39*, 779–786. [CrossRef]
78. Pulido-Moran, M.; Moreno-Fernandez, J.; Ramirez-Tortosa, C.; Ramirez-Tortosa, M. Curcumin and Health. *Molecules (Basel, Switz.)* **2016**, *21*, 264. [CrossRef]
79. Schubert, D.; Currais, A.; Goldberg, J.; Finley, K.; Petrascheck, M.; Maher, P. Geroneuroprotectors: Effective Geroprotectors for the Brain. *Trends Pharm. Sci* **2018**, *39*, 1004–1007. [CrossRef]
80. Ono, K.; Hasegawa, K.; Naiki, H.; Yamada, M. Curcumin has potent anti-amyloidogenic effects for Alzheimer's beta-amyloid fibrils in vitro. *J. Neurosci. Res.* **2004**, *75*, 742–750. [CrossRef]
81. Yang, F.; Lim, G.P.; Begum, A.N.; Ubeda, O.J.; Simmons, M.R.; Ambegaokar, S.S.; Chen, P.P.; Kayed, R.; Glabe, C.G.; Frautschy, S.A.; et al. Curcumin inhibits formation of amyloid beta oligomers and fibrils, binds plaques, and reduces amyloid in vivo. *J. Biol. Chem.* **2005**, *280*, 5892–5901. [CrossRef] [PubMed]
82. Garcia-Alloza, M.; Borrelli, L.A.; Rozkalne, A.; Hyman, B.T.; Bacskai, B.J. Curcumin labels amyloid pathology in vivo, disrupts existing plaques, and partially restores distorted neurites in an Alzheimer mouse model. *J. Neurochem.* **2007**, *102*, 1095–1104. [CrossRef] [PubMed]
83. Baell, J.; Walters, M.A. Chemistry: Chemical con artists foil drug discovery. *Nature* **2014**, *513*, 481–483. [CrossRef] [PubMed]
84. Nelson, K.M.; Dahlin, J.L.; Bisson, J.; Graham, J.; Pauli, G.F.; Walters, M.A. The Essential Medicinal Chemistry of Curcumin. *J. Med. Chem.* **2017**, *60*, 1620–1637. [CrossRef]
85. Malik, P.; Mukherjee, T.K. Structure-Function Elucidation of Antioxidative and Prooxidative Activities of the Polyphenolic Compound Curcumin. *Chin. J. Biol.* **2014**, *2014*, 1–8. [CrossRef]
86. Alzahrani, A.M.; Rajendran, P. The Multifarious Link between Cytochrome P450s and Cancer. *Oxidative Med. Cell. Longev.* **2020**, *2020*, 3028387. [CrossRef]
87. Wang, W.; Sun, C.; Mao, L.; Ma, P.; Liu, F.; Yang, J.; Gao, Y. The biological activities, chemical stability, metabolism and delivery systems of quercetin: A review. *Trends Food Sci. Technol.* **2016**, *56*, 21–38. [CrossRef]

88. Xu, D.; Hu, M.J.; Wang, Y.Q.; Cui, Y.L. Antioxidant Activities of Quercetin and Its Complexes for Medicinal Application. *Molecules (Basel, Switz.)* **2019**, *24*, 1123. [CrossRef]
89. Li, Y.; Yao, J.; Han, C.; Yang, J.; Chaudhry, M.T.; Wang, S.; Liu, H.; Yin, Y. Quercetin, Inflammation and Immunity. *Nutrients* **2016**, *8*, 167. [CrossRef]
90. Khan, H.; Ullah, H.; Aschner, M.; Cheang, W.S.; Akkol, E.K. Neuroprotective Effects of Quercetin in Alzheimer's Disease. *Biomolecules* **2019**, *10*, 59. [CrossRef]
91. Jimenez-Aliaga, K.; Bermejo-Bescos, P.; Benedi, J.; Martin-Aragon, S. Quercetin and rutin exhibit antiamyloidogenic and fibril-disaggregating effects in vitro and potent antioxidant activity in APPswe cells. *Life Sci.* **2011**, *89*, 939–945. [CrossRef] [PubMed]
92. Wang, D.-M.; Li, S.-Q.; Wu, W.-L.; Zhu, X.-Y.; Wang, Y.; Yuan, H.-Y. Effects of long-term treatment with quercetin on cognition and mitochondrial function in a mouse model of Alzheimer's disease. *Neurochem. Res.* **2014**, *39*, 1533–1543. [CrossRef] [PubMed]
93. Dajas, F.; Abin-Carriquiry, J.A.; Arredondo, F.; Blasina, F.; Echeverry, C.; Martinez, M.; Rivera, F.; Vaamonde, L. Quercetin in brain diseases: Potential and limits. *Neurochem. Int.* **2015**, *89*, 140–148. [CrossRef] [PubMed]
94. Parveen, N.; Shadab, G.G. The dual clastogenic and anti-clastogenic properties of quercetin is dose dependent. *Front. Biosci. (Sch. Ed.)* **2017**, *9*, 139–153.
95. Walle, T. Methylation of dietary flavones increases their metabolic stability and chemopreventive effects. *Int. J. Mol. Sci.* **2009**, *10*, 5002–5019. [CrossRef] [PubMed]
96. Chang, H.; Mi, M.; Ling, W.; Zhu, J.; Zhang, Q.; Wei, N.; Zhou, Y.; Tang, Y.; Yuan, J. Structurally related cytotoxic effects of flavonoids on human cancer cells in vitro. *Arch. Pharmacal Res.* **2008**, *31*, 1137–1144. [CrossRef]
97. Hara, Y.; Yang, C.S.; Isemura, M.; Tomita, I. *Health benefits of green tea: An evidence-based approach*; CABI: Wallingford, UK, 2017; ISBN 9781786392398.
98. Islam, M.A. Cardiovascular effects of green tea catechins: Progress and promise. *Recent Pat. Cardiovasc. Drug Discov.* **2012**, *7*, 88–99. [CrossRef]
99. Eng, Q.Y.; Thanikachalam, P.V.; Ramamurthy, S. Molecular understanding of Epigallocatechin gallate (EGCG) in cardiovascular and metabolic diseases. *J. Ethnopharmacol.* **2018**, *210*, 296–310. [CrossRef]
100. Singh, N.A.; Mandal, A.K.; Khan, Z.A. Potential neuroprotective properties of epigallocatechin-3-gallate (EGCG). *Nutr. J.* **2016**, *15*, 60. [CrossRef]
101. Chakrawarti, L.; Agrawal, R.; Dang, S.; Gupta, S.; Gabrani, R. Therapeutic effects of EGCG: A patent review. *Expert Opin. Ther. Pat.* **2016**, *26*, 907–916. [CrossRef]
102. Ngo, S.T.; Truong, D.T.; Tam, N.M.; Nguyen, M.T. EGCG inhibits the oligomerization of amyloid beta (16-22) hexamer: Theoretical studies. *J. Mol. Graph. Model.* **2017**, *76*, 1–10. [CrossRef] [PubMed]
103. Palhano, F.L.; Lee, J.; Grimster, N.P.; Kelly, J.W. Toward the molecular mechanism(s) by which EGCG treatment remodels mature amyloid fibrils. *J. Am. Chem. Soc.* **2013**, *135*, 7503–7510. [CrossRef] [PubMed]
104. Bieschke, J.; Russ, J.; Friedrich, R.P.; Ehrnhoefer, D.E.; Wobst, H.; Neugebauer, K.; Wanker, E.E. EGCG remodels mature alpha-synuclein and amyloid-beta fibrils and reduces cellular toxicity. *Proc. Natl. Acad. Sci. USA* **2010**, *107*, 7710–7715. [CrossRef] [PubMed]
105. Rezai-Zadeh, K.; Arendash, G.W.; Hou, H.; Fernandez, F.; Jensen, M.; Runfeldt, M.; Shytle, R.D.; Tan, J. Green tea epigallocatechin-3-gallate (EGCG) reduces beta-amyloid mediated cognitive impairment and modulates tau pathology in Alzheimer transgenic mice. *Brain Res.* **2008**, *1214*, 177–187. [CrossRef]
106. Rezai-Zadeh, K.; Shytle, D.; Sun, N.; Mori, T.; Hou, H.; Jeanniton, D.; Ehrhart, J.; Townsend, K.; Zeng, J.; Morgan, D.; et al. Green tea epigallocatechin-3-gallate (EGCG) modulates amyloid precursor protein cleavage and reduces cerebral amyloidosis in Alzheimer transgenic mice. *J. Neurosci. Off. J. Soc. Neurosci.* **2005**, *25*, 8807–8814. [CrossRef]
107. Younes, M.; Aggett, P.; Aguilar, F.; Crebelli, R.; Dusemund, B.; Filipič, M.; Frutos, M.J.; Galtier, P.; Gott, D.; Gundert-Remy, U.; et al. Scientific opinion on the safety of green tea catechins. *Efsa J.* **2018**, 16.
108. Dekant, W.; Fujii, K.; Shibata, E.; Morita, O.; Shimotoyodome, A. Safety assessment of green tea based beverages and dried green tea extracts as nutritional supplements. *Toxicol. Lett.* **2017**, *277*, 104–108. [CrossRef]
109. Bertram, B.; Bollow, U.; Rajaee-Behbahani, N.; Burkle, A.; Schmezer, P. Induction of poly(ADP-ribosyl)ation and DNA damage in human peripheral lymphocytes after treatment with (-)-epigallocatechin-gallate. *Mutat. Res.* **2003**, *534*, 77–84. [CrossRef]

110. Wang, D.; Wei, Y.; Wang, T.; Wan, X.; Yang, C.S.; Reiter, R.J.; Zhang, J. Melatonin attenuates (-)-epigallocatehin-3-gallate-triggered hepatotoxicity without compromising its downregulation of hepatic gluconeogenic and lipogenic genes in mice. *J. Pineal Res.* **2015**, *59*, 497–507. [CrossRef]
111. Wang, D.; Wang, Y.; Wan, X.; Yang, C.S.; Zhang, J. Green tea polyphenol (-)-epigallocatechin-3-gallate triggered hepatotoxicity in mice: Responses of major antioxidant enzymes and the Nrf2 rescue pathway. *Toxicol. Appl. Pharmacol.* **2015**, *283*, 65–74. [CrossRef]
112. Takumi-Kobayashi, A.; Ogura, R.; Morita, O.; Nishiyama, N.; Kasamatsu, T. Involvement of hydrogen peroxide in chromosomal aberrations induced by green tea catechins in vitro and implications for risk assessment. *Mutat. Res.* **2008**, *657*, 13–18. [CrossRef] [PubMed]
113. Kucera, O.; Mezera, V.; Moravcova, A.; Endlicher, R.; Lotkova, H.; Drahota, Z.; Cervinkova, Z. In vitro toxicity of epigallocatechin gallate in rat liver mitochondria and hepatocytes. *Oxidative Med. Cell. Longev.* **2015**, *2015*, 476180. [CrossRef] [PubMed]
114. Isbrucker, R.A.; Bausch, J.; Edwards, J.A.; Wolz, E. Safety studies on epigallocatechin gallate (EGCG) preparations. Part 1: Genotoxicity. *Food Chem. Toxicol. Int. J. Publ. Br. Ind. Biol. Res. Assoc.* **2006**, *44*, 626–635. [CrossRef] [PubMed]
115. Rho, T.; Choi, M.S.; Jung, M.; Kil, H.W.; Hong, Y.D.; Yoon, K.D. Identification of fermented tea (Camellia sinensis) polyphenols and their inhibitory activities against amyloid-beta aggregation. *Phytochemistry* **2019**, *160*, 11–18. [CrossRef]
116. Xie, H.; Wang, J.R.; Yau, L.F.; Liu, Y.; Liu, L.; Han, Q.B.; Zhao, Z.; Jiang, Z.H. Catechins and procyanidins of Ginkgo biloba show potent activities towards the inhibition of beta-amyloid peptide aggregation and destabilization of preformed fibrils. *Molecules (Basel, Switz.)* **2014**, *19*, 5119–5134. [CrossRef]
117. Chan, S.; Kantham, S.; Rao, V.M.; Palanivelu, M.K.; Pham, H.L.; Shaw, P.N.; McGeary, R.P.; Ross, B.P. Metal chelation, radical scavenging and inhibition of Abeta(4)(2) fibrillation by food constituents in relation to Alzheimer's disease. *Food Chem.* **2016**, *199*, 185–194. [CrossRef]
118. Verma, S.; Singh, A.; Mishra, A. Gallic acid: Molecular rival of cancer. *Environ. Toxicol. Pharmacol.* **2013**, *35*, 473–485. [CrossRef]
119. Badhani, B.; Sharma, N.; Kakkar, R. Gallic acid: A versatile antioxidant with promising therapeutic and industrial applications. *Rsc Adv.* **2015**, *5*, 27540–27557. [CrossRef]
120. Bastianetto, S.; Yao, Z.X.; Papadopoulos, V.; Quirion, R. Neuroprotective effects of green and black teas and their catechin gallate esters against beta-amyloid-induced toxicity. *Eur. J. Neurosci.* **2006**, *23*, 55–64. [CrossRef]
121. Liu, Y.; Pukala, T.L.; Musgrave, I.F.; Williams, D.M.; Dehle, F.C.; Carver, J.A. Gallic acid is the major component of grape seed extract that inhibits amyloid fibril formation. *Bioorganic Med. Chem. Lett.* **2013**, *23*, 6336–6340. [CrossRef]
122. LeVine, H.; Lampe, L.; Abdelmoti, L.; Augelli-Szafran, C.E. Dihydroxybenzoic Acid Isomers Differentially Dissociate Soluble Biotinyl-Aβ(1–42) Oligomers. *Biochemistry* **2012**, *51*, 307–315. [CrossRef] [PubMed]
123. National Toxicology Program. NTP Carcinogenesis Bioassay of Propyl Gallate (CAS No. 121-79-9) in F344/N Rats and B6C3F1 Mice (Feed Study). *Natl. Toxicol. Program Tech. Rep. Ser.* **1982**, *240*, 1–152.
124. Locatelli, C.; Filippin-Monteiro, F.B.; Creczynski-Pasa, T.B. Alkyl esters of gallic acid as anticancer agents: A review. *Eur. J. Med. Chem.* **2013**, *60*, 233–239. [CrossRef] [PubMed]
125. Hirakawa, K.; Yoshida, M.; Oikawa, S.; Kawanishi, S. Base oxidation at 5′ site of GG sequence in double-stranded DNA induced by UVA in the presence of xanthone analogues: Relationship between the DNA-damaging abilities of photosensitizers and their HOMO energies. *Photochem. Photobiol.* **2003**, *77*, 349–355. [CrossRef]
126. Ishii, T.; Mori, T.; Tanaka, T.; Mizuno, D.; Yamaji, R.; Kumazawa, S.; Nakayama, T.; Akagawa, M. Covalent modification of proteins by green tea polyphenol (–)-epigallocatechin-3-gallate through autoxidation. *Free Radic. Biol. Med.* **2008**, *45*, 1384–1394. [CrossRef]
127. Xiao, Y.; Ma, B.; McElheny, D.; Parthasarathy, S.; Long, F.; Hoshi, M.; Nussinov, R.; Ishii, Y. Aβ (1–42) fibril structure illuminates self-recognition and replication of amyloid in Alzheimer's disease. *Nat. Struct. Mol. Biol.* **2015**, *22*, 499–505. [CrossRef]
128. Sato, M.; Murakami, K.; Uno, M.; Nakagawa, Y.; Katayama, S.; Akagi, K.-i.; Masuda, Y.; Takegoshi, K.; Irie, K. Site-specific inhibitory mechanism for amyloid β42 aggregation by catechol-type flavonoids targeting the Lys residues. *J. Biol. Chem.* **2013**, *288*, 23212–23224. [CrossRef]

129. Oikawa, S.; Furukawa, A.; Asada, H.; Hirakawa, K.; Kawanishi, S. Catechins induce oxidative damage to cellular and isolated DNA through the generation of reactive oxygen species. *Free Radic. Res.* **2003**, *37*, 881–890. [CrossRef]
130. Nedumpully-Govindan, P.; Kakinen, A.; Pilkington, E.H.; Davis, T.P.; Chun Ke, P.; Ding, F. Stabilizing Off-pathway Oligomers by Polyphenol Nanoassemblies for IAPP Aggregation Inhibition. *Sci. Rep.* **2016**, *6*, 19463. [CrossRef]
131. Przygonska, K.; Pacewicz, M.; Sadowska, W.; Poznanski, J.; Bal, W.; Dadlez, M. His6, His13, and His14 residues in Abeta 1-40 peptide significantly and specifically affect oligomeric equilibria. *Sci. Rep.* **2019**, *9*, 9449. [CrossRef]
132. Kong, X.; Zhao, Z.; Lei, X.; Zhang, B.; Dai, D.; Jiang, L. Interaction of metal ions with the His13-His14 sequence relevant to Alzheimer's disease. *J. Phys. Chem. A* **2015**, *119*, 3528–3534. [CrossRef] [PubMed]
133. Vaz, F.N.C.; Fermino, B.L.; Haskel, M.V.L.; Wouk, J.; de Freitas, G.B.L.; Fabbri, R.; Montagna, E.; Rocha, J.B.T.; Bonini, J.S. The Relationship Between Copper, Iron, and Selenium Levels and Alzheimer Disease. *Biol. Trace Elem. Res.* **2018**, *181*, 185–191. [CrossRef] [PubMed]
134. James, S.A.; Churches, Q.I.; de Jonge, M.D.; Birchall, I.E.; Streltsov, V.; McColl, G.; Adlard, P.A.; Hare, D.J. Iron, Copper, and Zinc Concentration in Abeta Plaques in the APP/PS1 Mouse Model of Alzheimer's Disease Correlates with Metal Levels in the Surrounding Neuropil. *Acs Chem. Neurosci.* **2017**, *8*, 629–637. [CrossRef] [PubMed]
135. Yugay, D.; Goronzy, D.P.; Kawakami, L.M.; Claridge, S.A.; Song, T.B.; Yan, Z.; Xie, Y.H.; Gilles, J.; Yang, Y.; Weiss, P.S. Copper Ion Binding Site in beta-Amyloid Peptide. *Nano Lett.* **2016**, *16*, 6282–6289. [CrossRef] [PubMed]
136. Mira, L.; Fernandez, M.T.; Santos, M.; Rocha, R.; Florencio, M.H.; Jennings, K.R. Interactions of flavonoids with iron and copper ions: A mechanism for their antioxidant activity. *Free Radic. Res.* **2002**, *36*, 1199–1208. [CrossRef]
137. Vestergaard, M.; Kerman, K.; Tamiya, E. An electrochemical approach for detecting copper-chelating properties of flavonoids using disposable pencil graphite electrodes: Possible implications in copper-mediated illnesses. *Anal. Chim. Acta* **2005**, *538*, 273–281. [CrossRef]
138. Teng, Y.; Zhao, J.; Ding, L.; Ding, Y.; Zhou, P. Complex of EGCG with Cu(II) Suppresses Amyloid Aggregation and Cu(II)-Induced Cytotoxicity of alpha-Synuclein. *Molecules (Basel, Switz.)* **2019**, *24*.

International Journal of
Molecular Sciences

MDPI

Review

Prevention of Cognitive Decline in Alzheimer's Disease by Novel Antioxidative Supplements

Koh Tadokoro [1], Yasuyuki Ohta [1], Haruhiko Inufusa [2], Alan Foo Nyuk Loon [3] and Koji Abe [1,*]

1 Department of Neurology, Graduate School of Medicine, Dentistry and Pharmaceutical Sciences, Okayama University, 2-5-1 Shikata-cho, Kita-ku, Okayama 700-8558, Japan; p86m6fma@okayama-u.ac.jp (K.T.); yomdhot@hotmail.co.jp (Y.O.)
2 Division of Anti-Oxidant Research, Life Science Research Center, Gifu University, 1-1 Yanagido, Gifu 501-1193, Japan; inufusah@antioxidantres.jp
3 Hovid Berhad, 121, Jalan Tunku Abdul Rahman, Ipoh 30010, Perak, Malaysia; alanfoo@hovid.com
* Correspondence: me20048@s.okayama-u.ac.jp

Received: 18 February 2020; Accepted: 10 March 2020; Published: 13 March 2020

Abstract: Oxidative stress plays a crucial role in Alzheimer's disease (AD) from its prodromal stage of mild cognitive impairment. There is an interplay between oxidative stress and the amyloid β (Aβ) cascade via various mechanisms including mitochondrial dysfunction, lipid peroxidation, protein oxidation, glycoxidation, deoxyribonucleotide acid damage, altered antioxidant defense, impaired amyloid clearance, inflammation and chronic cerebral hypoperfusion. Based on findings that indicate that oxidative stress plays a major role in AD, oxidative stress has been considered as a therapeutic target of AD. In spite of favorable preclinical study outcomes, previous antioxidative components, including a single antioxidative supplement such as vitamin C, vitamin E or their mixtures, did not clearly show any therapeutic effect on cognitive decline in AD. However, novel antioxidative supplements can be beneficial for AD patients. In this review, we summarize the interplay between oxidative stress and the Aβ cascade, and introduce novel antioxidative supplements expected to prevent cognitive decline in AD.

Keywords: Alzheimer's disease; oxidative stress; supplement

1. Introduction

The number of dementia patients is rapidly increasing in aging societies. Over 46 million people lived with dementia worldwide in 2015, and this number is estimated to increase to 131.5 million by 2050 [1]. Alzheimer's disease (AD) is the most common cause of dementia. Of all dementia patients in our clinic, 62% had AD, and 69% of patients were in the late elderly (≥75 years old) subgroup [2]. There is a great demand for effective interventions to prevent cognitive decline in AD, but no effective drugs or supplemental therapies have yet been established.

AD is pathologically characterized by the presence of hallmark lesions such as neuronal loss, and senile plaque consisting of amyloid β (Aβ) and neurofibrillary tangles (NFT). In the Aβ hypothesis, Aβ is widely regarded as a primary cause of cognitive decline. Aβ peptides are cleaved from amyloid precursor protein (APP), a transmembrane protein associated with neuronal development, neurite outgrowth and axonal transport, and released outside the cell, where they are rapidly degraded or removed. Aβ monomers aggregate into oligomers, protofibrils and amyloid fibrils. Although Aβ is rapidly degraded or removed in normal subjects, under pathological conditions, Aβ peptides can accumulate to produce Aβ oligomers, protofibrils or fibrils [3]. Soluble oligomers or protofibrils are supposed to cause neuronal dysfunction including synaptic impairment/spine changes, dendritic simplification, axonopathy/neuronal loss and subsequent memory impairment in AD rather than end-stage amyloid fibrils [4,5]. In patients carrying a mutation for autosomal dominant AD (described below), changes of pathophysiological conditions preceded their cognitive decline; $A\beta_{42}$ in the cerebrospinal fluid (CSF) appeared to decline 25 years before the onset of symptoms, followed by fibrillar Aβ in positron emission tomography (PET), increased tau in the CSF, hippocampal atrophy and hypometabolism [6].

In familial cases of early-onset AD, autosomal dominant mutations of *APP*, *PSEN1* and *PSEN2* have been identified, and the global prevalence for autosomal dominant forms of early-onset AD is 5–10% [7]. These mutations lead to the accumulation of Aβ and subsequent development of AD. On the other hand, in sporadic cases of AD, the mechanism of Aβ accumulation in the brain remains unclear. There are several genetic and non-genetic risk factors of late-onset AD. The apolipoprotein E (ApoE) ε4 allele is a well-known genetic risk factor of AD [8]. In addition, genome-wide association studies identified susceptibility loci such as *CLU*, *PICALM*, *CR1* and *BIN1*, mainly clustered according to their immune response, APP processing and lipid metabolism and endocytosis [9]. Among non-genetic factors, cerebrovascular disease, hypertension, diabetes mellitus, both low and high body weight, dyslipidemia, metabolic syndrome, smoking and traumatic brain injury increase the risk of AD [10].

2. Oxidative Stress and Alzheimer's Disease

2.1. Oxidative Stress

Oxidative stress is a disturbance in the balance between the production of reactive oxygen species (ROS) or reactive nitrogen species (RNS) and antioxidant systems in the body [11]. ROS is a type of unstable molecule that contains oxygen and easily reacts with other molecules, including the superoxide anion ($O_2^{\bullet -}$), hydrogen peroxide (H_2O_2) and the hydroxyl radical ($^{\bullet}OH$). The sources of ROS are mitochondria, oxidases (such as nicotinamide adenine dinucleotide phosphate (NADPH) oxidase and xanthine oxidase), and autoxidation of different small molecules of endogenous and exogenous origin [12]. RNS are a family of molecules derived from nitric oxide ($NO^{\bullet}$) and $O_2^{\bullet -}$ produced via $NO^{\bullet}$ synthase and several enzymes including NADPH oxidase, xanthine oxidase, lipoxygenase and cyclooxygenase [13]. Antioxidants serve to counterbalance the effect of oxidants, and can be classified into enzymatic and non-enzymatic groups. Enzymatic antioxidants include superoxide dismutase (SOD), catalase, glutathione peroxidase (GPx), thioredoxin, thioredoxin, peroxiredoxin and glutathione-*S*-transferase (GST), commonly requiring NADPH as a reducing equivalent. Non-enzymatic antioxidants include low-molecular weight compounds such as vitamins C and E, β-carotene, uric acid and glutathione (GSH). Several studies proved that ROS modulate intracellular transduction pathways and transcriptional factors involved in cell proliferation, differentiation and maturation [14]. However, when ROS accumulation exceeds antioxidant defense, it is referred to as oxidative stress, and is related to pathological conditions [15].

The nervous system is vulnerable to oxidative stress because of its high consumption of oxygen, a large amount of polyunsaturated fatty acids and high iron content resulting in an increased generation of ROS. Therefore, oxidative stress plays a crucial role in various diseases of the nervous system including ischemic stroke, as well as neurodegenerative diseases including amyotrophic lateral sclerosis (ALS), Parkinson's disease, the prodromal stage of mild cognitive impairment (MCI) and AD [16]. Many previous studies revealed increased free radical production, lipid peroxidation, oxidative protein damage, decreased adenosine triphosphate (ATP) production and reduced cell viability in postmortem AD brains. Praticò et al. noted that individuals with MCI had increased oxidative damage before the onset of symptomatic dementia by measuring 8,12-iso-$iPF_{2\alpha}$-VI levels in urine, plasma and cerebrospinal fluid using gas chromatography–mass spectrometry [16]. Arimon et al. demonstrated that local infusion of oxidizing agents into the hippocampus of wild-type mice increased local $A\beta_{42}$ levels in the interstitial fluid, suggesting that oxidative stress is located upstream of Aβ pathology in AD [17]. Baldeiras et al. conducted a longitudinal study on 70 MCI patients and demonstrated that the accumulation of oxidative damage may start in presymptomatic phases of AD pathology and that progression to AD might be related to depletion of antioxidant defenses such as the oxidized/reduced GSH ratio and vitamin E [18].

2.2. *Amyloid, Neurofibrillary Tangle and Oxidative Stress*

Extracellular formation of senile plaques composed of Aβ is one of the hallmarks of AD pathology. Aβ is generated from APP though sequential cleavage by β and γ-secretases. Metal ions such as copper and zinc in the synaptic cleft of some neurons are supposed to play an important role in Aβ aggregation. Amyloid plaques also have a high content of these, and other, metal ions. Since these metal ions are involved in ROS production, aggregated Aβ takes part in ROS production [16]. Felica et al. showed that Aβ oligomers stimulated excessive formation of ROS through a mechanism requiring the activation of the N-methyl-D-aspartate (NMDA) receptor by using hippocampal neuronal cultures [19]. On the contrary, ROS themselves trigger Aβ generation by enhancing the amyloidogenic pathway [20]. Mitochondrial dysfunction and subsequent ROS production in a cell model induced by using rotenone and antimycin increased Aβ production [21]. In animal models, Aβ levels were enhanced by inhibiting complex I [21]. These findings suggest that there is bidirectional interplay between ROS and Aβ.

NFTs are another pathological hallmark of AD, and the formation of NFTs is thought to be linked closely to neuronal dysfunction in AD. NFTs are composed of a highly phosphorylated form of microtubule-associated protein tau [22]. Oxidative stress contributes to phosphorylation and the formation of NFTs [23]. By using an in vitro model of chronic oxidative stress through inhibition of glutathione synthesis with buthionine sulfoximine, Su et al. demonstrated that chronic oxidative stress increased levels of tau phosphorylated at the PHF 1 epitope in a time-dependent manner [24]. They also reported that a fragment of tau protein possessed copper reduction activity and initiated the copper-mediated generation of H_2O_2 [25].

2.3. *Mitochondrial Dysfunction*

The mitochondrion is an essential organelle that produces ATP through aerobic oxidative phosphorylation for sustaining cellular functions and survival, also serves as a regulator of cellular calcium concentration, and is a major generator of ROS. Therefore, dysfunctional mitochondria cause the loss of ATP, cellular calcium dysregulation, apoptosis and oxidative stress. $O_2^{\bullet-}$, which is the proximal mitochondrial ROS, is produced by the one-electron reduction of molecular oxygen. Most of the $O_2^{\bullet-}$ generated by intact mammalian mitochondria is produced by complex I, followed by complex III [26]. The production of $O_2^{\bullet-}$ increases when the mitochondria are not making ATP and consequently have a high proton-motive force and a reduced coenzyme Q pool, and when the NADH/NAD^+ ratio in the mitochondrial matrix is high [27].

Mitochondrial impairment is a common feature of the aging process [28] and AD [20]. Several mitochondrial functions decline with age, causing increased ROS production, mtDNA damage, changes in membranes and electrolytes and decreased recovery of damaged mitochondria [28]. Fluorodeoxyglucose positron emission tomography revealed reduced glucose metabolism in living AD patients, suggesting mitochondrial dysfunction [29]. In postmortem brains from AD patients, mitochondrial deoxyribonucleic acid (DNA) was damaged compared with the age-matched healthy control. Mitochondrial enzyme complexes are reduced in AD including cytochrome c oxidase, the pyruvate dehydrogenase complex, and the α-ketodehydrogenase complex, possibly explained by depletion of the mtDNA encoding subunits of such enzymes [30]. Manczak et al. reported that in APP transgenic AD model mice, expression of the mitochondrial fission genes *Drp1* and *Fis1* increased, expression of mitochondrial fusion genes *Mfn1*, *Mfn2*, *Opa1* and *Tomm40* decreased, and that Drp1 interacted with the Aβ monomer and oligomer, suggesting that increased production of Aβ and the interaction of Aβ with Drp1 are crucial factors in mitochondrial fragmentation, abnormal mitochondrial dynamics and synaptic damage [31].

2.4. Lipid Peroxidation

Lipid peroxidation occurs in the AD brain and is most prominent where degenerative changes are most pronounced [32]. A meta-analysis performed by Schrag et al. provided evidence of increased oxidative stress in serum, erythrocytes and circulating lymphocytes in AD, particularly in the lipid compartment [33]. Lipid peroxidation consists of a cascade of reactions, which causes the degradation of lipids mediated by free radicals. Free radicals abstract an allylic H from a methylene group in the acryl chain of phospholipids, followed by rearrangement of the double bonds to the conjugate diene form, producing a carbon-centered alkyl radical. When the alkyl radical reacts with paramagnetic molecular oxygen, a peroxyl radical is produced, which abstracts another allylic H atom to initiate a self-perpetuating chain reaction that ultimately leads to a variety of cyclic peroxides and hydroperoxides. Hydroperoxides can be further degraded to produce malondialdehyde (MDA), 4-hydroxynonenal (4-HNE) and acrolein, which can cause irreversible modification of phospholipids. Peroxidation of membrane lipids affects a variety of functions resulting in increased rigidity, decreased activity of membrane-bound enzymes, impairment of membrane receptors and altered permeability. 4-HNE binds to both nicastrin and beta-site amyloid precursor protein cleaving enzyme (BACE), differentially affecting γ- and β-secretase activity, suggesting that this naturally occurring product of lipid peroxidation may trigger the generation of toxic Aβ species [17]. Markers of lipid peroxidation are elevated in AD patients [34]. Membrane-associated oxidative stress occurs in association with the alterations in lipids, and exposure of the hippocampus to Aβ induces membrane oxidative stress and the accumulation of ceramide species and cholesterol [35].

2.5. Protein Oxidation

Protein oxidation also plays an important role in AD. Protein carbonyls generated by the oxidation of amino acids increased in AD brains [36]. Oxidative modification of proteins such as unfolding, conformational changes, protein–protein cross linking due to dityrosine formation, tyrosine halogenation and nitration and protein carbonylation can cause the loss of protein function, resulting in cell death [37]. Markers of protein oxidation such as carbonyls, dityrosine and 3-nitrotyrosine were elevated in the hippocampus and inferior parietal lobule of AD patients compared with age-matched controls [38].

Using a proteomics approach, Castegna et al. demonstrated that creatine kinase (CK) BB, ubiquitin carboxy-terminal hydrolase L-1, glutamine synthetase (GS), dihydropyrimidinase-related protein 2,α-enolase and heat shock cognate 71 were specifically oxidized in AD [39,40]. One consequence of oxidized CK is decreased availability of ATP in synaptic terminals, areas of the neuron that are probably most vulnerable and involved early in oxidative neurodegeneration in AD [40]. Impaired GS could reduce astrocyte protection against glutamate excitotoxicity to neurons [41].

2.6. Glycoxidation

Diabetes mellitus increases the risk of AD [42,43] through several mechanisms such as decreased Aβ clearance [44], cerebrovascular changes [45] and oxidative stress [46]. Advanced glycation endproducts (AGEs) are a group of heterogeneous compounds increasingly formed non-enzymatically by the Maillard reaction under hyperglycemic conditions [47]. Intracellular deposits of AGEs increased in both neurons and astrocytes of AD patients, and many neurons showed the co-localization of AGEs with hyperphosphorylated tau and nNOS [48].

The receptor for AGE (RAGE) is highly expressed in diabetes and the increased expression of RAGE was associated with increased oxidative and inflammatory stress [49]. RAGE expression levels in AD brains were higher than in control brains, and RAGE was present in neurons, glia and microglia in the hippocampus and cortex [50]. Wautier et al. demonstrated that enhanced oxidative stress by the AGE-RAGE signaling pathway is, at least in part, contributed by NADPH oxidase activation [51]. In addition, ligation of RAGE by AGEs resulted in the suppression of antioxidants such as GSH and ascorbic acid [52]. Askarova et al. demonstrated that Aβ binding to RAGE activated NADPH oxidase in endothelial cells and astrocytes, causing oxidative stress [53].

2.7. DNA Damage

ROS, especially •OH, reacts with DNA by adding double bonds to the DNA base and by abstraction of an H atom from the methyl group of thymine and each C-H bond of 2′-deoxyribose [54], leading to strand breaks, as well as DNA–DNA and DNA–protein cross-linking [55]. 8-Hydroxy-2′-deoxyguanosine (8-OHdG) is a major form of DNA damage induced by ROS and is regarded as a marker of DNA oxidation. Nunomura et al. analyzed autopsy brains to demonstrate that the levels of neuronal 8-OHdG decreased exponentially as the Aβ burden in AD increased, noting a similar pattern of exponential decrease in neuronal 8-OHdG with increasing disease duration, suggesting that oxidative DNA damage is the earliest event and that AD is associated with compensatory changes that reduce damage from ROS [56].

Although the brain is the most affected in AD, Mecocci et al. demonstrated that the level of 8-OHdG was elevated in peripheral lymphocytes of AD patients [57]. Using comet assay analysis, Migliore et al. revealed a significantly higher level of primary DNA damage in leukocytes of AD and also of MCI patients compared to control individuals, suggesting that DNA damage is an earlier event in the pathogenesis of AD [58]. Moslemnezhad et al. also demonstrated that the plasma level of 8-OHdG was significantly higher in AD than in the control, while the amount of total antioxidants was significantly lower in patients compared to controls [59]. Isobe et al. revealed that the concentration of 8-OHdG in the CSF of AD patients was significantly higher than in the CSF of controls, and was positively correlated with the percentage of coenzyme Q10 and the duration of the illness [60].

2.8. Altered Antioxidant Defense

Progressive changes in oxidative stress defense mechanisms during the progression from MCI towards severe AD have been reported [61]. Antioxidants such as glutathione, GPx, GST and SOD significantly declined in the mitochondrial and synaptosomal fractions in the postmortem frontal cortex of MCI and AD patients [62]. A reduction of GSH was also demonstrated in the hippocampus and frontal cortex of living MCI and AD patients by using proton-magnetic resonance spectroscopy [63]. By using APP23 transgenic mice, Bayer et al. noted that chronic APP overexpression per se reduced SOD1 activity in the transgenic mouse brain, which could be restored to normal levels after Cu treatment [64].

Nrf2 is an antioxidant transcription factor. In unstressed conditions, Nrf2 in the cytoplasm is negatively regulated by Keap1. In oxidative stress, Nrf2 is stabilized and accumulates in the nucleus and activates its target genes such as GST and Heme oxygenase 1 (HO-1) [65]. Ramsey et al. reported that Nfr2 is predominantly cytoplasmic in hippocampal neurons in hippocampal neurons in AD, while it is expressed in both the nucleus and the cytoplasm in normal hippocampi with predominant expression in the nucleus, suggesting that Nrf2 does not respond properly to oxidative stress in AD neurons [66]. Using AD model $App^{NL\text{-}G\text{-}F/NL\text{-}G\text{-}F}$ knock-in mice and a natural compound, 6-(methylsulfinyl)hexyl isothiocyanate, which mildly activated Nrf2 signaling, the induction of Nrf2 ameliorated cognitive impairment in the AD model mouse by suppressing oxidative stress and neuroinflammation [67].

2.9. Amyloid Clearance and Oxidative Stress

Ultrastructural studies demonstrated characteristic and extensive angioarchitectural distortion of cerebral capillaries in AD [68,69]. Low-density lipoprotein receptor-related protein 1 (LRP-1) is the primary moiety responsible for the efflux of Aβ from the brain to the blood across the blood–brain barrier (BBB). In an autopsy of AD brains, the levels of 4-HNE bound to transmembrane LRP-1 had significantly increased in the hippocampus, while the levels of LRP-1-3-nitrotyrosine had not, suggesting that Aβ impaired its own efflux from the brain by oxidation of its transporter LRP-1, leading to increased Aβ deposition [70]. On the contrary, RAGE not only causes oxidative stress described above, but also promotes influx of circulating Aβ across the BBB [71]. FPS-ZM1, a specific RAGE inhibitor, downregulated Aβ influx across the BBB, decreased hippocampal Aβ, inhibited NF-κB signaling and reduced apoptosis in db/db mice [72].

2.10. Inflammation

The interaction between oxidative stress and inflammation contributes to AD pathology [73]. Damaged neurons, insoluble Aβ peptide deposits and NFTs stimulate inflammation in the AD brain [74]. Microglia are resident macrophages in the central nervous system [75] and are principle immune effectors [76]. Inflammatory responses are mediated by the activation of microglia [77]. In AD brains, microglia are activated and are attracted to and surround senile plaque [78]. Quantitative in-vivo measurements of glial activation with PET and carbon-11-labelled (*R*)-PK11195 demonstrated that AD patients showed a significant increase of microglial activation in entorhinal, temporal and cingulate cortexes [79]. Fibrillar Aβ-stimulated microglia release ROS [74]. The primary source of ROS and the source of widespread oxidative damage found in AD brains is microglial NADPH oxidase [76]. Astrocytes are also activated in the AD brain [80]. Similar to microglia, astrocytes release cytokines, interleukins, $NO^{\bullet}$ and other potentially cytotoxic molecules upon exposure to Aβ [81]. Aβ upregulated both pro- and anti-inflammatory cytokines including IL-1β, IL-6, transforming growth factor-β and IL-10 [82]. In turn, neuro-inflammation-induced oxidative stress increases the expression of Aβ [73].

2.11. Chronic Cerebral Hypoperfusion

Chronic cerebral hypoperfusion (CCH) is ubiquitous in elderly AD patients, and can play pivotal roles in triggering and exacerbating the pathophysiological progress of AD. Our previous studies revealed that white matter hyperintensity (WMH) was observed in more than 88% of AD patients by magnetic resonance imaging (MRI) [83], and that high grade WMH was a risk factor for MCI conversion to AD as well as low educational attainment, a low baseline mini-mental state examination (MMSE) score and parahippocampal gyrus atrophy [83]. Analysis of the Alzheimer's Disease Neuroimaging Initiative public database revealed that Pittsburg compound B (PIB) positivity increased total WMH volume independently of the predicted AD diagnosis, that those diagnosed as having AD had greater WMH volume among PIB-positive subjects than normal control subjects and that both WMH and PIB status at the baseline conferred risk for future diagnosis of AD, suggesting that WMH contributes to the presentation of AD and may provide a second hit necessary for the clinical manifestation of the disease [84]. We revealed that neural oxidative stress and neuroinflammation were enhanced in AD model mice with CCH and that edaravone, a free radical scavenger, significantly improved motor and cognitive deficits, attenuated neuronal loss, reduced Aβ/phosphorylated tau (pTau) accumulation and alleviated neural oxidative stress and neuroinflammation in the AD mouse model with CCH [85]. Additionally, CCH greatly enhanced the number of Aβ oligomer-positive/pTau cells, the expression of peroxidation products (4-HNE and 8-OHdG), mitochondrial fission proteins (Drp1 and Fis1), and decreased the expression of mitochondrial fusion proteins (Opa1 and Mfn1) in the CTX and thalamus (TH) of AD model mice at 12 month of age, demonstrating that CCH shifted the balance in mitochondrial morphology from fusion to fission [86]. Furthermore, we investigated expressive changes of two main Aβ transport receptors, LRP-1 and RAGE, and revealed that CCH increased LRP-1 and RAGE expression in brain parenchyma, while a decrease of LRP1 and increase of RAGE were observed in vascular endothelial cells, suggesting double imbalances of Aβ efflux and influx transport-related proteins in the cortical blood vessel of AD mice. These neuropathological abnormalities were greatly ameliorated by edaravone [87].

3. Therapeutic Approach for Alzheimer's Disease

There have not been any effective treatments to prevent, halt or reverse AD [88]. Currently available therapies with cholinesterase inhibitors such as donepezil hydrochloride, galantamine and rivastigmine or NMDA receptor antagonist memantine offer little more than short-term palliative effects [89]. Although several trials of amyloid-targeting therapy have been performed recently, they have yet to show satisfactory results.

Based on findings that oxidative stress plays a major role in AD, oxidative stress has been considered as a therapeutic target of AD [89]. The free radical scavenger edaravone, which was approved as a treatment of acute ischemic stroke [90] and amyotrophic lateral sclerosis [91], inhibited Aβ aggregation and Aβ-induced oxidation in vitro, and improved AD pathology and cognitive behavioral deficits of AD model mice [92]. Treatment with coenzyme Q10, a component of the mitochondrial electron transport chain, decreased brain levels of carbonyls, plaque area and number in the hippocampus and in the overlying cortex immunostained with an $A\beta_{42}$-specific antibody, brain $A\beta_{42}$ levels and levels of Aβ protein precursor β-carboxyterminal fragments, and improved performance in the Tg19959 mouse model of AD [93]. Melatonin has inhibitory effects on the formation of secondary β-sheet structures and amyloid fibril formation in vitro [94,95], and administering melatonin into Tg2576 AD model mice partially inhibited the time-dependent elevation of Aβ, reduced abnormal nitration of proteins, and increased survival [96]. Overexpression of SOD-2 reduced hippocampal $O_2^{\bullet -}$ and prevented memory deficits in the Tg2576 mouse model of AD [97].

In spite of such favorable preclinical study outcomes, previous antioxidative components including a single antioxidative supplement such as vitamin C, vitamin E or their mixtures [98–100] did not clearly show a therapeutic effect on cognitive decline in AD [101–104], even though some of the clinical studies suggest a possible therapeutic effect of these compounds. However, we recently reported that novel supplements such as Twendee X® and tocotrienols can be beneficial for AD patients.

4. Twendee X

Twendee X® (TwX; TIMA Japan, Osaka, Japan) is a patented supplement containing coenzyme Q10, niacin amide, L-cystine, ascorbic acid, succinic acid, fumaric acid, L-glutamine and riboflavin, having stronger antioxidant and anti-inflammatory effects than single antioxidant vitamins [105]. Inufusa et al. characterized the antioxidant properties of TwX, assessing parameters of the redox state following the induction of oxidative stress by H_2O_2 in HepG2 cells. In HepG2 cells, H_2O_2 exposure increased ROS at the mitochondrial (69%) and cellular level (68%), reduced natural antioxidant enzymatic activity with reduced Mn (32%) and Cu/ZnSOD1 (31%) activities and increased H_2O_2 scavengers with a 31% increase in total glutathione activity. TwX modulated H_2O_2 oxidative activity by reducing the level of ROS produced in the two compartments, increasing direct antioxidant defense at the mitochondrial and cellular levels, and reducing H_2O_2-induced scavenging activity by reducing GSH activity [106]. The therapeutic effect of TwX was also examined in nude mice that underwent inoculation of colon or gastric cancer cells: TwX reduced tumor growth, increased antioxidants measured by the d-ROMs test, and reduced natural killer cell activity, suggesting that TwX significantly reduced tumor growth and inhibited metastasis by reducing oxidative stress [107]. In ischemic stroke model mice, we demonstrated that pretreatment of TwX (20 mg/kg/d for 14 days) reduced infarct size as well as the expression of both oxidative stress markers such as 8-OHdG, 4-HNE and Nε-(carboxymethyl) lysine (an advanced glycation end product) and inflammatory markers such as Iba-1, tumor necrosis factor α (TNF-α) and monocyte chemotactic protein [108,109].

We also investigated the therapeutic effect of TwX on cognitive function, Aβ pathology, oxidative stress and inflammation in a novel AD mouse model with CCH. TwX treatment (20 mg/kg/d, from 4.5 to 12 months) significantly ameliorated cognitive deficit, amyloid-β, tau and α-synuclein pathology, neuronal loss and neurovascular dysfunction, and this was accompanied by the attenuation of both oxidative stress (4-HNE, 8-OHdG) and inflammatory markers (NACHT, LRP and PYD domains-containing protein 3 (NLRP3), caspase-1, IL-1β, Iba-1 and TNF-α) [110].

In a multicenter, randomized, double-blind and placebo-controlled prospective interventional study, TwX showed a significantly higher score of the mini-mental state examination at 6 months compared with the placebo, and also a significant improvement of the Hasegawa dementia scale-revised score from baseline at 6 months, suggesting that a strong antioxidative therapy might be a useful way to prevent the conversion of MCI to AD [111].

5. Tocotrienols

Vitamin E is a lipid component of biological membranes and a potent antioxidant consisting of two categories: tocopherols and tocotrienols, each with 4 α, β, γ and δ analogs. Both tocopherols and tocotrienols consist of a chromanol ring and a 15-carbon tail, but tocotrienols differ from tocopherols by the presence of three unsaturated bonds in the hydrocarbon tail (Figure 1) [112]. Tocopherols are found in lipid-rich regions of cells such as mitochondrial membranes, fat depots and lipoproteins such as low-density lipoprotein cholesterol [113]. Vitamin E detected in the brain is virtually only α-tocopherol [114]. Orally supplemented α-tocotrienols were effectively delivered to most tissues, including the brain, in mice with α-tocopherol deficiency [115]. Epidemiological studies revealed that AD and MCI had lower levels of total tocopherols, total tocotrienols and total vitamin E compared with cognitively normal subjects [116,117].

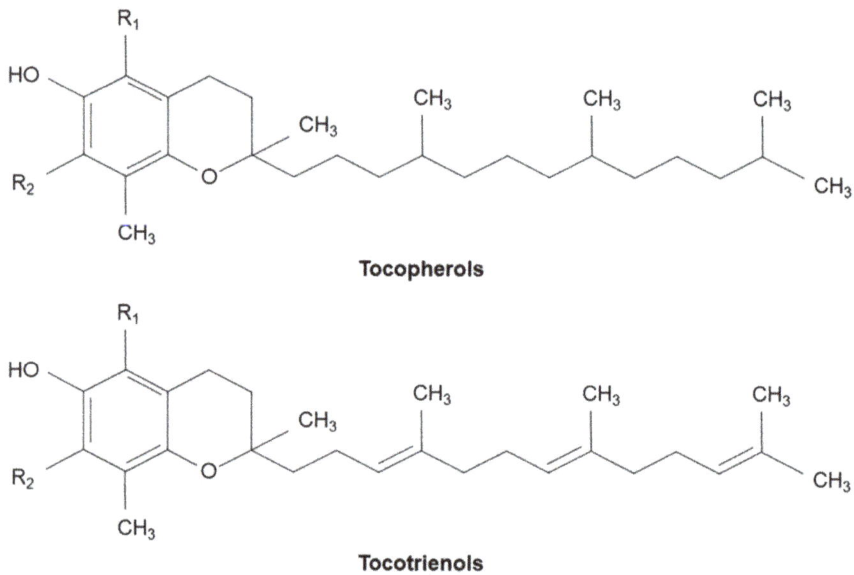

Figure 1. Molecular structure of tocopherols and tocotrienols.

Oral vitamin E supplement mainly consists of α-tocopherol, which was unable to prevent cognitive decline [98–100]. However, the antioxidant activity of α-tocotrienols is higher than that of α-tocopherols [114]. The antioxidative activity of tocopherols is related to the scavenging of free radicals in unsaturated lipid [118]. Compared to tocopherols, tocotrienols are widely distributed in the phospholipid bilayer and easily interact with lipids due to the unsaturated bonds of the hydrocarbon tail (Figure 1) [112]. An increasing number of studies have shown that tocotrienols possess additional beneficial pharmacological actions such as inhibited platelet aggregation [119], monocytic adhesion and cholesterol-lowering activity, which are independent of their antioxidant properties. Khanna et al. demonstrated that tocotrienols blocked glutamate-induced death by suppressing early activation of c-Src kinase and 12-Lox [120]. Gopalan et al. demonstrated that mixed tocotrienols attenuated the progression of white matter lesions, indicating cerebral small vessel disease [121]. Ibrahim et al. demonstrated that treatment with a tocotrienol-rich fraction (TRF) dose-dependently inhibited the formation of Aβ formation fibrils and Aβ oligomers in vitro, and that daily TRF supplementation to AβPPswe/PS1dE9 double transgenic mice for 10 months attenuated Aβ immunoreactive depositions and thioflavin-S-positive fibrillar type plaques in the brain and eventually improved cognitive function [122,123].

Tocovid Suprabio™ (Hovid, Perak, Malaysia) is a patented supplement that consists of 61.52 mg α-tocotrienol, 112.80 mg γ-tocotrienol, 25.68 mg δ-tocotrienol and 91.60 IU α-tocopherol in a capsule with a self-emulsifying system (Suprabio™), providing higher and more consistent absorption of tocotrienols. We reported that the neuroprotective effects of Tocovid Suprabio™ in the ischemic stroke model mice were accompanied by amelioration of motor dysfunction and infarct volumes [124,125]. Tocovid Suprabio™ significantly decreased the expression of oxidative stress markers (4-HNE, nitrotyrosine and 8-OHdG), advanced glycation markers (RAGE, carboxymethyl arginine (CMA) and carboxymethyl lysine (CML)) and apoptotic and autophagy markers (cleaved caspase-3 and LC3-II), and enhanced the expression of Nrf2 and multidrug resistance protein 1 (MRP1) accompanied by a decrease of the glutathione disulfide (GSSG)/GSH ratio [124]. In addition, Tocovid Suprabio™ decreased the expression of inflammatory markers such as TNF-α, monocyte chemotactic marker-1 (MCP-1) and Iba-1, and improved the damage of neurovascular units including matrix metalloproteinase 9 (MMP9), IgG and collagen IV [125]. These studies obviously demonstrated that Tocovid Suprabio™ treatment showed neuroprotective effects through antioxidative stress, antiapoptotic/autophagic and anti-inflammatory effects in the ischemic mouse brain.

A clinical trial reported that mixed tocotrienols attenuated the progression of white matter lesions in the human brain after 2 years [121] and that this was related with cognitive decline [126], suggesting that tocotrienols might prevent cognitive decline in AD.

6. Mitochondria-Targeted Antioxidants and Polyphenols

Plastoquinonyl-decyltriphenylphosphonium (SkQ1), mitoquinone mesylate (MitoQ) and astaxanthin are mitochondria-targeted antioxidants [127,128]. SkQ1 increased behavioral activity, and reduced destructive changes in mitochondria, pathological accumulation of AβPP, Aβ, hyperphosphorylation of tau-protein and hippocampal $A\beta_{40}$ and $A\beta_{42}$ protein levels in AD model rats [129,130]. MitoQ attenuated Aβ-neurotoxicity in the cortical neuron and prevented increased production of ROS, loss of mitochondrial membrane potential, cognitive decline, Aβ accumulation, astrogliosis, synaptic loss and caspase activation in AD model mice [131]. In a randomized, placebo-controlled, double-blind, crossover design study, MitoQ improved vascular function in healthy older adults [132]. Astaxanthin reduced cognitive impairment, soluble $A\beta_{42}$, insulin receptor substrate-1 S307 phosphorylation, glycogen synthase kinase-3β phosphorylation, in AD model rats [133]. A composite supplement containing astaxanthin and sesamin improved permeability psychomotor speed and processing speed of MCI subjects in a randomized, double-blind, placebo-control trial [134].

Polyphenols are also expected to be beneficial for AD patients [135]. Curcumin is a polyphenol from *Curcuma longa*, and has an antioxidant property as well as anti-inflammatory and antiamyloid effects [136]. Curcumin treatment improved the behavioral symptoms in AD patients [137]. Maiti et al. reported that solid lipid curcumin particles (SLCP) provide more antiamyloid, anti-inflammatory and neuroprotective outcomes than natural curcumin, and intraperitoneal injection of SLCP decreased Aβ plaque loads, pyknotic or tangle-like neurons, and reduced glial fibrillary acidic protein and Iba-1 immunoreactivity more strongly than natural curcumin [138]. Resveratrol is a polyphenol found in red grapes, red wine and other plant foods. Resveratrol treatment significantly prevented memory loss, reduced the amyloid burden and increased mitochondrial complex IV protein levels in the mouse brain mainly through Sirtuin 1 and adenosine monophosphate-activated protein kinase pathways in AD model mice [139]. In a randomized, double-blind, placebo-controlled trial, oral resveratrol supplementation stabilized CSF $A\beta_{40}$ and plasma $A\beta_{40}$ levels, which declined significantly in the placebo group [140].

7. Conclusions

There is considerable demand for effective interventions to prevent cognitive decline in AD, and oxidative stress can be a therapeutic target of AD. Novel antioxidative supplements might be hopeful antioxidative supplements for preventing dementia (Figure 2).

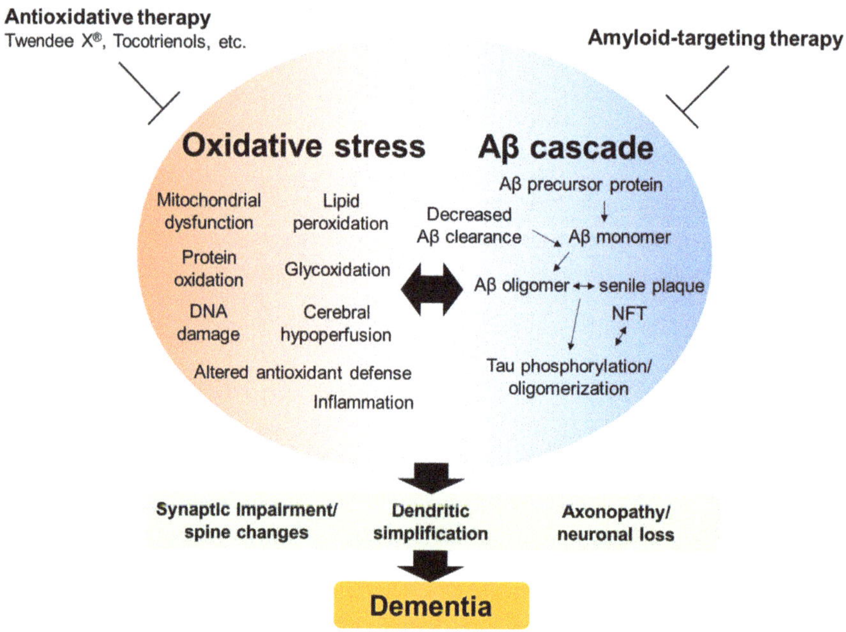

Figure 2. Schematic illustration of oxidative stress and amyloid β (Aβ) cascade in Alzheimer's disease. There is interplay between oxidative stress and the Aβ cascade, resulting in neuronal dysfunction and death. Antioxidative supplements such as Twendee X® and tocotrienols might be hopeful for preventing dementia.

Funding: This work was partly supported by a Grant-in-Aid for Scientific Research (B) 17H0419611, (C) 15K0931607, 17H0975609 and 17K1082709, and by Grants-in-Aid from the Research Committees (Kaji R, Toba K, and Tsuji S) from the Japan Agency for Medical Research and Development 7211800049, 7211800130 and 7211700121.

Acknowledgments: We thank members of our laboratory for helpful discussions.

Conflicts of Interest: The authors declare no conflict of interest.

Abbreviations

AD	Alzheimer's disease
AGE	advanced glycation end-product
ALS	amyotrophic lateral sclerosis
ApoE	apolipoprotein E
APP	amyloid precursor protein
ATP	adenosine triphosphate
Aβ	amyloid β
BACE	beta-site amyloid precursor protein cleaving enzyme
BBB	blood-brain barrier
CCH	chronic cerebral hypoperfusion
CK	creatine kinase
CMA	carboxymethyl arginine

CML	carboxymethyl lysine
CSF	cerebrospinal fluid
CTX	Cortex
GPx	glutathione peroxidase
GS	glutamine synthetase
GSH	Glutathione
GSSG	glutathione disulfide
GST	glutathione-S-transferase
HNE	Hydroxynonenal
H_2O_2	hydrogen peroxide
LRP-1	low-density lipoprotein receptor-related protein 1
MCI	mild cognitive impairment
MCP-1	monocyte chemotactic protein-1
MDA	Malondialdehyde
MitoQ	mitoquinone mesylate
MMP9	matrix metalloproteinase 9
MMSE	mini-mental state examination
MRI	magnetic resonance imaging
MRP1	multidrug resistance protein 1
NADPH	nicotinamide adenine dinucleotide phosphate
NFT	neurofibrillary tangles
NLRP3	NACHT, LRP and PYD domains-containing protein 3
NMDA	N-methyl-D-aspartate
$O_2^{\bullet -}$	superoxide anion
$^{\bullet}OH$	hydroxyl radical
PET	positron emission tomography
PIB	Pittsburg compound B
pTau	phosphorylated tau
RAGE	receptor for advanced glycation end-product
RNS	reactive nitrogen species
ROS	reactive oxygen species
SkQ1	plastoquinonyl-decyltriphenylphosphonium
SLCP	solid lipid curcumin particles
SOD	superoxide dismutase
TH	Thalamus
TNF-α	tumor necrosis factor-α
TRF	tocotrienol-rich fraction
TwX	Twendee X
WMH	white matter hyperintensity
8-OHdG	8-hydroxy-2′-deoxyguanosine

References

1. Prince, M.J.; Wimo, A.; Guerchet, M.M.; Ali, G.C.; Wu, Y.-T.; Prina, M. *World Alzheimer Report 2015—The Global Impact of Dementia*; Alzheimer's Disease International: London, UK, 2015.
2. Hishikawa, N.; Fukui, Y.; Sato, K.; Kono, S.; Yamashita, T.; Ohta, Y.; Deguchi, K.; Abe, K. Characteristic features of cognitive, affective and daily living functions of late-elderly dementia. *Geriatr. Gerontol. Int.* **2016**, *16*, 458–465. [CrossRef] [PubMed]
3. Kametani, F.; Hasegawa, M. Reconsideration of Amyloid Hypothesis and Tau Hypothesis in Alzheimer's Disease. *Front. Neurosci.* **2018**, *12*, 25. [CrossRef] [PubMed]
4. Bieschke, J.; Herbst, M.; Wiglenda, T.; Friedrich, R.P.; Boeddrich, A.; Schiele, F.; Kleckers, D.; Lopez del Amo, J.M.; Grüning, B.A.; Wang, Q.; et al. Small-molecule conversion of toxic oligomers to nontoxic β-sheet–rich amyloid fibrils. *Nat. Chem. Biol.* **2012**, *8*, 93–101. [CrossRef] [PubMed]
5. Brandt, R.; Bakota, L. Microtubule dynamics and the neurodegenerative triad of Alzheimer's disease: The hidden connection. *J. Neurochem.* **2017**, *143*, 409–417. [CrossRef]

6. Bateman, R.J.; Xiong, C.; Benzinger, T.L.; Fagan, A.M.; Goate, A.; Fox, N.C.; Marcus, D.S.; Cairns, N.J.; Xie, X.; Blazey, T.M.; et al. Clinical and biomarker changes in dominantly inherited Alzheimer's disease. *N. Engl. J. Med.* **2012**, *367*, 795–804. [CrossRef]
7. Van Cauwenberghe, C.; Van Broeckhoven, C.; Sleegers, K. The genetic landscape of Alzheimer disease: Clinical implications and perspectives. *Genet. Med.* **2016**, *18*, 421–430. [CrossRef]
8. Bertram, L.; McQueen, M.B.; Mullin, K.; Blacker, D.; Tanzi, R.E. Systematic meta-analyses of Alzheimer disease genetic association studies: The AlzGene database. *Nat. Genet.* **2007**, *39*, 17–23. [CrossRef]
9. Tosto, G.; Reitz, C. Genome-wide association studies in Alzheimer's disease: A review. *Curr. Neurol. Neurosci. Rep.* **2013**, *13*, 381. [CrossRef]
10. Reitz, C.; Mayeux, R. Alzheimer disease: Epidemiology, diagnostic criteria, risk factors and biomarkers. *Biochem. Pharmacol.* **2014**, *88*, 640–651. [CrossRef]
11. Betteridge, D.J. What is oxidative stress? *Metabolism* **2000**, *49*, 3–8. [CrossRef]
12. Lushchak, V.I. Free radicals, reactive oxygen species, oxidative stress and its classification. *Chem. Biol. Interact.* **2014**, *224*, 164–175. [CrossRef]
13. Iovine, N.M.; Pursnani, S.; Voldman, A.; Wasserman, G.; Blaser, M.J.; Weinrauch, Y. Reactive nitrogen species contribute to innate host defense against Campylobacter jejuni. *Infect. Immun.* **2008**, *76*, 986–993. [CrossRef]
14. Beckhauser, T.F.; Francis-Oliveira, J.; De Pasquale, R. Reactive Oxygen Species: Physiological and Physiopathological Effects on Synaptic Plasticity. *J. Exp. Neurosci.* **2016**, *10*, 23–48. [CrossRef] [PubMed]
15. Birben, E.; Sahiner, U.M.; Sackesen, C.; Erzurum, S.; Kalayci, O. Oxidative stress and antioxidant defense. *World Allergy Organ. J.* **2012**, *5*, 9–19. [CrossRef] [PubMed]
16. Pratico, D.; Clark, C.M.; Liun, F.; Rokach, J.; Lee, V.Y.; Trojanowski, J.Q. Increase of brain oxidative stress in mild cognitive impairment: A possible predictor of Alzheimer disease. *Arch. Neurol.* **2002**, *59*, 972–976. [CrossRef] [PubMed]
17. Arimon, M.; Takeda, S.; Post, K.L.; Svirsky, S.; Hyman, B.T.; Berezovska, O. Oxidative stress and lipid peroxidation are upstream of amyloid pathology. *Neurobiol. Dis.* **2015**, *84*, 109–119. [CrossRef]
18. Baldeiras, I.; Santana, I.; Proenca, M.T.; Garrucho, M.H.; Pascoal, R.; Rodrigues, A.; Duro, D.; Oliveira, C.R. Oxidative damage and progression to Alzheimer's disease in patients with mild cognitive impairment. *J. Alzheimers Dis.* **2010**, *21*, 1165–1177. [CrossRef]
19. De Felice, F.G.; Velasco, P.T.; Lambert, M.P.; Viola, K.; Fernandez, S.J.; Ferreira, S.T.; Klein, W.L. Abeta oligomers induce neuronal oxidative stress through an N-methyl-D-aspartate receptor-dependent mechanism that is blocked by the Alzheimer drug memantine. *J. Biol. Chem.* **2007**, *282*, 11590–11601. [CrossRef]
20. Grimm, A.; Friedland, K.; Eckert, A. Mitochondrial dysfunction: The missing link between aging and sporadic Alzheimer's disease. *Biogerontology* **2016**, *17*, 281–296. [CrossRef]
21. Leuner, K.; Schutt, T.; Kurz, C.; Eckert, S.H.; Schiller, C.; Occhipinti, A.; Mai, S.; Jendrach, M.; Eckert, G.P.; Kruse, S.E.; et al. Mitochondrion-derived reactive oxygen species lead to enhanced amyloid beta formation. *Antioxid Redox Signal.* **2012**, *16*, 1421–1433. [CrossRef]
22. Brion, J.P.; Anderton, B.H.; Authelet, M.; Dayanandan, R.; Leroy, K.; Lovestone, S.; Octave, J.N.; Pradier, L.; Touchet, N.; Tremp, G. Neurofibrillary tangles and tau phosphorylation. *Biochem. Soc. Symp.* **2001**, *67*, 81–88. [CrossRef]
23. Liu, Z.; Li, T.; Li, P.; Wei, N.; Zhao, Z.; Liang, H.; Ji, X.; Chen, W.; Xue, M.; Wei, J. The Ambiguous Relationship of Oxidative Stress, Tau Hyperphosphorylation, and Autophagy Dysfunction in Alzheimer's Disease. *Oxid. Med. Cell Longev.* **2015**, *2015*, 352723. [CrossRef] [PubMed]
24. Su, B.; Wang, X.; Lee, H.G.; Tabaton, M.; Perry, G.; Smith, M.A.; Zhu, X. Chronic oxidative stress causes increased tau phosphorylation in M17 neuroblastoma cells. *Neurosci. Lett.* **2010**, *468*, 267–271. [CrossRef] [PubMed]
25. Su, X.Y.; Wu, W.H.; Huang, Z.P.; Hu, J.; Lei, P.; Yu, C.H.; Zhao, Y.F.; Li, Y.M. Hydrogen peroxide can be generated by tau in the presence of Cu(II). *Biochem. Biophys. Res. Commun.* **2007**, *358*, 661–665. [CrossRef]
26. Ott, M.; Gogvadze, V.; Orrenius, S.; Zhivotovsky, B. Mitochondria, oxidative stress and cell death. *Apoptosis* **2007**, *12*, 913–922. [CrossRef]
27. Murphy, M.P. How mitochondria produce reactive oxygen species. *Biochem. J.* **2009**, *417*, 1–13. [CrossRef]
28. Balaban, R.S.; Nemoto, S.; Finkel, T. Mitochondria, oxidants, and aging. *Cell* **2005**, *120*, 483–495. [CrossRef]

29. Landau, S.M.; Harvey, D.; Madison, C.M.; Koeppe, R.A.; Reiman, E.M.; Foster, N.L.; Weiner, M.W.; Jagust, W.J.; Alzheimer's Disease Neuroimaging Initiative. Associations between cognitive, functional, and FDG-PET measures of decline in AD and MCI. *Neurobiol. Aging* **2011**, *32*, 1207–1218. [CrossRef]
30. Zhu, X.; Raina, A.K.; Lee, H.G.; Casadesus, G.; Smith, M.A.; Perry, G. Oxidative stress signalling in Alzheimer's disease. *Brain Res.* **2004**, *1000*, 32–39. [CrossRef]
31. Manczak, M.; Calkins, M.J.; Reddy, P.H. Impaired mitochondrial dynamics and abnormal interaction of amyloid beta with mitochondrial protein Drp1 in neurons from patients with Alzheimer's disease: Implications for neuronal damage. *Hum. Mol. Genet.* **2011**, *20*, 2495–2509. [CrossRef]
32. Markesbery, W.R. The role of oxidative stress in Alzheimer disease. *Arch. Neurol.* **1999**, *56*, 1449–1452. [CrossRef] [PubMed]
33. Schrag, M.; Mueller, C.; Zabel, M.; Crofton, A.; Kirsch, W.M.; Ghribi, O.; Squitti, R.; Perry, G. Oxidative stress in blood in Alzheimer's disease and mild cognitive impairment: A meta-analysis. *Neurobiol. Dis.* **2013**, *59*, 100–110. [CrossRef] [PubMed]
34. Sultana, R.; Perluigi, M.; Butterfield, D.A. Lipid peroxidation triggers neurodegeneration: A redox proteomics view into the Alzheimer disease brain. *Free Radic. Biol. Med.* **2013**, *62*, 157–169. [CrossRef] [PubMed]
35. Cutler, R.G.; Kelly, J.; Storie, K.; Pedersen, W.A.; Tammara, A.; Hatanpaa, K.; Troncoso, J.C.; Mattson, M.P. Involvement of oxidative stress-induced abnormalities in ceramide and cholesterol metabolism in brain aging and Alzheimer's disease. *Proc. Natl. Acad. Sci. USA* **2004**, *101*, 2070–2075. [CrossRef] [PubMed]
36. Ahmad, W.; Ijaz, B.; Shabbiri, K.; Ahmed, F.; Rehman, S. Oxidative toxicity in diabetes and Alzheimer's disease: Mechanisms behind ROS/RNS generation. *J. Biomed. Sci.* **2017**, *24*, 76. [CrossRef]
37. Butterfield, D.A.; Dalle-Donne, I. Redox proteomics: From protein modifications to cellular dysfunction and disease. *Mass Spectrom. Rev.* **2014**, *33*, 1–6. [CrossRef]
38. Di Domenico, F.; Coccia, R.; Butterfield, D.A.; Perluigi, M. Circulating biomarkers of protein oxidation for Alzheimer disease: Expectations within limits. *Biochim. Biophys. Acta* **2011**, *1814*, 1785–1795. [CrossRef]
39. Castegna, A.; Aksenov, M.; Aksenova, M.; Thongboonkerd, V.; Klein, J.B.; Pierce, W.M.; Booze, R.; Markesbery, W.R.; Butterfield, D.A. Proteomic identification of oxidatively modified proteins in alzheimer's disease brain. part I: Creatine kinase BB, glutamine synthase, and ubiquitin carboxy-terminal hydrolase L-1. *Free Radic. Biol. Med.* **2002**, *33*, 562–571. [CrossRef]
40. Butterfield, D.A.; Lauderback, C.M. Lipid peroxidation and protein oxidation in Alzheimer's disease brain: Potential causes and consequences involving amyloid β-peptide-associated free radical oxidative stress 1,2. *Free Radic. Biol. Med.* **2002**, *32*, 1050–1060. [CrossRef]
41. Zou, J.; Wang, Y.X.; Dou, F.F.; Lu, H.Z.; Ma, Z.W.; Lu, P.H.; Xu, X.M. Glutamine synthetase down-regulation reduces astrocyte protection against glutamate excitotoxicity to neurons. *Neurochem. Int.* **2010**, *56*, 577–584. [CrossRef]
42. Ott, A.; Stolk, R.P.; van Harskamp, F.; Pols, H.A.; Hofman, A.; Breteler, M.M. Diabetes mellitus and the risk of dementia: The Rotterdam Study. *Neurology* **1999**, *53*, 1937–1942. [CrossRef] [PubMed]
43. Ohara, T.; Doi, Y.; Ninomiya, T.; Hirakawa, Y.; Hata, J.; Iwaki, T.; Kanba, S.; Kiyohara, Y. Glucose tolerance status and risk of dementia in the community: The Hisayama study. *Neurology* **2011**, *77*, 1126–1134. [CrossRef] [PubMed]
44. Wakabayashi, T.; Yamaguchi, K.; Matsui, K.; Sano, T.; Kubota, T.; Hashimoto, T.; Mano, A.; Yamada, K.; Matsuo, Y.; Kubota, N.; et al. Differential effects of diet- and genetically-induced brain insulin resistance on amyloid pathology in a mouse model of Alzheimer's disease. *Mol. Neurodegener.* **2019**, *14*, 15. [CrossRef] [PubMed]
45. Takeda, S.; Sato, N.; Uchio-Yamada, K.; Sawada, K.; Kunieda, T.; Takeuchi, D.; Kurinami, H.; Shinohara, M.; Rakugi, H.; Morishita, R. Diabetes-accelerated memory dysfunction via cerebrovascular inflammation and Aβ deposition in an Alzheimer mouse model with diabetes. *Proc. Natl. Acad. Sci. USA* **2010**, *107*, 7036–7041. [CrossRef] [PubMed]
46. Butterfield, D.A.; Di Domenico, F.; Barone, E. Elevated risk of type 2 diabetes for development of Alzheimer disease: A key role for oxidative stress in brain. *Biochim. Biophys. Acta* **2014**, *1842*, 1693–1706. [CrossRef] [PubMed]
47. Nowotny, K.; Jung, T.; Hohn, A.; Weber, D.; Grune, T. Advanced glycation end products and oxidative stress in type 2 diabetes mellitus. *Biomolecules* **2015**, *5*, 194–222. [CrossRef] [PubMed]

48. Luth, H.J.; Ogunlade, V.; Kuhla, B.; Kientsch-Engel, R.; Stahl, P.; Webster, J.; Arendt, T.; Munch, G. Age- and stage-dependent accumulation of advanced glycation end products in intracellular deposits in normal and Alzheimer's disease brains. *Cereb Cortex* **2005**, *15*, 211–220. [CrossRef]
49. Yan, S.F.; Ramasamy, R.; Schmidt, A.M. The RAGE axis: A fundamental mechanism signaling danger to the vulnerable vasculature. *Circ. Res.* **2010**, *106*, 842–853. [CrossRef]
50. Choi, B.R.; Cho, W.H.; Kim, J.; Lee, H.J.; Chung, C.; Jeon, W.K.; Han, J.S. Increased expression of the receptor for advanced glycation end products in neurons and astrocytes in a triple transgenic mouse model of Alzheimer's disease. *Exp. Mol. Med.* **2014**, *46*, e75. [CrossRef]
51. Wautier, M.P.; Chappey, O.; Corda, S.; Stern, D.M.; Schmidt, A.M.; Wautier, J.L. Activation of NADPH oxidase by AGE links oxidant stress to altered gene expression via RAGE. *Am. J. Physiol. Endocrinol. Metab.* **2001**, *280*, E685–E694. [CrossRef]
52. Bierhaus, A.; Humpert, P.M.; Morcos, M.; Wendt, T.; Chavakis, T.; Arnold, B.; Stern, D.M.; Nawroth, P.P. Understanding RAGE, the receptor for advanced glycation end products. *J. Mol. Med.* **2005**, *83*, 876–886. [CrossRef] [PubMed]
53. Askarova, S.; Yang, X.; Sheng, W.; Sun, G.Y.; Lee, J.C. Role of Abeta-receptor for advanced glycation endproducts interaction in oxidative stress and cytosolic phospholipase A(2) activation in astrocytes and cerebral endothelial cells. *Neuroscience* **2011**, *199*, 375–385. [CrossRef] [PubMed]
54. Cooke, M.S.; Evans, M.D.; Dizdaroglu, M.; Lunec, J. Oxidative DNA damage: Mechanisms, mutation, and disease. *FASEB J.* **2003**, *17*, 1195–1214. [CrossRef] [PubMed]
55. Lovell, M.A.; Markesbery, W.R. Oxidative DNA damage in mild cognitive impairment and late-stage Alzheimer's disease. *Nucleic Acids Res.* **2007**, *35*, 7497–7504. [CrossRef] [PubMed]
56. Nunomura, A.; Perry, G.; Aliev, G.; Hirai, K.; Takeda, A.; Balraj, E.K.; Jones, P.K.; Ghanbari, H.; Wataya, T.; Shimohama, S.; et al. Oxidative damage is the earliest event in Alzheimer disease. *J. Neuropathol. Exp. Neurol.* **2001**, *60*, 759–767. [CrossRef]
57. Mecocci, P.; Polidori, M.C.; Cherubini, A.; Ingegni, T.; Mattioli, P.; Catani, M.; Rinaldi, P.; Cecchetti, R.; Stahl, W.; Senin, U.; et al. Lymphocyte oxidative DNA damage and plasma antioxidants in Alzheimer disease. *Arch. Neurol.* **2002**, *59*, 794–798. [CrossRef]
58. Migliore, L.; Fontana, I.; Trippi, F.; Colognato, R.; Coppede, F.; Tognoni, G.; Nucciarone, B.; Siciliano, G. Oxidative DNA damage in peripheral leukocytes of mild cognitive impairment and AD patients. *Neurobiol. Aging* **2005**, *26*, 567–573. [CrossRef]
59. Moslemnezhad, A.; Mahjoub, S.; Moghadasi, M. Altered plasma marker of oxidative DNA damage and total antioxidant capacity in patients with Alzheimer's disease. *Caspian J. Intern. Med.* **2016**, *7*, 88–92.
60. Isobe, C.; Abe, T.; Terayama, Y. Levels of reduced and oxidized coenzyme Q-10 and 8-hydroxy-2'-deoxyguanosine in the CSF of patients with Alzheimer's disease demonstrate that mitochondrial oxidative damage and/or oxidative DNA damage contributes to the neurodegenerative process. *J. Neurol.* **2010**, *257*, 399–404. [CrossRef]
61. Wojsiat, J.; Zoltowska, K.M.; Laskowska-Kaszub, K.; Wojda, U. Oxidant/Antioxidant Imbalance in Alzheimer's Disease: Therapeutic and Diagnostic Prospects. *Oxid. Med. Cell Longev.* **2018**, *2018*, 6435861. [CrossRef]
62. Ansari, M.A.; Scheff, S.W. Oxidative stress in the progression of Alzheimer disease in the frontal cortex. *J. Neuropathol. Exp. Neurol.* **2010**, *69*, 155–167. [CrossRef]
63. Mandal, P.K.; Saharan, S.; Tripathi, M.; Murari, G. Brain glutathione levels–a novel biomarker for mild cognitive impairment and Alzheimer's disease. *Biol. Psychiatry* **2015**, *78*, 702–710. [CrossRef]
64. Bayer, T.A.; Schafer, S.; Simons, A.; Kemmling, A.; Kamer, T.; Tepest, R.; Eckert, A.; Schussel, K.; Eikenberg, O.; Sturchler-Pierrat, C.; et al. Dietary Cu stabilizes brain superoxide dismutase 1 activity and reduces amyloid Abeta production in APP23 transgenic mice. *Proc. Natl. Acad. Sci. USA* **2003**, *100*, 14187–14192. [CrossRef] [PubMed]
65. Suzuki, T.; Yamamoto, M. Molecular basis of the Keap1-Nrf2 system. *Free Radic. Biol. Med.* **2015**, *88*, 93–100. [CrossRef] [PubMed]
66. Ramsey, C.P.; Glass, C.A.; Montgomery, M.B.; Lindl, K.A.; Ritson, G.P.; Chia, L.A.; Hamilton, R.L.; Chu, C.T.; Jordan-Sciutto, K.L. Expression of Nrf2 in neurodegenerative diseases. *J. Neuropathol. Exp. Neurol.* **2007**, *66*, 75–85. [CrossRef] [PubMed]

67. Uruno, A.; Matsumaru, D.; Ryoke, R.; Saito, R.; Kadoguchi, S.; Saigusa, D.; Saito, T.; Saido, T.C.; Kawashima, R.; Yamamoto, M. Nrf2 Suppresses Oxidative Stress and Inflammation in App Knock-In Alzheimer's Disease Model Mice. *Mol. Cell Biol.* **2020**, *40*, e00467-00419. [CrossRef]
68. de la Torre, J.C.; Mussivand, T. Can disturbed brain microcirculation cause Alzheimer's disease? *Neurol. Res.* **1993**, *15*, 146–153. [CrossRef]
69. Miyakawa, T. Vascular pathology in Alzheimer's disease. *Psychogeriatrics* **2010**, *10*, 39–44. [CrossRef]
70. Owen, J.B.; Sultana, R.; Aluise, C.D.; Erickson, M.A.; Price, T.O.; Bu, G.; Banks, W.A.; Butterfield, D.A. Oxidative modification to LDL receptor-related protein 1 in hippocampus from subjects with Alzheimer disease: Implications for Abeta accumulation in AD brain. *Free Radic. Biol. Med.* **2010**, *49*, 1798–1803. [CrossRef]
71. Deane, R.J. Is RAGE still a therapeutic target for Alzheimer's disease? *Future Med. Chem.* **2012**, *4*, 915–925. [CrossRef]
72. Wang, H.; Chen, F.; Du, Y.F.; Long, Y.; Reed, M.N.; Hu, M.; Suppiramaniam, V.; Hong, H.; Tang, S.S. Targeted inhibition of RAGE reduces amyloid-beta influx across the blood-brain barrier and improves cognitive deficits in db/db mice. *Neuropharmacology* **2018**, *131*, 143–153. [CrossRef] [PubMed]
73. Cai, Z.; Zhao, B.; Ratka, A. Oxidative stress and beta-amyloid protein in Alzheimer's disease. *Neuromol. Med.* **2011**, *13*, 223–250. [CrossRef] [PubMed]
74. Akiyama, H.; Barger, S.; Barnum, S.; Bradt, B.; Bauer, J.; Cole, G.M.; Cooper, N.R.; Eikelenboom, P.; Emmerling, M.; Fiebich, B.L.; et al. Inflammation and Alzheimer's disease. *Neurobiol. Aging* **2000**, *21*, 383–421. [CrossRef]
75. Kettenmann, H.; Hanisch, U.K.; Noda, M.; Verkhratsky, A. Physiology of microglia. *Physiol. Rev.* **2011**, *91*, 461–553. [CrossRef]
76. Wilkinson, B.L.; Landreth, G.E. The microglial NADPH oxidase complex as a source of oxidative stress in Alzheimer's disease. *J. Neuroinflamm.* **2006**, *3*, 30. [CrossRef]
77. Dheen, S.T.; Kaur, C.; Ling, E.A. Microglial activation and its implications in the brain diseases. *Curr. Med. Chem.* **2007**, *14*, 1189–1197. [CrossRef]
78. Simard, A.R.; Soulet, D.; Gowing, G.; Julien, J.P.; Rivest, S. Bone marrow-derived microglia play a critical role in restricting senile plaque formation in Alzheimer's disease. *Neuron* **2006**, *49*, 489–502. [CrossRef]
79. Cagnin, A.; Brooks, D.J.; Kennedy, A.M.; Gunn, R.N.; Myers, R.; Turkheimer, F.E.; Jones, T.; Banati, R.B. In-vivo measurement of activated microglia in dementia. *Lancet* **2001**, *358*, 461–467. [CrossRef]
80. Zhao, J.; O'Connor, T.; Vassar, R. The contribution of activated astrocy to Aβ production: Implications for Alzheimer's disease pathogenesis. *J. Neuroinflamm.* **2011**, *8*, 150. [CrossRef]
81. Heneka, M.T.; Carson, M.J.; El Khoury, J.; Landreth, G.E.; Brosseron, F.; Feinstein, D.L.; Jacobs, A.H.; Wyss-Coray, T.; Vitorica, J.; Ransohoff, R.M.; et al. Neuroinflammation in Alzheimer's disease. *Lancet Neurol.* **2015**, *14*, 388–405. [CrossRef]
82. Apelt, J.; Schliebs, R. β-Amyloid-induced glial expression of both pro- and anti-inflammatory cytokines in cerebral cortex of aged transgenic Tg2576 mice with Alzheimer plaque pathology. *Brain Res.* **2001**, *894*, 21–30. [CrossRef]
83. Tokuchi, R.; Hishikawa, N.; Sato, K.; Hatanaka, N.; Fukui, Y.; Takemoto, M.; Ohta, Y.; Yamashita, T.; Abe, K. Age-dependent cognitive and affective differences in Alzheimer's and Parkinson's diseases in relation to MRI findings. *J. Neurol. Sci.* **2016**, *365*, 3–8. [CrossRef] [PubMed]
84. Provenzano, F.A.; Muraskin, J.; Tosto, G.; Narkhede, A.; Wasserman, B.T.; Griffith, E.Y.; Guzman, V.A.; Meier, I.B.; Zimmerman, M.E.; Brickman, A.M. White matter hyperintensities and cerebral amyloidosis: Necessary and sufficient for clinical expression of Alzheimer disease? *JAMA Neurol.* **2013**, *70*, 455–461. [CrossRef] [PubMed]
85. Feng, T.; Yamashita, T.; Shang, J.; Shi, X.; Nakano, Y.; Morihara, R.; Tsunoda, K.; Nomura, E.; Sasaki, R.; Tadokoro, K.; et al. Clinical and Pathological Benefits of Edaravone for Alzheimer's Disease with Chronic Cerebral Hypoperfusion in a Novel Mouse Model. *J. Alzheimers Dis.* **2019**, *71*, 327–339. [CrossRef]
86. Feng, T.; Yamashita, T.; Zhai, Y.; Shang, J.; Nakano, Y.; Morihara, R.; Fukui, Y.; Hishikawa, N.; Ohta, Y.; Abe, K. Chronic cerebral hypoperfusion accelerates Alzheimer's disease pathology with the change of mitochondrial fission and fusion proteins expression in a novel mouse model. *Brain Res.* **2018**, *1696*, 63–70. [CrossRef]

87. Shang, J.; Yamashita, T.; Tian, F.; Li, X.; Liu, X.; Shi, X.; Nakano, Y.; Tsunoda, K.; Nomura, E.; Sasaki, R.; et al. Chronic cerebral hypoperfusion alters amyloid-beta transport related proteins in the cortical blood vessels of Alzheimer's disease model mouse. *Brain Res.* **2019**, *1723*, 146379. [CrossRef]
88. Huang, Y.; Mucke, L. Alzheimer mechanisms and therapeutic strategies. *Cell* **2012**, *148*, 1204–1222. [CrossRef]
89. Aliev, G.; Obrenovich, M.E.; Reddy, V.P.; Shenk, J.C.; Moreira, P.I.; Nunomura, A.; Zhu, X.; Smith, M.A.; Perry, G. Antioxidant therapy in Alzheimer's disease: Theory and practice. *Mini Rev. Med. Chem.* **2008**, *8*, 1395–1406. [CrossRef]
90. Edaravone Acute Infarction Study Group. Effect of a novel free radical scavenger, edaravone (MCI-186), on acute brain infarction. Randomized, placebo-controlled, double-blind study at multicenters. *Cerebrovasc. Dis.* **2003**, *15*, 222–229. [CrossRef]
91. Abe, K.; Itoyama, Y.; Sobue, G.; Tsuji, S.; Aoki, M.; Doyu, M.; Hamada, C.; Kondo, K.; Yoneoka, T.; Akimoto, M.; et al. Confirmatory double-blind, parallel-group, placebo-controlled study of efficacy and safety of edaravone (MCI-186) in amyotrophic lateral sclerosis patients. *Amyotroph. Lateral Scler. Frontotemporal Degener.* **2014**, *15*, 610–617. [CrossRef]
92. Jiao, S.S.; Yao, X.Q.; Liu, Y.H.; Wang, Q.H.; Zeng, F.; Lu, J.J.; Liu, J.; Zhu, C.; Shen, L.L.; Liu, C.H.; et al. Edaravone alleviates Alzheimer's disease-type pathologies and cognitive deficits. *Proc. Natl. Acad. Sci. USA* **2015**, *112*, 5225–5230. [CrossRef] [PubMed]
93. Dumont, M.; Kipiani, K.; Yu, F.; Wille, E.; Katz, M.; Calingasan, N.Y.; Gouras, G.K.; Lin, M.T.; Beal, M.F. Coenzyme Q10 decreases amyloid pathology and improves behavior in a transgenic mouse model of Alzheimer's disease. *J. Alzheimers Dis.* **2011**, *27*, 211–223. [CrossRef] [PubMed]
94. Pappolla, M.; Bozner, P.; Soto, C.; Shao, H.; Robakis, N.K.; Zagorski, M.; Frangione, B.; Ghiso, J. Inhibition of Alzheimer beta-fibrillogenesis by melatonin. *J. Biol. Chem.* **1998**, *273*, 7185–7188. [CrossRef] [PubMed]
95. Poeggeler, B.; Miravalle, L.; Zagorski, M.G.; Wisniewski, T.; Chyan, Y.J.; Zhang, Y.; Shao, H.; Bryant-Thomas, T.; Vidal, R.; Frangione, B.; et al. Melatonin reverses the profibrillogenic activity of apolipoprotein E4 on the Alzheimer amyloid Abeta peptide. *Biochemistry* **2001**, *40*, 14995–15001. [CrossRef]
96. Matsubara, E.; Bryant-Thomas, T.; Pacheco Quinto, J.; Henry, T.L.; Poeggeler, B.; Herbert, D.; Cruz-Sanchez, F.; Chyan, Y.J.; Smith, M.A.; Perry, G.; et al. Melatonin increases survival and inhibits oxidative and amyloid pathology in a transgenic model of Alzheimer's disease. *J. Neurochem.* **2003**, *85*, 1101–1108. [CrossRef]
97. Massaad, C.A.; Washington, T.M.; Pautler, R.G.; Klann, E. Overexpression of SOD-2 reduces hippocampal superoxide and prevents memory deficits in a mouse model of Alzheimer's disease. *Proc. Natl. Acad. Sci. USA* **2009**, *106*, 13576–13581. [CrossRef]
98. Jia, X.; McNeill, G.; Avenell, A. Does taking vitamin, mineral and fatty acid supplements prevent cognitive decline? A systematic review of randomized controlled trials. *J. Hum. Nutr. Diet.* **2008**, *21*, 317–336. [CrossRef]
99. Forbes, S.C.; Holroyd-Leduc, J.M.; Poulin, M.J.; Hogan, D.B. Effect of Nutrients, Dietary Supplements and Vitamins on Cognition: A Systematic Review and Meta-Analysis of Randomized Controlled Trials. *Can. Geriatr. J.* **2015**, *18*, 231–245. [CrossRef]
100. Butler, M.; Nelson, V.A.; Davila, H.; Ratner, E.; Fink, H.A.; Hemmy, L.S.; McCarten, J.R.; Barclay, T.R.; Brasure, M.; Kane, R.L. Over-the-Counter Supplement Interventions to Prevent Cognitive Decline, Mild Cognitive Impairment, and Clinical Alzheimer-Type Dementia: A Systematic Review. *Ann. Intern. Med.* **2018**, *168*, 52–62. [CrossRef]
101. Petersen, R.C.; Thomas, R.G.; Grundman, M.; Bennett, D.; Doody, R.; Ferris, S.; Galasko, D.; Jin, S.; Kaye, J.; Levey, A.; et al. Vitamin E and donepezil for the treatment of mild cognitive impairment. *N. Engl. J. Med.* **2005**, *352*, 2379–2388. [CrossRef]
102. Lee, L.K.; Shahar, S.; Chin, A.V.; Yusoff, N.A. Docosahexaenoic acid-concentrated fish oil supplementation in subjects with mild cognitive impairment (MCI): A 12-month randomised, double-blind, placebo-controlled trial. *Psychopharmacology* **2013**, *225*, 605–612. [CrossRef] [PubMed]
103. van der Zwaluw, N.L.; Dhonukshe-Rutten, R.A.; van Wijngaarden, J.P.; Brouwer-Brolsma, E.M.; van de Rest, O.; In't Veld, P.H.; Enneman, A.W.; van Dijk, S.C.; Ham, A.C.; Swart, K.M.; et al. Results of 2-year vitamin B treatment on cognitive performance: Secondary data from an RCT. *Neurology* **2014**, *83*, 2158–2166. [CrossRef] [PubMed]
104. de Jager, C.A.; Oulhaj, A.; Jacoby, R.; Refsum, H.; Smith, A.D. Cognitive and clinical outcomes of homocysteine-lowering B-vitamin treatment in mild cognitive impairment: A randomized controlled trial. *Int. J. Geriatr. Psychiatry* **2012**, *27*, 592–600. [CrossRef]

105. Inufusa, H. Composition for Protection Against Cell-Damaging Effects. U.S. Patent US9089548B2, 28 November 2015.
106. Inufusa, H. Abstract 3490: Mechanism of action for antioxidant Twendee X on hepatic cancer cell line HepG2. *Cancer Res.* **2018**, *78*, 3490. [CrossRef]
107. Inufusa, H. Abstract 2812: Targeting reactive oxygen species in human cancer cell growth and metastasis in nude mouse. *Cancer Res.* **2016**, *76*, 2812. [CrossRef]
108. Liu, X.; Yamashita, T.; Shang, J.; Shi, X.; Morihara, R.; Huang, Y.; Sato, K.; Takemoto, M.; Hishikawa, N.; Ohta, Y.; et al. Clinical and Pathological Benefit of Twendee X in Alzheimer's Disease Transgenic Mice with Chronic Cerebral Hypoperfusion. *J. Stroke Cerebrovasc. Dis.* **2019**, *28*, 1993–2002. [CrossRef] [PubMed]
109. Kusaki, M.; Ohta, Y.; Inufusa, H.; Yamashita, T.; Morihara, R.; Nakano, Y.; Liu, X.; Shang, J.; Tian, F.; Fukui, Y.; et al. Neuroprotective Effects of a Novel Antioxidant Mixture Twendee X in Mouse Stroke Model. *J. Stroke Cerebrovasc. Dis.* **2017**, *26*, 1191–1196. [CrossRef] [PubMed]
110. Liu, X.; Yamashita, T.; Shang, J.; Shi, X.; Morihara, R.; Huang, Y.; Sato, K.; Takemoto, M.; Hishikawa, N.; Ohta, Y.; et al. Twendee X Ameliorates Phosphorylated Tau, alpha-Synuclein and Neurovascular Dysfunction in Alzheimer's Disease Transgenic Mice With Chronic Cerebral Hypoperfusion. *J. Stroke Cerebrovasc. Dis.* **2019**, *28*, 104310. [CrossRef]
111. Tadokoro, K.; Morihara, R.; Ohta, Y.; Hishikawa, N.; Kawano, S.; Sasaki, R.; Matsumoto, N.; Nomura, E.; Nakano, Y.; Takahashi, Y.; et al. Clinical Benefits of Antioxidative Supplement Twendee X for Mild Cognitive Impairment: A Multicenter, Randomized, Double-Blind, and Placebo-Controlled Prospective Interventional Study. *J. Alzheimers Dis.* **2019**, *71*, 1063–1069. [CrossRef]
112. Sen, C.K.; Khanna, S.; Roy, S. Tocotrienols: Vitamin E beyond tocopherols. *Life Sci.* **2006**, *78*, 2088–2098. [CrossRef]
113. Shahidi, F.; de Camargo, A.C. Tocopherols and Tocotrienols in Common and Emerging Dietary Sources: Occurrence, Applications, and Health Benefits. *Int. J. Mol. Sci.* **2016**, *17*, 1745. [CrossRef] [PubMed]
114. Packer, L.; Weber, S.U.; Rimbach, G. Molecular aspects of alpha-tocotrienol antioxidant action and cell signalling. *J. Nutr.* **2001**, *131*, 369S–373S. [CrossRef] [PubMed]
115. Khanna, S.; Patel, V.; Rink, C.; Roy, S.; Sen, C.K. Delivery of orally supplemented alpha-tocotrienol to vital organs of rats and tocopherol-transport protein deficient mice. *Free Radic. Biol. Med.* **2005**, *39*, 1310–1319. [CrossRef] [PubMed]
116. Mangialasche, F.; Xu, W.; Kivipelto, M.; Costanzi, E.; Ercolani, S.; Pigliautile, M.; Cecchetti, R.; Baglioni, M.; Simmons, A.; Soininen, H.; et al. Tocopherols and tocotrienols plasma levels are associated with cognitive impairment. *Neurobiol. Aging* **2012**, *33*, 2282–2290. [CrossRef]
117. Mangialasche, F.; Solomon, A.; Kareholt, I.; Hooshmand, B.; Cecchetti, R.; Fratiglioni, L.; Soininen, H.; Laatikainen, T.; Mecocci, P.; Kivipelto, M. Serum levels of vitamin E forms and risk of cognitive impairment in a Finnish cohort of older adults. *Exp. Gerontol.* **2013**, *48*, 1428–1435. [CrossRef]
118. Ahsan, H.; Ahad, A.; Iqbal, J.; Siddiqui, W.A. Pharmacological potential of tocotrienols: A review. *Nutr. Metab. (Lond.)* **2014**, *11*, 52. [CrossRef]
119. Qureshi, A.A.; Karpen, C.W.; Qureshi, N.; Papasian, C.J.; Morrison, D.C.; Folts, J.D. Tocotrienols-induced inhibition of platelet thrombus formation and platelet aggregation in stenosed canine coronary arteries. *Lipids Health Dis.* **2011**, *10*, 58. [CrossRef]
120. Khanna, S.; Roy, S.; Slivka, A.; Craft, T.K.; Chaki, S.; Rink, C.; Notestine, M.A.; DeVries, A.C.; Parinandi, N.L.; Sen, C.K. Neuroprotective properties of the natural vitamin E alpha-tocotrienol. *Stroke* **2005**, *36*, 2258–2264. [CrossRef]
121. Gopalan, Y.; Shuaib, I.L.; Magosso, E.; Ansari, M.A.; Abu Bakar, M.R.; Wong, J.W.; Khan, N.A.; Liong, W.C.; Sundram, K.; Ng, B.H.; et al. Clinical investigation of the protective effects of palm vitamin E tocotrienols on brain white matter. *Stroke* **2014**, *45*, 1422–1428. [CrossRef]
122. Ibrahim, N.F.; Yanagisawa, D.; Durani, L.W.; Hamezah, H.S.; Damanhuri, H.A.; Wan Ngah, W.Z.; Tsuji, M.; Kiuchi, Y.; Ono, K.; Tooyama, I. Tocotrienol-Rich Fraction Modulates Amyloid Pathology and Improves Cognitive Function in AβPP/PS1 Mice. *J. Alzheimers Dis.* **2017**, *55*, 597–612. [CrossRef]
123. Durani, L.W.; Hamezah, H.S.; Ibrahim, N.F.; Yanagisawa, D.; Nasaruddin, M.L.; Mori, M.; Azizan, K.A.; Damanhuri, H.A.; Makpol, S.; Wan Ngah, W.Z.; et al. Tocotrienol-Rich Fraction of Palm Oil Improves Behavioral Impairments and Regulates Metabolic Pathways in AbetaPP/PS1 Mice. *J. Alzheimers Dis.* **2018**, *64*, 249–267. [CrossRef] [PubMed]

124. Shang, J.; Yan, H.; Jiao, Y.; Ohta, Y.; Liu, X.; Li, X.; Morihara, R.; Nakano, Y.; Fukui, Y.; Shi, X.; et al. Therapeutic Effects of Pretreatment with Tocovid on Oxidative Stress in Postischemic Mice Brain. *J. Stroke Cerebrovasc. Dis.* **2018**, *27*, 2096–2105. [CrossRef] [PubMed]
125. Jiao, Y.; Shang, J.; Ohta, Y.; Yan, H.; Liu, X.; Li, X.; Morihara, R.; Nakano, Y.; Fukui, Y.; Shi, X.; et al. Neuroprotective Effects of Tocovid Pretreatment in a Mouse Stroke Model. *J. Stroke Cerebrovasc. Dis.* **2018**, *27*, 2166–2174. [CrossRef] [PubMed]
126. Fukui, Y.; Hishikawa, N.; Ichinose, J.; Sato, K.; Nakano, Y.; Morihara, R.; Ohta, Y.; Yamashita, T.; Abe, K. Different clinical effect of four antidementia drugs for Alzheimer's disease patients depending on white matter severity. *Geriatr. Gerontol. Int.* **2017**, *17*, 1991–1999. [CrossRef]
127. Nazarov, P.A.; Osterman, I.A.; Tokarchuk, A.V.; Karakozova, M.V.; Korshunova, G.A.; Lyamzaev, K.G.; Skulachev, M.V.; Kotova, E.A.; Skulachev, V.P.; Antonenko, Y.N. Mitochondria-targeted antioxidants as highly effective antibiotics. *Sci. Rep.* **2017**, *7*, 1394. [CrossRef]
128. Cenini, G.; Voos, W. Mitochondria as Potential Targets in Alzheimer Disease Therapy: An Update. *Front. Pharmacol.* **2019**, *10*, 902. [CrossRef]
129. Kolosova, N.G.; Tyumentsev, M.A.; Muraleva, N.A.; Kiseleva, E.; Vitovtov, A.O.; Stefanova, N.A. Antioxidant SkQ1 Alleviates Signs of Alzheimer's Disease-like Pathology in Old OXYS Rats by Reversing Mitochondrial Deterioration. *Curr. Alzheimer Res.* **2017**, *14*, 1283–1292. [CrossRef]
130. Stefanova, N.A.; Muraleva, N.A.; Skulachev, V.P.; Kolosova, N.G. Alzheimer's disease-like pathology in senescence-accelerated OXYS rats can be partially retarded with mitochondria-targeted antioxidant SkQ1. *J. Alzheimers Dis.* **2014**, *38*, 681–694. [CrossRef]
131. McManus, M.J.; Murphy, M.P.; Franklin, J.L. The mitochondria-targeted antioxidant MitoQ prevents loss of spatial memory retention and early neuropathology in a transgenic mouse model of Alzheimer's disease. *J. Neurosci.* **2011**, *31*, 15703–15715. [CrossRef]
132. Rossman, M.J.; Santos-Parker, J.R.; Steward, C.A.C.; Bispham, N.Z.; Cuevas, L.M.; Rosenberg, H.L.; Woodward, K.A.; Chonchol, M.; Gioscia-Ryan, R.A.; Murphy, M.P.; et al. Chronic Supplementation With a Mitochondrial Antioxidant (MitoQ) Improves Vascular Function in Healthy Older Adults. *Hypertension* **2018**, *71*, 1056–1063. [CrossRef]
133. Rahman, S.O.; Panda, B.P.; Parvez, S.; Kaundal, M.; Hussain, S.; Akhtar, M.; Najmi, A.K. Neuroprotective role of astaxanthin in hippocampal insulin resistance induced by Abeta peptides in animal model of Alzheimer's disease. *Biomed. Pharmacother.* **2019**, *110*, 47–58. [CrossRef] [PubMed]
134. Ito, N.; Saito, H.; Seki, S.; Ueda, F.; Asada, T. Effects of Composite Supplement Containing Astaxanthin and Sesamin on Cognitive Functions in People with Mild Cognitive Impairment: A Randomized, Double-Blind, Placebo-Controlled Trial. *J. Alzheimers Dis.* **2018**, *62*, 1767–1775. [CrossRef] [PubMed]
135. Amato, A.; Terzo, S.; Mule, F. Natural Compounds as Beneficial Antioxidant Agents in Neurodegenerative Disorders: A Focus on Alzheimer's Disease. *Antioxidants (Basel)* **2019**, *8*, 608. [CrossRef] [PubMed]
136. Hewlings, S.J.; Kalman, D.S. Curcumin: A Review of Its' Effects on Human Health. *Foods* **2017**, *6*, 92. [CrossRef]
137. Hishikawa, N.; Takahashi, Y.; Amakusa, Y.; Tanno, Y.; Tuji, Y.; Niwa, H.; Murakami, N.; Krishna, U.K. Effects of turmeric on Alzheimer's disease with behavioral and psychological symptoms of dementia. *Ayu* **2012**, *33*, 499–504. [CrossRef]
138. Maiti, P.; Paladugu, L.; Dunbar, G.L. Solid lipid curcumin particles provide greater anti-amyloid, anti-inflammatory and neuroprotective effects than curcumin in the 5xFAD mouse model of Alzheimer's disease. *BMC Neurosci.* **2018**, *19*, 7. [CrossRef]
139. Porquet, D.; Grinan-Ferre, C.; Ferrer, I.; Camins, A.; Sanfeliu, C.; Del Valle, J.; Pallas, M. Neuroprotective role of trans-resveratrol in a murine model of familial Alzheimer's disease. *J. Alzheimers Dis.* **2014**, *42*, 1209–1220. [CrossRef]
140. Turner, R.S.; Thomas, R.G.; Craft, S.; van Dyck, C.H.; Mintzer, J.; Reynolds, B.A.; Brewer, J.B.; Rissman, R.A.; Raman, R.; Aisen, P.S.; et al. A randomized, double-blind, placebo-controlled trial of resveratrol for Alzheimer disease. *Neurology* **2015**, *85*, 1383–1391. [CrossRef]

International Journal of
Molecular Sciences

MDPI

Review

APP Osaka Mutation in Familial Alzheimer's Disease—Its Discovery, Phenotypes, and Mechanism of Recessive Inheritance

Takami Tomiyama [1,*] and Hiroyuki Shimada [2]

1 Department of Translational Neuroscience, Osaka City University Graduate School of Medicine, Osaka 545-8585, Japan

2 Clinical Research Center for Dementia, Osaka City University Graduate School of Medicine, Osaka 545-8585, Japan; h.shimada@med.osaka-cu.ac.jp

* Correspondence: tomi@med.osaka-cu.ac.jp

Received: 2 February 2020; Accepted: 17 February 2020; Published: 19 February 2020

Abstract: Alzheimer's disease is believed to begin with synaptic dysfunction caused by soluble Aβ oligomers. When this oligomer hypothesis was proposed in 2002, there was no direct evidence that Aβ oligomers actually disrupt synaptic function to cause cognitive impairment in humans. In patient brains, both soluble and insoluble Aβ species always coexist, and therefore it is difficult to determine which pathologies are caused by Aβ oligomers and which are caused by amyloid fibrils. Thus, no validity of the oligomer hypothesis was available until the Osaka mutation was discovered. This mutation, which was found in a Japanese pedigree of familial Alzheimer's disease, is the deletion of codon 693 of APP gene, resulting in mutant Aβ lacking the 22nd glutamate. Only homozygous carriers suffer from dementia. In vitro studies revealed that this mutation has a very unique character that accelerates Aβ oligomerization but does not form amyloid fibrils. Model mice expressing this mutation demonstrated that all pathologies of Alzheimer's disease can be induced by Aβ oligomers alone. In this review, we describe the story behind the discovery of the Osaka mutation, summarize the mutant's phenotypes, and propose a mechanism of its recessive inheritance.

Keywords: APP mutation; recessive inheritance; familial Alzheimer's disease; Aβ oligomers; amyloid imaging

---

1. Introduction

Alzheimer's disease (AD) is believed to begin with synaptic dysfunction caused by soluble Aβ oligomers. This idea, the so-called oligomer hypothesis, was proposed in 2002 [1] and has been supported by growing amounts of evidence from animal and organotypic experiments [2–8]. However, there was no direct evidence that Aβ oligomers actually disrupt synaptic function to cause cognitive impairment in humans. In the brains of AD patients, both soluble and insoluble Aβ species always coexist from the disease onset, with the latter usually observed as senile plaques in the brain parenchyma. The existence of senile plaques, along with neurofibrillary tangles (NFTs) that are composed of hyperphosphorylated tau and brain atrophy that comes from neuron loss, is an absolute requirement for the diagnosis of AD. In such circumstances, the determination of the pathological roles of Aβ oligomers separately from those of amyloid fibrils is nearly impossible, and therefore no validity of the oligomer hypothesis was available. To prove this hypothesis in humans, we needed clinical cases that developed AD without forming senile plaques, even though the feature of no senile plaques is contradictory to the definition of AD. Aside from this discrepancy, if we could find such patients and if we could identify novel etiologic mutation(s) in those patients, we could create new disease models to investigate the pathological and physiological roles of Aβ oligomers at the molecular, cellular, and individual levels for basic and translational research. However, we never expected such clinical cases to exist.

The Arctic mutation (E693G) in amyloid precursor protein (APP), which was discovered in a Swedish family showing clinical symptoms of early-onset AD [9], has been shown to accelerate Aβ protofibril formation in vitro, a soluble intermediate oligomer of Aβ fibrillization, and intraneuronal Aβ aggregation in model mice [10,11]. In those mice, intracellular Aβ aggregation occurred concomitantly with cognitive impairment before the onset of amyloid plaques. These features imply that the Arctic mutation might be favorable to study the pathological effects of Aβ oligomers, but the patients displayed AD pathologies including severe congophilic angiopathy, region-specific NFTs, and abundant amyloid plaques [12,13]. Interestingly, those plaques had a characteristic ring-like architecture and were negative for Congo red staining [12,13] with a low retention in Pittsburgh compound-B (PIB)-PET scans [14].

Meanwhile, in 2001, we had an opportunity to come across a very unusual case of hereditary dementia. The proband (female) showed typical clinical symptoms of AD, but almost no signals of senile plaques in PIB-PET scans [15]. From her pedigree, we identified a novel APP mutation at the same position as the Arctic mutation. This 'Osaka' mutation (E693Δ) was found to have a very unique character that accelerates Aβ oligomerization but does not form amyloid fibrils. This was the very mutation we had eagerly sought, but its discovery was quite accidental.

2. Discovery of the Osaka Mutation

In 2001, a Japanese woman who recognized memory disturbance visited the Osaka City University Hospital. She was 57 years old at the time and was diagnosed with mild cognitive impairment (MCI). The doctor in charge (the second author of this review) noticed that her pedigree had many patients with dementia and suspected familial AD. The doctor asked the first author's group to conduct a genetic test. We examined the patient's APP, presenilin 1, and presenilin 2 genes, and in September 2002, we discovered a new mutation on APP. This mutation was the first deletion-type mutation found in APP; the codon GAA encoding the 693rd glutamate, which corresponds to the 22nd amino acid in the Aβ sequence, had disappeared [15] (Figure 1). Interestingly, the woman had this mutation in both APP alleles, i.e., she was homozygous (Figure 2A,B). She showed progressive cognitive decline and was diagnosed as having AD at the age of 59. Under informed consent, we carried out gene examination for other family members. One of her younger sisters was found to also have this mutation in both APP alleles, while her elder sister and another younger sister and her two daughters possessed this mutation only in one allele. Their cognitive function was normal at the time, but the homozygous sister soon developed dementia at the same age (59 years old), as did the proband. ApoE genotype was ε3/ε3 for all sisters including the proband, and ε3/ε4 for the two daughters. These findings suggest that the inheritance mode of this mutation is recessive, which was the first case in familial AD.

This mutation, which we named the Osaka mutation, is located within the Aβ sequence, and therefore we considered that this mutation might affect the production and/or aggregation of Aβ like other pathological APP mutations (https://www.alzforum.org/mutations/app). Thus, we initially studied the effects of this mutation on Aβ production in cultured cells. HEK293 cells were transfected with mutant APP, and Aβ secretion was compared with wild-type APP. The ratio of Aβ40 and Aβ42 was almost the same as wild-type, but Aβ secretion was nearly half the level of wild-type [15]. Subsequently, we tested Aβ aggregation and toxicity using synthetic peptides. In PBS solution, neither mutant Aβ40 nor Aβ42 peptides showed any increase in thioflavin T fluorescence, which reflects the level of β-sheet conformation, even after 7 days incubation [15]. The aged peptide solutions were subjected to electron microscopy, and no amyloid fibrils were observed. Under the same condition, wild-type Aβ peptides showed a rapid increase in thioflavin T fluorescence and abundant amyloid fibrils. When added to human and mouse neuronal cells (IMR-32 and Neuro2a, respectively), wild-type, but not mutant Aβ42 peptide, showed toxicity in the MTT assay [16]. Soon after the discovery of the Osaka mutation, we started to generate transgenic (Tg) mice expressing human APP695 with this mutation under the mouse prion protein promoter. In the next year (2003), three mutant lines with different levels of APP expression were established. The expression level of human APP was almost the same as that of endogenous mouse APP in line 1 and was lower in lines 2 and 3. We examined brain sections from adult mice of line 1 for amyloid deposition, but no plaque-like structures were observed [17]. These results, for example, lowered Aβ production and aggregation in vitro and no amyloid plaques in vivo, which appeared to be against our expectation that the Osaka mutation plays an etiological role in AD. Rather, this mutation seemed protective, although the mutant peptides were found more resistant to enzymatic degradation than wild-type peptides [15]. We began to suspect that the Osaka mutation might be just a harmless genetic polymorphism that happened to occur in the pedigree of familial dementia.

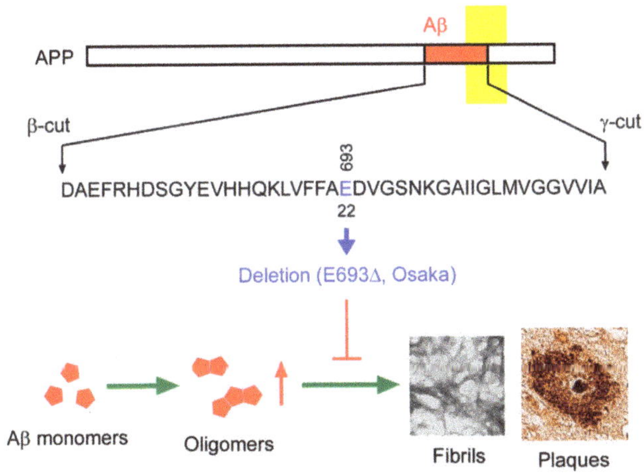

Figure 1. Identity of the Osaka mutation. The Osaka mutation is the deletion of codon 693 of APP gene that produces mutant Aβ lacking the 22nd glutamate. This mutation has a very unique character that accelerates Aβ oligomerization (red arrow) but does not form amyloid fibrils (T-shaped red line). Green arrows indicate Aβ aggregation process.

In the summer of 2003, when looking at the results of Western blots, we unexpectedly noticed that the mutant Aβ peptide formed more abundant oligomers than wild-type peptide [15,16]. These oligomers (dimers, trimers, and tetramers) appeared to be very stable and did not proceed into larger aggregates of amyloid fibrils. These findings that the mutant Aβ peptides tend to form oligomers but never fibrils seemed very unusual, because we thought that once Aβ starts to aggregate, the process does not stop until all monomers and oligomers are spent for fibril formation. In the same period, we studied the effects of the Osaka mutation on APP processing in transfected cells. This mutation suppressed neither β- nor γ-cut of APP, indicating no inhibition of Aβ production [18]. Nevertheless, Aβ secretion was decreased. Soon after, we found that mutant APP-expressing cells contained more abundant intracellular Aβ than wild-type APP-expressing cells [18]. In addition, the intracellular Aβ was accumulated as monomers, dimers, and perhaps trimers. This finding was later supported by immunostaining of the cells with Aβ oligomer-selective antibodies. If the Osaka mutation really does promote Aβ oligomerization, it could be a cause of AD by exacerbating oligomer-induced synaptic alteration regardless of the lack of amyloid plaques. In fact, when added to mouse hippocampal slice culture, the mutant Aβ42 peptide caused a decrease of synaptophysin, a marker of pre-synapses, more potently than wild-type peptide [16]. This was the first time we realized that the Osaka mutation could prove the oligomer hypothesis in humans.

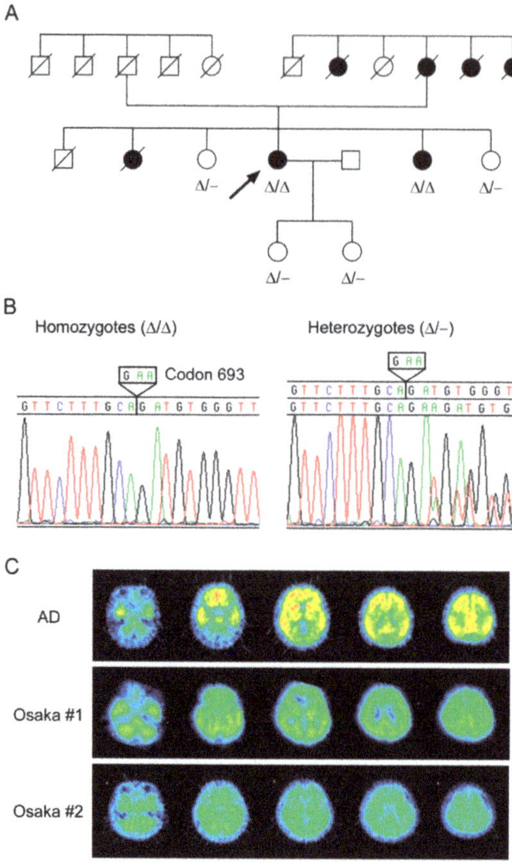

Figure 2. Recessive inheritance of the Osaka mutation. (**A**) Pedigree of familial Alzheimer's disease (AD) with the Osaka mutation. The proband is indicated with an arrow. Both of the affected members were homozygous (Δ/Δ) for the mutation, while unaffected family members were all heterozygous (Δ/-). Square, male; circle, female. (**B**) DNA sequence of the Osaka mutation. Homozygous and heterozygous deletion of codon 693 of APP gene were detected in the pedigree. (**C**) No amyloid plaques in the Osaka mutation. Pittsburgh compound-B (PIB-PET) images of AD (71-year-old woman), the proband (62-year-old woman), and her affected sister (61-year-old woman) are shown.

This notion led us to evaluate the toxic effects of mutant Aβ on synaptic plasticity by in vivo electrophysiology. Synthetic Aβ42 peptide was injected into rat cerebral ventricles at a dose of 10 ng. Ten minutes after the injection, high frequency stimulation (HFS) was delivered to the Schaffer collateral/commissural pathway, and field excitatory postsynaptic potential (fEPSP) was measured in the hippocampal CA1 region. Compared with wild-type Aβ, mutant Aβ suppressed long-term potentiation (LTP) more potently [15]. This result appeared to support our hypothesis that the Osaka mutation aggravates Aβ-induced synaptic dysfunction by accelerating its oligomerization.

From these results, we decided to answer a bigger question: Do patients with the Osaka mutation really have no senile plaques in their brains? In 2006, we commissioned the attending doctor (the second author), who was an expert on amyloid imaging, to perform PIB-PET analysis for the proband who was 62 years old at the time. Only slight signals were detected in the temporal, parietal, and occipital lobes and in the cerebellum, but not in the frontal lobe [15] (Figure 2C). The signals were far less than those in typical AD, indicating that there were almost no senile plaques in the patient. We next examined Aβ levels in her cerebrospinal fluid (CSF). The amounts of Aβ40 and Aβ42 and the ratio of Aβ42/Aβ40 were considerably lower than those in control patients with AD or other neurological disorders, whereas the ratio of Aβ oligomers to monomers was markedly higher than those in control patients [15]. The low levels of Aβ40 and Aβ42 in the CSF presumably reflect their accumulation in the brain parenchyma, and the high ratio of oligomers to monomers confirms the oligomer-prone aggregation property of mutant Aβ in vivo. Based on these results, we concluded that the Osaka mutation causes AD by accelerating the formation of synaptotoxic Aβ oligomers without forming amyloid plaques.

In the fall of 2007, we presented our findings at domestic and international conferences without disclosing the identity of this mutation. Our discovery of an oligomer-prone mutation garnered great interest from the audience (https://www.alzforum.org/news/conference-coverage/san-diego-oligomers-live-bad-reputation-part-2). In addition, our claim that AD develops by Aβ oligomers alone and that senile plaques are not necessary for disease onset was controversial, because it opposed the definition of AD, in that the existence of senile plaques is an absolute requirement for the disease. In the next year, the first report of the Osaka mutation was finally published [15] after it had been rejected by several journals, partly due to the lack of autopsy data and the small number of patients. However, since its publication, the Osaka mutation has been appreciated as the first evidence showing that the mechanism of the oligomer hypothesis actually applies to humans.

3. Phenotypes of the Osaka Mutation

3.1. Patients

Two pedigrees of familial AD with the Osaka mutation have so far been reported. One is from Osaka City University, Osaka, Japan [15,19], and the other is from Kawasaki Medical School, Okayama, Japan [20]. Probably both pedigrees originated from the same island in the Inland Sea of Japan. Two patients in the first pedigree (Osaka) and three patients in the second pedigree (Okayama) have been described. They were all homozygous for the Osaka mutation, while their unaffected family members were all heterozygous. In the first report of the Osaka mutation, we screened 5310 Japanese people recruited for the Japanese Genetic Study Consortium for AD for this mutation and found three independent additional carriers [15]. One was homozygous and had AD at 36 years old. The other two were heterozygous; one had MCI at 81 years old and the other was cognitively normal at 64 years old. These results clearly indicate that this mutation is recessive. In PIB-PET scans, all patients tested in the two pedigrees were almost negative, supporting our conclusion that this mutation causes disease without forming amyloid plaques.

The proband of the first pedigree experienced memory disturbance at 55 years old [15,19]. Her MMSE scores and MRI and fluorodeoxyglucose (FDG)-PET images were rather normal at 57 years old. However, three-dimensional stereotactic surface projection (3D-SSP) analysis showed a hypometabolism of FDG in the posterior cingulate cortex, which is similar to typical AD. At 59 years old, her MMSE score decreased to 22. According to the Diagnostic and Manual Statistical of Mental Disorders, Third Edition (DSM-IIIR), and the criteria of the National Institute of Neurological and Communication Disorders-Alzheimer's Disease and Related Disorders Association (NINC DS-ADRDA), she was diagnosed with AD. Two years later, at 61 years old, her MMSE score decreased to 18. She developed cerebellar ataxia, gait disturbance, ideomotor apraxia, and pyramidal signs. These symptoms are unusual in AD. At 62 years old, her MMSE score further decreased to 5. Nevertheless, MRI scans displayed only mild parietal lobe atrophy. Furthermore, [^{11}C]PIB-PET scans revealed almost no amyloid accumulation (Figure 2C). FDG-PET scans showed decreased glucose metabolism throughout the brain except the motor and sensory cortices and the cerebellum. At 63 years old, she became unable to walk. At 65 years old, MRI scans revealed moderate atrophy of the hippocampus, mild atrophy of the cerebrum, and mild dilatation of the third and lateral ventricles. The levels of Aβ42 and Aβ40 in her CSF were 4.1 and 242.8 pg/mL, respectively, and those of total tau and phosphorylated tau were 628 and 87.2 pg/mL, respectively. These values alone suggest that she was in an advanced phase of AD. We decided to perform tau imaging analysis for her at 70 years old with the [^{11}C]PBB3 probe. PET scans revealed massive NFT formation in the cerebellum and cerebral cortex (Shimada et al. unpublished observations). In addition, MRI scans revealed severe brain atrophy. These observations demonstrate that the pathological cascade of AD proceeds even in the absence of amyloid plaques.

The second patient in the first pedigree, the younger sister of the proband, experienced memory disturbance at 59 years old [19]. Her MMSE score was 27 at the time. Her memory gradually worsened, and at 61 years old, her MMSE score was 15. Similar to the proband, she exhibited cerebellar ataxia, gait disturbance, and pyramidal signs at this age. MRI scans displayed only mild cortical atrophy. PIB-PET scans revealed no amyloid accumulation (Figure 2C). The levels of Aβ42 and Aβ40 in her CSF were 9.5 and 285.2 pg/mL, respectively, and those of total tau and phosphorylated tau were 856 and 152 pg/mL, respectively. Again, these values alone suggest that she was in an advanced phase of AD. At 62 years old, she also became unable to walk.

Taken together, these two patients in the same pedigree developed very similar symptoms and showed a similar progression of the disease. In the early stages, they exhibited only memory disturbance without evident brain atrophy or evident amyloid deposition. In the late stages, they displayed mild to moderate brain atrophy and unexpected motor dysfunction. These clinical features are unique and distinctive from typical AD. To elucidate the underlying mechanism, neuropathological examination by brain autopsy is required for these patients. However, both patients are already deceased at 70 and 66 years old, respectively, and unfortunately, we could not obtain autopsy samples from them.

Compared with the first pedigree, the proband (female) in the second pedigree showed an earlier onset of dementia. According to one report [20], she experienced memory disturbance at 35 years old. She was diagnosed with AD at 42 years old based on progressive cognitive impairment and prominent spatial disorientation. At 48 years old, she had difficulty walking and became bedridden by 50 years old. She had spastic paraparesis and mild dysphagia at 56 years old. The serum level of Aβ and CSF levels of total tau and phosphorylated tau were normal. MRI scans showed remarkable brain atrophy, and FDG-PET scans displayed a greater reduction in glucose uptake in the cerebral cortex compared with typical AD. Nevertheless, PIB-PET scans revealed no amyloid deposition at 56 years old. Using Western blots, the authors detected high-molecular weight Aβ oligomers in her CSF under the non-denaturing condition, while the total level of Aβ under the denaturing condition was less than control level. Her elder brother and sister had memory disturbance at 59 and 44 years old, respectively. They also had difficulty walking due to spasticity of the lower limbs at 66 and 58 years old, respectively. Thus, similar to the first pedigree, the second pedigree also exhibited motor dysfunction, which is unusual in AD. The authors stated that compared with the first pedigree, motor impairment was more profound and brain atrophy was more severe in the second pedigree.

3.2. Animal Models

As mentioned above, we established Tg mice expressing Osaka-mutant APP in 2003. This model, which we named APP_{OSK} mice, did not show any amyloid plaques in our first examination. However, because we had learned the unique character of the Osaka mutation, we re-examined brain sections of the mice. Using Aβ oligomer-selective antibodies [21,22], we found that APP_{OSK} mice start to accumulate Aβ oligomers within neurons in the hippocampus, cerebral cortex, and cerebellum at 8 months [17] (Figure 3A). This accumulation was age-dependent, but amyloid plaque deposition was not detected even at 24 months. In accordance with Aβ accumulation, the level of synaptophysin in the hippocampus began to decrease at 8 months. We measured the synaptic plasticity of 8-month-old APP_{OSK} mice by in vivo electrophysiology in comparison with wild-type APP Tg mice (referred to as APP_{WT} mice) [23], which show almost the same level of APP expression as APP_{OSK} mice. HFS was delivered to the perforant path, and population spikes were recorded in the granular cell body layer of the dentate gyrus. Compared with non-Tg mice, basal synaptic transmission was not affected in either APP_{WT} or APP_{OSK} mice [17]. On the other hand, paired-pulse facilitation (PPF), a measure of short-term synaptic plasticity, and LTP were significantly inhibited in both Tg mice, with more severe deterioration in APP_{OSK} mice. Then we examined the cognitive function of APP_{OSK} mice at 8 months by the Morris water maze test. Compared with non-Tg mice, APP_{WT} mice showed only slight impairment in memory, but the memory of APP_{OSK} mice was markedly disturbed [17]. These results were in agreement with the hypothesis that Aβ oligomers cause early synaptic pathology in AD.

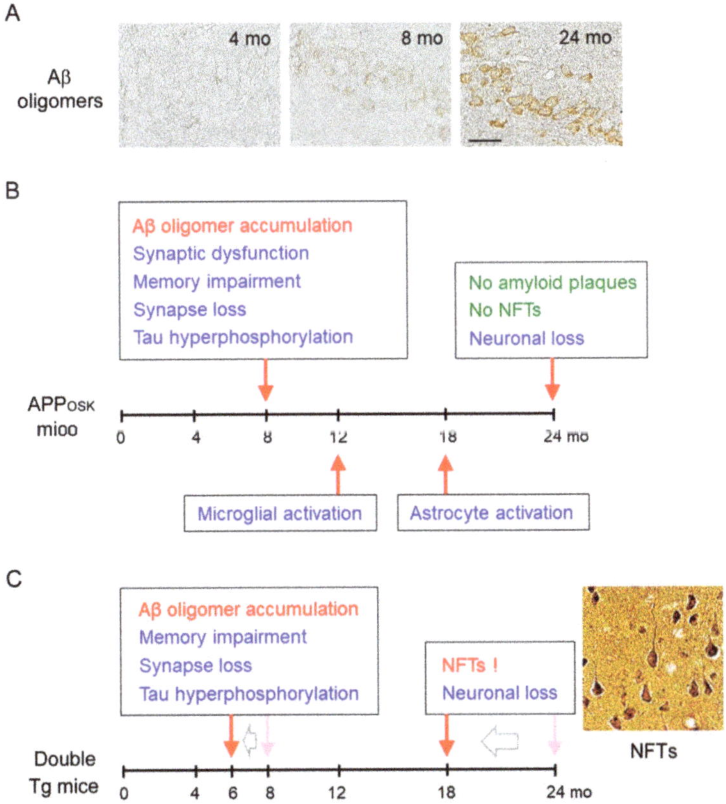

Figure 3. Phenotypes of APP_{OSK} mice. **(A)** Intraneuronal accumulation of Aβ oligomers in APP_{OSK} mice. The mice started to accumulate intracellular Aβ oligomers at 8 months. Images of the hippocampus are shown. Scale bar = 30 μm. **(B)** Aβ oligomers alone can induce most AD pathologies except for neurofibrillary tangles (NFT) formation. **(C)** The double Tg mice expressing both the Osaka mutation and wild-type human tau displayed NFTs at 18 months. Intracellular Aβ oligomers and tau might interact to accelerate each other's pathologies.

In AD brain, not only synaptic alterations but also many other pathologies are induced in parallel with Aβ accumulation, including tau hyperphosphorylation, NFT formation, glial activation, and neuron loss. It had been unclear which pathologies are attributable to Aβ oligomers and which are not. To address this question, we examined brain sections of APP_{OSK} mice for these pathologies [17]. Tau hyperphosphorylation occurred at 8 months in the hippocampal mossy fibers and at 12 months in the cerebral cortex, but NFTs were not observed even at 24 months. Microglial activation was first detected at 12 months, while astrocyte activation was at 18 months in both the hippocampus and cerebral cortex. Finally, neuron loss was observed at 24 months in the pyramidal cell layer of the hippocampal CA3 region, but not in the cerebral cortex at that age. These results clearly indicate that Aβ oligomers alone can induce most AD pathologies except for NFT formation (Figure 3B).

We speculated that NFT formation did not occur in APP_{OSK} mice because the mice express only mouse tau, which is different from human tau in amino acid sequence and isoform expression at adult age. If those mice expressed human tau with the same pattern of isoform expression, NFTs might be formed by the action of Aβ oligomers. Thus, we crossed APP_{OSK} mice with wild-type tau Tg mice. The latter were generated in our laboratory so that they express the longest 3-repeat and 4-repeat human tau isoforms under the mouse calcium/calmodulin-dependent kinase IIa promoter and were found not to show any pathologies, even at 24 months [24]. On the other hand, the double Tg mice displayed NFTs at 18 months in the hippocampus and cerebral cortex [25]. Interestingly, Aβ oligomer accumulation, tau hyperphosphorylation, synapse loss, and memory impairment were all accelerated in double Tg mice beginning at 6 months, and neuron loss was also started at 18 months. These findings demonstrate that Aβ oligomers can initiate the pathological cascade of AD including NFT formation and that intracellular Aβ oligomers and tau presumably interact to accelerate each other's pathologies (Figure 3C).

Our finding that Tg mice expressing the Osaka mutation exhibit an intraneuronal accumulation of Aβ oligomers but not extracellular deposition of amyloid plaques was supported by another research group. Kulic et al. [26] created a new mouse model that expresses human APP695 with both the Osaka and Swedish (K670N/M671L) mutations under the same mouse prion protein promoter as APP_{OSK} mice. These double mutant Tg mice displayed an intraneuronal accumulation of Aβ oligomers at 3 months, but no amyloid plaque deposition up to 15 months. Instead, the mice exhibited vascular amyloid deposits in the leptomeningeal cerebellar and cortical vessels at 24 months. These deposits were positive for thioflavin S and Congo red staining, indicating the existence of fibrillar aggregates of Aβ. In our APP_{OSK} mice, intracellular Aβ oligomers were only faintly thioflavin S-positive [17]. Nevertheless, they were mostly collected in SDS-insoluble fractions, implying that Osaka-mutant Aβ oligomers may possess insoluble fibrillar nature with the β-sheet conformation, as shown in an in vitro study [27].

Several groups have observed the opposite phenomena to our findings in experiments using synthetic peptides, showing that their own Osaka-mutant peptides promptly aggregated into amyloid fibrils [28–35]. However, we have often experienced that the aggregation property of synthetic peptides show remarkable lot-to-lot variations and that peptide aggregation largely depends on the experimental conditions, including the environment surrounding the peptides. In vitro aggregation studies are usually performed in test tubes only containing the peptides in a pure solution like PBS, which is completely different from the circumstances in the brain. We therefore consider that in vitro results using synthetic peptides are less reliable than in vivo observations in patients and animal models. Our initial in vitro finding that Osaka-mutant peptides form abundant oligomers but no amyloid fibrils have been confirmed in mouse models and patients [15,17,19,20].

There is an interesting animal model to study the pathological roles of Aβ oligomers [36]. A comparison of Aβ sequences of several animal species shows that chimpanzee and dog have the identical sequence as humans, while cat sequence differs by 1 amino acid (D7E) and mice sequence differs by 3 amino acids (R5G, Y10F, and H13R). Chimpanzee and dog display amyloid plaques in their brains, whereas cats accumulate only intraneuronal Aβ oligomers without amyloid plaques, and mice have no amyloid pathology. On the other hand, the amino acid sequence of tau is identical to human only in chimpanzees, and has homologies of 92% in dog, 93% in cat, and 89% in mice. All species but mice express 3-repeat and 4-repeat tau isoforms at an adult age like humans, whereas a mouse expresses only 4-repeat tau isoforms. Nonetheless, NFTs and neurodegeneration were observed only in aged cats. These results suggest that not amyloid plaques, but intracellular Aβ oligomers, can initiate the pathological cascade of AD, leading to NFT formation and neuron death. This finding appears to support our notion that Aβ oligomers can induce NFTs.

3.3. Cellular Models

Cell culture system has long been used to investigate pathological and physiological roles of Aβ. Using this system, several mechanisms for the toxicity of extracellular Aβ oligomers have been proposed. Aβ oligomers are shown to bind multiple cell surface receptors, including N-methyl-D-aspartate receptor (NMDAR), insulin receptor, cellular prion protein, and others, and disturb their normal function (reviewed in [8,37–39]). Aβ oligomers also bind the plasma membrane directly to form ion channel-like amyloid pores and disrupt cellular calcium homeostasis [37]. These mechanisms account for Aβ oligomer-induced synaptic dysfunction and neurodegeneration and implicate therapeutic strategies for AD.

On the other hand, in the Osaka mutation, Aβ oligomers were detected predominantly within neurons. Additionally, their subcellular localization and influence on cellular functions had remained unclear. To address these questions, we generated cellular models by transfecting COS-7 cells with mutant APP. Cells were double stained with Aβ42 or Aβ oligomers and organelle markers for ER (calnexin), Golgi apparatus (furin), early endosomes (EEA1), late endosomes (M6PR), lysosomes (LAMP2), and autophagosomes (LC3). Intracellular Aβ42 was detected in all organelles tested with preferential localization in late endosomes and found to predominantly form oligomers [18]. It is known that misfolded proteins in the ER often cause ER stress and apoptosis. We found ER stress markers, such as the molecular chaperone Grp78 and phosphorylated (i.e., down-regulated) translation initiation factor eIF2a, and apoptosis markers, including cleaved caspase-3 and -4 and DNA fragmentation, which was detected by the TUNEL method, were all positive in mutant APP-transfected cells [18]. We also detected in these cells an increase of the E3 ubiquitin ligase HRD1, a marker of ER-associated degradation, which is another cellular response to ER stress, and the cell surface expression of annexin V, an indicator of early apoptosis, together with positive signals for propidium iodide, a probe for the loss of membrane integrity in late apoptosis and necrosis [40]. Furthermore, cells expressing mutant APP displayed endosomal/lysosomal membrane damage, which was demonstrated by the leakage of a pinocytic tracer, Lucifer yellow, and a lysosomal enzyme, cathepsin D, from these organelles into the cytoplasm [40]. This might be caused by Aβ oligomer-derived amyloid pore formation in the endosomal/lysosomal membrane. In addition to the above organelle, intracellular Aβ oligomers also localized into the mitochondria with a marker for Tom20, and caused mitochondrial dysfunction, which was indicated by the aberrant fluorescence of JC-1, a reporter dye of the mitochondrial membrane potential, and the release of cytochrome c from the mitochondria into the cytoplasm, which is a sign of mitochondria-dependent apoptosis [40] (Figure 4A). These results collectively indicate that intracellular Aβ oligomers cause ER stress, endosomal/lysosomal damage, and mitochondrial dysfunction, all of which eventually lead to apoptosis (Figure 4B).

Following these findings, we examined whether these pathological events actually occur in the neurons of the Tg mouse brain. In 18-month-old APP_{OSK} mice, ER stress, lysosomal leakage, mitochondrial dysfunction, and apoptosis were shown to correlate with Aβ oligomer accumulation [40]. Our finding that the Osaka mutation induces the intracellular accumulation of Aβ oligomers followed by ER stress was confirmed by others using iPS cells derived from a patient with the Osaka mutation and from one sporadic AD patient [41]. Based on their results, the authors proposed that AD could be classified into two categories: extracellular Aβ type and intracellular Aβ type [41]. This is an interesting idea. However, in the brain of patients and model mice, the intracellular accumulation of Aβ is always followed by its extracellular deposition and these two events do not occur independently [11,42–47]. In some cases of AD, including the Osaka mutation, the excretion of Aβ aggregates from cells into the extracellular space might be impaired, resulting in the prolonged intracellular Aβ accumulation.

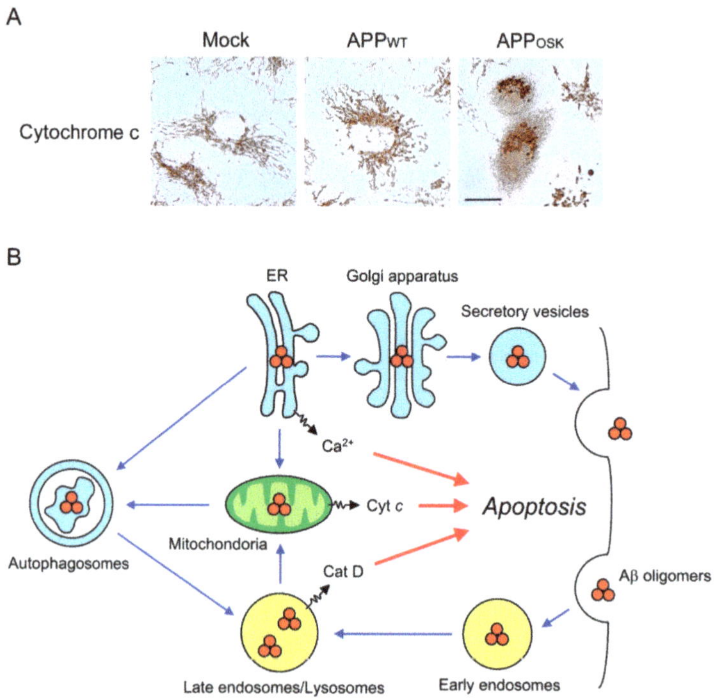

Figure 4. Cytotoxic effects of the Osaka mutation. (**A**) Disruption of mitochondria network by the Osaka mutation. Cytochrome c release from the mitochondria into the cytoplasm was detected in cells expressing the Osaka mutation. Scale bar = 20 μm. (**B**) Proposed mechanism underlying intracellular Aβ oligomer-induced cell death. Accumulation of Aβ oligomers caused ER stress, endosomal/lysosomal damage, and mitochondrial dysfunction, all of which eventually lead to apoptosis (red arrow). Blue arrows indicate the intracellular transport pathways of Aβ oligomers.

The above findings may explain the mechanism by which intracellular Aβ oligomers cause cell death. However, synaptic dysfunction by intracellular Aβ occurs long before neurodegeneration in Tg mice. In vitro studies have shown that extracellular Aβ oligomers alter synaptic spine density and morphology and disrupt axonal transport [48–52] and that these toxic effects are mediated by cell surface receptors, such as NMDAR [49,50], an ephrine receptor EphA4 [39], and a Nogo receptor NgR1 [53]. On the other hand, the effects of intracellular Aβ oligomers on synaptic spines and axonal transport were unclear. Thus, we transfected mouse or rat primary neurons with mutant and wild-type human APP and compared their effects. Initially, we confirmed that Aβ oligomers accumulated in neurons expressing Osaka-mutant APP, but not in those expressing wild-type APP [54]. The Osaka mutation induced a significant decrease in dendritic spine density with a dominant decrease of mushroom-type mature spine [54]. This toxic effect was tau-independent and unlikely to be caused by extracellular Aβ. Furthermore, this mutation disrupted the axonal and dendritic transport of BDNF, which was also tau-independent, and mitochondria. Spine formation and growth requires membrane trafficking mediated by recycling endosomes. The Osaka mutation was shown to also inhibit dendritic transport of recycling endosomes. These results could account for the synaptic alterations in the Osaka mutation.

The secretion of Aβ from cells was remarkably suppressed by the presence of the Osaka mutation. This may occur because the protein quality control system in the ER recognizes Osaka-mutant Aβ oligomers as a misfolded protein and thereby inhibits their transport from the ER to the cell surface.

This begs the question: Is Aβ secretion necessary for neural function? In other words, what is the physiological role of Aβ secretion? We had proposed that Aβ may play a role in cholesterol efflux from cells and from the brain [55]. During its secretion, Aβ assembled lipoprotein-like particles with cellular excess cholesterol, an assembly mediated by the ATP-binding cassette transporter A1. Soon after the secretion, the particles may be fused with apoE-containing nascent high-density lipoprotein (HDL), forming mature HDL. If this hypothesis is true, the Osaka mutation may have a defect in cholesterol efflux and thereby induce the intracellular accumulation of cholesterol. This would be unfavorable for the cells because, for example, cholesterol accumulation in mitochondria has been shown to increase the susceptibility of neurons to Aβ-induced reactive oxygen species (ROS) generation [56]. To test this possibility, we transfected COS-7 cells with wild-type and mutant APP and compared their cholesterol content after cholesterol loading. It is known that after its internalization, cholesterol is transported from the endosomes to the ER and Golgi apparatus, which is mediated by the endosomal cholesterol transporters Niemann–Pick type C1 (NPC1) and type C2 (NPC2) and cytosolic cholesterol transporters such as oxysterol-binding protein-related proteins [57]. Cholesterol is then transported to the plasma membrane and excreted from the cells as a lipoprotein. Compared with mock-transfected cells, wild-type APP-expressing cells exhibited an immediate clearance of cholesterol from the cells, whereas Osaka-mutant APP-expressing cells showed a prolonged accumulation of cholesterol [58]. In the Osaka mutation, intracellular cholesterol dominantly localized to the ER and less to the Golgi apparatus and endosomes. This subcellular localization was similar to that of Aβ, suggesting that intracellular cholesterol transport largely depends on Aβ trafficking. The impaired cholesterol transport observed in the Osaka mutation resembles the phenotype observed in cellular models of Niemann–Pick disease type C [59] which is caused by pathological mutations in NPC1 and/or NPC2 genes and shows neural dysfunction, including dementia. The Osaka mutation also caused an accumulation of Aβ and cholesterol in the mitochondria and increased ROS generation within the cells [58]. These results indicate that the cytotoxic effects of the Osaka mutation are, at least in part, displayed via an impaired ability to mediate cholesterol transport and efflux. This finding implies that high levels of cholesterol may aggravate the pathological phenotype of the Osaka mutation. Our finding that hypercholesterolemia accelerated the intraneuronal accumulation of Aβ oligomers and memory loss in APP_{OSK} mice [60] appears to support this notion.

4. Mechanism of Recessive Inheritance of the Osaka Mutation

As mentioned above, the Osaka mutation causes disease only in homozygotes, implying that it is recessive. When we found the Osaka mutation, all other APP mutations identified previously in familial AD were dominant (https://www.alzforum.org/mutations/app). Why is the Osaka mutation recessive even though it makes a high amount of toxic Aβ oligomers?

Whether a mutation shows a dominant or recessive inheritance is determined by one of the following four conditions. (1) The mutant protein acquires a new, often toxic, function (gain of function). (2) The mutant protein loses its function (loss of function) and simultaneously disturbs the action of its normal counterpart (dominant negative effect). (3) The mutant protein loses its function and the amount of normal counterpart (only half the usual level) is not enough to maintain physiological function (haploinsufficiency). In cases 1–3, pathological phenotypes appear even in heterozygotes, and the mutation is recognized as dominant. Almost all APP mutations cause gain-of-toxic function, increasing Aβ (particularly Aβ42) production and/or enhancing Aβ aggregation into fibrils, and thus are classified as case 1. Finally, in the last condition, (4) the mutant protein loses its function, but the normal counterpart can maintain physiological function even at half the usual level. In this case, pathological phenotypes can be observed only in homozygotes, and the mutation is regarded as recessive.

According to this rule, the Osaka mutation was assumed to cause a loss of certain APP function. The next question we asked was which function of APP is lost by the mutation? APP has been proposed to have several functions as a receptor, cell adhesion molecule, and growth factor [61–63], but the exact physiological role of APP is not well known. Thus, we created a new mouse model harboring the Osaka mutation. Using the homologous recombination technique, the codon CAA was deleted in endogenous mouse APP exon 17 to knock-in the Osaka mutation. Different from APP_{OSK} mice, this knock-in mouse model, named OSK-KI mice, produces mouse Aβ under the original mouse APP promotor [64]. In 2004, homozygous KI mice were established. This model is considered suitable for studying the influence of the mutation on APP function, because it lacks wild-type APP, which may compensate for the function lost in mutant APP, and avoids the artifacts by APP overexpression that is often observed in Tg mice.

We compared the memory and Aβ oligomer pathology between homo-, hetero-, and non-KI mice at 4, 6, and 8 months by the Morris water maze and immunohistochemical and biochemical analyses. Only homo-KI mice displayed memory impairment and intraneuronal accumulation of Aβ oligomers followed by synapse loss, tau hyperphosphorylation, glial activation, and neuron loss [64]. This result represents the recessive inheritance of the Osaka mutation and furthermore demonstrates that this mutation promotes Aβ oligomerization even in mouse Aβ. However, unlike Tg mice, the timing of the Aβ oligomer accumulation and of the memory impairment do not match in KI mice; the former was observed at 8 months and the latter at 4 months. In contrast, in APP_{OSK} mice, memory impairment was detected soon after the intraneuronal accumulation of Aβ oligomers.

We then examined synaptic plasticity by electrophysiology in hippocampal slices prepared at 4 and 8 months. HFS was delivered to the molecular layer of the dentate gyrus, and fEPSP was recorded in the same region. In the presence of a $GABA_A$ receptor antagonist, picrotoxin, HFS induced LTP in all groups (homo-, hetero-, and non-KI mice) at 4 months [64]. However, at 8 months, LTP was elicited only in the hetero- and non-KI mice. The LTP impairment in homo-KI mice was probably caused by the intraneuronal accumulation of Aβ oligomers. Interestingly, when the experiment was carried out without picrotoxin, something strange happened. Under the picrotoxin-free condition, GABAergic inhibitory inputs sufficiently prevented neuronal activation and thereby disturbed LTP induction as seen in hetero- and non-KI mice. On the contrary, in homo-KI mice, LTP was abnormally induced at 4 months, although it had diminished by 8 months again. These results suggest that in homo-KI mice, GABAergic neurotransmission is attenuated. Reduced inhibitory signaling would result in abnormal neural activation to impair normal cognition.

While we observed this phenomenon in 2014, we came across a paper showing that APP is highly expressed in GABA interneurons in the dentate gyrus and plays an essential role in GABAergic synapse formation [65]. This information led us to speculate that the Osaka mutation spoils APP function for the maintenance of dentate GABA neurons. To test this hypothesis, we counted the number of parvalbumin-positive GABA neurons in the dentate gyrus of OSK-KI mice. As we expected, compared with non-KI mice, only homo-KI mice showed a decrease of GABA neurons at 4 months [64] (Figure 5A). It has been shown that Aβ production depends on neuronal activity [66,67] and that the secreted Aβ, in turn, regulates synaptic activity [68]. Insufficient GABAergic inputs will lead to enhanced Aβ production via an abnormal activation of neurons. This may cause an eventual accumulation of Aβ oligomers when Aβ has an oligomer-prone mutation, as seen in homo-KI mice at 8 months. If this is the case, early treatment of homo-KI mice with a GABAergic supplement could normalize cognition and prevent Aβ oligomer accumulation. Thus, to activate GABAergic neurotransmission, we orally administered diazepam, a positive allosteric modulator of $GABA_A$ receptor, to 6-month-old homo-KI mice for 2 months and examined their memory and Aβ oligomers at 8 months. While the number of GABA neurons in the dentate gyrus remained lower, cognitive function was recovered and Aβ oligomers were not detected in the treated mice [64].

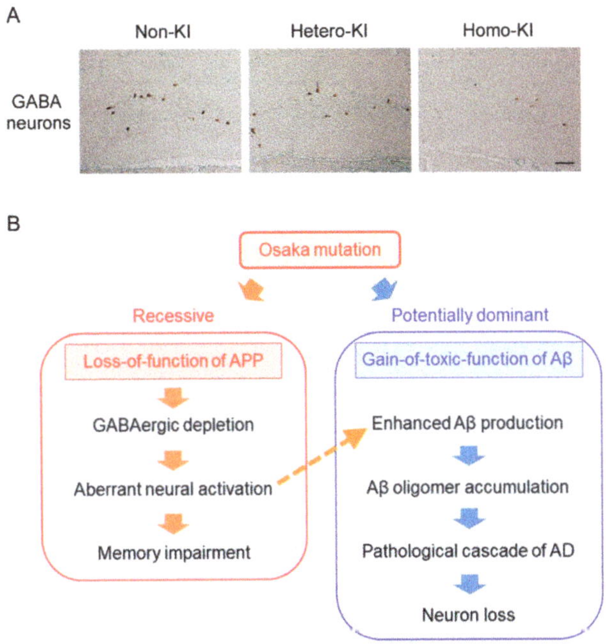

Figure 5. Phenotypes of OSK-KI mice. (**A**) GABAergic depletion by the Osaka mutation. Parvalbumin-positive GABA neurons in the dentate gyrus were decreased only in homo-KI mice. Scale bar = 30 μm. (**B**) Proposed mechanism for the recessive inheritance of the Osaka mutation. The Osaka mutation has dual effects; a loss-of-function of APP and gain-of-toxic-function of Aβ. The former causes GABAergic depletion, leading to abnormal neural activation and impaired cognition. The extensive neural activation results in enhanced Aβ production (dashed arrow). Then, the latter effect accelerates Aβ oligomerization that initiates the pathological cascade of AD leading to neuron loss.

Taken together, our findings indicate that the Osaka mutation primarily causes a loss of APP function that is essential for GABA neurons (Figure 5B). The resultant imbalance between excitatory and inhibitory inputs will lead to impaired cognition [69] and also enhanced Aβ production via abnormal neural activation. In the presence of the Osaka mutation, Aβ promptly forms oligomers and accumulates within neurons, which initiates the pathological cascade of AD. Thus, in patients and KI mice with this mutation, Aβ oligomer accumulation can be regarded as a secondary phenomenon that follows the loss of APP function. This is the mechanism we propose to account for the recessive inheritance of the Osaka mutation. In APP_{OSK} mice, mutant Aβ is produced under the powerful mouse prion protein promotor, and therefore extensive neural activation is not necessary for Aβ oligomer accumulation.

As described above, early treatment with diazepam, which is an antiepileptic drug, prevented disease onset in OSK-KI mice. Epileptic seizures, which are caused by excessive neural activation, are frequently observed in the early stages of AD and are likely responsible, in part, for the progression of AD [70]. A recent study has shown that extracellular Aβ oligomers mediate neuronal hyperactivation through blocking glutamate reuptake by astrocytes [71]. This mechanism implies the existence of a vicious cycle between Aβ production and neuronal activation. In AD mouse models, certain classes of antiepileptic drugs exhibited disease-modifying properties [72–74]. These findings suggest a new therapeutic strategy for AD targeting neural network hyperexcitability.

5. Concluding Remarks

The Osaka mutation is the first recessive mutation found in familial AD. Interestingly, it has dual effects: a loss-of-function of APP and a gain-of-toxic-function of Aβ. The former causes GABAergic depletion, leading to abnormal neural activation and impaired cognition. The latter causes accelerated Aβ oligomerization that initiates the pathological cascade of AD including tau hyperphosphorylation, synapse loss, glial activation, NFT formation, and eventual neuron loss, although the effect never appears until enhanced Aβ production is established by the former effect. The most striking feature of the Osaka mutation is no fibrillization of Aβ. Our finding that AD develops only by Aβ oligomers without amyloid plaque formation strongly supports the hypothesis that Aβ oligomers play a critical role in the onset and progression of AD [1]. Furthermore, it raises an important question on the definition of AD: Is the existence of amyloid plaques an absolute requirement for AD? We propose that AD should be redefined as a neurodegenerative dementia caused by Aβ accumulation regardless of the Aβ aggregation state.

In the Osaka mutation, Aβ oligomers accumulate within neurons, localizing into the ER, Golgi apparatus, endosomes, lysosomes, autophagosomes, and mitochondria. This accumulation disrupts the axonal and dendritic transport of mitochondria, BDNF, and recycling endosomes, resulting in synaptic alterations. It also causes ER stress, endosomal/lysosomal leakage, and mitochondrial dysfunction, which lead to cell death.

Finally, the Osaka mutation provides a useful means to investigate the pathological and physiological roles of Aβ oligomers without the influence of amyloid fibrils or plaques (for example, [75–78]). Additionally, our model mice harboring this mutation could be a good tool for developing effective approaches in the prevention, treatment, and diagnosis of AD that specifically target Aβ oligomers (for example, [79,80]).

Author Contributions: T.T. wrote all sections except 3.1 and organized the manuscript; H.S. wrote Section 3.1. All authors have read and agreed to the published version of the manuscript.

Funding: This study was supported by the Japan Ministry of Education, Culture, Sports, Science and Technology.

Acknowledgments: The authors would like to thank Hiroshi Mori, a past professor of the Neuroscience Department, Osaka City University, for supporting the research of the Osaka mutation. The authors also thank Peter Karagiannis for editing the manuscript.

Conflicts of Interest: The authors declare no conflict of interest.

References

1. Selkoe, D.J. Alzheimer's disease is a synaptic failure. *Science* **2002**, *298*, 789–791. [CrossRef] [PubMed]
2. Lambert, M.P.; Barlow, A.K.; Chromy, B.A.; Edwards, C.; Freed, R.; Liosatos, M.; Morgan, T.E.; Rozovsky, I.; Trommer, B.; Viola, K.L.; et al. Diffusible, nonfibrillar ligands derived from Abeta1-42 are potent central nervous system neurotoxins. *Proc. Natl. Acad. Sci. USA* **1998**, *95*, 6448–6453. [CrossRef] [PubMed]
3. Walsh, D.M.; Klyubin, I.; Fadeeva, J.V.; Cullen, W.K.; Anwyl, R.; Wolfe, M.S.; Rowan, M.J.; Selkoe, D.J. Naturally secreted oligomers of amyloid beta protein potently inhibit hippocampal long-term potentiation in vivo. *Nature* **2002**, *416*, 535–539. [CrossRef] [PubMed]
4. Cleary, J.P.; Walsh, D.M.; Hofmeister, J.J.; Shankar, G.M.; Kuskowski, M.A.; Selkoe, D.J.; Ashe, K.H. Natural oligomers of the amyloid-beta protein specifically disrupt cognitive function. *Nat. Neurosci.* **2005**, *8*, 79–84. [CrossRef]
5. Lesné, S.; Koh, M.T.; Kotilinek, L.; Kayed, R.; Glabe, C.G.; Yang, A.; Gallagher, M.; Ashe, K.H. A specific amyloid-beta protein assembly in the brain impairs memory. *Nature* **2006**, *440*, 352–357. [CrossRef]
6. Shankar, G.M.; Li, S.; Mehta, T.H.; Garcia-Munoz, A.; Shepardson, N.E.; Smith, I.; Brett, F.M.; Farrell, M.A.; Rowan, M.J.; Lemere, C.A.; et al. Amyloid-beta protein dimers isolated directly from Alzheimer's brains impair synaptic plasticity and memory. *Nat. Med.* **2008**, *14*, 837–842. [CrossRef]

7. Reed, M.N.; Hofmeister, J.J.; Jungbauer, L.; Welzel, A.T.; Yu, C.; Sherman, M.A.; Lesné, S.; LaDu, M.J.; Walsh, D.M.; Ashe, K.H.; et al. Cognitive effects of cell-derived and synthetically derived Aβ oligomers. *Neurobiol. Aging* **2011**, *32*, 1784–1794. [CrossRef]
8. Cline, E.N.; Bicca, M.A.; Viola, K.L.; Klein, W.L. The Amyloid-β Oligomer Hypothesis: Beginning of the Third Decade. *J. Alzheimers Dis.* **2018**, *64*, S567–S610. [CrossRef]
9. Nilsberth, C.; Westlind-Danielsson, A.; Eckman, C.B.; Condron, M.M.; Axelman, K.; Forsell, C.; Stenh, C.; Luthman, J.; Teplow, D.B.; Younkin, S.G.; et al. The 'Arctic' APP mutation (E693G) causes Alzheimer's disease by enhanced Abeta protofibril formation. *Nat. Neurosci.* **2001**, *4*, 887–893. [CrossRef]
10. Lord, A.; Kalimo, H.; Eckman, C.; Zhang, X.Q.; Lannfelt, L.; Nilsson, L.N. The Arctic Alzheimer mutation facilitates early intraneuronal Abeta aggregation and senile plaque formation in transgenic mice. *Neurobiol. Aging* **2006**, *27*, 67–77. [CrossRef]
11. Knobloch, M.; Konietzko, U.; Krebs, D.C.; Nitsch, R.M. Intracellular Abeta and cognitive deficits precede beta-amyloid deposition in transgenic arcAbeta mice. *Neurobiol. Aging* **2007**, *28*, 1297–1306. [CrossRef] [PubMed]
12. Basun, H.; Bogdanovic, N.; Ingelsson, M.; Almkvist, O.; Näslund, J.; Axelman, K.; Bird, T.D.; Nochlin, D.; Schellenberg, G.D.; Wahlund, L.O.; et al. Clinical and neuropathological features of the arctic APP gene mutation causing early-onset Alzheimer disease. *Arch. Neurol.* **2008**, *65*, 499–505. [CrossRef] [PubMed]
13. Kalimo, H.; Lalowski, M.; Bogdanovic, N.; Philipson, O.; Bird, T.D.; Nochlin, D.; Schellenberg, G.D.; Brundin, R.; Olofsson, T.; Soliymani, R.; et al. The Arctic AβPP mutation leads to Alzheimer's disease pathology with highly variable topographic deposition of differentially truncated Aβ. *Acta Neuropathol. Commun.* **2013**, *1*, 60. [CrossRef] [PubMed]
14. Schöll, M.; Wall, A.; Thordardottir, S.; Ferreira, D.; Bogdanovic, N.; Långström, B.; Almkvist, O.; Graff, C.; Nordberg, A. Low PiB PET retention in presence of pathologic CSF biomarkers in Arctic APP mutation carriers. *Neurology* **2012**, *79*, 229–236. [CrossRef] [PubMed]
15. Tomiyama, T.; Nagata, T.; Shimada, H.; Teraoka, R.; Fukushima, A.; Kanemitsu, H.; Takuma, H.; Kuwano, R.; Imagawa, M.; Ataka, S.; et al. A new amyloid beta variant favoring oligomerization in Alzheimer's-type dementia. *Ann. Neurol.* **2008**, *63*, 377–387. [CrossRef] [PubMed]
16. Takuma, H.; Teraoka, R.; Mori, H.; Tomiyama, T. Amyloid-beta E22Delta variant induces synaptic alteration in mouse hippocampal slices. *Neuroreport* **2008**, *19*, 615–619. [CrossRef]
17. Tomiyama, T.; Matsuyama, S.; Iso, H.; Umeda, T.; Takuma, H.; Ohnishi, K.; Ishibashi, K.; Teraoka, R.; Sakama, N.; Yamashita, T.; et al. A mouse model of amyloid beta oligomers: Their contribution to synaptic alteration, abnormal tau phosphorylation, glial activation, and neuronal loss in vivo. *J. Neurosci.* **2010**, *30*, 4845–4856. [CrossRef]
18. Nishitsuji, K.; Tomiyama, T.; Ishibashi, K.; Ito, K.; Teraoka, R.; Lambert, M.P.; Klein, W.L.; Mori, H. The E693Delta mutation in amyloid precursor protein increases intracellular accumulation of amyloid beta oligomers and causes endoplasmic reticulum stress-induced apoptosis in cultured cells. *Am. J. Pathol.* **2009**, *174*, 957–969. [CrossRef]
19. Shimada, H.; Ataka, S.; Tomiyama, T.; Takechi, H.; Mori, H.; Miki, T. Clinical course of patients with familial early-onset Alzheimer's disease potentially lacking senile plaques bearing the E693Δ mutation in amyloid precursor protein. *Dement. Geriatr. Cogn. Disord.* **2011**, *32*, 45–54. [CrossRef]
20. Kutoku, Y.; Ohsawa, Y.; Kuwano, R.; Ikeuchi, T.; Inoue, H.; Ataka, S.; Shimada, H.; Mori, H.; Sunada, Y. A second pedigree with amyloid-less familial Alzheimer's disease harboring an identical mutation in the amyloid precursor protein gene (E693delta). *Intern. Med.* **2015**, *54*, 205–208. [CrossRef]
21. Ishibashi, K.; Tomiyama, T.; Nishitsuji, K.; Hara, M.; Mori, H. Absence of synaptophysin near cortical neurons containing oligomer Abeta in Alzheimer's disease brain. *J. Neurosci. Res.* **2006**, *84*, 632–636. [CrossRef] [PubMed]
22. Lambert, M.P.; Velasco, P.T.; Chang, L.; Viola, K.L.; Fernandez, S.; Lacor, P.N.; Khuon, D.; Gong, Y.; Bigio, E.H.; Shaw, P.; et al. Monoclonal antibodies that target pathological assemblies of Abeta. *J. Neurochem.* **2007**, *100*, 23–35. [CrossRef] [PubMed]
23. Matsuyama, S.; Teraoka, R.; Mori, H.; Tomiyama, T. Inverse correlation between amyloid precursor protein and synaptic plasticity in transgenic mice. *Neuroreport* **2007**, *18*, 1083–1087. [CrossRef] [PubMed]

24. Umeda, T.; Yamashita, T.; Kimura, T.; Ohnishi, K.; Takuma, H.; Ozeki, T.; Takashima, A.; Tomiyama, T.; Mori, H. Neurodegenerative disorder FTDP-17-related tau intron 10 +16C → T mutation increases tau exon 10 splicing and causes tauopathy in transgenic mice. *Am. J. Pathol.* **2013**, *183*, 211–225. [CrossRef]
25. Umeda, T.; Maekawa, S.; Kimura, T.; Takashima, A.; Tomiyama, T.; Mori, H. Neurofibrillary tangle formation by introducing wild-type human tau into APP transgenic mice. *Acta Neuropathol.* **2014**, *127*, 685–698. [CrossRef]
26. Kulic, L.; McAfoose, J.; Welt, T.; Tackenberg, C.; Späni, C.; Wirth, F.; Finder, V.; Konietzko, U.; Giese, M.; Eckert, A.; et al. Early accumulation of intracellular fibrillar oligomers and late congophilic amyloid angiopathy in mice expressing the Osaka intra-Aβ APP mutation. *Transl. Psychiatry* **2012**, *2*, e183. [CrossRef]
27. Suzuki, T.; Murakami, K.; Izuo, N.; Kume, T.; Akaike, A.; Nagata, T.; Nishizaki, T.; Tomiyama, T.; Takuma, H.; Mori, H.; et al. E22Δ Mutation in Amyloid β-Protein Promotes β-Sheet Transformation, Radical Production, and Synaptotoxicity, But Not Neurotoxicity. *Int. J. Alzheimers Dis.* **2010**, *2011*, 431320.
28. Cloe, A.L.; Orgel, J.P.; Sachleben, J.R.; Tycko, R.; Meredith, S.C. The Japanese mutant Aβ (ΔE22-Aβ(1-39)) forms fibrils instantaneously, with low-thioflavin T fluorescence: Seeding of wild-type Aβ(1-40) into atypical fibrils by ΔE22-Aβ(1-39). *Biochemistry* **2011**, *50*, 2026–2039. [CrossRef]
29. Inayathullah, M.; Teplow, D.B. Structural dynamics of the ΔE22 (Osaka) familial Alzheimer's disease-linked amyloid β-protein. *Amyloid* **2011**, *18*, 98–107. [CrossRef]
30. Ovchinnikova, O.Y.; Finder, V.H.; Vodopivec, I.; Nitsch, R.M.; Glockshuber, R. The Osaka FAD mutation E22Δ leads to the formation of a previously unknown type of amyloid β fibrils and modulates Aβ neurotoxicity. *J. Mol. Biol.* **2011**, *408*, 780–791. [CrossRef]
31. Poduslo, J.F.; Howell, K.G. Unique molecular signatures of Alzheimer's disease amyloid β peptide mutations and deletion during aggregate/oligomer/fibril formation. *J. Neurosci. Res.* **2015**, *93*, 410–423. [CrossRef] [PubMed]
32. Jang, H.; Arce, F.T.; Ramachandran, S.; Kagan, B.L.; Lal, R.; Nussinov, R. Familial Alzheimer's disease Osaka mutant (ΔE22) β-barrels suggest an explanation for the different Aβ1-40/42 preferred conformational states observed by experiment. *J. Phys. Chem. B* **2013**, *117*, 11518–11529. [CrossRef] [PubMed]
33. Huber, M.; Ovchinnikova, O.Y.; Schütz, A.K.; Glockshuber, R.; Meier, B.H.; Böckmann, A. Solid-state NMR sequential assignment of Osaka-mutant amyloid-beta (Aβ1-40 E22Δ) fibrils. *Biomol. NMR Assign.* **2015**, *9*, 7–14. [CrossRef] [PubMed]
34. Schütz, A.K.; Vagt, T.; Huber, M.; Ovchinnikova, O.Y.; Cadalbert, R.; Wall, J.; Güntert, P.; Böckmann, A.; Glockshuber, R.; Meier, B.H. Atomic-resolution three-dimensional structure of amyloid β fibrils bearing the Osaka mutation. *Angew. Chem. Int. Ed. Engl.* **2015**, *54*, 331–335. [CrossRef] [PubMed]
35. Elkins, M.R.; Wang, T.; Nick, M.; Jo, H.; Lemmin, T.; Prusiner, S.B.; DeGrado, W.F.; Stöhr, J.; Hong, M. Structural Polymorphism of Alzheimer's β-Amyloid Fibrils as Controlled by an E22 Switch: A Solid-State NMR Study. *J. Am. Chem. Soc.* **2016**, *138*, 9840–9852. [CrossRef] [PubMed]
36. Chambers, J.K.; Tokuda, T.; Uchida, K.; Ishii, R.; Tatebe, H.; Takahashi, E.; Tomiyama, T.; Une, Y.; Nakayama, H. The domestic cat as a natural animal model of Alzheimer's disease. *Acta Neuropathol. Commun.* **2015**, *3*, 78. [CrossRef]
37. Kayed, R.; Lasagna-Reeves, C.A. Molecular mechanisms of amyloid oligomers toxicity. *J. Alzheimers Dis.* **2013**, *33*, S67–S78. [CrossRef]
38. Smith, L.M.; Strittmatter, S.M. Binding Sites for Amyloid-β Oligomers and Synaptic Toxicity. *Cold Spring Harb. Perspect. Med.* **2017**, *7*, a024075. [CrossRef]
39. Vargas, L.M.; Cerpa, W.; Muñoz, F.J.; Zanlungo, S.; Alvarez, A.R. Amyloid-β oligomers synaptotoxicity: The emerging role of EphA4/c-Abl signaling in Alzheimer's disease. *Biochim. Biophys. Acta Mol. Basis Dis.* **2018**, *1864*, 1148–1159. [CrossRef]
40. Umeda, T.; Tomiyama, T.; Sakama, N.; Tanaka, S.; Lambert, M.P.; Klein, W.L.; Mori, H. Intraneuronal amyloid β oligomers cause cell death via endoplasmic reticulum stress, endosomal/lysosomal leakage, and mitochondrial dysfunction in vivo. *J. Neurosci Res.* **2011**, *89*, 1031–1042. [CrossRef]
41. Kondo, T.; Asai, M.; Tsukita, K.; Kutoku, Y.; Ohsawa, Y.; Sunada, Y.; Imamura, K.; Egawa, N.; Yahata, N.; Okita, K.; et al. Modeling Alzheimer's disease with iPSCs reveals stress phenotypes associated with intracellular Aβ and differential drug responsiveness. *Cell Stem Cell* **2013**, *12*, 487–496. [CrossRef] [PubMed]

42. Gouras, G.K.; Tsai, J.; Naslund, J.; Vincent, B.; Edgar, M.; Checler, F.; Greenfield, J.P.; Haroutunian, V.; Buxbaum, J.D.; Xu, H.; et al. Intraneuronal Abeta42 accumulation in human brain. *Am. J. Pathol.* **2000**, *156*, 15–20. [CrossRef]
43. Wirths, O.; Multhaup, G.; Czech, C.; Blanchard, V.; Moussaoui, S.; Tremp, G.; Pradier, L.; Beyreuther, K.; Bayer, T.A. Intraneuronal Abeta accumulation precedes plaque formation in beta-amyloid precursor protein and presenilin-1 double-transgenic mice. *Neurosci. Lett.* **2001**, *306*, 116–120. [CrossRef]
44. Oddo, S.; Caccamo, A.; Shepherd, J.D.; Murphy, M.P.; Golde, T.E.; Kayed, R.; Metherate, R.; Mattson, M.P.; Akbari, Y.; LaFerla, F.M. Triple-transgenic model of Alzheimer's disease with plaques and tangles: Intracellular Abeta and synaptic dysfunction. *Neuron* **2003**, *39*, 409–421. [CrossRef]
45. Oddo, S.; Caccamo, A.; Smith, I.F.; Green, K.N.; LaFerla, F.M. A dynamic relationship between intracellular and extracellular pools of Abeta. *Am. J. Pathol.* **2006**, *168*, 184–194. [CrossRef]
46. Bayer, T.A.; Wirths, O. Intraneuronal Aβ as a trigger for neuron loss: Can this be translated into human pathology? *Biochem. Soc. Trans.* **2011**, *39*, 857–861. [CrossRef]
47. Gouras, G.K.; Tampellini, D.; Takahashi, R.H.; Capetillo-Zarate, E. Intraneuronal beta-amyloid accumulation and synapse pathology in Alzheimer's disease. *Acta Neuropathol.* **2010**, *119*, 523–541. [CrossRef]
48. Lacor, P.N.; Buniel, M.C.; Furlow, P.W.; Clemente, A.S.; Velasco, P.T.; Wood, M.; Viola, K.L.; Klein, W.L. Abeta oligomer-induced aberrations in synapse composition, shape, and density provide a molecular basis for loss of connectivity in Alzheimer's disease. *J. Neurosci.* **2007**, *27*, 796–807. [CrossRef]
49. Shankar, G.M.; Bloodgood, B.L.; Townsend, M.; Walsh, D.M.; Selkoe, D.J.; Sabatini, B.L. Natural oligomers of the Alzheimer amyloid-beta protein induce reversible synapse loss by modulating an NMDA-type glutamate receptor-dependent signaling pathway. *J. Neurosci.* **2007**, *27*, 2866–2875. [CrossRef]
50. Decker, H.; Lo, K.Y.; Unger, S.M.; Ferreira, S.T.; Silverman, M.A. Amyloid-beta peptide oligomers disrupt axonal transport through an NMDA receptor-dependent mechanism that is mediated by glycogen synthase kinase 3beta in primary cultured hippocampal neurons. *J. Neurosci.* **2010**, *30*, 9166–9171. [CrossRef]
51. Vossel, K.A.; Zhang, K.; Brodbeck, J.; Daub, A.C.; Sharma, P.; Finkbeiner, S.; Cui, B.; Mucke, L. Tau reduction prevents Abeta-induced defects in axonal transport. *Science* **2010**, *330*, 198. [CrossRef] [PubMed]
52. Tang, Y.; Scott, D.A.; Das, U.; Edland, S.D.; Radomski, K.; Koo, E.H.; Roy, S. Early and selective impairments in axonal transport kinetics of synaptic cargoes induced by soluble amyloid β-protein oligomers. *Traffic* **2012**, *13*, 681–693. [CrossRef] [PubMed]
53. Zhao, Y.; Sivaji, S.; Chiang, M.C.; Ali, H.; Zukowski, M.; Ali, S.; Kennedy, B.; Sklyar, A.; Cheng, A.; Guo, Z.; et al. Amyloid Beta Peptides Block New Synapse Assembly by Nogo Receptor-Mediated Inhibition of T-Type Calcium Channels. *Neuron* **2017**, *96*, 355–372. [CrossRef] [PubMed]
54. Umeda, T.; Ramser, E.M.; Yamashita, M.; Nakajima, K.; Mori, H.; Silverman, M.A.; Tomiyama, T. Intracellular amyloid β oligomers impair organelle transport and induce dendritic spine loss in primary neurons. *Acta Neuropathol. Commun.* **2015**, *3*, 51. [CrossRef] [PubMed]
55. Umeda, T.; Mori, H.; Zheng, H.; Tomiyama, T. Regulation of cholesterol efflux by amyloid beta secretion. *J. Neurosci. Res.* **2010**, *88*, 1985–1994. [PubMed]
56. Fernández, A.; Llacuna, L.; Fernández-Checa, J.C.; Colell, A. Mitochondrial cholesterol loading exacerbates amyloid beta peptide-induced inflammation and neurotoxicity. *J. Neurosci.* **2009**, *29*, 6394–6405. [CrossRef] [PubMed]
57. Chang, T.Y.; Chang, C.C.; Ohgami, N.; Yamauchi, Y. Cholesterol sensing, trafficking, and esterification. *Annu. Rev. Cell Dev. Biol.* **2006**, *22*, 129–157. [CrossRef]
58. Nomura, S.; Umeda, T.; Tomiyama, T.; Mori, H. The E693Δ (Osaka) mutation in amyloid precursor protein potentiates cholesterol-mediated intracellular amyloid β toxicity via its impaired cholesterol efflux. *J. Neurosci. Res.* **2013**, *91*, 1541–1550. [CrossRef]
59. Koh, C.H.; Cheung, N.S. Cellular mechanism of U18666A-mediated apoptosis in cultured murine cortical neurons: Bridging Niemann-Pick disease type C and Alzheimer's disease. *Cell. Signal.* **2006**, *18*, 1844–1853. [CrossRef]
60. Umeda, T.; Tomiyama, T.; Kitajima, E.; Idomoto, T.; Nomura, S.; Lambert, M.P.; Klein, W.L.; Mori, H. Hypercholesterolemia accelerates intraneuronal accumulation of Aβ oligomers resulting in memory impairment in Alzheimer's disease model mice. *Life Sci.* **2012**, *91*, 1169–1176. [CrossRef]
61. Deyts, C.; Thinakaran, G.; Parent, A.T. APP Receptor? To Be or Not To Be. *Trends Pharmacol. Sci.* **2016**, *37*, 390–411. [CrossRef] [PubMed]

62. Sosa, L.J.; Cáceres, A.; Dupraz, S.; Oksdath, M.; Quiroga, S.; Lorenzo, A. The physiological role of the amyloid precursor protein as an adhesion molecule in the developing nervous system. *J. Neurochem.* **2017**, *143*, 11–29. [CrossRef] [PubMed]
63. Coronel, R.; Bernabeu-Zornoza, A.; Palmer, C.; Muñiz-Moreno, M.; Zambrano, A.; Cano, E.; Liste, I. Role of Amyloid Precursor Protein (APP) and Its Derivatives in the Biology and Cell Fate Specification of Neural Stem Cells. *Mol. Neurobiol.* **2018**, *55*, 7107–7117. [CrossRef] [PubMed]
64. Umeda, T.; Kimura, T.; Yoshida, K.; Takao, K.; Fujita, Y.; Matsuyama, S.; Sakai, A.; Yamashita, M.; Yamashita, Y.; Ohnishi, K.; et al. Mutation-induced loss of APP function causes GABAergic depletion in recessive familial Alzheimer's disease: Analysis of Osaka mutation-knockin mice. *Acta Neuropathol. Commun.* **2017**, *5*, 59. [CrossRef]
65. Wang, B.; Wang, Z.; Sun, L.; Yang, L.; Li, H.; Cole, A.L.; Rodriguez-Rivera, J.; Lu, H.C.; Zheng, H. The amyloid precursor protein controls adult hippocampal neurogenesis through GABAergic interneurons. *J. Neurosci.* **2014**, *34*, 13314–13325. [CrossRef]
66. Kamenetz, F.; Tomita, T.; Hsieh, H.; Seabrook, G.; Borchelt, D.; Iwatsubo, T.; Sisodia, S.; Malinow, R. APP processing and synaptic function. *Neuron* **2003**, *37*, 925–937. [CrossRef]
67. Cirrito, J.R.; Yamada, K.A.; Finn, M.B.; Sloviter, R.S.; Bales, K.R.; May, P.C.; Schoepp, D.D.; Paul, S.M.; Mennerick, S.; Holtzman, D.M. Synaptic activity regulates interstitial fluid amyloid-beta levels in vivo. *Neuron* **2005**, *48*, 913–922. [CrossRef]
68. Palop, J.J.; Mucke, L. Amyloid-beta-induced neuronal dysfunction in Alzheimer's disease: From synapses toward neural networks. *Nat. Neurosci.* **2010**, *13*, 812–818. [CrossRef]
69. Vico Varela, E.; Etter, G.; Williams, S. Excitatory-inhibitory imbalance in Alzheimer's disease and therapeutic significance. *Neurobiol. Dis.* **2019**, *127*, 605–615. [CrossRef]
70. Vossel, K.A.; Tartaglia, M.C.; Nygaard, H.B.; Zeman, A.Z.; Miller, B.L. Epileptic activity in Alzheimer's disease: Causes and clinical relevance. *Lancet Neurol.* **2017**, *16*, 311–322. [CrossRef]
71. Zott, B.; Simon, M.M.; Hong, W.; Unger, F.; Chen-Engerer, H.J.; Frosch, M.P.; Sakmann, B.; Walsh, D.M.; Konnerth, A. A vicious cycle of β amyloid-dependent neuronal hyperactivation. *Science* **2019**, *365*, 559–565. [CrossRef] [PubMed]
72. Qing, H.; He, G.; Ly, P.T.; Fox, C.J.; Staufenbiel, M.; Cai, F.; Zhang, Z.; Wei, S.; Sun, X.; Chen, C.H.; et al. Valproic acid inhibits Abeta production, neuritic plaque formation, and behavioral deficits in Alzheimer's disease mouse models. *J. Exp. Med.* **2008**, *205*, 2781–2789. [CrossRef] [PubMed]
73. Shi, J.Q.; Wang, B.R.; Tian, Y.Y.; Xu, J.; Gao, L.; Zhao, S.L.; Jiang, T.; Xie, H.G.; Zhang, Y.D. Antiepileptics topiramate and levetiracetam alleviate behavioral deficits and reduce neuropathology in APPswe/PS1dE9 transgenic mice. *CNS Neurosci. Ther.* **2013**, *19*, 871–881. [CrossRef] [PubMed]
74. Zhang, M.Y.; Zheng, C.Y.; Zou, M.M.; Zhu, J.W.; Zhang, Y.; Wang, J.; Liu, C.F.; Li, Q.F.; Xiao, Z.C.; Li, S.; et al. Lamotrigine attenuates deficits in synaptic plasticity and accumulation of amyloid plaques in APP/PS1 transgenic mice. *Neurobiol. Aging* **2014**, *35*, 2713–2725. [CrossRef] [PubMed]
75. Maeda, J.; Zhang, M.R.; Okauchi, T.; Ji, B.; Ono, M.; Hattori, S.; Kumata, K.; Iwata, N.; Saido, T.C.; Trojanowski, J.Q.; et al. In vivo positron emission tomographic imaging of glial responses to amyloid-beta and tau pathologies in mouse models of Alzheimer's disease and related disorders. *J. Neurosci.* **2011**, *31*, 4720–4730. [CrossRef] [PubMed]
76. Takano, M.; Yamashita, T.; Nagano, K.; Otani, M.; Maekura, K.; Kamada, H.; Tsunoda, S.; Tsutsumi, Y.; Tomiyama, T.; Mori, H.; et al. Proteomic analysis of the hippocampus in Alzheimer's disease model mice by using two-dimensional fluorescence difference in gel electrophoresis. *Neurosci. Lett.* **2013**, *534*, 85–89. [CrossRef]
77. Honjo, Y.; Ayaki, T.; Tomiyama, T.; Horibe, T.; Ito, H.; Mori, H.; Takahashi, R.; Kawakami, K. Increased GADD34 in oligodendrocytes in Alzheimer's disease. *Neurosci. Lett.* **2015**, *602*, 50–55. [CrossRef]
78. Palavicini, J.P.; Wang, C.; Chen, L.; Hosang, K.; Wang, J.; Tomiyama, T.; Mori, H.; Han, X. Oligomeric amyloid-beta induces MAPK-mediated activation of brain cytosolic and calcium-independent phospholipase A2 in a spatial-specific manner. *Acta Neuropathol. Commun.* **2017**, *5*, 56. [CrossRef]

79. Umeda, T.; Ono, K.; Sakai, A.; Yamashita, M.; Mizuguchi, M.; Klein, W.L.; Yamada, M.; Mori, H.; Tomiyama, T. Rifampicin is a candidate preventive medicine against amyloid-β and tau oligomers. *Brain* **2016**, *139*, 1568–1586. [CrossRef]
80. Umeda, T.; Tanaka, A.; Sakai, A.; Yamamoto, A.; Sakane, T.; Tomiyama, T. Intranasal rifampicin for Alzheimer's disease prevention. *Alzheimers Dement.* **2018**, *4*, 304–313. [CrossRef]

International Journal of
Molecular Sciences

MDPI

Review

Protofibrils of Amyloid-β are Important Targets of a Disease-Modifying Approach for Alzheimer's Disease

Kenjiro Ono [1,*] and Mayumi Tsuji [2]

1 Department of Internal Medicine, Division of Neurology, School of Medicine, Showa University, Tokyo 142-8666, Japan
2 Department of Pharmacology, School of Medicine, Showa University, Tokyo 142-8666, Japan; tsujim@med.showa-u.ac.jp
* Correspondence: onoken@med.showa-u.ac.jp; Tel.: +81-3-3784-8710

Received: 31 December 2019; Accepted: 29 January 2020; Published: 31 January 2020

Abstract: Worldwide, Alzheimer's disease (AD) is the most common age-related neurodegenerative disease and is characterized by unique pathological hallmarks in the brain, including plaques composed of amyloid β-protein (Aβ) and neurofibrillary tangles of tau protein. Genetic studies, biochemical data, and animal models have suggested that Aβ is responsible for the pathogenesis of AD (i.e., the amyloid hypothesis). Indeed, Aβ molecules tend to aggregate, forming oligomers, protofibrils, and mature fibrils. However, while these Aβ species form amyloid plaques of the type implicated in AD neurodegeneration, recent clinical trials designed to reduce the production of Aβ and/or the plaque burden have not demonstrated clinical efficacy. In addition, recent studies using synthetic Aβ peptides, cell culture models, Arctic transgenic mice, and human samples of AD brain tissues have suggested that the pre-fibrillar forms of Aβ, particularly Aβ protofibrils, may be the most critical species, compared with extracellular fibrillar forms. We recently reported that protofibrils of $A\beta_{1\text{-}42}$ disturbed membrane integrity by inducing reactive oxygen species generation and lipid peroxidation, resulting in decreased membrane fluidity, intracellular calcium dysregulation, depolarization, and synaptic toxicity. Therefore, the therapeutic reduction of protofibrils may prevent the progression of AD by ameliorating neuronal damage and cognitive dysfunction through multiple mechanisms.

Keywords: Alzheimer's disease; amyloid β-protein (Aβ); mAb158; oligomers; protofibrils

1. Introduction

Neurodegenerative diseases, such as Alzheimer's disease (AD), Parkinson's disease, and spinocerebellar ataxia, have characteristic abnormal protein aggregates in the brain. In AD, the two neuropathological characteristics are amyloid plaques composed of amyloid β-protein (Aβ) and neurofibrillary tangles of hyperphosphorylated tau protein [1].

Human genetic association studies, biochemical analyses of AD plaque content, and various animal models with altered Aβ or tau expression have strongly implicated Aβ and tau in AD pathogenesis [1]. Furthermore, many in vivo and in vitro studies have demonstrated the neurotoxicity of these amyloidogenic proteins. However, amyloid neurotoxicity depends strongly on Aβ's primary structure and aggregation state. For example, two predominant Aβ forms are produced in humans and are comprised of either 40 ($A\beta_{1\text{-}40}$) or 42 ($A\beta_{1\text{-}42}$) amino acid residues. The relative proportion of $A\beta_{1\text{-}42}$ appears to be particularly crucial for AD progression, as this longer form is more prone to aggregation and is inherently more toxic than $A\beta_{1\text{-}40}$ [2]. Aβ molecules form low molecular weight (LMW) oligomers, high molecular weight (HMW) oligomers such as protofibrils (PFs), and mature fibrils, which have been suggested to be primary agents of neuronal dysfunction in AD [3]. Although these Aβ aggregates may directly cause neuronal injury by acting on synapses or indirectly by activating astrocytes and microglia [2], evidence also supports the hypothesis that soluble oligomeric Aβ plays an important role in AD pathogenesis (i.e., the oligomer hypothesis) [1,3,4].

Many types of oligomeric Aβ species have been demonstrated in vitro, with PFs being commonly described. Aβ PFs are defined as curved linear structures >100 kDa that remain soluble upon centrifugation at 16,000–18,000× g [3,5–7]. The neurotoxicity of these Aβ PFs formed in vitro, as well as their ability to induce electrophysiological effects on neurons, has been demonstrated by several groups [8–11]. Arctic Aβ is the result of a mutation in the gene that encodes the amyloid precursor protein (APP) and leads to the production of a particular Aβ species, [Glu22Gly]Aβ, with a high propensity to form PFs [12]. We recently reported that PFs disturb membrane integrity by inducing reactive oxygen species' (ROS) generation and lipid peroxidation, resulting in decreased membrane fluidity, intracellular calcium dysregulation, depolarization, and impaired long-term potentiation (LTP). In addition, the damaging effects of PFs were found to be significantly greater than those of LMW-$A\beta_{1\text{-}42}$ [13].

Current treatments for AD are primarily aimed at mitigating symptoms, while disease-modifying approaches are aimed at halting or attenuating the progression of the disease, such as inhibiting Aβ production and aggregation or promoting $A\beta_{1\text{-}42}$ clearance [14]. However, despite many long and expensive trials, no disease-modifying drug for AD has been approved [15,16]. A recent failure in phase 3 involved the investigation of a β secretase in patients with mild-to-moderate AD [17]. Other large, phase 3 trials using anti-amyloid approaches including semagacestat [18], bapineuzumab [19], and solanezumab [20], have yielded disappointing results. However, it has been recently reported that BAN2401 (mAb158), an antibody developed for early AD with a unique target binding profile selective for Aβ PFs, significantly slowed cognitive decline by 30%, with a concomitant reduction in amyloid plaques, compared with placebo at 18 months [21].

In this review, we focus on recent developments from basic and clinical studies of PFs, including research findings from our laboratory.

2. PFs Are Primary Toxins in AD

2.1. The Discovery of PFs and Their Role in AD Pathogenesis

PFs were first described by Teplow and colleagues in 1997 [6]. Using a size exclusion chromatography (SEC) system and the synthetic $A\beta_{1\text{-}42}$ peptide, they found a peak representing a large (>100 kDa) soluble species before the peak of the LMW-Aβ (mainly monomer) [6]. Using electron microscopy (EM), they further revealed that this peak contained predominantly curved fibrils, with a diameter of ~5 nm and a length of up to 200 nm, which they termed PFs [6]. Subsequently, the authors elucidated that the PFs were composed primarily of β-sheets and partially random coils and α-helices in a secondary structure [6]. In the same year, using atomic force microscopy (AFM), Lansbury's group found the existence of a metastable intermediate species, which was termed Aβ PF [22]. Many data have shown that LMW-Aβ oligomers are on-pathway to fibril formation, while HMW-Aβ oligomers such as PFs are off-pathway [22–25]. Although the PF-to-fibril transition, characterized by PF elongation, was very slow, preformed fibrillar seeds greatly accelerated this conversion [22]. Recently, using a combination of high-speed AFM with thioflavin T assay, EM, and re-injection assays by SEC, we demonstrated that fibril formation from PFs is more difficult than that from LMW-Aβ, suggesting that mature fibrils of $A\beta_{1\text{-}42}$ are primarily formed from LMW-$A\beta_{1\text{-}42}$ and not from PFs [24]. Furthermore, we determined that PFs instead supplied precursors to LMW-$A\beta_{1\text{-}42}$ by their dissociation, suggesting that PFs may not always represent the "on-pathway" of $A\beta_{1\text{-}42}$ aggregation from the monomer to the mature fibrils [24]. Kodali and Wetzel mentioned that, although $A\beta_{1\text{-}40}$ PFs can grow by monomer addition, their rate of growth is lower than that of mature fibrils. Additionally, while $A\beta_{1\text{-}40}$ monomer was able to support the extension of mature fibrils at low concentrations of,

A$\beta_{1\text{-}40}$ PFs exhibited no extension [23]. They suggested another terminology, "curvilinear fibrils", for the description of off-pathway PFs instead of PFs as on-pathway precursors of fibrils [23]. It was recently revealed that curvilinear fibrils inhibit fibril formation not only by slowing fibril nucleation and elongation, but also by actively disrupting either process based on combined thioflavin kinetics and AFM imaging data [26]. On the other hand, Iwatsubo's group showed that A$\beta_{1\text{-}42}$ PF injection induced Aβ deposition in the brains of A7 mice overexpressing human APP695 and harboring the K670N, M671L, and T714I familial AD neuronal mutations, suggesting that Aβ PFs may act as a seed for Aβ aggregation in vivo [27]. The injection of Aβ PFs mixed with apoE3 significantly attenuated Aβ deposition, whereas apoE4 did not, suggesting that the suppressive effect of apoE3 on the structural conversion of Aβ PFs to fibrils is stronger than that of apoE4, thereby impeding Aβ deposition in vivo [27].

2.2. PFs Are Primary Toxins in AD

The solubility and diffusible nature of soluble oligomers may render them more effective in terms of intra- and extra-cellular interactions and engaging microglial receptors compared with mature insoluble fibrils. Indeed, it has been demonstrated that astrocytes engulf large amounts of accumulated, rather than digested, A$\beta_{1\text{-}42}$ PFs. This intracellular storage of A$\beta_{1\text{-}42}$ results in severe astrocytic endosomal/lysosomal defects and the secretion of extracellular vesicles containing N-truncated, neurotoxic Aβ [28]. A$\beta_{1\text{-}42}$ PFs have also been shown to induce an inflammatory process through microglial activation [29] and initiate Toll-like receptor (TLR) signaling (Figure 1) [30]. In addition, these PFs are preferentially internalized by microglia [31]. Furthermore, it has been reported that A$\beta_{1\text{-}42}$ PFs are more effective at inducing microglial tumor necrosis factor α (TNFα) production in BV-2 and primary murine microglia in vitro than monomers and mature fibrils. Moreover, PFs of A$\beta_{1\text{-}40}$ exhibit significantly less activity than concentration-matched A$\beta_{1\text{-}42}$ [29]. A$\beta_{1\text{-}42}$ PFs also have been shown to trigger a time- and myeloid differentiation protein (MyD) 88-dependent process that generates TNFα and interleukin-1β (IL-1β) mRNA, along with pro and mature forms of the intracellular IL-1β protein [30]. The accumulation of both IL-1β forms has indicated that A$\beta_{1\text{-}42}$ PFs are able to prime and activate the Nod-like receptor (NLR) P3 inflammasome. In this process, Aβ has been shown to elicit a quantized burst of secreted IL-1β which occurs prior to the Aβ priming of the microglia. The IL-1β secretion burst appears to be rapid and not sustained, yet it may be re-initiated with additional Aβ stimulation. These findings indicate multiple modes of IL-1β regulation by A$\beta_{1\text{-}42}$ PFs, including TLR/MyD88-mediated priming, NLRP3 inflammasome activation, and modulation of the IL-1β secretory process, suggesting wide-ranging effects of Aβ on the innate immune response [30].

Recent evidence has suggested that the neuronal cell membrane is the chief site of oligomer-mediated neuronal damage. We recently studied the cellular response to short exposures to PFs using multiple indices of membrane integrity, cytolysis, oxidative stress, and synaptic function. We found that cellular membrane and metabolic integrity were more severely disrupted by PFs of A$\beta_{1\text{-}42}$ than LMW-A$\beta_{1\text{-}42}$, as evidenced by various experimental systems, including cell viability and leakage assays, fluorometric measures of ROS generation, lipid peroxidation assays, and electrophysiological recordings [13]. While our results for lactate dehydrogenase (LDH) and calcein and ethidium homodimer-1 assays reflected cellular membrane damage by PFs of A$\beta_{1\text{-}42}$ to a greater extent than LMW-A$\beta_{1\text{-}42}$, 3-[4,5-dimethylthiazol-2-yl]-2,5-diphenyltetrazolium bromide metabolism (MTT) and water soluble tetrazolium (WST) assays reflecting mitochondrial enzyme activity, they demonstrated only small differences between the Aβs in different cellular models, including SH-SY5Y cells and a healthy, human-induced pluripotent stem line [13]. From these results, in terms of short-term A$\beta_{1\text{-}42}$ PF treatment, A$\beta_{1\text{-}42}$ PFs may first attack the cell membrane, followed by subsequent damage to the mitochondria, although A$\beta_{1\text{-}42}$ dimers might not be removed clearly in LMW-A$\beta_{1\text{-}42}$ preparation using the above-mentioned SEC method [6]. Next, we found that exposure to PFs of A$\beta_{1\text{-}42}$ in SH-SY5Y cells induces more severe oxidative stress, including greater levels of ROS production and membrane lipid peroxidation, than LMW-A$\beta_{1\text{-}42}$. Indeed, many studies have reported that oxidative stress, which occurs in the presence of a physiological imbalance between ROS generation and antioxidant capacity, is a critical pathogenic mechanism in AD progression [32]. Along with the direct destruction/modification of lipids, DNA, and proteins, the byproducts of lipid peroxidation produced during oxidative stress cause damage to the mitochondria and upregulate tau phosphorylation, which appears essential for NFT formation [33]. In addition, the generation of superoxide by Aβ aggregates may lead to mitochondrial impairment and further induce ROS generation, thereby establishing a positive feedback pathway that ultimately results in cell death [34]. Moreover, Aβ aggregates may directly interact with the mitochondrial respiratory chain, causing metabolic dysfunction and increased ROS production [35]. In our study, the PFs of $A\beta_{1\text{-}42}$ also reduced neuronal membrane fluidity to a significantly greater extent than LMW-$A\beta_{1\text{-}42}$. Thus, we consider the possibility that the effects on membrane fluidity, and the resulting neuronal damage, depend on the specific Aβ conformation [13].

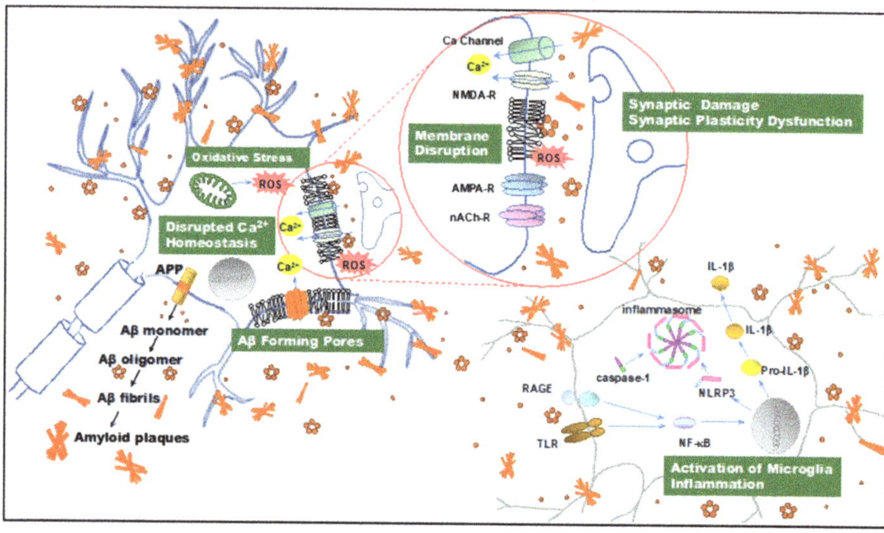

Figure 1. Illustration summarizing amyloid β-protein (Aβ) neurotoxicity. Aβ aggregates induce disruption of cellular homeostasis, which may be the result of inducing or exacerbating membrane disruption, oxidative stress, calcium dysregulation, synaptic plasticity dysfunction, and inflammation. APP: amyloid precursor protein; Aβ: amyloid β-protein; ROS: reactive oxygen species; NMDAR, *N*-methyl-D-aspartate receptor. AMPAR, α-amino-3-hydroxy-5-methyl-4-isoxazolepropionic acid receptor; nAChR, nicotinic acetylcholine receptor; TLR: toll-like receptor; RAGE: receptor for advanced glycation endproducts; NF-κB, nuclear factor κB; NLRP3: NOD-, LRR- and pyrin domain-containing protein 3; IL-1β: interleukin-1β.

We further demonstrated that short exposures to $A\beta_{1\text{-}42}$ PFs induces higher concentrations of $[Ca^{2+}]_i$ than LMW-$A\beta_{1\text{-}42}$, whereas a reduced depolarization-induced $[Ca^{2+}]_i$ influx through voltage-dependent Ca^{2+} channels was observed following longer exposures to $A\beta_{1\text{-}42}$ PFs [13]. These results suggested that PFs may not only directly damage voltage-gated calcium channels for a short time, but also alter the cell membrane environment required for proper channel insertion or gating for longer periods, as evidenced by lipid peroxidation and membrane fluidity measurements [13].

Consistent with the changes observed in $[Ca^{2+}]_i$ and the loss of membrane integrity, the application of PFs of $A\beta_{1\text{-}42}$, but not those of LMW-$A\beta_{1\text{-}42}$, also has been shown to depolarize SH-SY5Y cells and significantly reduce membrane input resistance [13]. Bode et al. monitored transmembrane currents during Aβ exposure at the extracellular face of excised membranes from HEK293 cells, and found that annular $A\beta_{1\text{-}42}$ oligomers formed ion channels, whereas $A\beta_{1\text{-}40}$ oligomers and mature fibrils and monomers did not [36]. Drolle et al. used multi-component lipid models to mimic healthy and

AD states of neuronal membranes and posited that $A\beta_{1\text{-}42}$ increases lipid membrane roughness and membrane conductance, possibly through pore formation [37]. Taken together, the $[Ca^{2+}]_i$ increase evoked by PFs may be due to pore formation and oxidative damage, as well as the suppression of calcium egress and sequestration pathways secondary to metabolic disruption.

We also demonstrated that $A\beta_{1\text{-}42}$ PFs significantly inhibit LTP formation in the mouse hippocampal CA1 subfield [13]. Similarly, it has been reported that PFs induce electrophysiological changes, including rapid increases in the excitatory post synaptic and action potentials, membrane depolarizations in rat cortical neurons [8], and the inhibition of LTP in the rat hippocampus [38]. Excessive ROS accumulation and decreased membrane fluidity are associated with effects on LTP and learning [39,40]. Furthermore, membrane pore formation may also impair cellular and synaptic functions (Figure 1) [41,42].

The small (35kDa) and highly dispersible protein, secreted-frizzled-related protein 1 (SFRP1), regulates transmembrane metalloprotease ADAM10 activity and is essential for the development of tissue homeostasis and constitutive levels of α-secretase in the brain [43]. As a novel player in AD pathogenesis, SFRP1 has been shown to be significantly increased in the brain and cerebrospinal fluid of patients with AD. In addition, SFRP1 has been demonstrated in human AD cases and mouse models to prevent Aβ PF formation by binding to Aβ, suggesting it may be a promising AD therapeutic target [44].

2.3. Arctic Mutation Causes Aβ PF Formation

Arctic mutation is a pathogenic APP mutation located within the Aβ sequence at codon 693, at which point glutamic acid is substituted for glycine (E693G). In 2001, Lannfelt's group named the mutation the 'Arctic' mutation because the family in which it was detected was from northern Sweden [12]. Affected subjects have clinical features of early AD and plasma levels of both $A\beta_{1\text{-}40}$ and $A\beta_{1\text{-}42}$ are lower in mutation carriers compared with healthy family members. In addition, concentrations of $A\beta_{1\text{-}42}$ were found to be reduced in media from cells transfected with APP_{E693G} [12]. Furthermore, the authors reported that the Arctic Aβ mutation ($A\beta_{1\text{-}40}$ Arc) causes enhanced the formation of $A\beta_{1\text{-}40}$ PFs in vitro [12]. Subsequently, Lannfelt's group found that the Arctic mutation significantly accelerated $A\beta_{1\text{-}42}$ PF formation, as well as PF fibrillization [7].

It has been reported that $A\beta_{1\text{-}40}$ Arc inhibits LTP ~100-fold more potently than wild-type $A\beta_{1\text{-}40}$ when wild-type and $A\beta_{1\text{-}40}$ Arc peptides are injected into the CA1 area in rats intracerebroventricularly. In this study, the isolated soluble fraction that included the PFs of $A\beta_{1\text{-}40}$ Arc after high-speed centrifugation was shown to still retain full LTP inhibitory activity [38]. In a later study, Lord et al. demonstrated that the Arc mutation accelerates early intraneuronal Aβ aggregation and PF formation, followed by plaque formation, in APP transgenic mice with both the Arctic (E693G) and Swedish (K670N, M671L) mutations (tg-APP_{ArcSwe}) [45,46]. In addition, cognitive deficits have been shown to occur concomitantly with the formation of intracellular Aβ deposits, but before plaque formation, in transgenic mice [45]. In addition, the levels of PFs in the brain, but not those of total Aβ, have been correlated with spatial learning, which adds further evidence to the theory that soluble PFs are the toxic species [47]. The pool of toxic Aβ species reportedly consists of molecules in the size range of 80 to 500 kDa [48].

3. Therapeutic Approaches Targeting Aβ PFs

3.1. Small Molecules Inhibit the Formation of Aβ PFs

Small molecules with the potential to mitigate toxic AD species such as $A\beta_{1\text{-}42}$ PFs are promising preventive and therapeutic candidates. We previously demonstrated that a grape-seed-derived polyphenol was able to inhibit $A\beta_{1\text{-}42}$ aggregation by preventing PF formation, pre-protofibrillar oligomerization, and random coil-aggregation-prone α-helix/β-sheet secondary structure transitions using various analyses, including circular dichroism spectroscopy, thioflavin T fluorescence, SEC, and EM [49]. Importantly, this polyphenol demonstrated protective effects in cytotoxicity assays, in which it was mixed with Aβ_{1-42} aggregates and exposed to cells [49]. Furthermore, our in vivo studies using the Tg2576 AD mouse model showed that this grape seed polyphenolic extract significantly attenuated AD-type cognitive deterioration and reduced cerebral amyloid deposition [50].

Using multiple molecular dynamics (MD) simulations, Jin et al. reported that dihydrochalcone, a compound extracted from the daemonorops draco tree, could effectively inhibit Aβ_{1-42} fibrillization and reduce Aβ-induced cytotoxicity by destabilizing the Aβ PFs. In this process, dihydrochalcone was shown to bind to the cavity of the Aβ_{1-40}/Aβ_{1-42} PFs themselves and disrupt the D23-K28 salt bridge and inter-peptide β–sheet in the β1 region [51]. In addition, Zhou et al. reported that 1,2-(dimethoxymethano)fullerene (DMF), a water-soluble fullerene derivative, strongly inhibited Aβ_{1-42} aggregation by binding with Aβ PFs on three dominant binding sites, namely, the central hydrophobic core (17LVFFA21), the turn site (27NKGAI31), and the C-terminal β-sheet site comprised of glycine and hydrophobic residues (31IIGLMVGGVVI41), by MD stimulations [52]. In addition, the binding of DMF to the turn region served to disrupt the D23-K28 salt-bridge critical for PF Aβ fibril formation [52]. Another series of MD stimulations showed that wgx-50, a compound extracted from the Sichuan pepper (Zanthoxylum bungeanum), can destabilize Aβ_{1-42} PFs through three possible stable binding sites, including two sites in the hydrophobic grooves on the surface of the Aβ PFs, which resulted in no significant changes in Aβ structure, and one site in the interior that caused PF destabilization. At this site, wgx-50 was observed to be packed against the side chains of I32 and L34, disrupting the D23-K28 salt bridge and partially opening the two tightly compacted β-sheets [53]. Recently, Saini et al. reported that a resveratrol and clioquinol hybrid compound, (E)-5-(4-hydroxystyryl)quinolone-8-ol, inhibits Aβ_{1-42} aggregation by preventing the conformational transition of the Aβ_{1-42} monomer and causing destablization of the Aβ_{1-42} PF structure using MD simulation [54]. The destabilizing mechanisms of the Aβ_{1-42} PF structure may be due to the increasing interchain distance between chains A–B, disrupting the salt-bridge interaction between D23-K28 and decreasing the number of backbone hydrogen bonds between the chains [54]. In the same year, it was reported that β-sheet breaker peptides, particularly PPFFE pentapeptides, display strong destablizing effects that shift the energy minima toward the lowest value of sheet content and the lowest number of hydrogen bonds in Aβ_{1-42} PFs, using in silico methodologies including the molecular mechanics Poisson–Bolzmann surface area method and MD simulations [55].

3.2. *Aβ PF-Selective Antibody*

PFs have been identified in the human brain and the APP transgenic mouse brain [48,56]. mAb158 is a murine monoclonal antibody developed to selectively target HMW-Aβ_{1-42} assemblies [56]. Using an enzyme-linked immunosorbent assay (ELISA), it has been elucidated that mAb158 has an at least 1000-fold higher selectivity for PFs than monomeric Aβ and 10-15 times better binding affinity to PFs than to mature fibrils, thereby targeting the more toxic species of the peptide [57]. In immunohistochemistry, mAb158 also detects Aβ in plaques and the vasculature of AD brains because of the massive amount of Aβ in these structures [58]. In addition, Lord et al. reported that mAb158 inhibits in vitro Aβ_{1-42} fibril formation and protected cells from Aβ PF-induced cytotoxicity [59]. A co-culture study of astrocytes, neurons, and oligodendrocytes exposed to Aβ_{1-42} PFs in the presence or absence of mAb158 demonstrated that the presence of mAb158 almost entirely abolished Aβ accumulation in astrocytes, indicating an effect towards Aβ PF degradation. Consequently, mAb158 treatment was shown to rescue neurons from Aβ-induced cell death [60].

The treatment of tg-APP$_{ArcSwe}$ mice with mAb158 resulted in the prevention of plaque formation if the antibody was administered before the appearance of plaques in young mice. If the treatment was started later in this mouse model, levels of insoluble Aβ were unaffected in the brains of plaque-bearing older mice. However, in both cases, soluble Aβ PF levels were diminished, supporting the notion that mAb158 can selectively reduce PF levels [59]. Similarly, the authors found that PF levels were elevated in young tg-APP$_{ArcSwe}$ mice compared with several transgenic models lacking the Arctic mutation. In older tg-APP$_{ArcSwe}$ mice with plaque deposition, the levels of Aβ PFs were approximately 50% higher than in younger mice, whereas levels of total Aβ were exponentially increased. Young tg-APP_{ArcSwe} mice showed deficits in spatial learning, and individual performances in the Morris water maze were inversely correlated with levels of Aβ PF, but not with total Aβ levels. These findings indicated that Aβ PFs accumulated in an age-dependent manner, and increased levels of Aβ PFs may result in spatial learning impairments in tg-APP_{ArcSwe} mice [47]. Lannfelt et al. reported that the murine version of mAb158 reached the brain and reduced brain PF levels by 42% in an exposure-dependent manner both after long-term (13 weeks) and short-term (4 weeks) treatment in tg-APP_{ArcSwe} mice [14]. Notably, a 53% reduction in PFs/oligomers in the cerebrospinal fluid (CSF), found to be correlated with reduced brain PF levels, was observed after long-term treatment, suggesting that CSF PFs/oligomers may be used as potential biomarkers of AD [14].

Recently, Sehlin's group succeeded in facilitating the brain uptake of mAb158 by using transferrin receptor-mediated transcytosis across the blood–brain barrier in tg-APP_{ArcSwe} mice [61]. ELISA analysis of the brain extracts demonstrated a 40% reduction in soluble Aβ PFs in both ten-fold lower-dose modified mAb158 and high-dose mAb158-treated mice, whereas there was no Aβ PF reduction in mice treated with a low dose of mAb158 [61]. Furthermore, ex vivo autoradiography and PET imaging have revealed different brain distribution patterns of modified mAb158 (brain parenchyma) and mAb158 (central periventricular areas), suggesting that these antibodies may affect Aβ levels by different mechanisms. This strategy may allow for decreased antibody doses, thereby reducing the side effects and treatment costs [61].

3.3. Clinical Application of mAb158

BAN2401, a humanized IgG1 monoclonal form of mAb158, exhibits a strong binding preference for soluble Aβ PFs compared with monomers [14]. In addition, it has been confirmed that both mAb158 and BAN2401 efficiently immunoprecipitate soluble Aβ aggregates in human AD brain extracts.

The first clinical study of BAN2401 demonstrated that the compound was safe and well tolerated in mild to moderate AD [62]. The incidence of amyloid-related imaging abnormalities (ARIA-E for edema /H for hemorrhage) on brain MRI scans was comparable to that of the placebo. BAN2401 exposure was approximately dose-proportional, with a serum terminal elimination half-life of approximately seven days. Only a slight increase in plasma $Aβ_{1\text{-}40}$ was observed, but there were no measurable effects of BAN2401 on CSF biomarkers such as $Aβ_{1\text{-}42}$, total-tau, and phosphorylayed-tau (p-tau) [62]. A recent phase 2 randomized trial reported that BAN2401's highest dose (10 mg/kg) significantly slowed cognitive decline in early AD, with a concomitant reduction in amyloid plaques, as measured by amyloid PET compared with placebo at 18 months [21]. BAN2401 significantly reduced amyloid plaques in the brain at all five treatment doses used in the trial, which involved 856 patients with mild cognitive impairment. The 30% slowing of cognitive decline at 18 months was based on the Alzheimer's Disease Composite Score (ADCOMS) created by Eisai. On the more widely used Alzheimer's Disease Assessment Scale cognitive subscale (ADAS-Cog), the highest dose of BAN2401 slowed a cognitive decline of 47% compared with placebo. However, the trial was not large enough to definitively demonstrate efficacy in improving cognitive function according to an overall optimistic statement from the Alzheimer Association. The drug also did not achieve its primary efficacy endpoint, namely, a change from baseline on the ADCOMS at 12 months [21]. Currently, BAN2401 is a part of an ongoing phase 3 clinical trial. In contrast, other clinical trials of monoclonal antibodies targeting fibrillar Aβ, such as bapineuzumab [63], or soluble monomeric Aβ, such as solanezumab [20], have failed to produce clinical effects.

In the fall of 2019, after trials of the drug EMERGE (aducanumab; BIIB037) were previously discontinued following a phase III futility analysis, Biogen, the company that developed the drug, announced that subsequent analysis of a larger dataset instead showed that EMERGE had met its primary endpoint. Patients on the highest dose, 10 mg/kg, had a significant reduction in decline in terms of the primary endpoint using the Clinical Dementia Rating Scale-Sum of Boxes (CDR-SB). This group also declined less in terms of secondary endpoints, including the Mini-Mental State Examination

(MMSE), ADAS-Cog, and the Alzheimer's Disease Cooperative Study/Activities of Daily Living scale adapted for patients with mild cognitive impairment (ADCS-ADL-MCI). In a parallel clinical trial of aducanumab, termed the ENGAGE trial, aducanumab did not meet the primary endpoint; however, an exploratory analysis suggested that a subgroup of people who had received 10 or more 10 mg/kg doses declined more slowly, which is consistent with the EMERGE participants. In both trials, aducanumab caused a dose-dependent reduction in brain Aβ and CSF p-tau. Based on the updated data analysis, Biogen announced plans to apply for regulatory approval of aducanumab in the US in early 2020 [64]. Since aducanumab may also bind aggregates such as oligomers of Aβ [65], these results may be important for interpreting data from the phase 3 clinical trial of BAN2401.

4. PFs Are Present in Other Neurodegenerative Diseases

PFs are formed from proteins implicated in other neurodegenerative diseases, including tauopathy [66], Parkinson's disease [67,68], familial amyloid polyneuropathy [69], and Huntington's disease [70], indicating a common mechanism. Similar to Aβ, tau and α-synuclein (αS) also form PFs with annular, pore-like structures, thereby exerting membrane permeabilization activity [66,67]. Analyses of annular tau PFs in brain tissue from patients with progressive supranuclear palsy, as well as that from the P301L mouse model, indicated that the annular PFs of tau are preceded by tau oligomers and do not go on to form neurofibrillary tangles (mature fibrils) [66]. In addition, it was recently reported that the αS oligomer and PFs interconvert during polymerization reactions, using the thioflavin T assay combined with SEC and EM [68]. Similarly, Groenning et al. described a dynamic transthyretin (TTR) protofibril structure that exchanges protomers with highly unfolded monomers in solution, using a combination of primarily small-angle X-ray scattering and hydrogen exchange mass spectrometry analysis. The TTR PFs were shown to only grow to an approximate final size of 2900 kDa and a length of 70 nm [69]. In a recent micro electron diffraction study at 0.75Å resolution, ultrahigh-resolution cryo-EM revealed that prion PFs are stabilized by a dense three-dimensional network of stabilizing hydrogen bonds that link residues between and within its β strands through polar clasps [71].

5. Conclusions and Future Perspectives

Unlike current therapies limited to the treatment of AD symptoms, research on Aβ aggregation has rapidly advanced, with growing evidence that soluble pre-fibrillar aggregates (i.e., oligomers of Aβ) are proximate neurotoxins. Indeed, recent data from both in vitro and in vivo studies have suggested that HMW oligomers as PFs induce neuronal injury and cognitive deficits via multiple mechanisms, including not only increasing Aβ plaque accumulation but also increasing direct membrane and synaptic damage. Furthermore, additional projects to fully characterize the PFs actually present in the human brain have been undertaken. Aβ PFs may be the primary pathogenic species of Aβ-related cognitive deficits, particularly in the early stage of AD, although it remains to be established how Aβ PFs, alone or together with other soluble oligomeric Aβ species, cause the neurodegeneration leading to AD. Disease-modifying therapies targeting toxic PFs will reach the clinical stage in the near future, and may have the potential to delay or even halt the further progression of AD. Further clarification of the toxic PFs of brain Aβ should aid in the development of more effective and safe drugs, as well as in novel diagnostic assays.

Author Contributions: K.O. and M.T. wrote the paper. Authorship must be limited to those who have contributed substantially to the work reported. All authors have read and agreed to the published version of the manuscript.

Funding: This work was supported by a Grant-in-Aid for Scientific Research (C) (26461266) (K.O.) from the Ministry of Education, Culture, Sports, Science and Technology, Japan, and a Grant for Research and Development Grants for Dementia from the Japan Agency for Medical Research and Development (16dk0207021h0001) (K.O.).

Conflicts of Interest: The authors declare no conflicts of interest.

Abbreviations

Aβ	amyloid β-protein
AD	Alzheimer's disease
ADAS-Cog	Alzheimer's Disease Assessment Scale cognitive subscale
ADCOMS	Alzheimer's Disease Composite Score
AFM	atomic force microscopy
APP	amyloid precursor protein
ARIA	amyloid-related imaging abnormalities
αS	α-synuclein
CSF	cerebrospinal fluid
DMF	1,2-(dimethoxymethano)fullerene
ELISA	enzyme-linked immunosorbent assay
EM	electron microscopy
HMW	high molecular weight
IL-1β	interleukin-1β
LDH	lactate dehydrogenase
LMW	low molecular weight
LTPs	long-term potentiation
MD	molecular dynamics
MTT	3-[4,5-dimethylthiazol-2-yl]-2,5-diphenyltetrazolium bromide metabolism
MyD	myeloid differentiation protein
NLR	Nod-like receptor
PFs	protofibrils
p-tau	phosphorylayed-tau
ROS	reactive oxygen species
SEC	size exclusion chromatography
SFRP1	secreted-frizzled-related protein 1
TLR	Toll-like receptor
TNFα	tumor necrosis factor α
TTR	transthyretin
WST	water soluble tetrazolium

References

1. Selkoe, D.J.; Hardy, J. The amyloid hypothesis of Alzheimer's disease at 25 years. *EMBO Mol. Med.* **2016**, *8*, 595–608. [CrossRef] [PubMed]
2. Hardy, J.; Selkoe, D.J. The amyloid hypothesis of Alzheimer's disease: Progress and problems on the road to therapeutics. *Science* **2002**, *297*, 353–356. [CrossRef] [PubMed]
3. Ono, K. Alzheimer's disease as oligomeropathy. *Neurochem. Int.* **2018**, *119*, 57–70. [CrossRef] [PubMed]
4. Ono, K.; Condron, M.M.; Teplow, D.B. Structure-neurotoxicity relationships of amyloid β-protein oligomers. *Proc. Natl. Acad. Sci. USA* **2009**, *106*, 14745–14750. [CrossRef] [PubMed]
5. Harper, J.D.; Lansbury, P.T., Jr. Models of amyloid seeding in Alzheimer's disease and scrapie: Mechanistic truths and physiological consequences of the time-dependent solubility of amyloid proteins. *Annu. Rev. Biochem.* **1997**, *66*, 385–407. [CrossRef]
6. Walsh, D.M.; Lomakin, A.; Benedek, G.B.; Condron, M.M.; Teplow, D.B. Amyloid β-protein fibrillogenesis. Detection of a protofibrillar intermediate. *J. Biol. Chem.* **1997**, *272*, 22364–22372. [CrossRef]
7. Johansson, A.S.; Berglind-Dehlin, F.; Karlsson, G.; Edwards, K.; Gellerfors, P.; Lannfelt, L. Physiochemical characterization of the Alzheimer's disease-related peptides Aβ 1-42Arctic and Aβ 1-42wt. *FEBS J.* **2006**, *273*, 2618–2630. [CrossRef]
8. Hartley, D.M.; Walsh, D.M.; Ye, C.P.; Diehl, T.; Vasquez, S.; Vassilev, P.M.; Teplow, D.B.; Selkoe, D.J. Protofibrillar intermediates of amyloid β-protein induce acute electrophysiological changes and progressive neurotoxicity in cortical neurons. *J. Neurosci.* **1999**, *19*, 8876–8884. [CrossRef]

9. Walsh, D.M.; Hartley, D.M.; Kusumoto, Y.; Fezoui, Y.; Condron, M.M.; Lomakin, A.; Benedek, G.B.; Selkoe, D.J.; Teplow, D.B. Amyloid β-protein fibrillogenesis. Structure and biological activity of protofibrillar intermediates. *J. Biol. Chem.* **1999**, *274*, 25945–25952. [CrossRef]
10. Ward, R.V.; Jennings, K.H.; Jepras, R.; Neville, W.; Owen, D.E.; Hawkins, J.; Christie, G.; Davis, J.B.; George, A.; Karran, E.H.; et al. Fractionation and characterization of oligomeric, protofibrillar and fibrillar forms of β-amyloid peptide. *Biochem. J.* **2000**, *348*, 137–144. [CrossRef]
11. Johansson, A.S.; Garlind, A.; Berglind-Dehlin, F.; Karlsson, G.; Edwards, K.; Gellerfors, P.; Ekholm-Pettersson, F.; Palmblad, J.; Lannfelt, L. Docosahexaenoic acid stabilizes soluble amyloid-β protofibrils and sustains amyloid-β-induced neurotoxicity in vitro. *FEBS J.* **2007**, *274*, 990–1000. [CrossRef] [PubMed]
12. Nilsberth, C.; Westlind-Danielsson, A.; Eckman, C.B.; Condron, M.M.; Axelman, K.; Forsell, C.; Stenh, C.; Luthman, J.; Teplow, D.B.; Younkin, S.G.; et al. The 'Arctic' APP mutation (E693G) causes Alzheimer's disease by enhanced Aβ protofibril formation. *Nat. Neurosci.* **2001**, *4*, 887–893. [CrossRef] [PubMed]
13. Yasumoto, T.; Takamura, Y.; Tsuji, M.; Watanabe-Nakayama, T.; Imamura, K.; Inoue, H.; Nakamura, S.; Inoue, T.; Kimura, A.; Yano, S.; et al. High molecular weight amyloid β1-42 oligomers induce neurotoxicity via plasma membrane damage. *FASEB J.* **2019**, *33*, 9220–9234. [CrossRef] [PubMed]
14. Tucker, S.; Moller, C.; Tegerstedt, K.; Lord, A.; Laudon, H.; Sjodahl, J.; Soderberg, L.; Spens, E.; Sahlin, C.; Waara, E.R.; et al. The murine version of BAN2401 (mAb158) selectively reduces amyloid-β protofibrils in brain and cerebrospinal fluid of tg-ArcSwe mice. *J. Alzheimers Dis.* **2015**, *43*, 575–588. [CrossRef]
15. Gauthier, S.; Albert, M.; Fox, N.; Goedert, M.; Kivipelto, M.; Mestre-Ferrandiz, J.; Middleton, L.T. Why has therapy development for dementia failed in the last two decades? *Alzheimers Dement.* **2016**, *12*, 60–64. [CrossRef]
16. Cummings, J.; Aisen, P.S.; DuBois, B.; Frolich, L.; Jack, C.R., Jr.; Jones, R.W.; Morris, J.C.; Raskin, J.; Dowsett, S.A.; Scheltens, P. Drug development in Alzheimer's disease: The path to 2025. *Alzheimers Res. Ther.* **2016**, *8*, 39. [CrossRef]
17. Burki, T. Alzheimer's disease research: The future of BACE inhibitors. *Lancet* **2018**, *391*, 2486. [CrossRef]
18. Doody, R.S.; Raman, R.; Farlow, M.; Iwatsubo, T.; Vellas, B.; Joffe, S.; Kieburtz, K.; He, F.; Sun, X.; Thomas, R.G.; et al. A phase 3 trial of semagacestat for treatment of Alzheimer's disease. *N. Engl. J. Med.* **2013**, *369*, 341–350. [CrossRef]
19. Vandenberghe, R.; Rinne, J.O.; Boada, M.; Katayama, S.; Scheltens, P.; Vellas, B.; Tuchman, M.; Gass, A.; Fiebach, J.B.; Hill, D.; et al. Bapineuzumab for mild to moderate Alzheimer's disease in two global, randomized, phase 3 trials. *Alzheimers Res. Ther.* **2016**, *8*, 18. [CrossRef]
20. Honig, L.S.; Vellas, B.; Woodward, M.; Boada, M.; Bullock, R.; Borrie, M.; Hager, K.; Andreasen, N.; Scarpini, E.; Liu-Seifert, H.; et al. Trial of Solanezumab for Mild Dementia Due to Alzheimer's Disease. *N. Engl. J. Med.* **2018**, *378*, 321–330. [CrossRef]
21. Abbasi, J. Promising Results in 18-Month Analysis of Alzheimer Drug Candidate. *JAMA* **2018**, *320*, 965. [CrossRef] [PubMed]
22. Harper, J.D.; Wong, S.S.; Lieber, C.M.; Lansbury, P.T. Observation of metastable Aβ amyloid protofibrils by atomic force microscopy. *Chem. Biol.* **1997**, *4*, 119–125. [CrossRef]
23. Kodali, R.; Wetzel, R. Polymorphism in the intermediates and products of amyloid assembly. *Curr. Opin. Struct. Biol.* **2007**, *17*, 48–57. [CrossRef] [PubMed]
24. Watanabe-Nakayama, T.; Ono, K.; Itami, M.; Takahashi, R.; Teplow, D.B.; Yamada, M. High-speed atomic force microscopy reveals structural dynamics of amyloid β1-42 aggregates. *Proc. Natl. Acad. Sci. USA* **2016**, *113*, 5835–5840. [CrossRef] [PubMed]
25. Cline, E.N.; Bicca, M.A.; Viola, K.L.; Klein, W.L. The Amyloid-β Oligomer Hypothesis: Beginning of the Third Decade. *J. Alzheimers Dis.* **2018**, *64*, S567–S610. [CrossRef] [PubMed]
26. Hasecke, F.; Miti, T.; Perez, C.; Barton, J.; Scholzel, D.; Gremer, L.; Gruning, C.S.R.; Matthews, G.; Meisl, G.; Knowles, T.P.J.; et al. Origin of metastable oligomers and their effects on amyloid fibril self-assembly. *Chem. Sci.* **2018**, *9*, 5937–5948. [CrossRef] [PubMed]
27. Hori, Y.; Hashimoto, T.; Nomoto, H.; Hyman, B.T.; Iwatsubo, T. Role of Apolipoprotein E in β-Amyloidogenesis: Isoform-Specific Effects On Protofibril To Fibril Conversion Of Aβ In Vitro And Brain Aβ Deposition In Vivo. *J. Biol. Chem.* **2015**, *290*, 15163–15174. [CrossRef]

28. Sollvander, S.; Nikitidou, E.; Brolin, R.; Soderberg, L.; Sehlin, D.; Lannfelt, L.; Erlandsson, A. Accumulation of amyloid-β by astrocytes result in enlarged endosomes and microvesicle-induced apoptosis of neurons. *Mol. Neurodegener.* **2016**, *11*, 38. [CrossRef]
29. Paranjape, G.S.; Gouwens, L.K.; Osborn, D.C.; Nichols, M.R. Isolated amyloid-β(1-42) protofibrils, but not isolated fibrils, are robust stimulators of microglia. *ACS Chem. Neurosci.* **2012**, *3*, 302–311. [CrossRef]
30. Terrill-Usery, S.E.; Mohan, M.J.; Nichols, M.R. Amyloid-β(1-42) protofibrils stimulate a quantum of secreted IL-1β despite significant intracellular IL-1β accumulation in microglia. *Biochim. Biophys. Acta* **2014**, *1842*, 2276–2285. [CrossRef]
31. Gouwens, L.K.; Makoni, N.J.; Rogers, V.A.; Nichols, M.R. Amyloid-β42 protofibrils are internalized by microglia more extensively than monomers. *Brain Res.* **2016**, *(Pt A)*, 485–495. [CrossRef]
32. Rosini, M.; Simoni, E.; Milelli, A.; Minarini, A.; Melchiorre, C. Oxidative stress in Alzheimer's disease: Are we connecting the dots? *J. Med. Chem.* **2014**, *57*, 2821–2831. [CrossRef] [PubMed]
33. Ansari, M.A.; Scheff, S.W. Oxidative stress in the progression of Alzheimer disease in the frontal cortex. *J. Neuropathol. Exp. Neurol.* **2010**, *69*, 155–167. [CrossRef] [PubMed]
34. Pohanka, M. Alzheimer's disease and oxidative stress: A review. *Curr. Med. Chem.* **2014**, *21*, 356–364. [CrossRef] [PubMed]
35. Hernandez-Zimbron, L.F.; Luna-Munoz, J.; Mena, R.; Vazquez-Ramirez, R.; Kubli-Garfias, C.; Cribbs, D.H.; Manoutcharian, K.; Gevorkian, G. Amyloid-β peptide binds to cytochrome C oxidase subunit 1. *PLoS ONE* **2012**, *7*, e42344. [CrossRef]
36. Bode, D.C.; Baker, M.D.; Viles, J.H. Ion Channel Formation by Amyloid-β42 Oligomers but Not Amyloid-β40 in Cellular Membranes. *J. Biol. Chem.* **2017**, *292*, 1404–1413. [CrossRef] [PubMed]
37. Drolle, E.; Negoda, A.; Hammond, K.; Pavlov, E.; Leonenko, Z. Changes in lipid membranes may trigger amyloid toxicity in Alzheimer's disease. *PLoS ONE* **2017**, *12*, e0182194. [CrossRef]
38. Klyubin, I.; Walsh, D.M.; Cullen, W.K.; Fadeeva, J.V.; Anwyl, R.; Selkoe, D.J.; Rowan, M.J. Soluble Arctic amyloid β protein inhibits hippocampal long-term potentiation in vivo. *Eur. J. Neurosci.* **2004**, *19*, 2839–2846. [CrossRef]
39. Massaad, C.A.; Klann, E. Reactive oxygen species in the regulation of synaptic plasticity and memory. *Antioxid Redox Signal.* **2011**, *14*, 2013–2054. [CrossRef]
40. Schaeffer, E.L.; Bassi, F., Jr.; Gattaz, W.F. Inhibition of phospholipase A2 activity reduces membrane fluidity in rat hippocampus. *J. Neural. Transm. (Vienna)* **2005**, *112*, 641–647. [CrossRef]
41. Alzheimer's Association Calcium Hypothesis Workgroup. Calcium Hypothesis of Alzheimer's disease and brain aging: A framework for integrating new evidence into a comprehensive theory of pathogenesis. *Alzheimers Dement.* **2017**, *13*, 178–182. [CrossRef] [PubMed]
42. Berridge, M.J. Calcium signalling in health and disease. *Biochem. Biophys. Res. Commun.* **2017**, *485*, 5. [CrossRef] [PubMed]
43. Jorissen, E.; Prox, J.; Bernreuther, C.; Weber, S.; Schwanbeck, R.; Serneels, L.; Snellinx, A.; Craessaerts, K.; Thathiah, A.; Tesseur, I.; et al. The disintegrin/metalloproteinase ADAM10 is essential for the establishment of the brain cortex. *J. Neurosci.* **2010**, *30*, 4833–4844. [CrossRef]
44. Esteve, P.; Rueda-Carrasco, J.; Ines Mateo, M.; Martin-Bermejo, M.J.; Draffin, J.; Pereyra, G.; Sandonis, A.; Crespo, I.; Moreno, I.; Aso, E.; et al. Elevated levels of Secreted-Frizzled-Related-Protein 1 contribute to Alzheimer's disease pathogenesis. *Nat. Neurosci.* **2019**, *22*, 1258–1268. [CrossRef] [PubMed]
45. Knobloch, M.; Konietzko, U.; Krebs, D.C.; Nitsch, R.M. Intracellular Aβ and cognitive deficits precede β-amyloid deposition in transgenic arcAβ mice. *Neurobiol. Aging* **2007**, *28*, 1297–1306. [CrossRef]
46. Lord, A.; Kalimo, H.; Eckman, C.; Zhang, X.Q.; Lannfelt, L.; Nilsson, L.N. The Arctic Alzheimer mutation facilitates early intraneuronal Aβ aggregation and senile plaque formation in transgenic mice. *Neurobiol. Aging* **2006**, *27*, 67–77. [CrossRef]
47. Lord, A.; Englund, H.; Soderberg, L.; Tucker, S.; Clausen, F.; Hillered, L.; Gordon, M.; Morgan, D.; Lannfelt, L.; Pettersson, F.E.; et al. Amyloid-β protofibril levels correlate with spatial learning in Arctic Alzheimer's disease transgenic mice. *FEBS J.* **2009**, *276*, 995–1006. [CrossRef] [PubMed]
48. Sehlin, D.; Englund, H.; Simu, B.; Karlsson, M.; Ingelsson, M.; Nikolajeff, F.; Lannfelt, L.; Pettersson, F.E. Large aggregates are the major soluble Aβ species in AD brain fractionated with density gradient ultracentrifugation. *PLoS ONE* **2012**, *7*, e32014. [CrossRef] [PubMed]

49. Ono, K.; Condron, M.M.; Ho, L.; Wang, J.; Zhao, W.; Pasinetti, G.M.; Teplow, D.B. Effects of grape seed-derived polyphenols on amyloid β-protein self-assembly and cytotoxicity. *J. Biol. Chem.* **2008**, *283*, 32176–32187. [CrossRef] [PubMed]
50. Wang, J.; Ho, L.; Zhao, W.; Ono, K.; Rosensweig, C.; Chen, L.; Humala, N.; Teplow, D.B.; Pasinetti, G.M. Grape-derived polyphenolics prevent Aβ oligomerization and attenuate cognitive deterioration in a mouse model of Alzheimer's disease. *J. Neurosci.* **2008**, *28*, 6388–6392. [CrossRef]
51. Jin, Y.; Sun, Y.; Lei, J.; Wei, G. Dihydrochalcone molecules destabilize Alzheimer's amyloid-β protofibrils through binding to the protofibril cavity. *Phys. Chem. Chem. Phys.* **2018**, *20*, 17208–17217. [CrossRef] [PubMed]
52. Zhou, X.; Xi, W.; Luo, Y.; Cao, S.; Wei, G. Interactions of a water-soluble fullerene derivative with amyloid-β protofibrils: Dynamics, binding mechanism, and the resulting salt-bridge disruption. *J. Phys. Chem. B* **2014**, *118*, 6733–6741. [CrossRef] [PubMed]
53. Fan, H.M.; Gu, R.X.; Wang, Y.J.; Pi, Y.L.; Zhang, Y.H.; Xu, Q.; Wei, D.Q. Destabilization of Alzheimer's Aβ42 Protofibrils with a Novel Drug Candidate wgx-50 by Molecular Dynamics Simulations. *J. Phys. Chem. B* **2015**, *119*, 11196–11202. [CrossRef] [PubMed]
54. Saini, R.K.; Shuaib, S.; Goyal, D.; Goyal, B. Insights into the inhibitory mechanism of a resveratrol and clioquinol hybrid against Aβ42 aggregation and protofibril destabilization: A molecular dynamics simulation study. *J. Biomol. Struct. Dyn.* **2019**, *37*, 3183–3197. [CrossRef] [PubMed]
55. Shuaib, S.; Narang, S.S.; Goyal, D.; Goyal, B. Computational design and evaluation of β-sheet breaker peptides for destabilizing Alzheimer's amyloid-β42 protofibrils. *J. Cell Biochem.* **2019**, *120*, 17935–17950.
56. Englund, H.; Sehlin, D.; Johansson, A.S.; Nilsson, L.N.; Gellerfors, P.; Paulie, S.; Lannfelt, L.; Pettersson, F.E. Sensitive ELISA detection of amyloid-β protofibrils in biological samples. *J. Neurochem.* **2007**, *103*, 334–345. [CrossRef]
57. Sehlin, D.; Hedlund, M.; Lord, A.; Englund, H.; Gellerfors, P.; Paulie, S.; Lannfelt, L.; Pettersson, F.E. Heavy-chain complementarity-determining regions determine conformation selectivity of anti-Aβantibodies. *Neurodegener. Dis.* **2011**, *8*, 117–123. [CrossRef]
58. Lannfelt, L.; Moller, C.; Basun, H.; Osswald, G.; Sehlin, D.; Satlin, A.; Logovinsky, V.; Gellerfors, P. Perspectives on future Alzheimer therapies: Amyloid-β protofibrils - a new target for immunotherapy with BAN2401 in Alzheimer's disease. *Alzheimers Res. Ther.* **2014**, *6*, 16. [CrossRef]
59. Lord, A.; Gumucio, A.; Englund, H.; Sehlin, D.; Sundquist, V.S.; Soderberg, L.; Moller, C.; Gellerfors, P.; Lannfelt, L.; Pettersson, F.E.; et al. An amyloid-β protofibril-selective antibody prevents amyloid formation in a mouse model of Alzheimer's disease. *Neurobiol. Dis.* **2009**, *36*, 425–434. [CrossRef]
60. Sollvander, S.; Nikitidou, E.; Gallasch, L.; Zysk, M.; Soderberg, L.; Sehlin, D.; Lannfelt, L.; Erlandsson, A. The Aβ protofibril selective antibody mAb158 prevents accumulation of Aβ in astrocytes and rescues neurons from Aβ-induced cell death. *J. Neuroinflammation* **2018**, *15*, 98. [CrossRef]
61. Syvanen, S.; Hultqvist, G.; Gustavsson, T.; Gumucio, A.; Laudon, H.; Soderberg, L.; Ingelsson, M.; Lannfelt, L.; Sehlin, D. Efficient clearance of Aβ protofibrils in AβPP-transgenic mice treated with a brain-penetrating bifunctional antibody. *Alzheimers Res. Ther.* **2018**, *10*, 49. [CrossRef] [PubMed]
62. Logovinsky, V.; Satlin, A.; Lai, R.; Swanson, C.; Kaplow, J.; Osswald, G.; Basun, H.; Lannfelt, L. Safety and tolerability of BAN2401–a clinical study in Alzheimer's disease with a protofibril selective Aβ antibody. *Alzheimers Res. Ther.* **2016**, *8*, 14. [CrossRef] [PubMed]
63. Salloway, S.; Sperling, R.; Fox, N.C.; Blennow, K.; Klunk, W.; Raskind, M.; Sabbagh, M.; Honig, L.S.; Porsteinsson, A.P.; Ferris, S.; et al. Two phase 3 trials of bapineuzumab in mild-to-moderate Alzheimer's disease. *N. Engl. J. Med.* **2014**, *370*, 322–333. [CrossRef] [PubMed]
64. Aducanumab. Available online: https://www.alzforum.org/therapeutics/aducanumab (accessed on 25 October 2019).
65. Sevigny, J.; Chiao, P.; Bussiere, T.; Weinreb, P.H.; Williams, L.; Maier, M.; Dunstan, R.; Salloway, S.; Chen, T.; Ling, Y.; et al. The antibody aducanumab reduces Aβ plaques in Alzheimer's disease. *Nature* **2016**, *537*, 50–56. [CrossRef] [PubMed]
66. Lasagna-Reeves, C.A.; Sengupta, U.; Castillo-Carranza, D.; Gerson, J.E.; Guerrero-Munoz, M.; Troncoso, J.C.; Jackson, G.R.; Kayed, R. The formation of tau pore-like structures is prevalent and cell specific: Possible implications for the disease phenotypes. *Acta Neuropathol. Commun.* **2014**, *2*, 56. [CrossRef]

67. Lashuel, H.A.; Petre, B.M.; Wall, J.; Simon, M.; Nowak, R.J.; Walz, T.; Lansbury, P.T., Jr. Alpha-synuclein, especially the Parkinson's disease-associated mutants, forms pore-like annular and tubular protofibrils. *J. Mol. Biol.* **2002**, *322*, 1089–1102. [CrossRef]
68. De Oliveira, G.A.P.; Silva, J.L. Alpha-synuclein stepwise aggregation reveals features of an early onset mutation in Parkinson's disease. *Commun. Biol.* **2019**, *2*, 374. [CrossRef]
69. Groenning, M.; Campos, R.I.; Hirschberg, D.; Hammarstrom, P.; Vestergaard, B. Considerably Unfolded Transthyretin Monomers Preceed and Exchange with Dynamically Structured Amyloid Protofibrils. *Sci. Rep.* **2015**, *5*, 11443. [CrossRef]
70. Beasley, M.; Stonebraker, A.R.; Hasan, I.; Kapp, K.L.; Liang, B.J.; Agarwal, G.; Groover, S.; Sedighi, F.; Legleiter, J. Lipid Membranes Influence the Ability of Small Molecules To Inhibit Huntingtin Fibrillization. *Biochemistry* **2019**, *58*, 4361–4373. [CrossRef]
71. Gallagher-Jones, M.; Glynn, C.; Boyer, D.R.; Martynowycz, M.W.; Hernandez, E.; Miao, J.; Zee, C.T.; Novikova, I.V.; Goldschmidt, L.; McFarlane, H.T.; et al. Sub-angstrom cryo-EM structure of a prion protofibril reveals a polar clasp. *Nat. Struct. Mol. Biol.* **2018**, *25*, 131–134. [CrossRef]

International Journal of
Molecular Sciences

MDPI

Case Report

Heavy Tau Burden with Subtle Amyloid β Accumulation in the Cerebral Cortex and Cerebellum in a Case of Familial Alzheimer's Disease with APP Osaka Mutation

Hiroyuki Shimada [1], Shinobu Minatani [2], Jun Takeuchi [2], Akitoshi Takeda [2], Joji Kawabe [3], Yasuhiro Wada [4], Aya Mawatari [4], Yasuyoshi Watanabe [4], Hitoshi Shimada [5], Makoto Higuchi [5], Tetsuya Suhara [5], Takami Tomiyama [6] and Yoshiaki Itoh [2,*]

1 Department of Radiology, Osaka City University Graduate School of Medicine, Osaka 545-8585, Japan; h.shimada@med.osaka-cu.ac.jp

2 Department of Neurology, Osaka City University Graduate School of Medicine, Osaka 545-8585, Japan; m4518413@med.osaka-cu.ac.jp (S.M.); jun-t@med.osaka-cu.ac.jp (J.T.); a-taked@med.osaka-cu.ac.jp (A.T.)

3 Department of Nuclear Medicine, Osaka City University Graduate School of Medicine, Osaka 45-8585, Japan; kawabe@med.osaka-cu.ac.jp

4 RIKEN Center for Biosystems Dynamics Research, Kobe 650-0047, Japan; yasuwada@riken.jp (Y.W.); a.mawatari@riken.jp (A.M.); yywata@riken.jp (Y.W.)

5 Department of Functional Brain Imaging Research (DOFI), National Institutes for Quantum and Radiological Science and Technology (QST), Chiba 263-8555, Japan; shimada.hitoshi@qst.go.jp (H.S.); higuchi.makoto@qst.go.jp (M.H.); suhara.tetsuya@qst.go.jp (T.S.)

6 Department of Translational Neuroscience, Osaka City University Graduate School of Medicine, Osaka 545-8585, Japan; tomi@med.osaka-cu.ac.jp

* Correspondence: y-itoh@med.osaka-cu.ac.jp

Received: 27 May 2020; Accepted: 19 June 2020; Published: 22 June 2020

Abstract: We previously identified a novel mutation in amyloid precursor protein from a Japanese pedigree of familial Alzheimer's disease, FAD (Osaka). Our previous positron emission tomography (PET) study revealed that amyloid β (Aβ) accumulation was negligible in two sister cases of this pedigree, indicating a possibility that this mutation induces dementia without forming senile plaques. To further explore the relationship between Aβ, tau and neurodegeneration, we performed tau and Aβ PET imaging in the proband of FAD (Osaka) and in patients with sporadic Alzheimer's disease (SAD) and healthy controls (HCs). The FAD (Osaka) patient showed higher uptake of tau PET tracer in the frontal, lateral temporal, and parietal cortices, posterior cingulate gyrus and precuneus than the HCs (>2.5 SD) and in the lateral temporal and parietal cortices than the SAD patients (>2 SD). Most noticeably, heavy tau tracer accumulation in the cerebellum was found only in the FAD (Osaka) patient. Scatter plot analysis of the two tracers revealed that FAD (Osaka) exhibits a distinguishing pattern with a heavy tau burden and subtle Aβ accumulation in the cerebral cortex and cerebellum. These observations support our hypothesis that Aβ can induce tau accumulation and neuronal degeneration without forming senile plaques.

Keywords: amyloid PET; tau PET; amyloid precursor protein; mutation; familial Alzheimer's disease; cerebellum; PBB3; PiB

1. Introduction

Senile plaques and neurofibrillary tangles (NFTs) in the brain are hallmarks of Alzheimer's disease (AD). These pathological changes can be visualized and assessed clinically by positron emission tomography (PET) with radioisotope-labeled probes specific for fibrillar Aβ and tau, such as

^{11}C-Pittsburgh compound-B (PiB) for amyloid and ^{11}C-pyridinyl-butadienyl-benzothiazole 3 (PBB3) for NFTs [1–3]. Since the existence of senile plaques is a prerequisite for the pathological diagnosis of AD, only individuals with dementia who are shown to be positive for amyloid by PET fulfill the recent clinical criteria for typical AD, including the IWG-2 criteria [4]. However, the view that the true culprit that initiates AD is not senile plaques but pathologically invisible, small oligomeric aggregates of Aβ has been widely accepted [5,6]. Animal and organotypic experiments have suggested that Aβ oligomers cause the synaptic and cognitive dysfunction as well as early pathological changes in AD, including tau hyperphosphorylation [7–9]. However, in humans, it is still unclear whether AD develops only with Aβ oligomers, or whether Aβ oligomers can induce the later pathologies of AD, including NFTs, in the absence of senile plaques.

Previously, we identified a novel mutation in amyloid precursor protein (APP) from a pedigree of familial AD in Osaka, Japan, FAD (Osaka) [10]. This 'Osaka' mutation is the deletion of codon 693 of the APP gene, resulting in mutant Aβ that lacks the 22nd glutamate. Only homozygous carriers suffer from dementia, indicating that this mutation is recessive. In vitro studies revealed that this mutation has a very unique characteristic that accelerates Aβ oligomerization but does not form amyloid fibrils [10]. Transgenic (Tg) mice expressing human APP with this mutation (APP_{OSK} mice) displayed intraneuronal accumulation of Aβ oligomers followed by synaptic and cognitive impairment, tau hyperphosphorylation, glial activation, and neurodegeneration but not amyloid plaques [11]. Furthermore, double Tg mice expressing both APP_{OSK} and wild-type human tau demonstrated that NFTs are also induced by Aβ oligomers alone [12].

To confirm these findings in humans, we evaluated amyloid accumulation in two sister patients harboring FAD (Osaka) [13]. PiB-PET scans revealed almost negligible amounts of Aβ accumulation in both patients, supporting our speculation that this mutation causes disease without forming senile plaques. To further explore the relationship between Aβ, tau and neurodegeneration, we performed tau and Aβ imaging in the proband of FAD (Osaka) and in patients with sporadic AD (SAD) and healthy controls (HCs) using PET and MRI.

2. Results

2.1. Demographic Data

Table 1 shows the demographic data for the groups that included the FAD (Osaka) patient (n = 1), early sporadic AD (early SAD) patients (n = 6), an advanced SAD patient (n = 1) and healthy controls (HCs) (n = 12). The age of the FAD (Osaka) patient at the time of the imaging, 70 years old, was comparable to that of the early SAD patients (mean ± standard deviation (SD), 69.7 ± 12.4 years old). The HCs were chosen to match the age distribution of the patients with AD (71.8 ± 8.7 years old).

Table 1. Demographic Data.

	n	Age	Gender (M/F)	Disease Duration (y)	MMSE
FAD (Osaka)	1	70	0/1	14	0
early SAD	6	69.7 ± 12.4	4/2	3.1 ± 1.7	23.3 ± 3.7
advanced SAD	1	53	0/1	6	0
HCs	12	71.8 ± 8.7	7/5	n.a.	28.8 ± 1.3

FAD (Osaka): familial Alzheimer's disease in Osaka; SAD: sporadic Alzheimer's disease; HCs: healthy controls; MMSE: Mini-Mental State Examination; average ± standard deviation; n.a.: not available.

In contrast, the disease duration of the FAD (Osaka) patient at the time of tau imaging, 14 years, was much longer than that of the early SAD patients (3.1 ± 1.7 years) and, to a lesser extent, longer than that of the advanced SAD patient (6 years). At this time, the FAD (Osaka) patient was in an extremely advanced stage with a Mini-Mental State Examination (MMSE) score of 0 points, whereas the early SAD patients had MMSE scores just below the cutoff level for dementia. One patient with SAD was in an advanced stage and unable to communicate at all. Subjects in the HC group lost few or no points on the cognitive test.

2.2. MRI Study

The FAD (Osaka) patient had severely advanced brain atrophy, including most of the cerebral cortex and brain stem shown in a T1 weighted MRI (Figure 1). Parahippocampal atrophy and ventricular enlargement were prominent in the coronal section. The cerebellum and primary motor cortex were relatively spared. These changes in FAD (Osaka) were all noticeable by comparing FAD (Osaka) to HC (Figure 1).

In contrast, representative images of the early SAD group showed only minor hippocampal atrophy with no other cortical involvement on MRI. The patient with advanced SAD had diffuse cortical atrophy with ventricular enlargement.

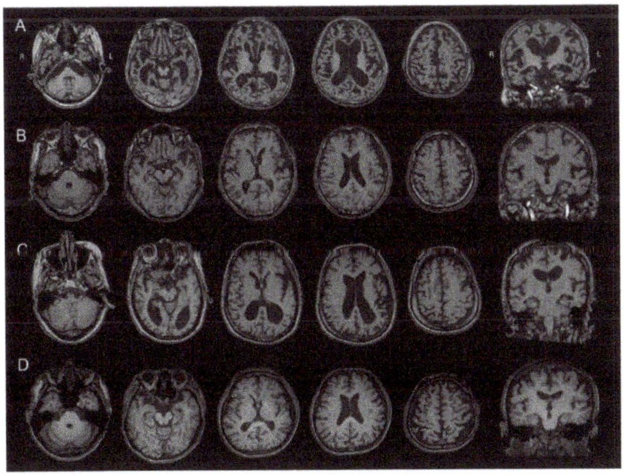

Figure 1. T1-weighted MRI scans of a patient with familial Alzheimer's disease with Osaka mutation (FAD (Osaka)) (**A**); a patient with early stage sporadic Alzheimer's disease (early SAD) (**B**); a patient with advanced stage of SAD (**C**); and a healthy control (HC) (**D**). The FAD (Osaka) patient had severely advanced brain atrophy including most of the cerebral cortex and brain stem. Parahippocampal atrophy and ventricular enlargement were prominent in the coronal section. The cerebellum and primary mortar cortex were relatively spared. R: right, L: left.

2.3. Tau PET Imaging

Representative tau PET images of the 4 groups are shown in Figure 2. The standard uptake value ratio (SUVR) values of accumulated PBB3 with reference to the midbrain were calculated. In the FAD (Osaka) patient, increased retention of ^{11}C-PBB3 was noticeable in most of the cerebral cortex except for the medial temporal cortex, including the hippocampus. The accumulation was also prominent in the cerebellar cortex. In contrast, elevated ^{11}C-PBB3 radio signals were small and limited to the frontal, parietal, and lateral temporal cortex, and the precuneus and the posterior cingulate gyrus in the early SAD patients. Furthermore, the advanced SAD patient showed diffusely increased ^{11}C-PBB3 signals in the cerebral cortex, but they seemed less increased than that of the FAD (Osaka) patient. No increase in the ^{11}C-PBB3 signals was found in HC.

The regional SUVR values with reference to the midbrain in each group are shown in Figure 3. In the cerebral cortex, ^{11}C-PBB3 accumulation in the FAD (Osaka) patient was higher than that in the HCs in all measured regions, including the frontal cortex, lateral temporal cortex, posterior cingulate gyrus, precuneus and parietal cortex (>2.5 SD). ^{11}C-PBB3 accumulation was also higher than that of both the early and advanced SAD groups, especially in the lateral temporal and parietal cortex (>2 SD and >2.5 SD, respectively, in early SAD). In contrast, remarkable uptake of ^{11}C-PBB3 in the cerebellum was found only in the FAD (Osaka) patient.

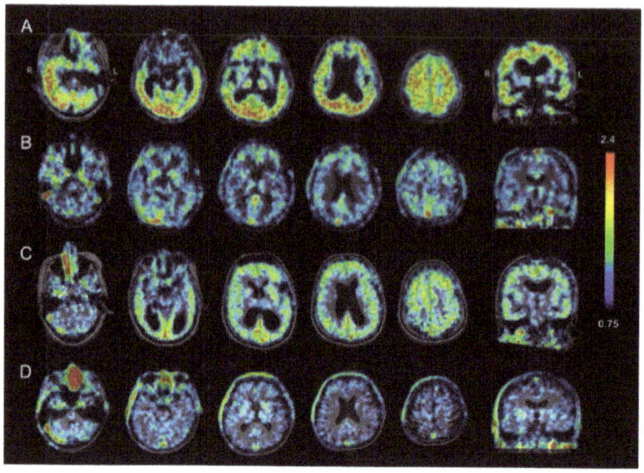

Figure 2. Tau PET using PBB3 in a patient with familial Alzheimer's disease with Osaka mutation (FAD (Osaka)) (**A**); a patient of early stage of sporadic Alzheimer's disease (early SAD) (**B**); a patient with advanced stage of SAD (**C**); and a healthy control (HC) (**D**). The heat map range (colored bar) of tau tracer uptake indicates standard uptake value ratio (SUVR) with reference to the midbrain. In the FAD (Osaka) patient, noticeable PBB3 accumulation was observed in the cerebral cortex and the cerebellar cortex, whereas the AD patient had much less tau accumulation that was more localized in the frontal, parietal, and lateral temporal cortices.

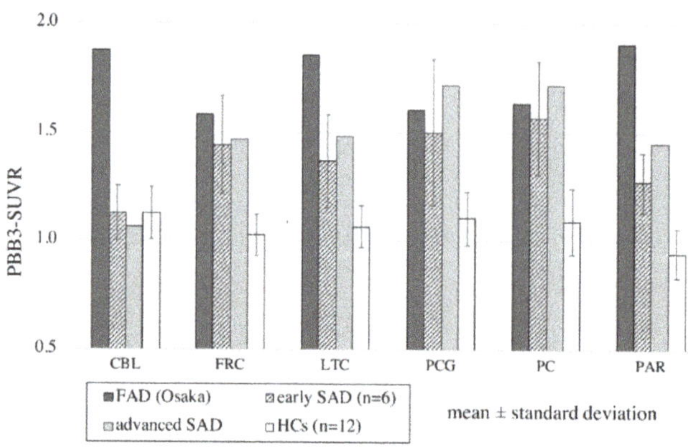

Figure 3. Regional PBB3 uptake with reference to the midbrain in familial Alzheimer's disease with Osaka mutation (FAD (Osaka)), in early and advanced stage patients with sporadic Alzheimer's disease (SAD) and healthy controls (HCs). Regions were set in the cerebellum (CBL), frontal cortex (FRC), lateral temporal cortex (LTC), posterior cingulate gyrus (PCG), precuneus (PC) and parietal cortex (PAR). In the cerebral cortex, PBB3 uptake in the FAD (Osaka) patient was higher than that in the HCs in all regions (>2.5 SD). PBB3 uptake was even higher than that in the early stage SAD patients and advanced stage SAD patients in the lateral temporal (>2 SD) and parietal cortices (>2.5 SD). Remarkably elevated PBB3 uptake in the cerebellum was found only in the FAD (Osaka) patient.

2.4. Amyloid β PET Imaging

Figure 4 shows representative Aβ PET images taken with ^{11}C-PiB in the 4 groups. The SUVR values of accumulated PiB with reference to the cerebellum were calculated and projected on the MRI images of each patient. In the FAD (Osaka) patient, the accumulation of ^{11}C-PiB was negligible in most of the cerebral cortex except for the temporal cortex and limited parts of the parietal and frontal cortices where very low accumulation could be seen. Relatively elevated ^{11}C-PiB uptake was found in the cerebellar cortex compared to the cerebral cortex, and the more severe atrophy in the cerebral cortex than in the cerebellum might affect the apparent ^{11}C-PiB accumulation.

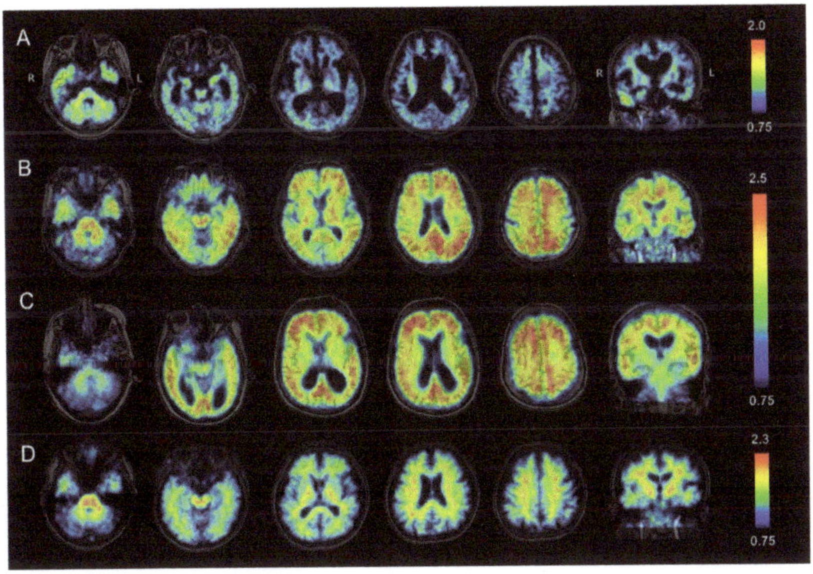

Figure 4. Amyloid PET using PiB in a patient with familial Alzheimer's disease with Osaka mutation (FAD (Osaka)) (**A**); a patient with early sporadic Alzheimer's disease (SAD) (**B**); a patient with advanced stage SAD (**C**); and a healthy control (HC) (**D**). Heat map range (colored bar) of amyloid tracer uptake defined by standard uptake value ratio (SUVR) with reference in the cerebellum. In the FAD (Osaka) patient, only negligible amounts of PiB retention were detected in any part of the cerebral cortex. Both the early and advanced SAD patients had highly elevated PiB uptake in the frontal, parietal and lateral cortices. No elevation in PiB accumulation was found in the HC.

In contrast, elevated ^{11}C-PiB uptake was remarkable in the frontal, parietal and lateral cortices in both the early and advanced SAD patients. The HC patients showed no evident increase in ^{11}C-PiB accumulation.

According to the J-ADNI PET core criteria [14], PiB uptake is regarded as positive when the cortical accumulation is higher than that in the white matter just below the cortex, as shown in the SAD patients in Figure 4. The negative patient had a reversed pattern of PiB, as in the HC patients. In the FAD (Osaka) patient, PiB appeared to be higher in the cortex compared to the white matter in the temporal, parietal and frontal cortices, but the severe cortical atrophy made it quite difficult to discern.

2.5. Scatter Plot Analysis of Tau vs. Aβ PET

Scatter plot analysis of the two tracers revealed that the FAD (Osaka) patient exhibited a distinguishing pattern with a high tau burden and subtle Aβ accumulation in all parts of the AD signature cortices (Figure 5). In contrast, the SAD patients exhibited high accumulation of both tau and Aβ, whereas both were negligible in the HCs.

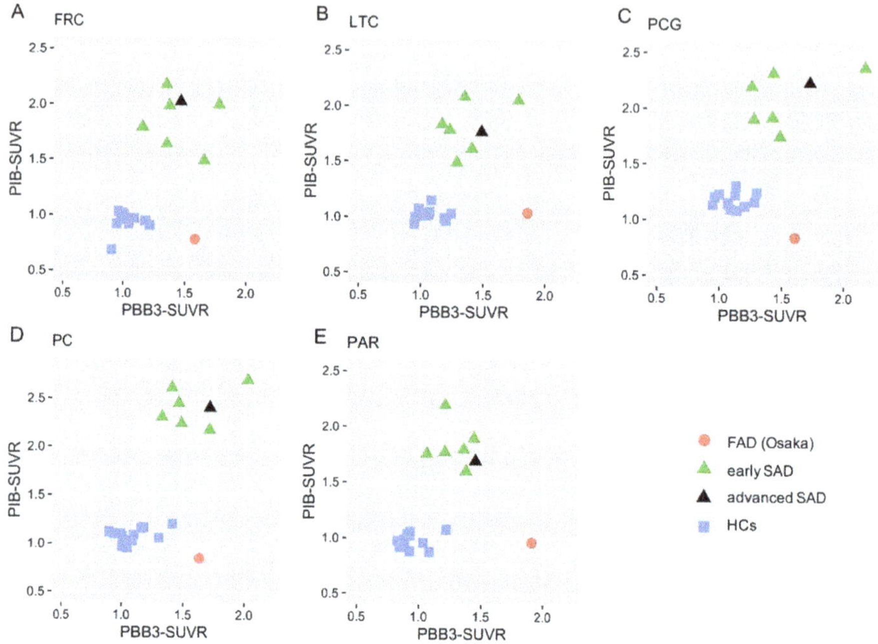

Figure 5. Scatter plot analysis between tau (PBB3-SUVR) and Aβ (PiB-SUVR) accumulation in the frontal cortex ((**A**): FRC); lateral temporal cortex ((**B**): LTC); posterior cingulate gyrus ((**C**): PCG); precuneus ((**D**): PC); and parietal cortex ((**E**): PAR) in a patient with familial Alzheimer's disease with Osaka mutation (FAD (Osaka)), patients with early and advanced sporadic Alzheimer's disease (SAD) and healthy controls (HCs). The FAD (Osaka) patient showed a distinguishing pattern with highly elevated PBB3 uptake and subtle uptake of PiB in all regions. In contrast, the SAD patients exhibited highly elevated accumulation of both PBB3 and PiB, whereas both were negligible in HCs.

3. Discussion

This report is the first report of tau PET imaging in the proband of familial AD with Osaka mutation, which revealed heavy tau burden in the cerebral cortex and cerebellum with only negligible Aβ in the cerebral cortex. The findings indicate that Aβ may induce tau accumulation and neuronal degeneration without forming senile plaques as previously reported in basic experiments [10–12].

We previously reported that in SAD, tau accumulation spreads from the parahippocampal gyrus to the cerebral cortex with advancing phases of AD, whereas Aβ distribution is already advanced in the clinically earliest stage [15]. Similar reports supporting our data have been recently published by Jack et al. [16]. The increase in tau accumulation in a patient with advanced SAD compared with early SAD patients was also confirmed in the present study. The FAD (Osaka) patient also exhibited higher tau accumulation than the HCs in all the AD signature ROIs in the cerebral cortex. It is noteworthy that the accumulation was even higher than early SAD by more than 2 SD in the lateral temporal cortex and parietal cortex. Although the difference between FAD (Osaka) and advanced SAD was also noticeable in these ROIs, whether this distribution pattern is specific to FAD (Osaka) or just reflecting the advanced stage of AD in general, requires further validation with more advanced SAD cases.

As is widely known, Aβ accumulates heavily and diffusely in the cerebral cortex even in the early stages of SAD and remains high in the advanced stage as in the present study. In contrast,

Aβ accumulation was negligibly low in earlier cases of FAD (Osaka) previously reported [1] and remained low in the advanced stage in the present study. These findings strongly support our hypothesis that accumulated Aβ that is usually found with senile plaques is not a prerequisite for the disease progression that includes tau accumulation and neurodegeneration in any phase of FAD (Osaka). Instead, small oligomeric aggregates of Aβ, not detectable with PiB, may play a key role in the onset and progression of the disease. These mechanisms were experimentally confirmed by our previous studies revealing that the Osaka mutation promotes Aβ oligomerization in the brain without forming senile plaques [10–12]. In addition to Aβ, more recent evidences have pointed to the pertinent role of soluble oligomeric tau in AD onset and progression. These oligomers may share a common mechanism of toxicity [17]. As the present case of FAD (Osaka) exhibited heavy tau burden on PBB3 PET, any other FAD mutation without tau accumulation, if found, might prove the toxicity of tau oligomer.

The possible non-harmful profile of accumulated Aβ is critical for the development of AD treatment. Previous clinical trials targeting accumulated Aβ were mostly successful in reducing Aβ but failed to improve the clinical outcomes and even worsened the outcomes in some studies [18]. In contrast, in the present and previous studies, we demonstrated that tau accumulates more widely as the disease progresses in SAD [15]. The burden of tau was also heavy in FAD (Osaka) without amyloid accumulation. These findings strongly warrant further development of novel treatments to reduce soluble Aβ oligomer or target tau, including aducanumab, which is a human monoclonal antibody that selectively binds to Aβ fibrils and soluble oligomers [19].

Cerebellar lesions with senile plaques and/or neurofibrillary tangles as well as Purkinje cell loss were pathologically reported in some FAD patients [20–23], whereas these pathological changes were usually not found until the end stage in SAD [24]. In the present study, Aβ and tau accumulation in the cerebellum were only found in FAD (Osaka). The present patient and other family members with dementia symptoms demonstrated cerebellar ataxia even in the early phase, which is quite rare in most patients with SAD. Cerebellar accumulation of Aβ or tau may be helpful in distinguishing FAD from SAD if familial history cannot be obtained.

In the advanced stage of AD, neuronal degeneration results in cerebral atrophy. The process may decompose and reduce the actual accumulation of Aβ and tau. In addition, atrophy reduces apparent tracer accumulation. The accumulation of PBB3 was even higher in the lateral temporal and parietal cortex in the FAD (Osaka) patient than in the advanced SAD patients with similar cerebral atrophy, indicating substantial tau accumulation in these regions. In contrast, PiB accumulation seemed much lower in the patient with FAD (Osaka) than in the patient with advanced SAD, suggesting much lower Aβ accumulation in FAD (Osaka), but the small amount of Aβ in the FAD (Osaka) patient compared to that in the HCs cannot be ruled out completely.

4. Materials and Methods

4.1. Ethical Approval

All procedures performed in studies involving human participants were approved by the Institutional Research Ethics Committee of Osaka City University Graduate School of Medicine (IRB# 3009 approved on 25 December 2014) and were conducted in accordance with the 1964 Helsinki Declaration and its later amendments.

4.2. Informed Consent

Written informed consent was obtained from all participants or from close family members when the participants were cognitively impaired.

4.3. Subjects

4.3.1. Familial Alzheimer's Disease with APP Mutation

Details of the FAD (Osaka) pedigree were described elsewhere [10]. We previously reported PET imaging of Aβ accumulation in two sister patients of the pedigree [13]. In the present study, the elder sister, who was the proband, was examined. She was in the advanced stages of dementia, bed-ridden and did not speak.

4.3.2. Sporadic AD

We recruited SAD patients from patients attending the memory clinic of Osaka City University Hospital. We evaluated medical history, neurological findings and general blood tests. Two qualified clinical psychologists (M.A. and N.K.) scored the MMSE. An MRI of the brain including coronal sections for hippocampal evaluation was taken.

AD was diagnosed based on the IWG-2 criteria for typical AD with amyloid PET as in vivo evidence of AD pathology [4]. Apparent familial AD was not included in the SAD diagnosis. The same exclusion criteria applied to the HCs were applied to the AD patients to exclude confounding diseases, such as diabetes, dyslipidemia and hypertension.

4.3.3. Healthy Controls

HCs without a history of brain disorders or subjective abnormalities were openly recruited as candidates. All candidates took physical examinations covering vital and cardiopulmonary systems. Tests that were administered to the AD patients were applied to the HCs.

The exclusion criteria were (1) a history of any brain diseases, brain surgery, or head trauma that requires hospitalization; (2) high risk for cerebrovascular diseases, including poorly controlled diabetes, severe dyslipidemia, and hypertension above the recommended level of standard guidelines; (3) any neurological findings suggesting brain diseases; (4) low cognitive test scores [MMSE, and Rivermead Behavioral Memory Test (RBMT)]; (5) significant MRI lesions including asymptomatic lacunas (less than 15 mm in diameter, high in T2, FLAIR and low in T1), severe white matter lesions (more than 1 grade in Fazekas score [25]), more than 4 cerebral microbleeds, and atrophy beyond average for their age by visual inspection; and (6) positive amyloid PET imaging. No other biomarkers were evaluated in the HC group.

4.4. PET Data Acquisition

^{11}C-PBB3 and ^{11}C-Pittsburgh compound-B (2-[4-(^{11}C-methylamino) phenyl]-1,3-benzothiazol-6-ol, ^{11}C-PiB) were produced following the methods previously reported [2,26,27]. ^{11}C-PBB3-and ^{11}C-PiB-PET images were obtained with a Siemens Biograph16 scanner (Siemens/CTI, Knoxville, TN, USA) and with an Eminence-B PET scanner (Shimadzu Co., Kyoto, Japan), respectively. ^{11}C-PBB3, a tau tracer, was intravenously injected in the range of 370 MBq (body weight $\leq$ 50 kg) to 555 MBq (body weight $\geq$ 70 kg) in a dimly lit room to avoid photoracemization. A 60-min PET scan was performed in list mode. The acquired data were sorted into dynamic data with 6×10 s, 3×20 s, 6×60 s, 4×180 s, and 8×300 s frames. To evaluate Aβ accumulation, each subject received 400 to 500 MBq of ^{11}C-PiB intravenously over 1 min. After the injection, a static scan image acquisition was performed for 50 to 70 min. Reconstruction of PET images for ^{11}C-PBB3 and ^{11}C-PiB was performed by filtered back projection using a 4-mm full width at half maximum (FWHM) Hanning filter and a 5-mm FWHM Gaussian filter with attenuation and scatter correction, respectively.

4.5. Criteria for Aβ Accumulation in SAD

Aβ accumulation was determined visually based on the Japanese Alzheimer's Disease Neuroimaging Initiative (J-ADNI) Visual Criteria for ^{11}C-PiB-PET (J-ADNI_PETQC_Ver1.1) modified from the ADNI PET core criteria [28]. The four regions selected for the assessment were the frontal lobe, lateral temporal lobe, lateral parietal lobe and the combined area of precuneus and posterior cingulate gyrus. The patient was considered positive when the accumulation in one of the four cerebral cortices was higher than that in the white matter just below the cortex and negative when none of the four cortices had higher accumulation than the white matter.

4.6. MRI Acquisition

We used a 1.5- or 3-Tesla magnetic resonance scanner (MAGNETOM Avanto, Siemens Healthcare, Erlangen, Germany or Ingenia, Philips Healthcare, Best, The Netherlands) with three-dimensional volumetric acquisition of a T1-weighted gradient echo sequence (repetition time range/echo time range, 6.5 ms/3.2 ms; field of view [frequency × phase], 240 mm × 240 mm; matrix, 256 × 256; contiguous axial slices of 1.5 mm thickness).

4.7. Image Processing

For all image processes, we used PMOD software version 3.7 (PMOD Technologies Ltd., Zurich, Switzerland) [29]. Acquired ^{11}C-PBB3 and ^{11}C-PiB images were transformed into standard brain and then SUVR images were reconstructed with the midbrain as the reference region using the frame summation of dynamics of the image for 30 to 50 min after ^{11}C-PBB3 injection and for 50 to 70 min after ^{11}C-PiB injection. The SUVR level of each volume of interest (VOI) was calculated in a manually set region of interest (ROI) as shown in Supplementary Figure S1.

4.8. Statistical Analysis

To quantitatively evaluate tau accumulation, SUVR values relative to the midbrain were calculated in each ROI. Statistically, the value of FAD (Osaka) was regarded as significantly high when it was higher than the average by more than 2 SD (standard deviation) in SAD patients and HCs.

5. Conclusions

Here, we report a patient with familial AD who had heavy tau burden in the cerebral cortex and cerebellum with only a negligible amount of Aβ. The unprecedented observations support our hypothesis that Aβ, probably in oligomers, induces tau accumulation and neuronal degeneration without forming senile plaques.

Supplementary Materials: The following are available online at http://www.mdpi.com/1422-0067/21/12/4443/s1, Figure S1: Region of interest (ROI).

Author Contributions: H.S. (Hiroyuki Shimada): Clinical management; S.M., J.T., A.T.: Imaging and statistical analysis; J.K., Y.W. (Yasuhiro Wada), A.M., Y.W. (Yasuyoshi Watanabe): Tracer preparation; H.S. (Hitoshi Shimada), M.H., T.S.: Tracer supply and funding; T.T.: Basic research to support this study; Y.I.: General management and writing the manuscript. All authors have read and agreed to the published version of the manuscript.

Funding: NIRS/QST grant (Japan Agency for Medical Research and Development 16768966).

Acknowledgments: The authors thank M. Ando and N. Kotani (Osaka City University) for psychological tests. The present study was financially supported by the NIRS/QST grant (Japan Agency for Medical Research and Development 16768966) in which Osaka City University was involved as a collaborating facility.

Conflicts of Interest: Authors H.S. (Hitoshi Shimada), M.H., and T.S. possess patent royalty of PBB3.

References

1. Shimada, H.; Kitamura, S.; Shinotoh, H.; Endo, H.; Niwa, F.; Hirano, S.; Kimura, Y.; Zhang, M.R.; Kuwabara, S.; Suhara, T.; et al. Association between Abeta and tau accumulations and their influence on clinical features in aging and Alzheimer's disease spectrum brains: A [11C]PBB3-PET study. *Alzheimers Dement.* **2017**, *6*, 11–20.
2. Maruyama, M.; Shimada, H.; Suhara, T.; Shinotoh, H.; Ji, B.; Maeda, J.; Zhang, M.R.; Trojanowski, J.Q.; Lee, V.M.; Ono, M.; et al. Imaging of tau pathology in a tauopathy mouse model and in Alzheimer patients compared to normal controls. *Neuron* **2013**, *79*, 1094–1108. [CrossRef]

3. Ono, M.; Sahara, N.; Kumata, K.; Ji, B.; Ni, R.; Koga, S.; Dickson, D.W.; Trojanowski, J.Q.; Lee, V.M.; Yoshida, M.; et al. Distinct binding of PET ligands PBB3 and AV-1451 to tau fibril strains in neurodegenerative tauopathies. *Brain* **2017**, *140*, 764–780. [CrossRef]
4. Dubois, B.; Feldman, H.H.; Jacova, C.; Hampel, H.; Molinuevo, J.L.; Blennow, K.; DeKosky, S.T.; Gauthier, S.; Selkoe, D.; Bateman, R.; et al. Advancing research diagnostic criteria for Alzheimer's disease: The IWG-2 criteria. *Lancet Neurol.* **2014**, *13*, 614–629. [CrossRef]
5. Walsh, D.M.; Klyubin, I.; Fadeeva, J.V.; Cullen, W.K.; Anwyl, R.; Wolfe, M.S.; Rowan, M.J.; Selkoe, D.J. Naturally secreted oligomers of amyloid beta protein potently inhibit hippocampal long-term potentiation in vivo. *Nature* **2002**, *416*, 535–539. [CrossRef] [PubMed]
6. Tomiyama, T.; Shimada, H. APP osaka mutation in familial Alzheimer's disease-its discovery, phenotypes, and mechanism of recessive inheritance. *Int. J. Mol. Sci.* **2020**, *21*, 1413. [CrossRef] [PubMed]
7. Cleary, J.P.; Walsh, D.M.; Hofmeister, J.J.; Shankar, G.M.; Kuskowski, M.A.; Selkoe, D.J.; Ashe, K.H. Natural oligomers of the amyloid-β protein specifically disrupt cognitive function. *Nat. Neurosci.* **2005**, *8*, 79–84. [CrossRef] [PubMed]
8. Lesné, S.; Koh, M.T.; Kotilinek, L.; Kayed, R.; Glabe, C.G.; Yang, A.; Gallagher, M.; Ashe, K.H. A specific amyloid-β protein assembly in the brain impairs memory. *Nature* **2006**, *440*, 352–357. [CrossRef]
9. Shankar, G.M.; Li, S.; Mehta, T.H.; Garcia-Munoz, A.; Shepardson, N.E.; Smith, I.; Brett, F.M.; Farrell, M.A.; Rowan, M.J.; Lemere, C.A.; et al. Amyloid-β protein dimers isolated directly from Alzheimer's brains impair synaptic plasticity and memory. *Nat. Med.* **2008**, *14*, 837–842. [CrossRef]
10. Tomiyama, T.; Nagata, T.; Shimada, H.; Teraoka, R.; Fukushima, A.; Kanemitsu, H.; Takuma, H.; Kuwano, R.; Imagawa, M.; Ataka, S.; et al. A new amyloid beta variant favoring oligomerization in Alzheimer's-type dementia. *Ann. Neurol.* **2008**, *63*, 377–387. [CrossRef]
11. Tomiyama, T.; Matsuyama, S.; Iso, H.; Umeda, T.; Takuma, H.; Ohnishi, K.; Ishibashi, K.; Teraoka, R.; Sakama, N.; Yamashita, T.; et al. A mouse model of amyloid beta oligomers: Their contribution to synaptic alteration, abnormal tau phosphorylation, glial activation, and neuronal loss in vivo. *J. Neurosci.* **2010**, *30*, 4845–4856. [CrossRef] [PubMed]
12. Umeda, T.; Maekawa, S.; Kimura, T.; Takashima, A.; Tomiyama, T.; Mori, H. Neurofibrillary tangle formation by introducing wild-type human tau into APP transgenic mice. *Acta Neuropathol.* **2014**, *127*, 685–698. [CrossRef] [PubMed]
13. Shimada, H.; Ataka, S.; Tomiyama, T.; Takechi, H.; Mori, H.; Miki, T. Clinical course of patients with familial early-onset Alzheimer's disease potentially lacking senile plaques bearing the E693Delta mutation in amyloid precursor protein. *Dement. Geriatr. Cogn. Disord.* **2011**, *32*, 45–54. [CrossRef]
14. Yamane, T.; Ishii, K.; Sakata, M.; Ikari, Y.; Nishio, T.; Ishii, K.; Kato, T.; Ito, K.; Senda, M. Inter-rater variability of visual interpretation and comparison with quantitative evaluation of (11)C-PiB PET amyloid images of the Japanese Alzheimer's Disease Neuroimaging Initiative (J-ADNI) multicenter study. *Eur. J. Nucl. Med. Mol. Imaging* **2017**, *44*, 850–857. [CrossRef]
15. Aohara, K.; Minatani, S.; Kimura, H.; Takeuchi, J.; Takeda, A.; Kawabe, J.; Wada, Y.; Mawatari, A.; Watanabe, Y.; Shimada, H.; et al. Staging of tau distribution by positron emission tomography may be useful in clinical staging of Alzheimer disease. *Neurol. Clin. Neurosci.* **2020**, *8*, 61–67. [CrossRef]
16. Jack, C.R.; Wiste, H.J.; Botha, H.; Weigand, S.D.; Therneau, T.M.; Knopman, D.S.; Graff-Radford, J.; Jones, D.T.; Ferman, T.J.; Boeve, B.F.; et al. The bivariate distribution of amyloid-beta and tau: Relationship with established neurocognitive clinical syndromes. *Brain* **2019**, *142*, 3230–3242. [CrossRef] [PubMed]
17. Kayed, R.; Head, E.; Thompson, J.L.; McIntire, T.M.; Milton, S.C.; Cotman, C.W.; Glabe, C.G. Common structure of soluble amyloid oligomers implies common mechanism of pathogenesis. *Science* **2003**, *300*, 486–489. [CrossRef]
18. Honig, L.S.; Vellas, B.; Woodward, M.; Boada, M.; Bullock, R.; Borrie, M.; Hager, K.; Andreasen, N.; Scarpini, E.; Liu-Seifert, H.; et al. Trial of solanezumab for mild dementia due to Alzheimer's disease. *N. Engl. J. Med.* **2018**, *378*, 321–330. [CrossRef]
19. Servick, K. Doubts persist for claimed Alzheimer's drug. *Science* **2019**, *366*, 1298. [CrossRef]
20. Lemere, C.A.; Lopera, F.; Kosik, K.S.; Lendon, C.L.; Ossa, J.; Saido, T.C.; Yamaguchi, H.; Ruiz, A.; Martinez, A.; Madrigal, L.; et al. The E280A presenilin 1 Alzheimer mutation produces increased A beta 42 deposition and severe cerebellar pathology. *Nat. Med.* **1996**, *2*, 1146–1150. [CrossRef]

21. Sepulveda-Falla, D.; Matschke, J.; Bernreuther, C.; Hagel, C.; Puig, B.; Villegas, A.; Garcia, G.; Zea, J.; Gomez-Mancilla, B.; Ferrer, I.; et al. Deposition of hyperphosphorylated tau in cerebellum of PS1 E280A Alzheimer's disease. *Brain Pathol.* **2011**, *21*, 452–463. [CrossRef] [PubMed]
22. Fukutani, Y.; Cairns, N.J.; Rossor, M.N.; Lantos, P.L. Cerebellar pathology in sporadic and familial Alzheimer's disease including APP 717 (Val–>Ile) mutation cases: A morphometric investigation. *J. Neurol. Sci.* **1997**, *149*, 177–184. [CrossRef]
23. Sutovsky, S.; Smolek, T.; Turcani, P.; Petrovic, R.; Brandoburova, P.; Jadhav, S.; Novak, P.; Attems, J.; Zilka, N. Neuropathology and biochemistry of early onset familial Alzheimer's disease caused by presenilin-1 missense mutation Thr116Asn. *J. Neural. Transm. (Vienna)* **2018**, *125*, 965–976. [CrossRef] [PubMed]
24. Jacobs, H.I.L.; Hopkins, D.A.; Mayrhofer, H.C.; Bruner, E.; van Leeuwen, F.W.; Raaijmakers, W.; Schmahmann, J.D. The cerebellum in Alzheimer's disease: Evaluating its role in cognitive decline. *Brain* **2017**, *141*, 37–47. [CrossRef]
25. Fazekas, F.; Chawluk, J.B.; Alavi, A.; Hurtig, H.I.; Zimmerman, R.A. MR signal abnormalities at 1.5 T in Alzheimer's dementia and normal aging. *AJR Am. J. Roentgenol.* **1987**, *149*, 351–356. [CrossRef]
26. Kimura, Y.; Ichise, M.; Ito, H.; Shimada, H.; Ikoma, Y.; Seki, C.; Takano, H.; Kitamura, S.; Shinotoh, H.; Kawamura, K.; et al. PET Quantification of Tau Pathology in Human Brain with 11C-PBB3. *J. Nucl. Med.* **2015**, *56*, 1359–1365. [CrossRef]
27. Hashimoto, H.; Kawamura, K.; Igarashi, N.; Takei, M.; Fujishiro, T.; Aihara, Y.; Shiomi, S.; Muto, M.; Ito, T.; Furutsuka, K.; et al. Radiosynthesis, photoisomerization, biodistribution, and metabolite analysis of 11C-PBB3 as a clinically useful PET probe for imaging of tau pathology. *J. Nucl. Med.* **2014**, *55*, 1532–1538. [CrossRef]
28. Jagust, W.J.; Bandy, D.; Chen, K.; Foster, N.L.; Landau, S.M.; Mathis, C.A.; Price, J.C.; Reiman, E.M.; Skovronsky, D.; Koeppe, R.A.; et al. The ADNI PET Core. *Alzheimers Dement.* **2010**, *6*, 221–229. [CrossRef]
29. Choi, W.H.; Um, Y.H.; Jung, W.S.; Kim, S.H. Automated quantification of amyloid positron emission tomography: A comparison of PMOD and MIMneuro. *Ann. Nucl. Med.* **2016**, *30*, 682–689. [CrossRef]

International Journal of
Molecular Sciences

MDPI

Commentary

What Are the Molecular Mechanisms by Which Functional Bacterial Amyloids Influence Amyloid Beta Deposition and Neuroinflammation in Neurodegenerative Disorders?

Robert P. Friedland [1,*], Joseph D. McMillan [2] and Zimple Kurlawala [3]

1 Department of Neurology, University of Louisville School of Medicine, Louisville, KY 40202, USA
2 Department, University of South Florida, Tampa, FL 33620, USA; jmcmillan3@health.usf.edu
3 National Institute on Aging, Bethesda, MD 33620, USA; zimple.kurlawala@louisville.edu
* Correspondence: Robert.Friedland@Louisville.edu

Received: 3 February 2020; Accepted: 26 February 2020; Published: 28 February 2020

Abstract: Despite the enormous literature documenting the importance of amyloid beta (Aβ) protein in Alzheimer's disease, we do not know how Aβ aggregation is initiated and why it has its unique distribution in the brain. In vivo and in vitro evidence has been developed to suggest that functional microbial amyloid proteins produced in the gut may cross-seed Aβ aggregation and prime the innate immune system to have an enhanced and pathogenic response to neuronal amyloids. In this commentary, we summarize the molecular mechanisms by which the microbiota may initiate and sustain the pathogenic processes of neurodegeneration in aging.

Keywords: Alzheimer's disease; microbiota; bacterial amyloid; FUBA; curli; CsgA; amyloid beta; neuroinflammation

---

1. Introduction

The original descriptions of cortical plaques in subjects with dementia was provided by several scientists, including Emil Redlich (1898), Koichi Miyake (1906), Alois Alzheimer (1906), Oskar Fischer (1907), and Soloman Carter Fuller (1907) [1]. However, the molecular analysis of the material in the plaque was not clarified until Glenner and Wong (1984) described the amino acid sequence of the amyloid which they found to have no homology to any known protein. Subsequently, it has been determined that the protein, now called amyloid Beta (Aβ), is a breakdown product of a larger molecule, the amyloid precursor protein (APP), coded by a gene on chromosome 21. Mutations in APP have been found to be a genetic cause of early-onset Alzheimer's disease (AD). This led to the "amyloid hypothesis", which posited that accumulation of the Aβ protein in the brain is responsible for the pathogenesis of AD [2,3].

There is now overwhelming evidence that Aβ buildup in the brain is certainly an important part of the AD pathogenic process. But it remains to be determined if it is the critical factor responsible. Furthermore, the amyloid hypothesis does not explain how the process starts. That is, what is the initiating factor responsible for Aβ accumulation, and why do some people get the disease and others remain unaffected? This is a critical question in the ~99% of cases of AD that are sporadic, and do not have a causative gene.

Prusiner has suggested that the key mechanism in Alzheimer's etiology is stochastic—an unfortunate misfolding of the Aβ protein causing a non-catalytic, prion-like self-replicating pathogenic process [4]. We find the stochastic explanation unsatisfactory and prefer to look for the opportunity for environmental factors to be involved. Our largest environmental exposure is to the myriad organisms (bacteria, viruses, archaea, yeasts, parasites, fungi) which reside on our body surfaces and in our body cavities. It has now been extensively documented that more than half of a person's cells are of microbial origin, and there are 100 times more nucleotide sequences in our bodies coding for bacterial, rather than human proteins [5]. In this commentary, we will examine the potential molecular mechanisms by which the microbiota may influence cortical Aβ deposition. The mechanisms noted below may operate independently, or more likely, work in unison (i.e, neuroinflammation triggers Aβ aggregation [6]).

2. Functional Bacterial Amyloid Proteins (FUBA) May Cross-Seed the Aggregation of Aβ as Well as Other Neuronal Amyloids

Chapman and colleagues originally described amyloid proteins made by bacteria in 2002 [7]. It has now been demonstrated that many organisms present inside our bodies are capable of producing FUBAs, including Streptococcus, Staphylococcus, Salmonella, Mycobacteria, Klebsiella, Citrobacter, and Bacillus species [8]. These bacterial amyloids are adaptive, help the organisms stick together, and protect bacterial communities by resisting destruction by viruses and other agents. Bacterial amyloids have been linked to infections as well to autoimmunity [9]. In 2015, Friedland proposed that bacterial amyloid may cross-seed the aggregation of neuronal proteins such as Aβ, alpha synuclein (AS), tau and others to initiate a prion-like propagation [10]. This process may begin with enteroendocrine and M cells in the gut epithelium and be transmitted to the brain through the autonomic nervous system via a bidirectional pathway (the gut-brain axis, [11,12]). It was subsequently demonstrated by Chen et al. that feeding bacterial amyloid to aged rats accelerates AS deposition in the gut, as well as in the brain [13].

It has been observed that protein folding in an amyloid configuration is a highly conserved process throughout evolution [14]. Recently, a series of studies have shown that FUBAs may interact with neuronal proteins in the manner predicted. That is, a bacterial amyloid protein may influence the aggregation of neuronal proteins. Perov et al. discovered structural similarity between the fibers of the best studied bacterial amyloid protein curli, and neuronal amyloids [14]. They noted that the curli protein CsgA cross-seeds the fibrillation of Aβ, as was previously proposed [10]. Curli cross-seeding with Aβ was also observed to have a concentration dependent effect, suggesting that this cross-seeding may be of widespread impact [15]. It has also been shown that the functional amyloid protein FapC, produced by Pseudomonas, may affect fibrillation of AS [16]. It should be noted that these studies were done in vitro with nonbiological buffers, suggesting a need for in vivo studies using simple systems such as yeast or *C. elegans* to demonstrate this process.

3. FUBAs May Enhance Neuro-Inflammation through Molecular Mimicry

Gut bacteria have been shown to remarkably influence microglial function [17] and bacterial amyloid is recognized by the innate immune system as a pathogen associated molecular pattern (PAMP) involving toll-like receptor 2 (TLR2), cluster of differentiation 14 (CD14), nuclear factor kappa light chain enhancer of activated B cells (NFkB), and inducible nitric oxide synthase (iNOS) [10]. Remarkably, this pathway is also involved in the recognition of misfolded neuronal proteins such as Aβ and AS [18]. The presence of Aβ aggregates in the brain, often found with healthy aging, may be recognized by the innate immune system as a bacterial product, due to its structural similarity to bacterial proteins. Bacterial amyloid also causes activation of the NLRP3 inflammasome, causing downstream release of pro-inflammatory interleukin-1β in microglia, eventually leading to Aβ and tau aggregation [6,19,20].

4. Microbiota May Influence Inflammation in the Brain through Effects on Circulating Immune Cells

Over 70% of our lymphocytes reside in the gut—a sign of the enormous influence of the microbiota on the immune system. Bacteria produce short-chain fatty acids such as propionate, butyrate and acetate that epigenetically promote the function of anti-inflammatory regulatory lymphocytes (Tregs) [21]. Conversely, some groups of bacteria enhanced by a low fiber diet do not produce short chain fatty acids and upregulate the production of pro-inflammatory CD4+ Th17 cells [21]. As we have discussed above, there is an inflammatory component to all neurodegenerations.

5. Microbiota May Influence the Effects of Apolipoprotein E (Apo E) Genotype on Disease Risk

It was proposed in 2015 that the effect of Apo E on Alzheimer risk may be caused by the influence of Apo E genotype on the microbiota [10]. Recently, Tran et al. provided evidence that in humans and transgenic mice the microbial populations and metabolite panels are different between ApoE genotypes [22]. This relationship may provide a valuable clue to development of microbiota-based therapies. It is important to recall that susceptibility to Creutzfeldt Jakob disease and bovine spongiform encephalopathy is determined by the genotype at codon 129 of the prion protein gene (people who are heterozygous MV, or valine homozygotes VV are resistant).

6. Microbiota May Influence Neurodegeneration through Production of Vitamins and Other Nutrients

The microbiome produce important metabolic products that influence health and disease. It was recently demonstrated that gut bacteria are involved in amyotrophic lateral sclerosis (ALS). Transgenic ALS hSOD1 G93A mice were found to have bacterial dysbiosis and mice lacking gut bacteria had accelerated disease. Preliminary studies suggest that it may be due to a deficiency of nicotinamide (vitamin B3) [23]. Nicotinamide is a NAD+ precursor crucial for energy and redox metabolism [24].

7. Further Considerations

Current microbiota studies have focused largely on bacterial populations in the intestines involving studies of stool. The potential role of oral, nasal, laryngeal and pharyngeal organisms has not been extensively studied. This is especially important in neurodegeneration, as hyposmia is a feature of both AD and Parkinson's disease (PD). Furthermore, the proximity of the brain to the nose (which has its own microbiota) demonstrates the opportunity for interactions between olfactory receptors, microbes and microbial metabolites [25,26]. The mouth is also a critical ecological niche for hundreds of different bacterial species in humans. Also, the common oral pathogen *Streptococcus mutans* may produce an amyloid product. Studies of stool do not assess the intestinal microbiota that reside in the mucus layer. In addition to the description of the populations present in the gut, it is imperative to consider metabolic products generated by the microbiota, which undoubtedly influence health and disease [27,28].

A central question which remains unanswered is: what FUBAs are present in the body, and how much? Current work has been devoted largely to bacteria while the role of viruses, fungi, parasites, and other agents have not been considered. It is estimated that there may be twice as many phagesin the gut as bacteria—their contribution to bacterial homeostasis is just beginning to be addressed. Furthermore, fungi can act as commensals in the gut and a few fungal amyloid proteins have been described [29]. Additional variables that need to be considered are sex-specific microbiota changes, as recently demonstrated in antibiotic treated transgenic mouse models of Aβ amyloidosis [30], and their potential effect of sex on FUBAs. These outstanding questions are addressed in the Table 1.

One potentially exciting aspect to microbiota studies is the relative ease with which microbial populations can be adjusted. This can be called "gene therapy in the kitchen", because it is possible to change bacterial populations in the gut by changing diet in as little as two weeks [31]. The potential for therapeutic interventions based on FUBA is illustrated by a recent paper by Sampson et al. [32] that elegantly reports the role of the curli (CsgA) protein in producing neurodegeneration in alpha synuclein overexpressing transgenic PD model mice. Oral treatment of wild type *E. coli* producing the bacterial amyloid protein CsgA with epigallocatechin gallate (EGCG) reduced biofilm formation and inhibited the production of aggregated alpha synuclein. EGCG is a potent polyphenol found in plants, which impairs the production of bacterial CsgA. As EGCG is not well-absorbed from the gut to the bloodstream, it is likely that these influences are localized to the intestinal microbiota. This work supports the exciting opportunity to treat neurodegenerative diseases with agents that act primarily on the microbiota, and not directly on the brain. The potential for influencing pathogenic processes in AD through therapies targeting the gut shows great promise and is now receiving academic and commercial attention worldwide.

Table 1. Outstanding questions concerning the role of functional bacterial amyloid proteins (FUBA) and neurodegeneration.

• Which organisms are producing these FUBA?
• Is there a basal *normal* level of FUBA in the gut?
• Are there unique niches for the production of FUBA? (nose, mouth, pharynx, larynx, stomach, duodenum, Ilium, appendix, colon, skin, external ear, elsewhere)
• Are there strains of FUBA which interact with specific neurodegenerative disease proteins involved in AD, PD, ALS, progressive supranuclear palsy, cortical basal degeneration, multiple system atrophy?
• Can the production of FUBA be regulated by diet and drugs?
• Do phages, fungi and archaea influence the production of FUBA?
• What are the effects of FUBA on metabolic products of the microbiota?
• How does sex influence FUBA and their effects?
• Are functional amyloids produced by non-bacterial species like fungi involved in human diseases?

Author Contributions: Conceptualization R.P.F.; methodology R.P.F.; software R.P.F.; validation R.P.F.; formal analysis R.P.F.; investigation R.P.F.; resources R.P.F., Z.K.; data creation, R.P.F.; Z.K., J.D.M.; writing—review and editing R.P.F., Z.K., J.D.M.; visualization R.P.F.; supervision R.P.F. project administration R.P.F.; funding acquisition R.P.F. All authors have read and agreed to the published version of the manuscript.

Funding: This research was funded by the Japan Society for the Promotion of Science, the University of Louisville, the Jewish Heritage Fund for Excellence, Walter Cowan, the National Institute on Aging, the family of E.A. Ford III and Axial Biotherapeutics.

Conflicts of Interest: The authors declare no conflict of interest.

References

1. Goedert, M. Oskar Fischer and the study of dementia. *Brain* **2009**, *132*, 1102–1111. [CrossRef] [PubMed]
2. Selkoe, D.J.; Hardy, J. The amyloid hypothesis of Alzheimer's disease at 25 years. *EMBO Mol. Med.* **2016**, *8*, 595–608. [CrossRef] [PubMed]
3. Hardy, J.; Selkoe, D.J. The amyloid hypothesis of Alzheimer's disease: progress and problems on the road to therapeutics. *Science* **2002**, *297*, 353–356. [CrossRef] [PubMed]
4. Prusiner, S.B. Cell biology. A unifying role for prions in neurodegenerative diseases. *Science* **2012**, *336*, 1511–1513. [CrossRef]
5. Cryan, J.F.; O'Riordan, K.J.; Sandhu, K.; Peterson, V.; Dinan, T.G. The gut microbiome in neurological disorders. *Lancet. Neurol.* **2019**, *19*, 179–194. [CrossRef]
6. Venegas, C.; Kumar, S.; Franklin, B.S.; Dierkes, T.; Brinkschulte, R.; Tejera, D.; Vieira-Saecker, A.; Schwartz, S.; Santarelli, F.; Kummer, M.P.; et al. Microglia-derived ASC specks cross-seed amyloid-beta in Alzheimer's disease. *Nature* **2017**, *552*, 355–361. [CrossRef]
7. Chapman, M.R.; Robinson, L.S.; Pinkner, J.S.; Roth, R.; Heuser, J.; Hammar, M.; Normark, S.; Hultgren, S.J. Role of Escherichia coli curli operons in directing amyloid fiber formation. *Science* **2002**, *295*, 851–855. [CrossRef]
8. Bokranz, W.; Wang, X.; Tschape, H.; Romling, U. Expression of cellulose and curli fimbriae by Escherichia coli isolated from the gastrointestinal tract. *J. Med. Microbiol.* **2005**, *54*, 1171–1182. [CrossRef]
9. Nicastro, L.; Tukel, C. Bacterial Amyloids: The Link between Bacterial Infections and Autoimmunity. *Trends Microbiol.* **2019**, *27*, 954–963. [CrossRef]
10. Friedland, R.P. Mechanisms of molecular mimicry involving the microbiota in neurodegeneration. *J. Alzheimers Dis.* **2015**, *45*, 349–362. [CrossRef] [PubMed]

11. Ulusoy, A.; Phillips, R.J.; Helwig, M.; Klinkenberg, M.; Powley, T.L.; Di Monte, D.A. Brain-to-stomach transfer of alpha-synuclein via vagal preganglionic projections. *Acta Neuropathol.* **2017**, *133*, 381–393. [CrossRef] [PubMed]
12. Kim, S.; Kwon, S.H.; Kam, T.I.; Panicker, N.; Karuppagounder, S.S.; Lee, S.; Lee, J.H.; Kim, W.R.; Kook, M.; Foss, C.A.; et al. Transneuronal Propagation of Pathologic alpha-Synuclein from the Gut to the Brain Models Parkinson's Disease. *Neuron* **2019**, *103*, 627–641. [CrossRef] [PubMed]
13. Chen, S.G.; Stribinskis, V.; Rane, M.J.; Demuth, D.R.; Gozal, E.; Roberts, A.M.; Jagadapillai, R.; Liu, R.; Choe, K.; Shivakumar, B.; et al. Exposure to the Functional Bacterial Amyloid Protein Curli Enhances Alpha-Synuclein Aggregation in Aged Fischer 344 Rats and Caenorhabditis elegans. *Sci. Rep.* **2016**, *6*, 34477. [CrossRef] [PubMed]
14. Perov, S.; Lidor, O.; Salinas, N.; Golan, N.; Tayeb-Fligelman, E.; Deshmukh, M.; Willbold, D.; Landau, M. Structural Insights into Curli CsgA Cross-beta Fibril Architecture Inspire Repurposing of Anti-amyloid Compounds as Anti-biofilm Agents. *PLoS Pathog.* **2019**, *15*, e1007978. [CrossRef] [PubMed]
15. Hartman, K.; Brender, J.R.; Monde, K.; Ono, A.; Evans, M.L.; Popovych, N.; Chapman, M.R.; Ramamoorthy, A. Bacterial curli protein promotes the conversion of PAP248-286 into the amyloid SEVI: cross-seeding of dissimilar amyloid sequences. *PeerJ* **2013**, *1*, e5. [CrossRef]
16. Christensen, L.F.B.; Jensen, K.F.; Nielsen, J.; Vad, B.S.; Christiansen, G.; Otzen, D.E. Reducing the Amyloidogenicity of Functional Amyloid Protein FapC Increases Its Ability to Inhibit alpha-Synuclein Fibrillation. *ACS Omega* **2019**, *4*, 4029–4039. [CrossRef]
17. Erny, D.; Hrabe de Angelis, A.L.; Jaitin, D.; Wieghofer, P.; Staszewski, O.; David, E.; Keren-Shaul, H.; Mahlakoiv, T.; Jakobshagen, K.; Buch, T.; et al. Host microbiota constantly control maturation and function of microglia in the CNS. *Nat. Neurosci.* **2015**, *18*, 965–977. [CrossRef]
18. Friedland, R.P.; Chapman, M.R. The role of microbial amyloid in neurodegeneration. *PLoS Pathog.* **2017**, *13*, e1006654. [CrossRef]
19. Tejera, D.; Mercan, D.; Sanchez-Caro, J.M.; Hanan, M.; Greenberg, D.; Soreq, H.; Latz, E.; Golenbock, D.; Heneka, M.T. Systemic inflammation impairs microglial Abeta clearance through NLRP3 inflammasome. *Embo J.* **2019**, *38*, e101064. [CrossRef]
20. Ising, C.; Venegas, C.; Zhang, S.; Scheiblich, H.; Schmidt, S.V.; Vieira-Saecker, A.; Schwartz, S.; Albasset, S.; McManus, R.M.; Tejera, D.; et al. NLRP3 inflammasome activation drives tau pathology. *Nature* **2019**, *575*, 669–673. [CrossRef]
21. Makki, K.; Deehan, E.C.; Walter, J.; Backhed, F. The Impact of Dietary Fiber on Gut Microbiota in Host Health and Disease. *Cell Host Microbe* **2018**, *23*, 705–715. [CrossRef] [PubMed]
22. Tran, T.T.T.; Corsini, S.; Kellingray, L.; Hegarty, C.; Le Gall, G.; Narbad, A.; Muller, M.; Tejera, N.; O'Toole, P.W.; Minihane, A.M.; et al. APOE genotype influences the gut microbiome structure and function in humans and mice: relevance for Alzheimer's disease pathophysiology. *FASEB J.* **2019**, *33*, 8221–8231. [CrossRef] [PubMed]
23. Blacher, E.; Bashiardes, S.; Shapiro, H.; Rothschild, D.; Mor, U.; Dori-Bachash, M.; Kleimeyer, C.; Moresi, C.; Harnik, Y.; Zur, M.; et al. Potential roles of gut microbiome and metabolites in modulating ALS in mice. *Nature* **2019**, *572*, 474–480. [CrossRef] [PubMed]
24. Williams, P.A.; Harder, J.M.; Foxworth, N.E.; Cochran, K.E.; Philip, V.M.; Porciatti, V.; Smithies, O.; John, S.W. Vitamin B3 modulates mitochondrial vulnerability and prevents glaucoma in aged mice. *Science* **2017**, *355*, 756–760. [CrossRef]
25. Rawls, M.; Ellis, A.K. The microbiome of the nose. *Ann. Allergy Asthma Immunol.* **2019**, *122*, 17–24. [CrossRef]
26. Proctor, D.M.; Relman, D.A. The Landscape Ecology and Microbiota of the Human Nose, Mouth, and Throat. *Cell Host Microbe* **2017**, *21*, 421–432. [CrossRef]
27. Dorrestein, P.C.; Mazmanian, S.K.; Knight, R. Finding the missing links among metabolites, microbes, and the host. *Immunity* **2014**, *40*, 824–832. [CrossRef]
28. Sharon, G.; Garg, N.; Debelius, J.; Knight, R.; Dorrestein, P.C.; Mazmanian, S.K. Specialized metabolites from the microbiome in health and disease. *Cell Metab.* **2014**, *20*, 719–730. [CrossRef]
29. Saupe, S.J.; Jarosz, D.F.; True, H.L. Amyloid Prions in Fungi. *Microbiol. Spectr.* **2016**, *4*, 1–13. [CrossRef]
30. Dodiya, H.B.; Kuntz, T.; Shaik, S.M.; Baufeld, C.; Leibowitz, J.; Zhang, X.; Gottel, N.; Zhang, X.; Butovsky, O.; Gilbert, J.A.; et al. Sex-specific effects of microbiome perturbations on cerebral Abeta amyloidosis and microglia phenotypes. *J. Exp. Med.* **2019**, *216*, 1542–1560. [CrossRef]

31. O'Keefe, S.J.; Li, J.V.; Lahti, L.; Ou, J.; Carbonero, F.; Mohammed, K.; Posma, J.M.; Kinross, J.; Wahl, E.; Ruder, E.; et al. Fat, fibre and cancer risk in African Americans and rural Africans. *Nat. Commun.* **2015**, *6*, 6342. [CrossRef] [PubMed]
32. Sampson, T.R.; Challis, C.; Jain, N.; Moiseyenko, A.; Ladinsky, M.S.; Shastri, G.G.; Thron, T.; Needham, B.D.; Horvath, I.; Debelius, J.W.; et al. A gut bacterial amyloid promotes alpha-synuclein aggregation and motor impairment in mice. *eLife* **2020**, *9*, 1–19. [CrossRef] [PubMed]

MDPI
St. Alban-Anlage 66
4052 Basel
Switzerland
Tel. +41 61 683 77 34
Fax +41 61 302 89 18
www.mdpi.com

International Journal of Molecular Sciences Editorial Office
E-mail: ijms@mdpi.com
www.mdpi.com/journal/ijms

www.ingramcontent.com/pod-product-compliance
Lightning Source LLC
Chambersburg PA
CBHW061831210926
51850CB00006B/856
* 9 7 8 3 0 3 6 5 6 7 1 2 9 *